T0192747

Phytopharmacy

Phytopharmacy

An evidence-based guide to herbal medicinal products

Sarah E. Edwards
Inês da Costa Rocha
Elizabeth M. Williamson
Michael Heinrich

WILEY Blackwell

Registered office: John Wiley & Sons, Ltd, The Atrium, Southern Gate, Chichester, West Sussex, PO19 8SQ, UK
Editorial offices: 9600 Garsington Road, Oxford, OX4 2DQ, UK
The Atrium, Southern Gate, Chichester, West Sussex, PO19 8SQ, UK
111 River Street, Hoboken, NJ 07030-5774, USA

For details of our global editorial offices, for customer services and for information about how to apply for permission to reuse the copyright material in this book please see our website at www.wiley.com/wiley-blackwell.

Library of Congress Cataloging-in-Publication Data

Edwards, Sarah E., author.
  Phytopharmacy : an evidence-based guide to herbal medical products / Sarah E. Edwards, Michael Heinrich, Ines Rocha, Elizabeth M. Williamson.
    p. ; cm.
  Includes bibliographical references.
  ISBN 978-1-118-54356-6 (pbk.)
  I. Heinrich, Michael, 1957 July 4-, author. II. Rocha, Ines, author. III. Williamson, Elizabeth M., author. IV. Title.
  [DNLM: 1. Plant Preparations–pharmacology. 2. Evidence-Based Practice.
  3. Phytotherapy. 4. Plant Preparations–therapeutic use. QV 766]
  RM666.H33
  615.3′21–dc23
                                                                              2014033180

A catalogue record for this book is available from the British Library.

Wiley also publishes its books in a variety of electronic formats. Some content that appears in print may not be available in electronic books.

The cover image is of *Ginkgo biloba* L., which is used in the treatment of diseases associated with milder forms of memory disorders and to enhance cognition. Photograph courtesy of Michael Heinrich

Typeset in 9/10pt TimesLTStd by Laserwords Private Limited, Chennai, India

1 2015

Contents

Preface

The increasing use of herbal medicines and botanical food supplements, either taken alone or in addition to orthodox treatment, presents a conundrum for the conventionally trained healthcare professional. Patients nowadays often prefer – and are encouraged – to take some responsibility for their own health and treatment options, and frequently purchase herbal medicines/botanical food supplements that they have read about in the popular press. This may be in an effort to generally improve their health or 'boost the immune system' or, as they see it, for the adjunctive treatment of specific disorders, regardless of whether expert advice was sought. *Doctors* may be asked whether it is safe for patients to take these products alongside their prescribed medicines; *pharmacists* who dispense those prescribed medicines and also sell herbal products, are asked which they can recommend; and *nurses* who look after patients on a long-term practical basis are asked for advice on all kinds of health issues. Numerous studies have shown that these practitioners consider their knowledge in the area of herbal medicines to be generally weak, especially regarding the potential therapeutic benefits, adverse effects, or possible interactions with prescribed or over-the-counter medicines. In addition to the healthcare professional, many patients are well-informed about their own health issues and medication (the 'expert patient') and are quite capable of making safe decisions about their own use of herbal medicines if they have access to the relevant information.

In the European Union, many herbal medicines are now regulated as medicines under the Traditional Herbal Medicinal Products Directive (http://www.mhra .gov.uk/Howweregulate/Medicines/Herbalmedicinesregulation/), placing new responsibilities on healthcare professionals. These traditional herbal registered (THR) products guarantee safety and quality, but in place of clinical trials (which may not have been carried out for economic reasons), a documented history of use in Europe is used instead. Herbal substances that are major ingredients in UK THR products are indicated as such in the presented monographs.

This book provides relevant information in a practical and useful way for the busy pharmacist, nurse or doctor, as well as the 'expert patient'. It gives a summary of the properties and uses of the most important herbal medicinal products and botanical food supplements, including an assessment of the available scientific evidence. It has been compiled on the premise that healthcare professionals (regardless of their own personal opinions) recognise patient and consumer demands for these products and need to be knowledgeable about them. The evidence available, both clinical and pre-clinical, is summarised to enable an evidence-based decision to be made as to whether the use of a particular nutritional or herbal medicinal product is advisable and safe.

We gratefully acknowledge funding through the UK government's matched funding scheme provided by Fa. Schwabe Pharmaceuticals and Bionorica (Germany) to the School of Pharmacy, Univ. London (M.Heinrich), which funded SE's and ICR's positions. The donors had no influence on the writing of the book.

Sarah Edwards,
UCL School of Pharmacy & Royal Botanic Gardens, Kew

Inês da Costa Rocha,
UCL School of Pharmacy

Elizabeth M. Williamson,
School of Pharmacy, University of Reading

Michael Heinrich,
UCL School of Pharmacy

Introduction

A Handbook of Herbal Medicines for the Practitioner and the Expert Patient

Herbal medicines are used increasingly in the United Kingdom, either alone or in addition to conventional treatment, which presents difficulties for the conventionally trained pharmacist, doctor, nurse, dentist, and so on. Herbal and nutritional products tend to be ignored by these practitioners, despite the fact that they are also used by some of these same professionals! It is important to be able to advise patients on the safe use of such products, including any possible interactions with prescribed or over-the-counter (OTC) medicines. However, although based on studies that are sparse, small and/or restricted to a particular setting, it can be concluded that generally health care professionals feel that their knowledge in this area is weak.

Doctors tend to know little about any aspect of herbal medicines, and often do not ask the patient if they are taking any (e.g. Lisk 2012), whereas pharmacists are more likely to answer correctly about the use of herbs, rather than about cautions, adverse effects, and interactions (e.g. Cuzzolin and Benoni 2009). Nurses also feel that their knowledge of herbal medicines is lacking (e.g. Temple et al. 2005) and in all surveys reported, respondents felt that health care professionals should know more, and that they themselves would benefit from training in this area. This book is an attempt to redress that lack of knowledge and provide useful practical information for the busy practitioner. It is intended to provide an overview of the most important medicinal and health food plants and products commonly used in the British Isles, including an assessment of the scientific evidence available for these 'herbal medicines'. It is based on the premise that health care professionals must recognise patient and consumer demands for these products, regardless of their own personal opinions, and therefore be knowledgeable about them - especially as many are now regulated as medicines.

A recent concept in medical treatment is that of the 'expert patient': someone who usually has a long-term health condition but who is able to take more control over their health by understanding and managing the condition, leading to an improved quality of life. Many patients who wish to take herbal and nutritional supplements do their own research, often over the Internet, and so are at risk of receiving biased information by vested interests, or politically or philosophically motivated groups. The information given in this book is taken only from peer-reviewed resources and written in language that the expert patient can normally understand, in an attempt to provide an informative and safe resource for the patient, as well as the practitioner.

The introductory chapters are based on a series of articles by the authors in the *Pharmaceutical Journal* in 2012: 288:565-566; 288:627-628; 288: 685-686; 289:161-162; 289: 270-271.

Phytopharmacy: An evidence-based guide to herbal medicinal products, First Edition.
Sarah E. Edwards, Inês da Costa Rocha, Elizabeth M. Williamson and Michael Heinrich.
© 2015 John Wiley & Sons, Ltd. Published 2015 by John Wiley & Sons, Ltd.

Definitions, the market and the legal position

What are herbal medicines and who uses them? Many people in the United Kingdom regularly use complementary and alternative medicine (CAM), either alone or in addition to conventional treatment. A recent systematic review has concluded that the average 1-year prevalence of use of CAM in the United Kingdom is >40% and the average lifetime prevalence >50%, with the most popular type being herbal medicine, at 29.5% (Posadzki et al. 2013). In 2003–2004, £3.25 bn was spent on herbal treatments alone in Western Europe (WHO 2008), and the global herbal supplements market is forecast to reach about £70 bn by the year 2017 (Global Industry Analysts 2013). These products are not only used by the 'worried well': a UK study found that 20% of cancer patients have used herbal medicines (Damery et al. 2011) and elsewhere, usage is even higher (see Williamson et al. 2013 for more details). Despite this, there generally is a lack of understanding of what herbal medicines actually are – or are not (IPSOS-MORI 2008).

Historically, plants have yielded many of our most important drugs, including morphine, taxol and digoxin, which are highly potent natural product – but not 'herbal' – medicines. Isolated compounds from plants are, in effect, identical as far as formulation, quality control and regulatory issues are concerned, to synthetic drugs or 'single chemical entities'. Herbal medicines are different in that they are prepared from plant material, but with little or no chemical fractionation and thus contain a wide range of natural compounds, some of which are pharmacologically active, and some of which are not. They can be licensed in the same way as 'conventional' medicines (e.g. *Senna alexandrina* Mill. tablets, ispaghula husk preparations, capsaicin cream), and even regulated as controlled drugs (e.g. cannabis oromucosal spray; MHRA 2010) - but if so, they are not usually considered to be 'herbal' medicines. Most frequently, botanical 'drugs' are available as food supplements and herbal medicines (e.g. rhodiola and black cohosh preparations), and can be purchased from health food and general stores, as well as pharmacies.

CAM encompasses a wide range of therapies, based generally on philosophical and cultural traditions rather than clinical evidence, and may or may not have been investigated scientifically. Some herbal medicinal products (HMPs) are fully licensed (and therefore not part of CAM), whereas others are registered under the Traditional Herbal Registration (THR) scheme on the basis of traditional use only (see later). The licensed products, and even many of the THRs, have been demonstrated to be pharmacologically active medicines and should be treated as such by health care professionals, with all of the issues that entails. The regulatory framework of HMPs has, however, been interpreted in a variety of ways, and consequently some products remain unlicensed and are classified as 'food supplements'.

UK regulation is largely based on European Union legislation, Directive 2001/83/EC, the European Traditional Herbal Medicinal Products Directive (THMPD) and the 1968 Medicines Act (Heinrich et al. 2012). Under this legislation, manufacturers of all products (including herbal remedies) classified as medicinal products must hold a marketing authorisation (MA, or product licence, PL) for that product, unless it satisfies the criteria for exemption from the requirement for an MA. In essence, medicinal products are defined by presentation (the purpose of the product), *or* by function (the actual effect of the product). All new chemical entities, including isolated constituents from plant and other natural sources, must have MAs for those products, based on the full dossier of chemical, pharmaceutical, pharmacological, toxicological and clinical data.

The legal position of herbal medicines in the United Kingdom: Herbal medicines are classed as medicinal products by the MHRA (2012a). While globally between

40,000 and 50,000 plant species are used for medicinal purposes in both traditional and modern medical systems (Heywood 2011), only a few hundred are used more widely in the United Kingdom, other European countries, North America or Australia.

Herbal products are available on the UK market as:

- Licensed (herbal) medicines
- Traditional herbal medicinal products registered under the THMPD
- Herbal medicines exempt from licensing, which comprise three groups:
  a) Unlicensed herbal medicines supplied (and often made) by a practitioner following a one-to-one consultation
  b) Manufactured or imported herbal products for individual patients commissioned from a third party ('specials')
  c) Unprocessed, that is, dried and cut herbal medicines (produced by subjecting a plant or mixture of plants to drying, crushing, cutting or a simple process of extraction)
- Medical devices, that is, products used to diagnose, prevent, or treat disease, but *without* chemical effects on the body
- Products sold as food or dietary supplements, often over the Internet
- Prescription-only medicines (POMs): potentially hazardous plants may only be dispensed by order of a prescription by a registered doctor.
- Pharmacy-only medicines (P), supplied by a registered pharmacist; these may be subject to restrictions of dose (but not duration of treatment) and/or route of administration.

Terminology used for herbal medicines: In addition to '*herbal medicines*', the term '*herbal medicinal products*' is also used, and highlights the commercial nature of these preparations. Less commonly, they may be called '*phytopharmaceuticals*', '*phytomedicines*', or even '*traditional medicines*'. In the United States, they are often referred to as '*botanicals*'; but in the United Kingdom, that also includes nutritional and cosmetic products. Similarly, a range of terms is also applied to foods with acclaimed health benefits: '*food supplements*', '*nutraceuticals*', '*health foods*' or '*medicinal foods*'. In this book, we use the term '*herbal medicines*' or '*herbal medicinal products*' *(HMPs)* to describe those which have a clearly defined medicinal use, and '*food supplements*' for those which are derived from foods and intended to supplement the diet or maintain health, rather than treat disease.

The production of herbal medicines: From a pharmaceutical perspective, herbal medicines may be extracts (usually aqueous, ethanolic or hydroalcoholic) or unprocessed (but usually powdered) dried plant material. 'Herbal drugs' are products that are either:

- derived from a plant: it may be the whole plant, or part of the plant such as the leaf, fruit, root, and so on, and prepared by simply drying and packaging (e.g. as a tea bag);
- obtained from a plant, but no longer retaining any recognisable structure of the plant: they still contain a complex mixture of compounds (e.g. essential oils, resins).

The chemical composition of individual plants is influenced by a combination of genetic and environmental factors, including soil, weather, season or time of day harvested, and use of any pesticides, herbicides and fertilisers. This has been demonstrated in various strains of *Sedum roseum* (better known by its synonym name, *Rhodiola rosea*) which were moved geographically and also grown under varying conditions (Peschel et al. 2013). Processing and extraction procedures affect the final chemical composition of HMPs, and also explain why the chemical profile of two HMPs derived from the same plant species may differ considerably. The variation is significant because not all constituents make an equal contribution to the pharmacological effects of the herb. Herbs used for registered HMPs are either grown/produced under controlled (cultivated) agricultural conditions or wild harvested in compliance with Good Agricultural and Collection Practice (GACP), providing a high level of product quality, which is intrinsically linked with safety.

The next key step in the production of HMPs is the processing, including harvesting, of the relevant plant part. The processing (drying, cutting, storage, packaging and transport, etc.) of registered products must be in compliance with Good Manufacturing Practice (GMP). Human error and/or unscrupulous operators also influence the quality of the raw material. Accidental or intentional botanical substitution are far more likely to occur with unregistered products that don't comply with GMP, and the intentional adulteration with conventional drugs (e.g. corticosteroids) and contamination with microorganisms and pesticides continues to be of concern. An important benefit of registration under THMPD is the safeguarding of patients' health by implementing a number of stringent manufacturing and quality control requirements. There is therefore an ethical argument that health care professionals should only recommend registered or licensed products.

Specific extraction and processing techniques are available in both the British and European Pharmacopoeia (BP, Ph Eur) for processing crude plant material. These are tightly controlled by European and national legislation and the monographs provide legally binding quality assurance procedures for products available on the British market. The variability in content and concentrations of constituents of the plant material, together with the range of extraction techniques and processing steps used by different manufacturers, results in a marked variability in the content and quality of all herbal products. Both raw and processed materials therefore require monitoring in order to produce HMPs of consistent quality and to ensure bioequivalence (Loew and Kaszkin 2002). For registration under the THMPD, the applicant has to provide details about the production, processing, extraction and formulation process, as well as on the composition of the medicine, the dose per unit, and the daily dose.

The question of quality assurance is also linked to the concept of 'standardisation'. Although relatively new for HMPs, it is essential to ensure that patients are provided with *consistent,* high-quality, herbal products. Standardisation is only possible where the active constituents are known (which is not the case for many HMPs) and can be defined as the requirement for *a specified amount or range of one or several pharmacologically active compounds, or groups of compounds, in the extract.* Reproducibility of the chemical constituents in HMPs prevents accidental overdosing due to batch-to-batch variation as well as under-dosing, and therefore contributes to efficacy. Unfortunately, the term 'standardisation' is often misunderstood, if not misused, in herbal medicine promotion – but it is easy to comprehend if compared to blending coffee or even whisky, to make the consistent and familiar product that the customer expects! If the active constituents are not known, the extract cannot be standardised, although one or more 'marker' compounds characteristic of the botanical drug can be used to characterise the HMP chemically (Heinrich et al. 2012).

Quality issues for registered/licensed HMPs – a checklist of key parameters:

- Harvesting/collection of plants using GACP
- Full botanical authentication
- Test for contaminants (pesticide, herbicide residues; heavy metals; microbial contamination)
- Extraction methodology (quality assurance)
- Standardised extracts (active constituents (single or groups) are known)
- Quantified extracts (known therapeutic or pharmacological activity)

Combination effects and their importance: Single compounds that are derived and/or purified from plants are not HMPs and do not exhibit combination effects. Herbal medicines, however, even when prepared from only one plant, contain a large number of phytochemicals, rather than a single pharmacologically active substance. A principal tenet of herbal medicine is that this results in a unique activity profile, in which several compounds act on each other, either moderating, opposing, or enhancing an effect. An enhancement may be an '*additive*' or '*synergistic*' action, whereby the combination of constituents is greater than would have been expected from the sum of individual contributions. There is some evidence for this: in the case of *Ginkgo biloba* L., synergy in inhibiting platelet aggregation has been shown for ginkgolides, using the isobole method, and other components of cannabis are seen to enhance the activity of the CB1 agonist Δ-9 tetrahydrocannabinol in the extract (Williamson 2001). '*Antagonism*' is when the effect of a compound is inhibited by the presence of another, but this may of course be beneficial if the particular effect is unwanted. The term '*polyvalence*' is now often used to describe the full range of biological activities that contribute to the overall effects, and includes multi-target and well as multi-component effects (Wagner and Ulrich-Merzenich 2009). Polyvalence can also be shown by St. John's wort (*Hypericum perforatum* L.), used to treat mild-to-moderate depression, which contains a variety of compounds acting in different ways. For example, hyperforin inhibits serotonin reuptake, whereas hypericin inhibits binding to some subtypes of dopamine receptors, and the flavonoids also contribute to the activity (Russo et al. 2014).

The 'one target, one disease' (or 'silver bullet') concept is increasingly considered inadequate in many clinical situations (Wermuth 2004) and polypharmacy is routine in conditions such as cancer, hypertension and HIV infection. The use of multiple drugs increases the risk of adverse effects and drug interactions, and while HMPs and food supplements are generally not included in definitions of polypharmacy, they can also increase the risk of drug interactions, although a great deal of speculation and exaggeration surrounds this issue (Williamson et al. 2013).

Traditional Herbal Registration and why it is necessary: The Traditional Herbal Medicinal Products Directive stipulates that only *registered* herbal products may be sold as OTC medicines. THR medicines have known quality and safety, and documented traditional use. Only limited therapeutic claims can be made, and their use is only for minor self-limiting conditions. They may be administered via any route of administration (topical, oral, etc.) except for injectables, which are always POMs. They must be sold with a patient information leaflet (PIL) and can be identified by a THR number. They may also display the certification mark see Fig I.1 (which is not compulsory) on the packaging.

The implementation of the THMPD has resolved a number of safety issues surrounding the production of unregulated HMPs, by ensuring consistent quality

Figure 1   Traditional Herbal Registration certification mark

based on good manufacturing, agricultural and/or collection practices. The aim is to reduce the risk of problems caused by:

- contamination (e.g. with heavy metals, pesticides, insects or moulds);
- substitution (e.g. with other plant species, which may be toxic or ineffective)
- adulteration (both accidental and deliberate: this may be with other plant parts of the correct plant, such as stems and fruits in a leaf drug, or with other – usually inferior and cheaper – species, or with synthetic drugs such as corticosteroids).

A few potentially dangerous medicinal plants remain restricted to use as POMs. These include *Digitalis*, *Strychnos* and *Aconitum* species, with maximum doses and/or route of administration specified, but, in fact, are rarely found in practice in the United Kingdom. Some other herbal ingredients are prohibited, including *Aristolochia* species which are highly nephrotoxic (Heinrich et al. 2009).

Patients who use unlicensed herbal products have no guarantee that these comply with any regulations, or any definition of good practice, and so may be exposing themselves to risk. The MHRA's Yellow Card Scheme for pharmacovigilance applies to all *registered and licensed HMPs* in addition to conventional medicines, and should be used where there is concern that an adverse event or interaction has occurred as a result of their use.

In a few cases, a product may actually hold a product licence under the 'Well Established Use Directive'. This is where an HMP is supported by sufficient safety and efficacy data and consequently has a well-established medicinal use rather than just being based on 'traditional use'. It is a licensing route more commonly used in continental Europe than in the United Kingdom.

The importance of using THR products is very well illustrated by the case of butterbur (*Petasites hybridus*), which is used traditionally for migraine, asthma and hay fever. Products containing this herb have been linked to 40 cases of liver toxicity, including two cases of liver failure requiring transplantation. This plant is known to contain hepatotoxic pyrrolizidine alkaloids (PAs), but what is of real concern is that these cases involved the use of butterbur-containing products where the PAs had been removed, indicating that other constituents (possibly sesquiterpenes) were responsible for the toxicity. Butterbur is not found in any THRs registered in the United Kingdom but is an ingredient in a number of herbal products sold as food supplements (MHRA 2012c).

Herbal medicines as 'food supplements':   A vast number of medicinal plants are also used as foods or in cosmetic preparations. The MHRA is responsible for classifying which herbal products are primarily medicines, and, therefore, fall within the remit of the THMPD, whereas those classified as 'food supplements' must comply with regulations set out by the Department of Health (DH 2011). Food supplements may be almost indiscernible from HMPs in terms of physiological effects (in fact, the same herb may be sold as both a food supplement and an HMP), but they may not be sold with any therapeutic claims. Like HMPs, food supplements have the potential to interact with other medications; for example, garlic and cranberry may increase the risk of bleeding associated with antiplatelet or anticoagulant agents such as aspirin and warfarin (Williamson et al. 2013).

Herbal medicines as 'medical devices': In rare cases, a herbal product may be registered as a 'medical device'. For example, a 'fibre complex' from stems of the prickly pear cactus (*Opuntia ficus-indica*), has been registered as a medical device for weight loss. Since a number of adverse drug reactions have been reported for it (MHRA 2012b), this product, and the concept in general, remains controversial.

In summary, *there is a regulatory framework available for the control of the quality of herbal medicines, and for ethical reasons, health care professionals should only recommend registered or licensed products* (assuming they are available, of course). After all, no health care professional should prescribe, dispense or administer a product if they have no guarantee that it even contains what it says on the label.

Information given in this book is not a substitute for medical advice and no responsibility can be taken by the authors for adverse reactions.

References

Cuzzolin L, Benoni G. (2009) Attitudes and knowledge toward natural products safety in the pharmacy setting: an Italian study. *Phytotherapy Research.* 23(7) 1018–1023.

Damery S, Gratus C, Grieve R, Warmington S, Jones J, Routledge P, Greenfield S, Dowswell G, Sherriff J, Wilson S. (2011) The use of herbal medicines by people with cancer: a cross-sectional survey. *British Journal of Cancer* 104: 927–933.

Department of Health, UK. (2011) Food supplements - Summary information on legislation relating to the sale of food supplements, https://www.gov.uk/government/uploads/system /uploads/attachment_data/file/204303/Supplements_Summary__Jan_2012__DH_FINAL .doc.pdf (accessed July 2013).

Global Industry Analysts. (2013). Herbal Supplements and Remedies: A Global Strategic Business Report. http://www.companiesandmarkets.com/Market/Consumer-Goods /Market-Research/Herbal-Supplements-and-Remedies-A-Global-Strategic-Business -Report/RPT907407 (accessed July 2013).

Heinrich M, Barnes J, Gibbons S, Williamson EM. (2012) *Fundamentals of Pharmacognosy and Phytotherapy.* 2nd Edition. Churchill Livingston Elsevier, Edinburgh and London.

Heinrich M, Chan J, Wanke S, Neinhuis C, Simmonds MSJ. (2009) Local uses of *Aristolochia* species and content of Aristolochic acid 1 and 2 – a global assessment based on bibliographic sources. *Journal of Ethnopharmacology* 125(1): 108–144.

Heywood V. (2011) Ethnopharmacology, food production, nutrition and biodiversity conservation: towards a sustainable future for indigenous peoples. *Journal of Ethnopharmacology* 137(1): 1–15.

IPSOS – MORI 2008 *Public Perceptions of Herbal Medicines General Public Qualitative & Quantitative Research.* IPSOS-Mori, London, UK.

Lisk C. (2012) Food for thought: doctors' knowledge of herbal medicines needs to be better. *Acute Medicine* 11(3): 134–137.

Loew D, Kaszkin M. (2002) Approaching the problem of bioequivalence of herbal medicinal products. *Phytotherapy Research* 16(8): 705–711.

MHRA. (2010). Medicines and Healthcare products Regulatory Agency Public Assessment Report. Sativex Oromucosal Spray. http://www.mhra.gov.uk/home/groups/par/documents /websiteresources/con084961.pdf (accessed July 2013).

MHRA. (2012a). A guide to what is a medicinal product. MHRA Guidance Note No. 8. http://www.mhra.gov.uk/home/groups/is-lic/documents/publication/con007544.pdf (accessed July 2013).

MHRA. (2012b) Drug Analysis Print. *Opuntia ficus-indica.* http://www.mhra.gov.uk/home /groups/public/documents/sentineldocuments/dap_4869624197041.pdf (accessed July 2013).

MHRA. (2012c) Unlicensed butterbur (*Petasites hybridus*) herbal remedies http://www .mhra.gov.uk/home/groups/comms-ic/documents/websiteresources/con140851.pdf (accessed July 2013).

Peschel W, Prieto MJ, Karkour C, Williamson EM. (2013) Effect of provenance, plant part and processing on extract profiles from cultivated European *Rhodiola* for medicine use. *Phytochemistry* 86: 92–102.

Posadzki P, Watson LK, Alotaibi A, Ernst E. (2013) Prevalence of use of complementary and alternative medicine (CAM) by patients/consumers in the UK: systematic review of surveys. *Clinical Medicine* 13(2): 126–31.

Russo E, Scicchitano F, Whalley BJ, Mazzitello C, Ciriaco M, Esposito S, Patanè M, Upton R, Pugliese M, Chimirri S, Mammì M, Palleria C, De Sarro G. (2014) *Hypericum perforatum*: pharmacokinetic, mechanism of action, tolerability and clinical drug-drug interactions. *Phytotherapy Research* 28(5): 643–655.

Temple MD, Fagerlund K, Saewyc E. (2005) A national survey of certified registered nurse anesthetists' knowledge, beliefs, and assessment of herbal supplements in the anesthesia setting. *American Association of Nurse Anesthetists Journal* 73(5): 368–377.

Wagner H, Ulrich-Merzenich G. (2009) Synergy research: approaching a new generation of phytopharmaceuticals. *Phytomedicine* 16(2–3): 97–110.

Wermuth CG. (2004) Multitargeted drugs: the end of the "one-target-one-disease" philosophy? *Drug Discovery Today* 9(19): 826–827.

Williamson EM. (2001) Synergy and other interactions in phytomedicines. *Phytomedicine* 8(5): 401–440.

Williamson EM, Driver S, Baxter K. (Eds.) (2013) *Stockley's Herbal Medicines Interactions.* 2nd Edition. Pharmaceutical Press, London, UK.

WHO. (2008) *Traditional Medicine.* Fact Sheet No. 134. WHO, Geneva.

The Evidence Base for Herbal Medicines

Efficacy, clinical effectiveness and comparative effectiveness: The term 'efficacy' has a somewhat different definition according to context. In medicine, it is considered to be *the capacity for a health intervention to produce a therapeutic effect*; whereas within the discipline of pharmacology, it means the maximum response achievable from a drug.

To avoid confusion, the term 'clinical effectiveness' can be used to describe how well a medicine works in practice, as it encompasses both the biological effects of the constituents as well as any non-pharmacological influences (such as placebo or nocebo effects).

In recent years, the classical approach to evaluating medicines has been scrutinised from the perspective of how clinical studies can actually inform about treatment outcomes. For herbal medicines, 'comparative effectiveness' research has been proposed as a means of assessing their effectiveness in everyday practice settings, meaning the use of trials that compare 'real-world' situations rather than isolated interventions. Overall, this results in the capacity to inform specific clinical decisions (see Witt 2013 and references therein).

Comparative effectiveness research includes:

- the possibility of comparing two or more health interventions (a specific medication or therapy) to determine which option works best for which type of patient;
- the design of studies using 'typical patients' and in 'everyday' settings, that is, similar to those in which the intervention will be used.

The non-pharmacological functions of medicines are often overlooked, yet they play a significant role in eliciting the so-called 'meaning response', which refers to the 'physiological or psychological effects of meaning in the origins or treatment of illness', such as the production of endogenous opiates in response to an intervention. Placebo effects appear to be the result of a number of different mechanisms, including expectation, anxiety and reward, in addition to learning phenomena such as Pavlovian conditioning, and cognitive and social learning. With regard to traditional herbal medicines, the sociocultural aspects are even more likely to elicit a physiological response, in addition to any intrinsic pharmacological activity of the plant, since they often exist within religious and mythical traditions, creating a vivid associated meaning (Moerman and Jonas 2002).

The use of herbal medicines and similar products should therefore be considered within the context of their use, as this undoubtedly will have an impact on their clinical effectiveness. In traditional medicine, a healer may also use ritual and prayer along with herbal treatment, which may induce a meaning response, adding to the therapeutic effects of the plant extract. (This is not as far from modern medicine as

Phytopharmacy: An evidence-based guide to herbal medicinal products, First Edition.
Sarah E. Edwards, Inês da Costa Rocha, Elizabeth M. Williamson and Michael Heinrich.

might be supposed and most pharmacists have patients who only find a particular brand of drug to be effective: for example, 'only the blue tablets work on me'!) It is also essential to understand that the UK approach to herbal medicines is very cautious, and that some products, which are not considered to have evidence for clinical effectiveness, may be widely accepted in medical practice and as OTC products in other countries (e.g. see Schulz et al. 2004).

What evidence is available and how can it be interpreted? Evidence of clinical effectiveness is rated according to quality, with the highest levels of evidence ascribed to systematic reviews and meta-analyses of randomised controlled trials (RCTs). Key resources for such evidence include the Cochrane reviews, which are produced by an international collaboration of scientists. These reviews are globally recognised as '*the benchmark for high quality information about the effectiveness of health care*' and are freely available at http://www.cochrane.org/.

Figure 2 summarises the levels of evidence recognised today. This is largely a clinical perspective highlighting how detailed the clinical and pharmacological assessment has been, but of course it is intrinsically linked to the composition of a particular product and therefore to the production and processing of the herbal medicine. These levels of evidence are not as sharp as they appear in the diagram, and they are sometimes interpreted differently by the regulatory authorities of EU member states!

A criticism commonly levelled at HMPs is that for many of those on the market there is a lack of clinical data from good quality RCTs, usually, it is stated, due to limited funding. Although it is true that there is a paucity of data, Cochrane reviews do exist for a few herbal medicines. For example, a Cochrane review of RCTs on herbal medicine to treat low back pain, found that Devil's claw (*Harpagophytum procumbens* DC.) and white willow bark (*Salix alba* L.) seemed to reduce pain more than placebo in short-term trials, with the qualification that further trials were needed to clarify their equivalence against standard treatments in terms of efficacy, and that for long-term use there was no evidence that these substances are safe and useful (Gagnier et al. 2006). Another Cochrane review on St. John's

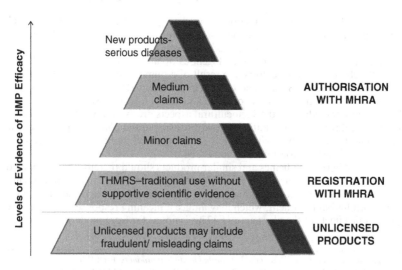

Figure 2   Levels of evidence (adapted from figure by Prof. Dr. W. Knoss, BfArM, Bonn, Germany)

wort (*Hypericum perforatum* L.), showed that it was equivalent to selective serotonin uptake inhibitors (Linde et al. 2008). Despite this, in the United Kingdom, health claims for all these drugs can only be based on 'traditional use'.

Due to the nature of herbal medicines, that is, their variability in phytochemical makeup according to genotype, plant part used, and environmental conditions, even well-designed clinical trials may potentially be flawed unless these factors are taken into consideration. A study of Echinacea, which showed no pharmacological effect beyond placebo for treating the common cold in children, was criticised because it used non-standardised pressed plant juice from aerial parts only, rather than standardised extracts from the entire plant (Firenzuoli and Gori 2004; Kim et al. 2004; Taylor et al. 2003). Furthermore, although Cochrane reviews are considered by many to be the gold standard for evaluating clinical effectiveness, in the case of herbal medicines, concerns have been raised. A study exploring the 11 most relevant Cochrane reviews on herbal medicine identified that frequently, the herbal medicines in the included studies had not been sufficiently well characterised. The plausibility of the medication for the specific indication needs to be considered in the light of the chemical composition and it has been suggested that the guidelines for preparing Cochrane reviews be revised for herbal products (Davidson et al. 2013).

Understanding the mechanisms of action responsible for the clinical effects of herbal products is challenging due to the presence of multiple constituents within one herbal ingredient, and thus pharmacokinetic/ pharmacodynamic data for these products are often unavailable. In Ayurvedic and traditional Chinese medicine (TCM) preparations, the complexity is compounded by the fact that each preparation usually contains multiple herbal (and sometimes non-herbal) components, each ingredient possibly containing a number of (as yet unknown) bioactive constituents. While pharmacological (and occasionally clinical) evidence for the properties of herbal extracts, mixtures and products is constantly being published, in journals such as the *Journal of Ethnopharmacology*, *Phytotherapy Research*, *Planta Medica*, *Fitoterapia* and *Phytomedicine*, there is a lack of a critical synthesis and applicability into everyday applications.

In this book we take a pragmatic approach and want to make clear that, while the level of evidence may be limited, these levels of evidence actually vary, in the manner shown in Figure 2. Note that this classification system is based on *available* evidence and should not be confused with clinical efficacy. We have defined five levels which highlight the level of evidence available for a specific botanical drug and the preparations derived from it.

Of course, 'lack of evidence for efficacy' is not the same 'as evidence for lack of efficacy', but from the perspective of the THMPD, the proven traditional use is the *only* relevant criterion (as well as the absence of toxicity reports) for registration, and we do not attempt to replace this classification. We do however highlight that different levels of evidence exist even within this group of HMPs.

In the Complementary and Alternative Medicine (CAM) sector, limited clinical evidence for herbal medicines is often not considered a serious problem by consumers (and often practitioners). It is assumed that a history of traditional use over many generations, without observed negative effects and in addition to being 'natural', implies that they are safe.

However, that implies that detailed knowledge of these products is available and that they have been produced using GACP, GMP and other Good Practice guidelines, which is often not the case, and it also depends hugely on what happens along the chain of supply from the grower to the buyer (Booker et al. 2012). The issue of quality has been discussed in more detail in the introduction.

Even with a long tradition of use for a particular herbal medicine, toxic effects may be delayed and a connection between cause and effect will not be made.

A well-known example is that of *Aristolochia*, a genus of plants used for perhaps thousands of years in traditional medical systems throughout the world. Awareness of the toxicity of *Aristolochia* only developed in the 1990s when a number of women in Belgium attending a slimming clinic developed kidney failure after being given a herbal medicine containing the herb *Aristolochia fangchi* (for the full story, see Heinrich et al. 2009). Today, despite a ban on this ingredient in most countries, cases of toxicity caused by the accidental or deliberate supply of products containing *Aristolochia* still occur, and the problem is exacerbated by the fact that other medicinal herbs have a similar appearance and similar Chinese common names.

There are other key concerns relating to the safety of use of TCM and Ayurvedic medicines, where the presence of high levels of heavy metals such as mercury, lead and arsenic have been found (MHRA 2013). Another complex example of toxicity is aconite (species of *Aconitum*), which is used in TCM for a wide range of indications, including many chronic conditions. The cardio- and neurotoxicity of this drug is potentially lethal, and the improper use of *Aconitum* in China, India, Japan and some other countries has led to cases of severe intoxication and death. According to claims made by some proponents of TCM, the tubers and roots can be detoxified by unique preparation methods, which are claimed to reduce the amount of toxic aconitine-type alkaloids present. While botanical drugs, which contain less than a threshold of aconitine, can be used medicinally in China, this position is not accepted in most Europe an countries, where aconite species may not be used under any circumstances (Singhuber et al. 2009). (In the United Kingdom, aconite may be found in licensed homeopathic remedies in which it is rigorously diluted. As an ingredient of an oral medicine its use is restricted and only available on prescription by a registered doctor or dentist.) TCM uses many toxic materials, especially in China (Liu et al. 2013). Despite this, the most commonly used TCM drugs in practice, in both China and the European Union, seem to be fairly safe. The most toxic drugs are used for serious diseases, a practice more likely in China (Williamson et al. 2013a). It is of course illegal to advertise non-licensed medicines for the treatment of conditions such as cancer, hypertension, and so on, in Europe.

Assessing the interaction potential of herbal medicines: Since the majority of HMPs are OTC medicines, it is of utmost importance that pharmacists, as well as manufacturers through patient information leaflets (PILs), raise awareness of the interaction potential and associated side effects of these products. Pharmacists in primary care encounter patients who take herbal medicines or supplements every day, and often sell HMPs in their stores, whereas pharmacists in secondary care should be aware that patients do bring their supplements into hospital with them, and sometimes continue to take them unbeknownst to the surgeon, anaesthetist, clinical pharmacists and nurses looking after them.

Patients should therefore be asked routinely about their use of herbal products, when dispensing medicines and taking a drug history on hospital admission, so that pharmacists can help the doctor and patient make a fully informed decision about their care. Special attention needs to be paid to long-term users and/or consumers of large amounts of HMPs, or patients who use many different medicinal products concomitantly, as they are more likely to suffer from adverse reactions. It is also rather worrying than many pregnant or nursing women take herbal medicines (Cuzzolin et al. 2011), and that these are also given to babies and children (Gottschling et al. 2013; Lim et al. 2011). Other groups at risk include the elderly, the malnourished or undernourished, as well as patients with serious medical conditions including heart disease, cancer, diabetes and asthma. As an example, patients with hypertension or congestive heart failure should not use *Ephedra*, often included in weight loss products, as it increases heart rate and blood pressure. *Ephedra* is legally restricted in the United Kingdom, but can be bought easily over the Internet. The herb can also be supplied following a one-to-one consultation with a practitioner at a specified dosage and route of administration, but otherwise can only be supplied under the supervision of a pharmacist.

Drug interactions are always complex and in the case of HMPs, assessing the situation is even more difficult. There is a great deal of misinformation surrounding herb–drug interactions (HDIs), and animal and cell-based studies, and individual case reports where causality has not been proved, have been cited as 'evidence' for clinical HDIs. The potential for interaction is variable, but most likely when patients are on cardiovascular, immunosuppressant and CNS drugs. All practitioners wisely err on the side of caution when warfarin, statins and digoxin are involved, as well as ciclosporin and tacrolimus, and midazolam and phenytoin, and this is equally true when combined with HMPs. St. John's wort has become notorious for its interaction profile, and should be avoided with the drugs mentioned, although it is in fact a very safe drug when used alone. In the monographs for the herbal drugs in this book, important HDIs are specified and warnings given when there may be a risk, even if not proved clinically; for more information on mechanisms and individual reports, see *Stockley's Herbal Medicines Interactions* (Williamson et al. 2013b).

The data available on clinically validated cases are limited and often do not allow an evidence-based decision. It is especially important that patients suffering from renal failure or liver disorders are assessed, but data is rarely available in these cases. If in doubt, the HMP should be avoided or discontinued, because the evidence in favour of the efficacy of the prescribed medicines will almost certainly be far greater. Despite the exaggerations, there is a real issue surrounding HDIs, and there is also likely to be an under-reporting of such cases, for example, via the Yellow Card Scheme (Da Costa Rocha et al. 2012). This is further explained later.

Some common safety issues to be considered during a consultation or when using HMPs are given in Table 1 using four important and commonly used HMPs as examples.

Further risks and pharmacovigilance: Patients may not read the PIL now supplied with each registered HMP, which includes warnings and information on contraindications, possible adverse reactions or interactions with other medicines, and could consume the product inappropriately. This again highlights the responsibilities of pharmacists and other health care professionals in advising patients.

Pharmacovigilance is a critical task in ensuring the safe use of HMPs. However, since the information on an HMP is often limited, this is even more challenging than it is for conventional medicines (Jordan et al. 2010). The UK Commission on Human Medicines (CHM) has extended its Yellow Card Scheme for adverse drug reaction (ADR) reporting to all hospital and community pharmacists, as well as to any member of the public. For the Yellow Card Scheme to be effective, patients need to be aware that they can report any suspected side effect associated with their medicines, including OTC and herbal products. *Thus, pharmacists have an important role to play in ADR reporting for herbal medicines. Community pharmacists should encourage patients to submit completed forms to the Medicines and Health care Products Regulatory Agency (MHRA).* As some patients may require assistance in completing these forms, the pharmacist is ideally placed to help patients with this process and, in particular, to decide whether a possible side effect is due to a medicine, or whether the MHRA criteria for reporting are met. Patient reporting of possible adverse effects has several advantages, including making the system faster, and by providing more detailed information of aspects such as how it has affected their quality of life. This should never exclude reporting by a health care

Table 1 Safety of HMPs: four examples of botanical materials and potential risks associated with their use

Herbal medicine	Contraindications	Adverse effects	Interactions with other drugs
St. John's wort (*Hypericum perforatum* L.)	Pregnancy, lactation	Gastrointestinal symptoms, allergic reactions	Can reduce the effects of warfarin and related drugs, digoxin, opioids, hormonal contraceptives, voriconazole, protease inhibitors (indinavir), statins (simvastatin and atorvastatin but not pravastatin). Serotonin syndrome has been reported in patients taking SNRIs, SSRIs and triptans.
Black cohosh (*Actaea racemosa* L.)	Pregnancy, lactation, hormone-dependent tumours	Occasionally upset stomach	Possible interactions with antineoplastic drugs such as cisplatin (experimental-based evidence only). Contains oestrogenic compounds. Might have an additive or antagonist effect in women receiving chemotherapy (tamoxifen) or hormone antagonists.
Ginseng (Asian) (*Panax ginseng* C.A.Mey)	Hypertension	Insomnia, hypertension and oedema as symptoms of overdose	Can increase the effect of antidiabetics. Might decrease the effect of tamoxifen and other oestrogen antagonists. Might reduce the effect of warfarin and related drugs (Asian and American ginseng)
Ginkgo (*Ginkgo biloba* L.)	Hypersensitivity to ginkgo preparations	Mild gastrointestinal symptoms, headaches, rare allergic reactions	Case reports describe seizure with antiepiletics (valproate and phenytoin). Might lead to haemorrhage when in combination with antiplatelet drugs (aspirin, clopidogrel and ticlopidine) and warfarin. May increase the effect of calcium channel blockers (nifedipine). Induces the metabolism of proton pump inhibitors (omeprazole). Coma development in an elderly patient taking trazodone.

Taken from Williamson et al. 2013b and Heinrich et al. 2012

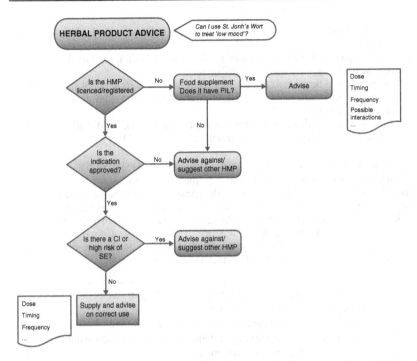

Figure 3 The key points to consider in consultations regarding HMPs. Abbreviations: HMP: herbal medicinal product; PIL: patient information leaflet; CI: contraindications; SE: side effects.

professional in parallel; pharmacists, for example, should not be concerned about the possibility of duplication of the reports and the information provided should be as detailed as possible to allow cross-checking. The MHRA is interested in any ADR reports, but specifically the ones by groups at increased risk, like children and the elderly.

Practical advice: The most important points to keep in mind during a consultation about HMPs are summarised in the flow chart above (Figure 3), which assumes that the patient is seeking help in the treatment of a self-limiting minor condition, and not one which requires medical attention.

The changes in the regulation of HMPs outlined in the introduction have brought new responsibilities for pharmacists. During a consultation with a patient, it is necessary to determine whether the herbal product is registered as herbal medicine (i.e. has a THR number on the packaging and includes a PIL), is a licensed medicine (has a product licence PL number or a marketing authorisation MA number), or is a food supplement.

Particular caution must be exercised if a patient is taking an unlicensed or unregistered product, which will not include a PIL. All health care professionals should be aware of the key points resulting from these changing responsibilities:

- HMPs and food supplements are increasingly popular for self-treatment.
- HMPs are pharmacologically active and should be treated as if they are conventional medicines; use the Yellow Card Scheme to report suspected adverse events.

- Patients should be advised to use only licensed/registered HMPs wherever possible.
- HMPs should only be used for *minor self-limiting diseases*, such as the common cold, or the temporary relief of mild anxiety.
- Patients should be advised not to use herbal medicines/food supplements alongside prescribed medicines due to risk of interactions, unless they are known to be safe to use together.
- Drugs used for cardiovascular, immunosuppressant and CNS disorders are especially liable to all types of drug interaction, including HDIs.

Within the next few years, the evidence base of many HMPs is likely to increase significantly (e.g. with the application of the 'omic'- technologies), enabling a more holistic understanding of how all the constituents in a herbal medicine work together and act on biological systems. For the time being, while evidence on efficacy or mode of action may be limited, community pharmacists can rely on licensed/registered HMPs meeting the levels of quality and safety that consumers require. Finally, an aide-memoire is provided to remind health care professionals of the steps to be taken in a consultation, as shown in Figure 4.

- Does the patient belong to any of the groups at higher risk for herb-drug interactions or idiosyncratic drug reactions?
- Do their medicines belong to a high risk therapeutic area?
- What unique risks are known about this specific herbal material and of extracts or products derived from them?
- Will the product be used for a shorter or a longer period?
- How detailed is the patient's understanding of such products, their health benefits and potential risks?

Figure 4   Some simple guiding principles for advising patients.

References

Booker A, Johnston D, Heinrich M. (2012) Value chains of herbal medicines – research needs and key challenges in the context of ethnopharmacology. *Journal of Ethnopharmacology* 140(3): 624–633.

Cuzzolin L, Francini-Pesenti F, Verlato G, Joppi M, Baldelli P, Benoni G. et al. (2011) Use of herbal products among 392 Italian pregnant women: focus on pregnancy outcome. *Pharmacoepidemiology and Drug Safety* 19(11): 1151–1158.

Da Costa Rocha I, Edwards SE, Lawrence MJ, Cable C, Heinrich M. (2012) Quality and safety of herbal products: Part 1 - New legislation and production. *Pharmaceutical Journal* 288(7708–7709): 685–686.

Davidson E, Vlachojannis J, Cameron M, Chrubasik S. (2013) Best available evidence in Cochrane reviews on herbal medicine? *Evidence-Based Complementary and Alternative Medicine* 2013: 163412.

Firenzuoli F, Gori L. (2004) Echinacea for Treating Colds in Children. *Journal of the American Medical Association* 291(11): 1323–1324.

Gagnier JJ, van Tulder M, Berman B, Bombardier C. (2006) Herbal medicines for low back pain (review). *Cochrane Database of Systematic Reviews* (2): CD004504.

Gottschling S, Gronwald B, Schmitt S, Schmitt C, Längler A, Leidig E, Meyer S, Baan A, Shamdeen MG, Berrang J, Graf N. (2013).Use of complementary and alternative medicine in healthy children and children with chronic medical conditions in Germany. *Complementery Therapies in Medicine* 21(Suppl 1): S61–S69.

Heinrich, M, Barnes J, Gibbons S and Williamson EM. (2012) *Fundamentals of Pharmacognosy and Phytotherapy*. 2nd Edition. Churchill Livingston Elsevier, Edinburgh and London.

Heinrich, M., Chan J, Wanke S, Neinhuis Ch., Simmonds MSS. (2009) Local uses of *Aristolochia* species and content of Aristolochic acid 1 and 2 – a global assessment based on bibliographic sources. *Journal of Ethnopharmacology* 125(1): 108–144.

Jordan SA, Cunningham DG, Marles RJ. (2010) Assessment of herbal medicinal products: challenges, and opportunities to increase the knowledge base for safety assessment. *Toxicology and Applied Pharmacology* 243(2): 198–216.

Kim L, Wollner D, Anderson P, Brammer D. (2004) Echinacea for Treating Colds in Children. *JAMA* 291(11): 1323–1324.

Lim A Cranswick N, South M. (2011) Adverse events associated with the use of complementary and alternative medicine in children. *Archives of Diseases in Childhood* 96(3): 297–300.

Linde K, Berner MM, Kriston L. (2008) St John's Wort for major depression. *Cochrane Database of Systematic Reviews* (4): CD000448.

Liu X, Wang Q, Song G, Zhang G, Ye Z, Williamson EM. (2013) The Classification and Application of Toxic Chinese *Materia Medica*. *Phytotherapy Research* 28(3): 334–347.

MHRA. (2013) Toxic poisoning alert for online Chinese medicines http://www.mhra.gov.uk /home/groups/comms-po/documents/news/con239404.pdf (accessed July 2013).

Moerman DE, Jonas WB. (2002) Deconstructing the placebo effect and finding the meaning response. *Annals of Internal Medicine* 136(6): 471–476.

Schulz V, Haensel R, Blumenthal M, Tyler V. (2004). *Rational Phytotherapy: A Reference Guide for Physicians and Pharmacists*. 5th Edition. Springer-Verlag, Berlin.

Singhuber J, Zhu M, Prinz S, Kopp B. (2009) Aconitum in Traditional Chinese Medicine - A valuable drug or an unpredictable risk? *Journal of Ethnopharmacology* 126(6): 18–30.

Taylor JA, Weber W, Standish L, Quinn H, Goesling J, McGann M, Calabrese C. (2003) Efficacy and Safety of *Echinacea* in Treating Upper Respiratory Tract Infections in Children: A Randomized Controlled Trial. *JAMA* 290(21): 2824–2830.

Williamson EM, Lorenc A, Booker A, Robinson N. (2013a).The rise of Traditional Chinese Medicine and its *materia medica*: a comparison of frequency and safety of materials and species used in Europe and China. *Journal of Ethnopharmacology* 149(2); 453–462.

Williamson EM, Driver S, Baxter K. (Eds.) (2013b) *Stockley's Herbal Medicines Interactions*. 2nd Edition. Pharmaceutical Press, London, UK.

Witt CM. (2013) Clinical research on traditional drugs and food items – the potential of comparative effectiveness research for interdisciplinary research. *Journal of Ethnopharmacology* 147(1): 254–8.

How to use

Some notes on how to use these monographs: The monographs in the following section are for drugs which most importantly are widely sold throughout the United Kingdom, either because they:

- are registered under the MHRA's scheme as a 'Traditional Herbal Medicinal Product' (THR);
- have a *product licence* (PL);
- are widely available as a *food supplement* with some specific claims;
- are a 'herbal health product' widely available via the Internet or in specialised shops;
- or, in a few cases, are products also sold as a *medical device*.

In general, the availability of PL or THR products are indicated (top right of the monograph), but this does not imply that *all* products on the market are regulated. Species sold as food supplements of an uncertain status are not indicated. These botanical drugs are generally also important in other European countries and in many other medical systems including the United States, Canada, Australia and New Zealand. If a botanical drug is important in traditional Chinese medicine (TCM) or Ayurveda this is stated. If major toxicological concerns have been raised, this is also indicated, but this will, of course, vary depending on the exact chemical composition of the preparation and there may be (generally registered) products on the market where the toxic metabolites have been largely removed (e.g. as in the case of butterbur, *Petasites hybridus*).

With the huge number of botanical drugs available, this has to be a selection of the most widely used species.

Excluded are all products which are not of plant or fungal origin, or which are simply sold as a food without any specific health claim.

In the top part of each monograph, key data about the drug are summarised:

- *Scientific* and (the most widely used) *common name*.
- *Synonyms*: these are Latin names under which information on these species may also have been published and are essential both for identifying which material may be in a product or for finding additional information.
- *Family*: The plant (or fungal) family that the species belongs to.
- *Other common name(s)*: Again, additional common names are included if they are frequently used in order to allow a tentative identification.
- *Drug name*: In international trade, for botanical drugs yielding licensed or registered medicines, and in many medical traditions, Latin drug names are used. They are generally found in pharmacopoeias but also in many scientific works. If a drug is included in a pharmacopeia (and, therefore, is seen as a medical substance) and has had a drug name assigned, we include it for ease of reference.

Phytopharmacy: An evidence-based guide to herbal medicinal products, First Edition.
Sarah E. Edwards, Inês da Costa Rocha, Elizabeth M. Williamson and Michael Heinrich.
© 2015 John Wiley & Sons, Ltd. Published 2015 by John Wiley & Sons, Ltd.

- *Botanical drug used*: The plant part commonly used and for which scientific evidence is relevant is listed.
- *Main chemical compounds*: In this section we list all main classes of compounds which are relevant for the botanical drug's pharmacological activity, or which may be used as a marker substance used in the pharmacognositc identification of the drug.

While (as pointed out in 'The Evidence Base for Herbal Medicines'), in most cases the evidence for the drug's use is based on a traditional use, the level of pre-clinical and clinical evidence varies. This classification system is based on available evidence and not limited to clinical evidence based on efficacy. Of course, the lack of evidence is not the same as lack of efficacy. We use a five-scale system in order to indicate this:

Rating system for available evidence

♣ ♣ ♣ ♣ ♣	Generally, there are products on the market which have a full marketing authorisation. Considerable clinical evidence available for benefits reported on its main uses, including systematic reviews/meta-analysis.
♣ ♣ ♣ ♣ ♣	Robust clinical evidence available for benefits reported on its main uses, including case-controlled studies, cohort studies and randomised controlled trial.
♣ ♣ ♣ ♣ ♣	Limited clinical evidence available for benefits on its main uses for products which are available on the UK and other European markets.
♣ ♣ ♣ ♣ ♣	No relevant clinical evidence reported but *in vivo* pharmacological data support the main uses or observations studies.
♣ ♣ ♣ ♣ ♣	Use is plausible but not based on sound pharmacological or clinical evidence. This group also includes botanical drugs which are on the market without any evidence for these products and for which no data on quality assurance could be found. Other preparations from the same species or from complex multi-component preparations may have been studied clinically, but these data are not transferable to the products commonly sold (often via the Internet).

In general, this provides an assessment of what level of pharmacological and clinical evidence is available, and it is intended to help heath care professionals and expert patients to decide what plausible scientific and clinical base there is for using this particular drug. It does not take into consideration the chemical composition of a specific product and, therefore, must be read in combination with the section where the chemical composition and pre-clinical/clinical evidence is discussed. Consequently, it cannot be used to assess a specific product, but *provides information on a botanical drug and the extracts derived from it*.

The ranking does highlight that in many cases considerable evidence is available, even though there are only very few cases where there is *sufficient clinical evidence* (cannabis, St. John's wort, capsicum, senna). There are many cases where robust data are available, like devil's claw, echinacea, ivy and pelargonium. This ranking is not necessarily an endorsement of the use for a certain drug, but recognises the level of evidence available. In a few cases, there is considerable evidence, but there may be concerns about the drug's safety. Most importantly, the ranking highlights that the evidence for botanical drugs commonly sold varies and that there are now many products on the market which can be recommended based on clinical and pre-clinical evidence.

Next, each monograph includes a box with a key summary on data relevant to everyday practice. It focuses on the following aspects:

- *Indications*: Here both approved (e.g. via the THR scheme) and popular uses are given. Therefore, this is not an endorsement of a certain use, but simply a recognition that it is used for these indications.
- *Evidence*: A short assessment summary on the evidence available or the lack thereof is given, which is discussed in more detail under *Clinical Evidence* and *Pre-clinical evidence and mechanisms of action*.
- *Safety*: Again, this is a short assessment and summary of data discussed in more detail under *Interactions, Contraindications, Adverse effects*.

The next part of the monograph provides more detailed information including references to some key works which include further data on *Clinical evidence* and *Pre-clinical evidence* as well as on the drug's safety – *Interactions, Contraindications, Adverse effects*. These sections are intended to allow a more detailed evaluation of the existing clinical and scientific information on each drug. In some cases, research has centered on one or two chemically well-defined extracts (e.g. St. John's wort, maritime pine bark) and if this is the case, we spell this out.

Some information on the dosage is included. This may either be based on the recommended doses, which have been developed as part of clinical or other studies or be based on empirical, but less well-substantiated data.

Also included are some additional relevant data of interest and key references.

Açaí   *Euterpe oleracea* Mart.

Family: Arecaceae (Palmae)

Other common names: Cabbage palm fruit

Botanical drug used: Fruit and pressed fruit juice

Indications/uses: Açaí products are marketed to produce rapid weight loss, improve digestion, prevent cardiovascular disease and other degenerative diseases and to 'delay the ageing process', as an antioxidant.

Evidence: Claims about alleged health benefits can only be substantiated to a very limited degree.

Safety: Overall, açaí fruit is considered to be safe although very limited data are available. High levels of manganese have been identified in the fruit, which may be above the daily recommended dose and thus might negatively impact on iron absorption (da Silva Santos et al. 2014).

Main chemical compounds: The main chemical compounds of açaí include a range of fatty acids and polyphenols, most notably anthocyanins and flavonoids. The main polyunsaturated, monounsaturated and saturated fatty acids include oleic (56.2%), palmitic (24.1%) and linoleic acid (12.5%) (Schauss et al. 2006). The main anthocyanins are cyanidin, delphinidin, malvidin, pelargonidin and peonidin and the other phenolics include catechin, ferulic acid, quercetin and resveratrol (Poulose et al. 2012).

Clinical evidence: Overall, the clinical evidence is weak (Heinrich et al. 2011). In an uncontrolled pilot study, consumption of açaí fruit pulp reduced levels of selected markers of metabolic disease risk in overweight adults, indicating that further studies are warranted (Udani et al. 2011). Antioxidant and lipid peroxidation levels in 12 volunteers in a randomised, placebo-controlled crossover experiment showed statistically significant positive changes both 1 hour and 2 hours after consumption. The same product was also tested for its effects in 14 volunteers with reduced range of motion (RoM) and produced a significant pain reduction and an improvement in measures of RoM and other activities of daily living (Jensen et al. 2011).

Phytopharmacy: An evidence-based guide to herbal medicinal products, First Edition.
Sarah E. Edwards, Inês da Costa Rocha, Elizabeth M. Williamson and Michael Heinrich.
© 2015 John Wiley & Sons, Ltd. Published 2015 by John Wiley & Sons, Ltd.

Pre-clinical evidence and mechanisms of action: In rabbits, an improvement of lipid profile and an attenuation of atherosclerosis were observed after açaí administration (Feio et al. 2012). Açaí pulp promoted a hypocholesterolaemic effect in a rat model of dietary-induced hypercholesterolaemia, through an increase in the expression of ATP-binding cassette, subfamily G transporters and LDL-R genes (de Souza et al. 2012). Açaí pulp extracts also produced a significant concentration-dependent reduction in cyclooxygenase-2, p38 mitogen-activated protein kinase, tumour necrosis factor-α (TNF-α) and nuclear factor-κB (NF-κB) in a mouse brain BV-2 microglial cells, suggesting a protective effect of açaí pulp fractions on brain cells, which could have implications for improved cognitive and motor functions (Poulose et al. 2012).

Interactions: No clinical reports of any interactions are available.

Contraindications: No specific contraindications are known. Based on da Silva Santos et al. (2014), a daily consumption of 300 ml of açaí pulp would lead to a six-fold excess of the recommended daily intake of manganese for an adult, which in turn would negatively impact on iron absorption.

Adverse effects: No data available.

Dosage: For products, follow manufacturers' instructions. No adequate dose-finding studies have been conducted. The use of 120 ml fruit juice per day has been suggested, but evidence for this dose is lacking.

General plant information: A native of northeastern Brazil, this palm is one of the most naturally abundant species in the Brazilian Amazon's estuary floodplains. It has become a 'poster child' of the power of the Internet to promote food supplements or HMPs (Heinrich et al. 2011).

In Brazil, as the fruits are commonly prepared using lukewarm water (i.e. not boiled), attention has been drawn to the potential risk of outbreaks of Chagas disease (*Trypanosoma cruzi* infections) transmitted by triatomines (especially *Rhodnius pictipes*), which have been found to contaminate açaí fruit with faeces or with dead insects (Nóbrega et al. 2009).

References

Feio CA, Izar MC, Ihara SS, Kasmas SH, Martins CM, Feio MN, Maues LA, Borges NC, Moreno RA, Povoa RM, Fonseca FA. (2012) *Euterpe oleracea* (açaí) modifies sterol metabolism and attenuates experimentally-induced atherosclerosis. *Journal of Atherosclerosis and Thrombosis* 19(3): 237–245.

Heinrich M, Dhanji T, Casselman I. (2011) Açai (*Euterpe oleracea* Mart.) – a phytochemical and pharmacological assessment of the species' health claims. *Phytochemistry Letters* 4 (1): 10–21.

Jensen G, Ager DM, Redman K. A., Mitzner MA, Benson KF, Schauss AG. (2011) Pain reduction and improvement in range of motion after daily consumption of an açaí (*Euterpe oleracea* Mart.) pulp-fortified polyphenolic-rich fruit and berry juice blend. *Journal of Medicinal Food* 14(7–8): 702–711.

Nóbrega AA, Garcia MH, Tatto E, Obara MT, Costa E, Sobel Jeremy, Araujo WN. (2009) Oral transmission of Chagas disease by consumption of Açaí palm fruit, Brazil. *Emerging Infectious Diseases* 15(4): 653–655.

Poulose SM, Fisher DR, Larson J, Bielinski DF, Rimando AM, Carey AN, Schauss AG, Shukitt-Hale B. (2012) Anthocyanin-rich açai (*Euterpe oleracea* Mart.) fruit pulp fractions attenuate inflammatory stress signaling in mouse brain BV-2 microglial cells. *Journal of Agricultural and Food Chemistry* 60(4): 1084–1093.

Schauss AG, Wu RL, Ou B, Patel D, Huang D, Kababick JP. 2006 Phytochemical and nutrient composition of the freeze-dried Amazonian palm berry *Euterpe oleracea* Mart. (açaí). *Journal of Agricultural and Food Chemistry* 54(22): 8598–8603.

da Silva Santos V, Henrique de Almeida Teixeira G, Barbosa Jr. F. (2014) Açaí (*Euterpe oleracea* Mart.): a tropical fruit with high levels of essential minerals – especially manganese – and its contribution as a source of natural mineral supplementation. *Journal of Toxicology and Environmental Health, Part A: Current Issues* 77 (1–3): 80–89.

de Souza MO, Souza E, Silva L, de Brito Magalhães CL, de Figueiredo BB, Costa DC, Silva ME, Pedrosa ML. (2012) The hypocholesterolemic activity of açaí (*Euterpe oleracea* Mart.) is mediated by the enhanced expression of the ATP-binding cassette, subfamily G transporters 5 and 8 and low-density lipoprotein receptor genes in the rat. *Nutrition Research* 32(12): 976–84.

Udani JK, Singh BB, Singh VJ, Barrett ML. (2011) Effects of Açaí (*Euterpe oleracea* Mart.) berry preparation on metabolic parameters in a healthy overweight population: a pilot study. *Nutrition Journal* 10:45.

Aloe Vera (Gel)

Aloe vera (L.) Burm.f., *A. arborescens* Mill. and other *Aloe spp.*

Synonym: *A. barbadensis* Mill.

Family: Xanthorrhoeaceae (Aloaceae); the genus Aloe has also been placed in the Liliaceae and Asphodelaceae

Other common names: *A. arborescens* is also known as krantz aloe

Drug name: Aloe vera; aloe vera 'juice'

Botanical drug used: Gel extracted from the internal tissues of the succulent leaf.

Indications/uses: Aloe vera gel is applied externally to treat skin irritation (such as from insect bites), burns, psoriasis, wounds, radiation dermatitis and frost-bite. It is also frequently found in cosmetic preparations, usually at low doses. Internally, aloe 'juice' is taken as a general tonic to enhance the immune system and to treat constipation. Aloe has also reported antidiabetic, anticancer and antibiotic properties.

N.B.: Aloe plants also yield a resinous exudate known as 'aloes' or 'bitter aloes', which is rich in anthraquinones and a strong laxative. These preparations are no longer recommended due to their potential genotoxic and mutagenic effects.

Evidence: Most clinical uses are based on anecdotal data. Data from small trials suggests effectiveness in healing first- and second-degree burns and increased survival of frost-bitten tissue. Aloe has been demonstrated to lower blood glucose and lipid levels in diabetic patients with hyperlipidaemia and patients with metabolic syndrome (see the subsequent sections).

Safety: External use of aloe vera gel appears to be safe, but there is limited data available on its safety for internal use. Large doses and long-term use should be avoided due to risk of nephritis (kidney inflammation), and studies in mice and rats have indicated genotoxicity and carcinogenicity (see the subsequent sections).

Phytopharmacy: An evidence-based guide to herbal medicinal products, First Edition.
Sarah E. Edwards, Inês da Costa Rocha, Elizabeth M. Williamson and Michael Heinrich.
© 2015 John Wiley & Sons, Ltd. Published 2015 by John Wiley & Sons, Ltd.

Main chemical compounds: Polysaccharides (e.g. acemannan in *A. vera* and aloemannan in *A. arborescens*); glycoproteins (lectins) such as aloctin A and B, enzymes such as carboxypeptidases (Harlev et al. 2012; Reynolds and Dweek 1999; Rodríguez et al. 2010); hexadecanoic acid and sterols including sitosterol and stigmasterol (Bawankar et al. 2013). Traces of anthraquinones, such as aloin and aloe emodin, from cells of the fibrovascular bundles of leaf are present in varying amounts (Reynolds and Dweck 1999).

Clinical evidence: There is a lack of evidence based on high quality clinical trials to support the topical use of aloe for treating acute or chronic wounds (Dat et al. 2012). Despite the paucity of evidence, data from a few other small trials suggest that topical *A. vera* may be effective in healing first- and second-degree burns, although these conclusions should be treated with caution (Maenthaisong et al. 2007).

In two randomized double-blind placebo-controlled clinical trials, aloe leaf gel demonstrated significant lowering of blood glucose and cholesterol levels, including low-density lipoprotein cholesterol (LDL) levels in hyperlipidimic type 2 diabetic patients and patients with pre-diabetes/metabolic syndrome (Deveraj et al. 2013; Huseini et al. 2012). In another trial, oral aloe gel produced a beneficial clinical effect greater than placebo in patients with active ulcerative colitis (Langmead et al. 2004).

Pre-clinical evidence and mechanisms of action: *In vivo* studies in animals have indicated that *A. vera* may accelerate wound healing and prevent progressive dermal ischaemia caused by burns, frostbite and electrical injury by inhibition of thromboxane A_2 (Heggers et al. 1993). Topical application of aloe has been shown to significantly enhance survival of frost-bitten tissue in both rabbit ears and in a clinical trial in humans (Reynolds and Dweck 1999). Antipsoriatic effects have been demonstrated in mice (Dhanabal et al. 2012) and the polysaccharides and glycoproteins have *in vitro* and *in vivo* immunomodulating and anticancer effects (Hevrev et al. 2012); hexadecanoic acid has antifungal effects (Bawankar et al. 2013).

Interactions: A report of an interaction suggests that taken internally, 'aloe vera tablets' may interact with the anaesthetic sevoflurane, increasing its antiplatelet effects (Lee et al. 2004). However, the composition of the preparation was not stated and no other interactions have been recorded.

Contraindications: Its contraindications include allergy to any of the constituents. While external use during pregnancy is not thought to be of concern, internal use of aloe vera gel and whole leaf aloe extracts should be avoided during pregnancy and breastfeeding, due to their gastrointestinal stimulant and genotoxic effects (Rodrigues et al. 2012).

Adverse effects: Taken internally in large doses, aloe preparations are known to cause nephritis. Two cases of acute renal failure have been reported following aloe administration. It may also precipitate allergic reactions and has been associated with gastrointestinal complications (colonic injury) (Evangelos et al. 2005). Two acute cases of hepatitis have been associated with internal use of *A. vera* (Kanat et al. 2006). An *in vivo* study in mice demonstrated a weak enhancing effect of aloe gel and aloe whole leaf on the photocarcinogenicity of simulated sun light (National Toxicology Program 2010). Other *in vivo* studies have indicated that *A. vera* whole leaf extract is a gastrointestinal irritant in mice and rats, and carcinogenic in the large intestine of rats (Boudreau et al. 2012).

Dosage: For external use, it can be applied liberally; for internal use, manufacturers' instructions should be followed and large doses or long-term use should be avoided.

General plant information: Succulent *Aloe* plants have been used to treat wounds since ancient times and are mentioned in the Ebers papyrus, an important medical text of ancient Egypt, dating from 1550 BCE (Dat et al. 2012).

References

Bawankar R, Deepti VC, Singh P, Subashkumar R, Vivekanandhan G, Babu S. (2013) Evaluation of bioactive potential of an *Aloe vera* sterol extract. *Phytotherapy Research* 27(6): 864–868.

Boudreau MD, Mellick P, Olsen G, Felton R, Thorn B, Beland F. (2012) Clear evidence of carcinogenic activity by a whole leaf extract of *Aloe barbadensis* Miller (*Aloe vera*) in F344/N rats. *Toxicological Sciences* 131(1): 26–39.

Dat AD, Poon F, Pharm KB, Doust J. (2012) *Aloe vera* for treating acute and chronic wounds. *Cochrane Database of Systematic Reviews* (2): CD008762.

Deveraj S, Yimam M, Brownell LA, Jialal I, Singh S, Jia Q. (2013) Effects of *Aloe vera* supplementation in subjects with pre-diabetes/metabolic syndrome. *Metabolic Syndrome and Related Disorders* 11(1): 35–40.

Dhanabal SP, Dwarampudi LP, Muruganantham N, Vadivelan R. (2012) Evaluation of the antipsoriatic activity of *Aloe vera* leaf extract using a mouse tail model of psoriasis. *Phytotherapy Research* 26(4): 617–619.

Evangelos C, Spyros K, Spyros D. (2005) Henoch-Schonlein purpura associated with *Aloe vera* administration. *European Journal of Internal Medicine* 16(1): 59–60.

Harlev E, Nevo E, Lansky EP, Ofir R, Bishayee A. (2012) Anticancer potential of Aloes: antioxidant, antiproliferative, and immunostimulatory attributes. *Planta Medica* 78(9): 843–852.

Heggers JP, Pelley RP, Robson MC. (1993) Beneficial effects of *Aloe* in wound healing. *Phytotherapy Research* 7(7): S48–S52.

Huseini HF, Kianbakht F, Hajiaghaee R, Dabaghian FH. (2012) Anti-hyperglycemic and anti-hypercholesterolemic effects of *Aloe vera* leaf gel in hyperlipidemic type 2 diabetic patients: a randomized double-blind placebo-controlled clinical trial. *Planta Medica* 78(4): 311–316.

Kanat O, Ozet A, Ataergin S. (2006) *Aloe vera*-induced acute toxic hepatitis in a healthy young man. *European Journal of Internal Medicine* 17(8): 589.

Langmead L, Feakins RM, Goldthorpe S, Holt H, Tsironi E, De Silva A, Jewell DP, Rampton DS. (2004) Randomized, double-blind, placebo-controlled trial of oral *Aloe vera* gel for active ulcerative colitis. *Alimentary Pharmacology & Therapeutics* 19(7): 739–747.

Lee A, Chui PT, Aun CS, Gin T, Lau AS. (2004) Possible interaction between sevoflurane and *Aloe vera*. *Annals of Pharmacotherapy* 38(10): 1651–1654.

Maenthaisong R, Chaiyakunapruk N, Niruntraporn S, Kongkaew C. (2007) The efficacy of Aloe vera used for burn wound healing: A systematic review. *Burns* 33(6): 713–718.

National Toxicology Program. (2010) Photocarcinogenesis study of *Aloe vera* [CAS NO. 481-72-1(Aloe-emodin)] in SKH-1 mice (simulated solar light and topical application study). *National Toxicology Program Technical Report Series* (553): 7–33; 35–97; 99–103.

Reynolds T, Dweck AC. (1999) *Aloe vera* gel: a review update. *Journal of Ethnopharmacology* 68(1–3): 3–37.

Rodríguez RE, Darias MJ, Díaz RC. (2010) *Aloe vera* as a functional ingredient in foods. *Critical Reviews in Food Science and Nutrition* 50(4): 305–326.

Arnica   *Arnica montana* L.

Synonyms: *A. alpina* Willd. ex Steud.; *A. helvetica* G.Don ex Loudon; *Cineraria cernua* Thore

Family: Asteraceae (Compositae)

Other common names: Leopard's bane; mountain tobacco; wolf's bane

Drug name: Arnicae flos

Botanical drug used: Flowers

Indications/uses: Used topically to treat sprains, bruises, inflammation, fracture oedema and muscle pain; may support treatment of rheumatic diseases.

*N.B.: Arnica is often taken orally **as a homoeopathic preparation** to treat or prevent bruising, including after surgery, but a clinical study for this use did not show efficacy.*

Evidence: There is some clinical evidence and a long history of use to support arnica for reducing bruising, but not for other indications. Its use is plausible based on the main active compounds (the sesquiterpene lactones), which are known anti-inflammatory agents and are absorbed through the skin.

Safety: If taken internally arnica is poisonous, and potentially fatal.
N.B.: Homoeopathic arnica tablets are not toxic.

Main chemical compounds: Sesquiterpene lactones (including helenalin and arnifolin), which are believed to be mainly responsible for anti-inflammatory attributes; flavonoids (including eupafolin, hispidulin, kaempferol and quercetin); essential oil (thymol and its derivatives); caffeic acids; pyrollizidine alkaloids (Pharmaceutical Press Editorial Team 2013; Williamson et al. 2013).

Clinical evidence: In a randomised, double-blind, placebo-controlled trial, laser-induced bruising was shown to be reduced more effectively than placebo by a topical 20% arnica ointment (Leu et al. 2010). Although arnica is reputed to decrease muscle pain, there is no clinical evidence to support this. In fact, in one randomised, double-blind, placebo-controlled trial investigating the effect of topical arnica on muscle pain, arnica was found to *increase* muscle pain after 24 h, although there was no significant difference at 48 h (Adkinson et al. 2010).

Phytopharmacy: An evidence-based guide to herbal medicinal products, First Edition.
Sarah E. Edwards, Inês da Costa Rocha, Elizabeth M. Williamson and Michael Heinrich.
© 2015 John Wiley & Sons, Ltd. Published 2015 by John Wiley & Sons, Ltd.

Pre-clinical evidence and mechanisms of action: Antioxidant activity and cyto-protective effects against oxidative damage have been demonstrated in a study using arnica extract on mice fibroblast-like cells, supporting the traditional use of arnica for skin disorders (Craciunescu et al. 2012). Sesquiterpene lactones in arnica inhibit transcription factors NF-kappaB and NF-AT at micromolar concentrations, an effect which is potent, but also comparatively non-specific (Bremner and Heinrich 2002; Merfort 2003). These transcription factors are responsible for regulation of genes that modulate inflammation.

Interactions: None known.

Contraindications: Arnica should not be taken internally (except in homeopathic dilutions), and should not be used on broken skin. As the active compounds are absorbed through the skin, use of arnica should also be avoided during pregnancy and lactation due to lack of safety data. Arnica may cause allergic reactions in sensitive individuals who should discontinue its use immediately.

Adverse effects: Arnica may cause contact dermatitis. It is an irritant to mucous membranes and may be fatal if taken internally.

Dosage: For external use only, see manufacturers' instructions.

General plant information: A native to mountainous regions of Central and Southern Europe, wild populations of *A. montana* are decreasing due to habitat loss and over-exploitation. In many places it is a protected species. It is cultivated in Germany for medicinal use.

References

Adkinson JD, Bauer DW, Chang T. (2010) The effect of topical arnica on muscle pain. *Annals of Pharmacotherapy* 44(10): 1579–1584.

Bremner P, Heinrich M. (2002) Natural products as targeted modulators of the nuclear factor-kappaB-pathway. *Journal of Pharmacy and Pharmacology* 54(4): 453–472.

Craciunescu O, Constantin D, Gaspar A, Toma L, Utioiu E, Moldovan L. (2012) Evaluation of antioxidant and cytoprotective activities of *Arnica montana* L. and *Artemisia absinthium* L. ethanolic extracts. *Chemistry Central Journal* 6(1): 97.

Leu S, Havey J, White LE, Martin N, Yoo SS, Rademaker AW, Alam M. (2010) Accelerated resolution of laser-induced bruising with topical 20% arnica: a rater-blinded randomized controlled trial. *British Journal of Dermatology* 163(3): 557–563.

Merfort I. (2003) Arnica: new insights on the molecular mode of action of a traditional medicinal plant. *Forschende Komplementärmedizin und klassische Naturheilkunde* 10(Suppl 1): 45–48.

Pharmaceutical Press Editorial Team. (2013) *Herbal Medicines*, 4th Edition. Pharmaceutical Press, London, UK.

Williamson EM, Driver S, Baxter K. (Eds.) (2013) *Stockley's Herbal Medicines Interactions*. 2nd Edition. Pharmaceutical Press, London, UK.

Artichoke   *Cynara cardunculus* L.

Synonym: *C. scolymus* L.

Family: Asteraceae (Compositae)

Other common names: Cynara; globe artichoke

Drug name: Cynarae folium

Botanical drug used: Leaf

Indications/uses: Used for the symptomatic relief of digestive disorders such as dyspepsia and also irritable bowel syndrome.

Evidence: There is some clinical evidence supporting its use in dyspepsia, irritable bowel syndrome and hypercholesterolaemia, although larger trials are needed before any conclusive recommendations can be drawn.

Safety: Overall it is considered to be safe, although it should be avoided in cases of allergenicity to the daisy family of plants.

Main chemical compounds: The main constituents of the leaf are phenolic acids, mainly chlorogenic acid, cynarin and caffeic acid. Sesquiterpene lactones such as cynaropicrin, flavonoids including scolymoside and luteolin, and phytosterols (e.g. taraxasterol) are also present along with sugars, inulin and a small amount of volatile oil (EMA 2011a).

Clinical evidence:

Functional dyspepsia and irritable bowel: A cohort study conducted in outpatients with a clinical diagnosis of functional dyspepsia showed that a commercially available product containing artichoke significantly reduced the symptom severity associated with dyspepsia, along with a modest decrease in total cholesterol, low-density lipoprotein (LDL) cholesterol, and triglyceride levels after 60 days (Sannia 2010). Similar results were found in a double-blind, randomised, controlled clinical trial (247 patients) in patients with functional dyspepsia. Patients were given a commercial artichoke leaf preparation (two capsules of 320 mg plant extract, three times a day) or a placebo. After 6 weeks, there was a greater improvement of the dyspeptic symptoms compared to placebo with a substantial impact on the disease-specific quality of life (Holtmann et al. 2003). Another open, dose-ranging study using artichoke leaf extract (320 or 640 mg daily for 2 months) in over 500 patients also showed a clear improvement of symptoms (gastro-intestinal symptoms and improved quality of life) during the treatment (Marakis et al. 2002).

The symptom overlap between dyspeptic syndrome and irritable bowel syndrome (IBS) suggests that artichoke extract may have potential for treating IBS as well. A subgroup of patients with IBS symptoms, identified from individuals with dyspeptic syndrome that were already being monitored in a post-marketing surveillance study of artichoke extract for 6 weeks, revealed significant reductions in the severity of symptoms as assessed by both physicians and patients. Furthermore, 96% of patients rated artichoke extract as 'better than' or 'at least equal to' their previous treatment for those symptoms, and the tolerability was very good (Walker et al. 2001). However, further, larger clinical studies are needed to conclusively support its use in dyspepsia and IBS.

Hypercholesterolaemia: A Cochrane review on the use of artichoke leaf extract (1280–1920 mg a day between 6 and 12 weeks) for treating hypercholesterolaemia included only randomised clinical trials (262 participants) on artichoke leaf extract as a single-herb preparation. The authors concluded that there is some evidence of a cholesterol-lowering effect, although the effect is modest (Wider et al. 2013). In an 8-week intervention trial (92 subjects), the use of artichoke leaf extract (250 mg twice a day) was shown to be efficacious in increasing the high-density lipoprotein (HDL) cholesterol levels, while decreasing the total cholesterol and LDL cholesterol levels, as well as being well tolerated compared to placebo in subjects with primary mild hypercholesterolaemia (Rondanelli et al. 2013a). Artichoke leaf extract also reduced fasting glucose levels and improved blood lipid patterns in overweight subjects (Rondanelli et al. 2013b).

Again, further studies are needed for developing clear recommendations.

Pre-clinical evidence and mechanisms of action: The mechanism of action is unclear but artichoke has antioxidant, hepato- and cardioprotective and choleretic effects (stimulation of bile secretion) as well as cholesterol-lowering effect (EMA 2011a). Animal studies with artichoke leaf extract decreased the levels of plasma cholesterol and triglycerides and prevented the development of atherosclerotic plaque. This mechanism of action might be due to its capacity of reducing the LDL oxidation and by its cholesterol synthesis inhibitory effect. Luteolin was identified as one major compound responsible (Wider et al. 2013). An ethanol extract of artichoke leaf showed beneficial effects on acute hemorrhagic gastritis in rats by gastric mucus glycoprotein-increasing action. Furthermore, its anti-gastritic action is mainly due to cynaropicrin and not chlorogenic acid (Ishida et al. 2010).

Interactions: None known.

Contraindications: Not recommended during pregnancy and lactation due to lack of safety data. Not recommended in children under 12 years of age. Should not be used in cases of occlusion of the bile ducts, cholangitis, gallstones and any other biliary diseases and hepatitis. A health care professional should be consulted if symptoms persist for longer than 2 weeks. Hypersensitivity to active substance or to plants from the Asteraceae (EMA 2011b). A health care professional should be sought before its use in patients with gallstones.

Adverse effects: Slight diarrhoea with abdominal spasm, epigastric complaints such as nausea and heartburn as well as allergic reactions have been reported (EMA 2011b).

Dosage: As recommended by the manufacturer. As comminuted dried leaves 1.5–3 g as a herbal tea one–four times a day up to 6 g a day. As an extract 600–2400 mg a day divided into doses (EMA 2011b). They are usually standardised to the caffeoylquinic acid derivative, chlorogenic acid (Williamson et al. 2013).

General plant information: Artichoke flowers are used as food while its extracts are used as flavouring agents. It has been cultivated in Europe and used as food and medicine since Greek and Roman times (EMA 2011a).

References

EMA. (2011a) Assessment report on *Cynara scolymus* L., folium. European Medicines Agency http://www.ema.europa.eu/docs/en_GB/document_library/Herbal_-_HMPC_assessment_report/2011/12/WC500119940.pdf (accessed August 2014).

EMA. (2011b) Community herbal monograph on *Cynara scolymus* L., folium. European Medicines Agency http://www.ema.europa.eu/docs/en_GB/document_library/Herbal_-_Community_herbal_monograph/2011/12/WC500119942.pdf (accessed August 2014).

Holtmann G, Adam B, Haag S, Collet W, Grunewald E, Windeck T. (2003) Efficacy of artichoke leaf extract in the treatment of patients with functional dyspepsia: a six-week placebo-controlled, double-blind, multicentre trial. *Alimentary Pharmacology & Therapeutics* 18(11–12): 1099–1105.

Ishida K, Kojima R, Tsuboi M, Tsuda Y, Ito M. (2010) Effects of artichoke leaf extract on acute gastric mucosal injury in rats. *Biological and Pharmaceutical Bulletin* 33(2): 223–229.

Marakis G, Walker AF, Middleton RW, Booth JC, Wright J, Pike DJ. (2002) Artichoke leaf extract reduces mild dyspepsia in an open study. *Phytomedicine* 9(8): 694–699.

Rondanelli M, Giacosa A, Faliva MA, Sala P, Perna S, Riva A, Morazzoni P, Bombardelli E. (2013a) Beneficial effects of artichoke leaf extract supplementation on increasing HDL-cholesterol in subjects with primary mild hypercholesterolaemia: a double-blind, randomized, placebo-controlled trial. *International Journal of Food Sciences and Nutrition* 64(1): 7–15.

Rondanelli M, Opizzi A, Faliva MA, Sala P, Perna S, Riva A, Morazzoni P, Bombardelli E, Giacosa A. (2013b) Metabolic management in overweight subjects with naive impaired fasting glycaemia by means of a highly standardized extract from *Cynara scolymus*: a double-blind placebo-controlled, randomized clinical trial. *Phytotherapy Research* 28(1): 33–41.

Sannia A. (2010) Phytotherapy with a mixture of dry extracts with hepato-protective effects containing artichoke leaves in the management of functional dyspepsia symptoms. *Minerva Gastroenterologica e Dietologica* 56(2): 93–99.

Walker AF, Middleton RW, Petrowicz O. (2001) Artichoke leaf extract reduces symptoms of irritable bowel syndrome in a post-marketing surveillance study. *Phytotherapy Research* 15(10): 58–61.

Wider B, Pittler MH, Thompson-Coon J, Ernst E. (2013) Artichoke leaf extract for treating hypercholesterolaemia. *Cochrane Database of Systematic Reviews* 3: CD003335.

Williamson EM, Driver S, Baxter K. (Eds.) (2013) *Stockley's Herbal Medicines Interactions.* 2nd Edition. Pharmaceutical Press, London, UK.

Ashwagandha *Withania somnifera* (L.) Dunal

Synonyms: *Physalis somnifera* L.; *W. kansuensis* K.Z.Kuang & A.M.Lu; *W. microphysalis* Suess.

Family: Solanaceae

Other common names: Asagandh; Indian ginseng; winter cherry

Drug name: Withania somnifera radix

Botanical drug used: Roots; more rarely leaf and berries

Indications/uses: Ashwagandha is considered to be an adaptogen in herbal medicine and a Rasayana in Ayurveda, that is to enhance the body's ability to respond to stress and have rejuvenating effects. It is used as a sedative to treat anxiety, enhance memory, treat neurodegenerative diseases, reduce inflammation (including arthritis), lower blood pressure, modulate the immune system and improve fertility. More recently, it has been used as an adjunctive therapy for serious disorders including cancer and infectious diseases.

Evidence: Clinical trials and animal studies support the use of *W. somnifera* for anxiety, cognitive, neurological and degenerative disorders and inflammation. Its chemopreventive properties make it a potential useful adjunct for patients undergoing radiation and chemotherapy, although more research is needed.

Safety: Ashwagandha appears to be safe when taken orally short-term, but there is a lack of safety data for long-term use. Ashwagandha is not recommended for use during pregnancy/lactation or by patients with autoimmune disease, due to lack of data.

Main chemical compounds: The main active constituents of the root are the steroidal lactones, the withanolides, which include withaferin A, and sitoindosides IX and X (glycowithanolides) and acylsteryl glucosides including sitoindosides VII and VIII. Phytosterols and alkaloids such as tropine, pseudotropine, isopelletierine and anaferine are also present (Chen et al. 2011; Engels and Brinckmann 2013; Williamson 2002). A series of long-chain amides, the withanamides, has been isolated from the fruit (Jayaprakasam et al. 2010).

Phytopharmacy: An evidence-based guide to herbal medicinal products, First Edition.
Sarah E. Edwards, Inês da Costa Rocha, Elizabeth M. Williamson and Michael Heinrich.
© 2015 John Wiley & Sons, Ltd. Published 2015 by John Wiley & Sons, Ltd.

Clinical evidence:

Numerous clinical trials have been carried out on ashwagandha, and although most have yielded positive results, many of the studies are of poor quality, use multiherb ingredient formulae, or are too small to make definite conclusions about the efficacy of the herb.

Sedative, anti-ageing and anti-stress effects: Clinical data from a few small trials supports its use to some extent, as a sedative, anti-stress agent and to help alleviate anxiety. A clinical trial in 50–59-year-olds showed that it significantly improved haemoglobin, red blood cell count and hair melanin; and decreased serum cholesterol; erythrocyte sedimentation rate decreased and 71.4% reported increase in sexual performance. Other studies have given similar results (Engels and Brinckmann 2013).

Effects on the immune system: Two human pilot studies found that administration of ashwagandha in a herbal tea combined with other herbs increased NK (natural killer) cell activity compared to 'regular' tea (*Camellia sinsensis*) (Bhat et al. 2010). In a small uncontrolled study, Ashwagandha root extract increased immune cell activation (Mikolai et al. 2009).

Adjunctive use with cancer chemotherapy: An open-label, prospective, non-randomised comparative trial on 100 patients with breast cancer in all stages evaluated the use of either a combination of chemotherapy with oral *W. somnifera* or chemotherapy alone. The chemotherapy regimens were either taxotere, adriamycin, and cyclophosphamide or 5-fluorouracil, epirubicin and cyclophosphamide. Ashwangandha root extract was administered at a dose of 2 g every 8 hours, throughout the course of chemotherapy. Quality-of-life and fatigue scores were evaluated before, during and on the last cycles of chemotherapy using the EORTC QLQ-C30, Piper Fatigue Scale (PFS) and Schwartz Cancer Fatigue Scale (SCFS-6). Patients in the control arm experienced statistically significant higher fatigue scores compared with the study group ($p < 0.001$ PFS, $p < 0.003$ SCFS-6) and other symptoms were lower in 7 out of 18 symptoms in the intervention group compared with the control group ($p < 0.001$) (Biswal et al. 2013).

Anti-inflammatory effects: A small double-blind, placebo-controlled clinical trial of a herbal formula containing ashwagandha as the main ingredient showed that it significantly reduced pain in arthritis (Engels and Brinckmann 2013).

Pre-clinical evidence and mechanisms of action:

There are many pharmacological papers showing useful potential activity for ashwagandha; some of these are described in the subsequent sections but full details are available from the review references Williamson (2003); Vanden Berghe et al. (2012) and Chen et al. (2011).

Sedative, anti-ageing and anti-stress effects: An aqueous extract of *W. somnifera* root inhibited amyloid beta fibril formation *in vitro* (Kumar et al. 2012) and the withanamides (in *W. somnifera* fruit) have been shown to protect PC-12 cells from beta-amyloid responsible for Alzheimer's disease (Jayaprakasam et al. 2010).

Anti-inflammatory and immunodulating effects: A number of *in vitro* and *in vivo* studies have demonstrated anti-inflammatory and immune enhancing effects of ashwagandha and particularly withaferin A, which is one of the most bioactive compounds, exerting anti-inflammatory, proapoptotic but also anti-invasive and anti-angiogenic effects (reviewed in Vanden Berghe et al. 2012). In a mouse model of lupus, *W. somnifera* exerted a prophylactic effect on progression of the disease (Minhas et al. 2012).

Cytotoxicity: Although no clinical trials have evaluated the use of ashwagandha in oncology except as an adjunctive therapy (see the earlier sections), and cancer chemotherapy is outside the scope of this book, *in vitro* and animal studies have demonstrated that *W. somnifera* can reduce tumour cell proliferation while increasing animal survival time, reduce the side effects of chemotherapy without interfering with the anti-tumour properties, and increase the effects of radiotherapy while mitigating undesirable side effects (see Winters 2006).

Interactions: Ashwagandha may potentiate the sedative effect of other drugs, although no clinical cases have been reported. There is limited evidence to suggest that it may increase thyroid hormone levels and may therefore interfere in control of hypo- and hyperthyroidism (Williamson et al. 2013).

Contraindications: Ashwagandha is not recommended for use during pregnancy or breastfeeding or for paediatric use due to lack of safety data. Due to its immunomodulatory effects, it has been suggested that ashwagandha should be avoided in autoimmune diseases such as lupus and multiple sclerosis (WHO 2009), although experimental evidence suggests that it may in fact be helpful (Minhas et al. 2012).

Adverse effects: Ashwagandha has been shown to be safe in several human studies, although large doses can cause gastrointestinal upset, vomiting and diarrhoea (Engels and Brinckmann 2013; WHO 2009). A case report of thyrotoxicosis was reported in a 32-year-old woman following ingestion of ashwagandha capsules for fatigue. On discontinuing use of ashwagandha, her symptoms resolved. In mice, ashwagandha increased serum levels of thyroid hormones (reported in Williamson et al. 2013).

Dosage: Powdered crude drug: 3–6 g of the dried powdered root or equivalent as extract (WHO 2009). For products, manufacturers' instructions should be followed.

General plant information: *W. somnifera* is one of the most important medicinal plants used in Asian medicine, it is found in over 200 formulations employed in Ayurvedic, Unani and Siddha medicine. It is also of economic importance and is cultivated in several regions of India (Mirjalili et al. 2009). *W. coagulans* (Stocks) Dunal is also used in Ayurveda: it contains similar compounds but less is known about it.

References

Ashwagandha is very well investigated compared to many herbal medicines. This is a selection of references and recent reviews to support the information included in this monograph.
Bhat J, Damle A, Vaishnav PP, Albers R, Joshi M, Banerjee G. (2010) *In vivo* enhancement of natural killer cell activity through tea fortified with Ayurvedic herbs. *Phytotherapy Research* 24(1): 129–135.
Biswal BM, Sulaiman SA, Ismail HC, Zakaria H, Musa KI. (2013) Effect of *Withania somnifera* (Ashwagandha) on the development of chemotherapy-induced fatigue and quality of life in breast cancer patients. *Integrative Cancer Therapy* 12(4): 312–322.
Chen LX, He H, Qiu F. (2011) Natural withanolides: an overview. *Natural Product Reports* 28(4): 705–40.
Engels G, Brinckmann J. (2013) Ashwagandha. *HerbalGram* 99: 1–7.

Jayaprakasam B, Padmanabhan K, Nair MG. (2010) Withanamides in *Withania somnifera* fruit protect PC-12 cells from beta-amyloid responsible for Alzheimer's disease. *Phytotherapy Research* 24(6): 859–863.

Kumar S, Harris RJ, Seal CJ, Okello EJ. (2012) An Aqueous Extract of *Withania somnifera* Root Inhibits Amyloid β Fibril Formation *in vitro*. *Phytotherapy Research* 26(1): 113–117.

Mikolai J, Erlandsen A, Murison A, Brown KA, Gregory WL, Raman-Caplan P, Zwickey HL. (2009) In vivo effects of Ashwagandha (*Withania somnifera*) extract on the activation of lymphocytes. *Journal of Alternative and Complementary Medicine* 15(4): 423–430.

Minhas U, Minz R, Das P, Bhatnagar A. (2012) Therapeutic effect of *Withania somnifera* on pristane-induced model of SLE. *Inflammopharmacology* 20(4): 195–205.

Mirjalili MH, Moyano E, Bonfill M, Cusido RM, Palazón J. (2009) Steroidal lactones from *Withania somnifera*, an ancient plant for modern medicine. *Molecules* 14: 2373–2393.

Vanden Berghe W, Sabbe L, Kaileh M, Haegeman G, Heyninck K. (2012) Molecular insight in the multifunctional activities of Withaferin A. *Biochemical Pharmacology* 84(10): 1282–1291.

WHO. (2009) *WHO Monographs on Selected Medicinal Plants* Vol. 4. WHO, Geneva, 456 pp.

Williamson EM. (Ed.) (2002) *Major Herbs of Ayurveda*. Dabur Research Foundation. Churchill Livingstone Elsevier, UK.

Williamson EM, Driver S, Baxter K. (Eds.) (2013) *Stockley's Herbal Medicines Interactions*. 2nd Edition. Pharmaceutical Press, London, UK.

Winters M. (2006) Ancient medicine, modern use: *Withania somnifera* and its potential role in integrative oncology. *Alternative Medicine Review* 11(4): 269–277.

Asparagus   *Asparagus officinalis* L.

Synonyms: *A. paragus* Gueldenst. ex Ledeb.; *A. polyphyllus* Steven ex Ledeb.; *A. vulgaris* Gueldenst. ex Ledeb.; and others

Family: Asparagaceae (previously assigned to Liliaceae)

Other common names: Garden asparagus; sparrowgrass; wild asparagus

Drug name: Asparagi rhizoma; Asparagi herba

Botanical drug used: Rhizomes with roots attached; aerial parts

---

Indications/uses: Traditionally, asparagus rhizome/root preparations are used to flush out the urinary tract and assist in minor urinary complaints. Asparagus root powder is found in UK registered (THR) products in combination with parsley herb, *Petroselinum crispum*. Asparagus herb preparations are used as diuretics, but there is little documented clinical evidence to support this application and no registered products are available. Asparagus is also used to prevent kidney stones.

Evidence: Clinical data to support the traditional uses of asparagus are lacking.

Safety: There is a lack of safety data for the use of asparagus root preparations during pregnancy and lactation or in paediatric use. Cases of hypersensitivity have been reported.

Main chemical compounds: A number of bioactive compounds have been isolated from asparagus roots, including furostanol-type and spirostanol-type steroidal saponins, including asparagosides A to I, asparasaponins I and II, officinalisnins I and II, asparanin A, smilagenin, sarsasapogenin O); and several fructo-oligosaccharides. Flavonoids have also been isolated from asparagus shoots, including rutin, quercetin and kaempferol (Huang and Kong 2006; Pharmaceutical Press Editorial Team 2013).

Clinical evidence: Clinical studies to support the THR-approved uses of *A. officinalis* are lacking. In a six week clinical surveillance of 163 patients receiving a proprietary preparation containing powdered *A. officinalis* roots with parsley herb

---

Phytopharmacy: An evidence-based guide to herbal medicinal products, First Edition.
Sarah E. Edwards, Inês da Costa Rocha, Elizabeth M. Williamson and Michael Heinrich.
© 2015 John Wiley & Sons, Ltd. Published 2015 by John Wiley & Sons, Ltd.

(200 mg each per tablet), administered at maximally tolerable doses of the product (12 tablets per day), no clinically useful antihypertensive effects were found. Due to kidney-related adverse events, seven patients left the surveillance early. In light of this, it raises questions about the use of such products to treat hypertension or to promote flushing of the urinary tract in renal inflammatory conditions or kidney stones (Chrubasik et al. 2006; Chrubasik et al. 2009).

Pre-clinical evidence and mechanisms of action: Pharmacological studies on *A. officinalis* constituents (predominantly extracted from the aerial parts) have demonstrated anti-inflammatory, anti-tumour, antimutagenic, hypolipidaemic and antifungal activities (Bousserouel et al. 2013; Huang et al. 2008; Pharmaceutical Press Editorial Team 2013; Zhao et al. 2012).

Cytotoxicity against cancer cell lines has been demonstrated *in vitro* for extracts from both *A. officinalis* roots and aerial parts. Eight steroids isolated from the roots demonstrated significant cytotoxicities against human A2780, HO-8910, Eca-109, MGC-803, CNE, LTEP-a-2, KB and mouse L1210 tumour cells (Huang et al. 2008). Cytotoxic activity against human hepatocellular carcinoma HepG2 cells was also found in the steroidal saponin asparanin A. The study indicated that asparanin A induces cell cycle arrest and triggers apoptosis in HepG2 cells independently of the tumour-suppressor gene p53 pathway, indicating that it may have potential therapeutic application against human hepatoma (Liu et al. 2009). Significant dose-dependent anti-tumour activity (against HeLa and Bel-7404 human hepatoma cells) and antioxidant capacities were also found in polysaccharide fractions of green asparagus (*A. officinalis* shoots) in an *in vitro* study (Zhao et al. 2012).

Interactions: None reported. Asparagus shoots contain moderate amounts of vitamin K, which if eaten in large quantities could reduce the effectiveness of anti-coagulants (Pharmaceutical Press Editorial Team 2013).

Contraindications: Its contraindications include hypersensitivity to asparagus plants and patients with impaired renal function, oedema secondary to heart disease, current or previous kidney disease and conditions where a reduced fluid intake is recommended, for example severe cardiac or renal diseases. It is not recommended during pregnancy and lactation or for children and adolescents under 18 years due to lack of safety data (German Commission E 1991; Pharmaceutical Press Editorial Team 2013).

Adverse effects: Skin reactions have been reported, although the frequency is unknown. Acute contact dermatitis and immunoglobulin E (IgE) mediated allergy to asparagus (the vegetable) is fairly common among asparagus production workers (Tabar et al. 2004). In an open surveillance study of a proprietary product containing *A. officinalis* rhizome with parsley herb, in 163 patients 89 adverse events were possibly attributable to the product. These included renal pain/and or peripheral oedema, exacerbation of acute gout and skin allergies (Chrubasik et al. 2006).

Dosage: Daily oral adult dosage: 45–60 g of rhizome or equivalent preparations. Note – sufficient fluids are required for use in irrigation therapy (German Commission E 1991). Traditionally, it is used for a duration of 2–4 weeks.

General plant information: Asparagus is native to Europe, Asia (as far east as Mongolia) and northwest Africa (Algeria, Morocco and Tunisia). Asparagus is also cultivated in North and South America, China and Europe. In areas of the world where it has been introduced, it is considered a persistent weed of cultivated ground. Young asparagus shoots ('spears') have been eaten as a vegetable at least since Roman times (Kew 2013).

References

Bousserouel S, Le Grandois J, Gossé F, Werner D, Barth SW, Marchioni E, Marescaux J, Raul F. (2013) Methanolic extract of white asparagus shoots activates TRAIL apoptotic death pathway in human cancer cells and inhibits colon carcinogenesis in a preclinical model. *International Journal of Oncology* 43(2): 394–404.

Chrubasik S, Droste C, Black A. (2009) Asparagus P® cannot compete with first-line diuretics in lowering the blood pressure in treatment-requiring antihypertensives. *Phytotherapy Research* 23(9): 1345–1346.

Chrubasik S, Droste C, Dragano N, Glimm E, Black A. (2006) Effectiveness and tolerability of the herbal mixture Asparagus P® on blood pressure in treatment-requiring antihypertensives. *Phytomedicine* 13(9–10): 740–742.

German Commission E. (1991) Asparagus root (Asparagi rhizoma) http://www.heilpflanzen -welt.de/1991-07-Asparagus-root-Asparagi-rhizoma/ (accessed August 2013).

Huang X, Kong L. (2006) Steroidal saponins from roots of *Asparagus officinalis*. *Steroids* 71(2): 171–176.

Huang XF, Lin YY, Kong LY. (2008) Steroids from roots of *Asparagus officinalis* and their cytotoxic activity. *Journal of Integrative Plant Biology* 50(6): 717–722.

Kew. (2013) *Asparagus officinalis* (garden asparagus) http://www.kew.org/plants-fungi /Asparagus-officinalis.htm (accessed August 2013).

Liu W, Huang XF, Qi Q, Dai QS, Yang L, Nie FF, Lu N, Gong DD, Kong LY, Guo QL. (2009) Asparanin A induces G2/M cell cycle arrest and apoptosis in human hepatocellular carcinoma HepG2 cells. *Biochemical and Biophysical Research Communications* 381(4): 700–705.

Pharmaceutical Press Editorial Team. (2013) *Herbal Medicines*. 4th Edition. Pharmaceutical Press, London, UK.

Tabar AI, Alvarez-Puebla MJ, Gomez B, Sanchez-Monge R, Garćia BE, Echechipia S. (2004) Diversity of asparagus allergy: clinical and immunological features. *Clinical and Experimental Allergy* 34(1): 131–136.

Zhao Q, Xie B, Yan J, Zhao F, Xiao J, Yao L, Zhao B, Huang Y. (2012) *In vitro* antioxidant and antitumor activities of polysaccharides extracted from *Asparagus officinalis*. *Carbohydate Polymers* 87(1): 392–396.

Astragalus

Astragalus mongholicus Bunge

Synonyms: *A. membranaceus* (Fisch.) Bunge; *A. membranaceus* var. *mongholicus* (Bunge) P.G.Xiao

Family: Fabaceae (Leguminosae)

Other common names: Huang Qi; membranous milk vetch

Drug name: Astragali radix

Botanical drug used: Root

Indications/uses: In Europe and North America, it is used as an 'adaptogen' and for colds and flu, often in combinations. It is used widely in traditional Chinese medicine (TCM) for similar indications, as a tonic to strengthen the immune system, and for viral infections, fatigue and loss of blood and also for its renal protective effect in diabetic nephropathy.

Evidence: There is anecdotal, but little clinical, evidence that astragalus alone or in combination may help in the treatment of the common cold or for impaired immunity, adjunctive cancer treatment and viral infections such as *Herpes simplex*.

Safety: Overall, it is considered to be safe and the usual precautions are required (see the subsequent sections).

Main chemical compounds: The main constituents are the triterpene saponins, which include the astragalosides, agroastragalosides and astromembranins, polysaccharides known as astraloglucans, isoflavonoids based mainly on formononetin and calycosin (WHO 1999; Williamson et al. 2013).

Clinical Evidence: In a randomised, double-blind, placebo controlled trial in 48 patients over 6 weeks, a herbal–mineral complex (HMC) containing 80 mg of astragalus root (the only herbal component) showed positive results in patients with allergic rhinitis. Efficacy was evaluated by the mean change in the symptom score, quality of life, serum IgE and IgG, nasal eosinophils and physicians' and patients' global evaluation. Compared to placebo, HMC significantly decreased the intensity of rhinorrhoea although other indicators did not differ. The investigators and

---

Phytopharmacy: An evidence-based guide to herbal medicinal products, First Edition.
Sarah E. Edwards, Inês da Costa Rocha, Elizabeth M. Williamson and Michael Heinrich.
© 2015 John Wiley & Sons, Ltd. Published 2015 by John Wiley & Sons, Ltd.

patients both judged the treatment with HMC as more efficacious, but further studies with a larger number of patients need to be carried out (Matkovic et al. 2010).

Pre-clinical evidence and mechanisms of action: Astragalus has been shown to possess anti-inflammatory properties via reducing the release of inflammatory mediators and inactivation of NF-κB through the MAPK signalling pathway (Lai et al. 2013). Several fractions have been isolated that inhibited nitric oxide release from lipopolysaccharide-stimulated mouse macrophage cells, and an active compound identified as formononetin (Lai et al. 2013). Astragalus has also been shown to stimulate the immune system in a variety of tests in mice and also in humans, including potentiation of phagocytic function, increase of thymus weight, proliferation of splenocytes, stimulation of NK-cell activity of peripheral blood lymphocytes and enhancement (Anon. 2003; Brush et al. 2006; Cho and Leung 2007). Astragalus is also effective in reducing fasting blood glucose and albuminuria levels, reversing glomerular hyperfiltration, and ameliorating pathological changes of early diabetic nephropathy in rat models (Zhang et al. 2009). *In vitro*, astragalus shows antibacterial activity against *Shigella dysenteriae, Streptococcus hemolyticus, Diplococcus pneumonia* and *Staphylococcus aureus* (Anon. 2003).

Interactions: Due to its immunomodulatory effect, preliminary evidence suggests that astragalus may have beneficial effects if it is given with antineoplastic drugs and interferon-α or interleukin-2 (IL-2). Clinical data showed that it improved the response to chemotherapy with the MVP chemotherapy regimen (mitomycin, vinca alkaloid and cisplatin). However, its use with immunosuppressant treatment for life-threatening conditions should be taken with caution as experimental data suggested that it may diminish the immunosuppressant effects of cyclophosphamide (Williamson et al. 2013). Since an extract showed remarkable inhibiting effects on the metabolism of CYP3A4 *in vitro* and *in vivo*, caution should be taken with drugs that also use this metabolic pathway (Pao et al. 2012).

Contraindications: It is not recommended during pregnancy and lactation as safety has not been established. People with autoimmune diseases and who have had transplant surgery should not use it as a precaution.

Adverse effects: Not known. Astragalus should be avoided in autoimmune diseases due to its immunomodulating activity. No adverse effects were observed in mice after oral administration of up to 100 g/kg, a dose that is much higher than the effective oral dose in humans (WHO 1999).

Dosage: In TCM, preparations equivalent to 9–30 g/per day dried, cut root prepared as a decoction (PPRC 2000). For maintenance of healthy immune system, preparations equivalent to 2–4.8 g/per day of dried root can be used.

General plant information: The species is a native to Asia and has been used in TCM for thousands of years. It is also used as a liver protectant, an adjunct in chemotherapy and impaired immunity, and for a variety of other conditions such as cardiovascular disease and diabetic complications (Williamson et al. 2013). Some *Astragalus* species can be toxic, but these are not found in standardised products.

References

Anon. (2003) *Astragalus membranaceus*. Monograph. *Alternative Medicine Review* 8(1): 72–77.

Brush J, Mendenhall E, Guggenheim A, Chan T, Connelly E, Soumyanath A, Buresh R, Barrett R, Zwickey H. (2006) The effect of *Echinacea purpurea, Astragalus membranaceus* and *Glycyrrhiza glabra* on CD69 expression and immune cell activation in humans. *Phytotherapy Research* 20(8): 687–695.

Cho WC, Leung KN. (2007) *In vitro* and *in vivo* immunomodulating and immunorestorative effects of *Astragalus membranaceus. Journal of Ethnopharmacology* 113(1): 132–141.

Lai PK, Chan JY, Wu S, Cheng L, Ho GK, Lau CP, Kennelly EJ, Leung PC, Fung KP, Lau CB. (2013) Anti-inflammatory activities of an active fraction isolated from the root of *Astragalus membranaceus* in RAW 264.7 *Macrophages, Phytotherapy Research* 28(3): 395–404.

Matkovic Z, Zivkovic V, Korica M, Plavec D, Pecanic S, Tudoric N. (2010) Efficacy and safety of *Astragalus membranaceus* in the treatment of patients with seasonal allergic rhinitis. *Phytotherapy Research* 24(2): 175–181.

Pao LH, Hu OY, Fan HY, Lin CC, Liu LC, Huang PW. (2012) Herb-drug interaction of 50 Chinese herbal medicines on CYP3A4 activity *in vitro* and *in vivo. American Journal of Chinese Medicine* 40(1): 57–73.

PPRC. (2000) *Pharmacopoeia of the People's Republic of China.* The State Pharmacopoeia Commission of the People's Republic of China, Beijing, China.

WHO. (1999) *WHO Monographs on Selected Medicinal Plants.* Vol. 1. WHO, Geneva, 295 pp.

Williamson EM, Driver S, Baxter K. (Eds.) (2013) *Stockley's Herbal Medicines Interactions.* 2nd Edition. Pharmaceutical Press, London, UK.

Zhang J, Xie X, Li C, Fu P. (2009) Systematic review of the renal protective effect of *Astragalus membranaceus* (root) on diabetic nephropathy in animal models. *Journal of Ethnopharmacology* 126(2): 189–196.

Baobab   *Adansonia digitata* L.

Synonyms: *A. bahobab* L.; *A. baobab* Gaertn.

Family: Bombacaceae (also sometimes included in Malvaceae)

Other common names: Cream-of-tartar tree; monkey-bread tree; upside-down tree

Botanical drug used: Fruit pulp (in Africa all plant parts are used in traditional medicine, including bark, roots, leaves, flowers and seeds, but these are not generally available in the European Union).

Indications/uses: The nutrient rich fruit pulp is being marketed in the US and EU as yet another 'super food'. In Africa baobab fruit pulp has many uses including in baking and ice cream, as a cream of tartar substitute and medicinally to treat fever. In Malawi, the fruit pulp has been used in traditional medicine as an antidysenteric and in the treatment of measles and smallpox (Bosch et al. 2004).

Evidence: There is limited clinical evidence, although baobab fruit pulp has high levels of antioxidants, which may account for its reputed uses.

Safety: The fruit pulp is generally regarded as safe (has US GRAS approval and EU novel foods authorisation). However, the use of other parts of the plant during pregnancy/lactation, and for long-term use, is not recommended due to lack of safety data.

Main chemical compounds: A number of aromatic compounds have been isolated from the fruit pulp of baobab, including isopropyl myristate and nonanal. The pulp also contains organic acids including citric, tartaric, malic, succinic and ascorbic acid (vitamin C), which is found at levels seven to ten times higher than in orange. It has high amounts of carbohydrates (ca 70%), crude fibre (ca 11%), a low amount of ash (ca 6%), protein (ca 2%) and traces of fat. Amino acids isolated from the pulp include alanine, arginine, glycine, lysine, methionine, proline, serine and valine. The fruit pulp and/or leaves have been shown to contain vitamins B1, B2, B3 and A. The minerals copper, iron, potassium, magnesium, manganese, sodium, phosphorus and zinc have also been isolated from the fruit pulp (Kamatou et al. 2011).

Clinical evidence: Evidence from clinical trials is sparse. One clinical trial compared the efficacy of a baobab fruit solution to the WHO standard solution in the

---

Phytopharmacy: An evidence-based guide to herbal medicinal products, First Edition.
Sarah E. Edwards, Inês da Costa Rocha, Elizabeth M. Williamson and Michael Heinrich.
© 2015 John Wiley & Sons, Ltd. Published 2015 by John Wiley & Sons, Ltd.

treatment of children ($n = 161$, age 6 months) with acute diarrhoea. Although the results showed that the WHO solution was superior in terms of duration of diarrhoea, weight gain and rehydration, these results were not statistically significant (Kamatou et al. 2011).

Pre-clinical evidence and mechanisms of action: A number of *in vitro* and *in vivo* studies have demonstrated analgesic, antipyretic, anti-inflammatory, antiviral and hepatoprotective activities of baobab fruit pulp extracts. Of interest may also be the significant reduction of the glycaemic response by 18.5 and 37 g in an aqueous drink in 250 ml of water, which, however, was not accompanied by significant effect on satiety or on energy expenditure (Coe et al. 2013).

An *in vivo* study in mice demonstrated that hot water extract of baobab had an analgesic effect 2 hours after administration. The reaction time was 15.4 minutes at concentrations of 800 mg/kg in comparison to 10.2 minutes in the negative control. Another *in vivo* study in rats demonstrated that baobab fruit extract at concentrations of 800 mg/kg exhibited antipyretic activity. In a rat paw oedema study, hot water extract of baobab was shown to exhibit anti-inflammatory effects. After 24 hours administration of the aqueous extract at concentrations of 400 and 800 mg/kg, the mean swelling of the foot was 1.81 and 1.75 mm respectively, compared with the negative control of 6.35 mm. The DMSO fruit pulp extract was shown to significantly inhibit the pro-inflammatory cytokine IL-8. Several *in vitro* and *in vivo* studies have investigated the antiviral activity of various plant parts of baobab, and although all extracts exhibited antiviral activity, the leaf extracts were far more effective than the fruit pulp. Aqueous extract of fruit pulp has been shown to be highly hepatoprotective in an *in vivo* study in rats treated with CCl_4 (76% alanine transferase, 77% aspartate transferase, 87% alkaline phosphatase activity when the extract was administered at the start of CCl_4 toxicity). Although the mechanism of action is unknown, hepatoprotective activity could be due to triterpenoids, β-sitosterol, β-amyrin palmitate and/or α-amyrin and ursolic acid present in the fruit.

The antioxidant activity of fresh ripe baobab fruit was shown to be 1000 mg AEAC/100 g (ascorbic acid equivalent antioxidant content). Phenolic content of baobab fruit pulp was found to be not statistically different from *Citrus sinensis* (orange), but the total antioxidant activity was shown to correspond to 6–7 mmol/g of Trolox® in comparison to 0.1 mmol/g in the fruit pulp of orange (Kamatou et al. 2011).

Interactions: None known.

Contraindications: None known. On April 11th 2012, the European Commission authorised the use of Baobab dried fruit pulp as a novel food ingredient, which was considered substantially equivalent to a food already on the market (pursuant to Article 5 of Regulation (EC) No 258/97).

Adverse effects: None known.

Dosage: Although the fruit pulp can be taken as desired, there is not enough data to determine the appropriate dose of baobab preparations for medicinal use. In lieu of lack of data, the manufacturers' instructions should be followed.

General plant information: *A. digitata* is endemic to, and widely-distributed in the savannahs and savannah woodlands of sub-Saharan Africa. It is considered to be the largest succulent plant in the world, with a diameter of 10–12 m and a height exceeding 23 m (Kamatou et al. 2011). In rural West Africa baobab is used extensively for subsistence, and increasing commercialisation and export of the fruits is putting pressure on the resource and could negatively influence livelihoods, including reduced nutritional intake (Buchmann et al. 2010).

References

Bosch CH, Sié K, Asafa BA. (2004) *Adansonia digitata L.* [Internet] Record from PROTA4U. Grubben GJH, Denton OA (Eds). *PROTA (Plant Resources of Tropical Africa / Resources végétales de l'Afrique tropicale)*. Wageningen, Netherlands. http://www.prota4u.org /search.asp (accessed June 2013).

Buchmann C, Prihsler S, Hartl A, Vogl C. (2010) The importance of baobab (*Adansonia digitata* L.) in rural West Africa subsistence – suggestion of a cautionary approach to international market export of baobab fruits. *Ecology of Food and Nutrition* 49(3): 145–172.

Coe SA, Clegg M, Armengol M, Ryan L. (2013) The polyphenol-rich baobab fruit (*Adansonia digitata* L.) reduces starch digestion and glycemic response in humans. *Nutrition Research* 33(11): 888–896.

Kamatou GPP, Vermaak I, Viljoen AM. (2011) An updated review of *Adansonia digitata*: a commercially important African tree. *South African Journal of Botany* 77(4): 908–919.

Bearberry   *Arctostaphylos uva-ursi* (L.) Spreng

Synonyms: *Arbutus uva-ursi* L.

Family: Ericaceae

Other common name: Uva ursi

Drug name: Uvae ursi folium

Botanical drug used: Leaf

Indications/uses: Bearberry preparations are intended to relieve symptoms of mild cases of water retention and early symptoms of mild urinary tract infections (UTI; burning sensations during urination and/or frequent urination).

Evidence: Limited clinical information is available to support its use but pre-clinical data in the short-term treatment of uncomplicated lower UTI can be considered plausible.

Safety: Overall, it is considered to be safe, but not recommended during pregnancy and lactation, for patients under 18-years old or those with kidney disorders. Bearberry should not be used for prolonged periods of time (more than 1 week).

Main chemical compounds: The main constituents are arbutin (hydroquinone beta-glucoside) and related hydroquinone derivatives such as hydroquinone and methylarbutin. Other compounds present include polyphenols (gallotannins, ellagictannin, corilagin, catechin and anthocyanidin derivatives), phenolic acids (gallic acid, *p*-coumaric and syringic acids), flavonoids (hyperoside, quercitrin) and triterpenes (ursolic acid and uvaol) and piceoside, an iridoid glucoside (EMA 2010b; WHO 2004; Williamson et al. 2013).

Clinical evidence: No clinical data on bearberry leaf extract alone are available. The prophylactic effect of a hydroalcoholic extract of bearberry leaves (standardised to arbutin and methylarbutin content) in combination with *Taraxacum officinale* (dandelion) in a double-blind, prospective, randomised study in women (three tablets, three times a day) suggested that this preparation is not the treatment of choice for treating acute cystitis but might be useful as a prophylactic in the treatment of recurrent UTI (Larsson et al. 1993) and for relief of mild symptoms (EMA 2010b).

Phytopharmacy: An evidence-based guide to herbal medicinal products, First Edition.
Sarah E. Edwards, Inês da Costa Rocha, Elizabeth M. Williamson and Michael Heinrich.
© 2015 John Wiley & Sons, Ltd. Published 2015 by John Wiley & Sons, Ltd.

Pre-clinical evidence and mechanisms of action: Pre-clinical data supports the use of bearberry leaf extract, as well as its main constituents arbutin or hydroquinone, as antibacterial agents for the treatment of mild infections of the lower urinary tract (EMA 2010b; ESCOP 2003). The mechanism of action is thought to be due to the antimicrobial and anti-inflammatory activities of the hydroquinone derivatives, mainly arbutin (which is hydrolysed to hydroquinone during renal excretion), and which has an astringent action on the urinary mucous membranes and may result in a more acidic urine (EMA 2010b; WHO 2004).

Interactions: It should not be administered with foods (like acidic fruits or fruit juice) or medicines that acidify the urine as concomitant acidification of the urine may result in a reduction of the antibacterial efficacy (EMA 2010b; ESCOP 2003; WHO 2004). There was one case report of a patient with lithium toxicity who took a diuretic herbal product containing bearberry among other ingredients (Williamson et al. 2013).

Contraindications: Bearberry is not recommended during pregnancy and lactation, as safety has not been established. Moreover, it is not recommended for children and adolescents under 18-years old or for patients with kidney disorders. Bearberry should not be used for prolonged periods of time (more than 1 week) and if fever, dysuria, spasm or blood in the urine occur, or if symptoms of urinary tract infection persist, a health-care professional should be consulted (EMA 2010b; ESCOP 2003; WHO 2004).

Adverse effects: Bearberry may cause nausea, vomiting and stomach-ache (EMA 2010b; WHO 2004).

Dosage: For specific products, manufacturers' indications should be followed. A dose of 1–4 g of the herbal medicine in 150 ml water, as an infusion or cold macerate, can be taken three or four times daily (equivalent to 400–800 mg hydroquinone derivatives) (EMA 2010b; WHO 2004).

General plant information: Bearberry is native to North America, Asia and northern Europe. The leaves have been used for centuries, being first documented by the Welsh 'Physicians of Myddfai' in the 13th century, where it was used as a tea for UTI and as a diuretic. Bearberry may cause a colour change in urine (to greenish-brown), which darkens when exposed to air due to the oxidation of hydroquinone. Extracts from the leaves have been used in cosmetic preparations as skin-whitening agents but may cause irritation and discolouration of the skin (EMA 2010b; WHO 2004).

References

EMA. (2010a) Community herbal monograph on *Arctostaphylos uva-ursi* (L.) Spreng, folium. European Medicines Agency http://www.ema.europa.eu/docs/en_GB/document_library/Herbal_-_Community_herbal_monograph/2010/08/WC500095779.pdf.

EMA. (2010b) Assessment report on *Arctostaphylos uva-ursi* (L.) Spreng., folium. European Medicines Agency http://www.ema.europa.eu/docs/en_GB/document_library/Herbal_-_HMPC_assessment_report/2011/07/WC500108750.pdf.

ESCOP. (2003) *ESCOP Monographs: The Scientific Foundation for Herbal Medicinal Products.* 2nd Edition. Thieme, Exeter and London, UK.

Larsson B, Jonasson A, Fianu S. (1993) Prophylactic effect of UVA-E in women with recurrent cystitis: a preliminary report. *Current Therapeutic Research* 53(4): 441–443.

WHO. (2004) *WHO Monographs on Selected Medicinal Plants.* Vol. 2. WHO, Geneva, 358 pp.

Williamson EM, Driver S, Baxter K. (Eds.) (2013) *Stockley's Herbal Medicines Interactions.* 2nd Edition. Pharmaceutical Press, London, UK.

Bilberry; Blueberry

Vaccinium myrtillus L.; *V. angustifolium* Aiton, *V. corymbosum* L.

Family: Ericaceae

Other common names: *V. myrtillus* – blaeberry; bog bilberry; European blueberry; huckleberry; whortleberry; wineberry; *V. angustifolium* – lowbush blueberry; *V. corymbosum* – highbush blueberry

Drug name: Myrtilli fructus; Myrtilli folium

Botanical drug used: The berries are used most frequently, but the leaves also have a traditional use.

Indications/uses: Preparations of bilberry fruit are used to improve visual acuity and atherosclerosis and to treat haemorrhoids, venous insufficiency, diarrhoea, gastrointestinal inflammation and urinary complaints. Modern research is focusing on memory-enhancement effects and improvement of metabolic syndrome. The leaf has traditionally been used to treat diabetes.

Evidence: The clinical evidence to support these uses is limited.

Safety: The berries are considered to be safe; they are widely eaten.

Main chemical compounds: The main constituents of the berries are anthocyanoside flavonoids (anthocyanins) mainly glucosides of cyanidin, delphinidin, malvidin, petunidin and peonidin, with flavonoids including catechins, quercetin-3-glucuronide and hyperoside vitamin C, sugars and pectins (Anon. 2001; Pharmaceutical Press Editorial Team 2013; Williamson et al. 2013).

Clinical evidence:

Effects on night vision: Uncontrolled trials have shown a beneficial effect of the extract on patients suffering from retinal degeneration, myopia and glaucoma; but a systematic review of randomised placebo-controlled trials of bilberry fruit extracts on night vision showed no evidence to support its use (Canter and Ernst 2004).

Cardiovascular effects: A randomised controlled trial assessing the effect of bilberry juice (330 ml daily for 4 weeks) on serum and plasma biomarkers of inflammation and antioxidant status, in subjects at increased risk of cardiovascular disease, found that the juice produced a significant decrease in plasma concentrations of

C-reactive protein, interleukin (IL)-6, IL-15 and other cytokines, thus warranting further studies (Karlsen et al. 2010). A study of 48 men and women found that blueberries may improve selected features of metabolic syndrome and related cardiovascular risk factors at dietary achievable doses (Basu et al. 2010).

Effects on insulin sensitivity: Daily dietary supplementation with bioactives from whole blueberries improved insulin sensitivity in obese, non-diabetic, and insulin-resistant men and women in a small study of 32 participants (Stull et al. 2010).

Effects on ulcerative colitis: An open label pilot study on the effect of a daily standardised anthocyanin-rich bilberry preparation (average anthocyanin dose of 840 mg per day for a total of 6 weeks) for ulcerative colitis (UC) showed a beneficial effect on inflammation in UC. However, larger, double-blind controlled trials are needed to support these results (Biedermann et al. 2013)

Effects on memory: Blueberry intake acutely improves vascular function in healthy men in a time- and intake-dependent manner. These benefits may be mechanistically linked to the actions of circulating phenolic metabolites on neutrophil NADPH oxidase activity (Rodriguez-Mateos et al. 2013a).

Pre-clinical evidence and mechanisms of action:

The anthocyanosides are considered the most important of the pharmacologically active components and show antioxidant and anti-inflammatory effects (Laplaud et al. 1997; Pharmaceutical Press Editorial Team 2013).

There is no evidence to support the traditional antidiabetic use of bilberry leaf (Helmstädter and Schuster 2010).

Cardiovascular effects: Bilberry anthocyanins have been shown to have beneficial effects on microvascular blood flow and ischaemia–reperfusion injury in rats. Oral administration of a bilberry anthocyanin complex reduced the increase in capillary permeability, decreased leucocyte adhesion and improved capillary perfusion compared with controls (Pharmaceutical Press Editorial Team 2013). Blueberry extracts also improve vascular reactivity and lower blood pressure in high-fat and high-cholesterol-fed rats (Rodriguez-Mateos et al. 2013b).

Effects on memory: Consumption of flavonoid-rich blueberries had a positive impact on spatial learning performance in young healthy rats and these improvements are linked to the activation of ERK-CREB-BDNF pathway in the hippocampus (Rendeiro et al. 2012).

Anticancer effects: Mechanisms by which blueberries have been shown to prevent carcinogenesis include inhibition of the production of pro-inflammatory mediators, oxidative stress and its products such as DNA damage, inhibition of cancer cell proliferation and increased apoptosis (Johnson and Arjmandi 2013).

Interactions: None known but may enhance the antiplatelet effect of some drugs (Williamson et al. 2013).

Contraindications: No contraindication is known for the berries. During pregnancy and lactation consumption of bilberry leaf should be avoided, as safety has not been established.

Adverse effects: None known.

Dosage: Standardised extracts containing not less than 0.3% of anthocyanins (expressed as cyanidin-3-glucoside chloride) or not less than 1% tannins (expressed as pyrogallol) are available (Williamson et al. 2013).

General plant information: Bilberry is widespread in areas of acidic, poor nutrient soils, such as heaths, moorland, mountains and forests of Europe, while the closely related blueberries are native to the United States. Bilberries are seldom cultivated, unlike blueberries, which are a popular food and sometimes described as a 'super-fruit'.

References

Anon. (2001) *Vaccinium myrtillus* (bilberry). Monograph. *Alternative Medicine Review* 6(5): 500–504.

Basu A, Du M, Leyva MJ, Sanchez K, Betts NM, Wu M, Aston CE, Lyons TJ. (2010) Blueberries decrease cardiovascular risk factors in obese men and women with metabolic syndrome. *Journal of Nutrition* 140(9): 1582–1587.

Biedermann L, Mwinyi J, Scharl M, Frei P, Zeitz J, Kullak-Ublick GA, Vavricka SR, Fried M, Weber A, Humpf HU, Peschke S, Jetter A, Krammer G, Rogler G. (2013) Bilberry ingestion improves disease activity in mild to moderate ulcerative colitis – an open pilot study. *Journal of Crohn's and Colitis* 7(4): 271–279.

Canter, PH, Ernst E. (2004) Anthocyanosides of *Vaccinium myrtillus* (bilberry) for night vision – a systematic review of placebo-controlled trials. *Survey of Ophthalmology* 49(1): 38–50.

Helmstädter A, Schuster N. (2010) *Vaccinium myrtillus* as an antidiabetic medicinal plant – research through the ages. *Pharmazie* 65(5): 315–321.

Johnson SA, Arjmandi BH. (2013) Evidence for anti-cancer properties of blueberries: a mini-review. *Anti-Cancer Agents in Medicinal Chemistry* 13(8): 1142–1148.

Karlsen A, Paur I, Bohn SK, Sakhi AK, Borge GI, Serafini M, Erlund I, Laake P, Tonstad S, Blomhoff R. (2010) Bilberry juice modulates plasma concentration of NF-kappaB related inflammatory markers in subjects at increased risk of CVD. *European Journal of Nutrition* 49(6): 345–355.

Laplaud PM, Lelubre A, Chapman MJ. (1997) Antioxidant action of *Vaccinium myrtillus* extract on human low density lipoproteins *in vitro*: initial observations. *Fundamental & Clinical Pharmacology*. 11(1): 35–40.

Pharmaceutical Press Editorial Team. (2013) *Herbal Medicines*. 4rd Edition. Pharmaceutical Press, London,UK.

Rendeiro C, Vauzour D, Kean RJ, Butler LT, Rattray M, Spencer JP, Williams CM. (2012) Blueberry supplementation induces spatial memory improvements and region-specific regulation of hippocampal BDNF mRNA expression in young rats. *Psychopharmacology*. 223(3): 319–330.

Rodriguez-Mateos A, Rendeiro C, Bergillos-Meca T, Tabatabaee S, George TW, Heiss C, Spencer JP. (2013a) Intake and time dependence of blueberry flavonoid-induced improvements in vascular function: a randomized, controlled, double-blind, crossover intervention study with mechanistic insights into biological activity. *American Journal of Clinical Nutrition* 98(5): 1179–1191.

Rodriguez-Mateos A, Ishisaka A, Mawatari K, Vidal-Diez A, Spencer JP, Terao J. (2013b) Blueberry intervention improves vascular reactivity and lowers blood pressure in high-fat-, high-cholesterol-fed rats. *British Journal of Nutrition* 109(10): 1746–1754.

Stull AJ, Cash KC, Johnson WD, Champagne CM, Cefalu WT. (2010) Bioactives in blueberries improve insulin sensitivity in obese, insulin-resistant men and women. *Journal of Nutrition* 140(10): 1764–1768.

Williamson EM, Driver S, Baxter K. (Eds.) (2013) *Stockley's Herbal Medicines Interactions*. 2nd Edition. Pharmaceutical Press, London, UK.

Birch, Silver and Downy

Betula pendula Roth; *B. pubescens* Ehrh. and hybrids

Synonyms: *B. verrucosa* Ehrh. (= *B. pendula*); *B. alba* L. (= *B. pubescens*)

Family: Betulaceae

Other common names: European white birch; moor birch (=*B. pubescens*)

Drug name: Betulae folium; Betulae cortex

Botanical drug used: Leaves; bark

Indications/uses: Topical preparations made from birch leaf or bark extracts are traditionally used to treat inflammation and in the alleviation of pain in rheumatism. Internally, birch leaf extract is used as a diuretic and as an adjuvant in minor urinary complaints. Birch bark extracts have been used for the topical treatment of skin diseases, especially eczema, and liver disorders. UK products (THRs) currently registered are made from leaf extracts in combination with *Arnica montana* for topical use.

Evidence: Clinical data to support the over-the-counter (OTC) or traditional uses of birch extracts are lacking. Preliminary studies show promise for use of birch bark extract in treating actinic keratosis and normalising liver function in hepatitis. However, these should only be treated by a medically qualified professional. Active constituents in the leaf extract have known anti-inflammatory and antioxidant properties.

Safety: Limited safety or toxicity data are available. Safety during pregnancy and lactation has not been established.

Main chemical compounds: Several classes of pharmacologically active compounds have been isolated from both leaf and bark extracts.

Leaf: The main constituents are polymeric proanthocyanidins; *B. pubescens* dried leaves contain slightly lower levels of flavonoids (e.g. hyperoside, avacularin, quercetrin) compared with *B. pendula*. Other phenolic compounds include 3,4′dihydroxypropiophenone 3-glucoside (DHPPG), caffeic acid and chlorogenic acid. The essential oil (0.04–0.05% in leaves) contains methyl salicylate.

Phytopharmacy: An evidence-based guide to herbal medicinal products, First Edition.
Sarah E. Edwards, Inês da Costa Rocha, Elizabeth M. Williamson and Michael Heinrich.
© 2015 John Wiley & Sons, Ltd. Published 2015 by John Wiley & Sons, Ltd.

Bark: The bark is rich in pentacyclic triterpenes including betulinic acid, betulin, lupeol and minor components including oleanolic acid, ursolic acid and betulinic aldehyde; fatty acids, e.g. linolenic and linoleic acids; with tannins (4–5%) and smaller amounts of lignans and diarylheptanoids (Cîntă-Pînzaru et al. 2012; Dadáková et al. 2010; Duric et al. 2013; EMA 2008a).

Clinical evidence: Clinical data are lacking for all indications. There is limited evidence to suggest that birch bark topical preparations may be a potential therapeutic option for actinic keratoses (squamous cell carcinomas *in situ*). In one non-randomised pilot study, 14 patients with actinic keratoses were treated with a standardised birch bark ointment, and 14 patients received a combination therapy with cryotherapy and birch bark ointment. After 2 months of treatment, 79% of the patients receiving the birch bark ointment monotherapy, and 93% of patients receiving the combination therapy had more than 75% clearance of lesions (Huyke et al. 2006).

Another uncontrolled clinical study investigated the hepatoprotective effect of birch bark extract in patients with chronic hepatitis C. Patients [$n = 42$] were treated for 12 weeks with an oral dose of 160 mg/day of standardised birch bark extract (confirmed with 75% betulin and 3.5% betulinic acid). After 12 weeks, levels of alanine amino-transferase (ALT) were significantly decreased and normalised, and hepatitis C virus RNA was significantly reduced. In addition, reports of fatigue and abdominal discomfort were significantly reduced (by sixfold and threefold, respectively), and dyspepsia was no longer reported (Shikov et al. 2011).

Pre-clinical evidence and mechanisms of action: Studies suggest that *B. pendula* extracts exert a diuretic effect by their anti-inflammatory capacity and inhibiting endopeptidases. In an *in vitro* study, an aqueous extract of *B. pendula* was shown to inhibit the growth and cell division of activated T-lymphocytes in a dose-dependent manner, induce apoptosis and decrease cell proliferation. The extract is registered in Germany and clinically used for rheumatoid arthritis as a subcutaneous injection to the inflamed joint area (Gründemann et al. 2011). Birch leaf extract has been shown to be a promising treatment for corneal inflammation (Wacker et al. 2012), and is antibacterial against uropathogenic *Escherichia coli* (Wojnicz et al. 2012).

Betulin and betulinic acid are well known anti-inflammatory agents, with documented anti-tumour properties in animal studies (Pisha et al. 1995). A number of *in vitro* and *in vivo* studies have shown betulin and betulinic acid to be active against different cancer cell lines (including colorectal, breast, prostate and lung cancer cells) by targeting mitochondrial pathways and inducing apoptosis, and as a result of anti-angiogenic effects (Şoica et al. 2012). Most research has been done on melanoma and other skin cancers but the evidence is mainly still at the preclinical stage.

Interactions: None reported, although it would be wise to avoid warfarin because of the salicylate content. Due to the reputed diuretic activity of birch extracts, orally administered preparations should be avoided if using diuretics.

Contraindications: Hypersensitivity to salicylates. Birch extracts should not be taken by children under 16 years of age (as with aspirin), or during pregnancy and lactation, due to general lack of safety data (EMA 2008b). Oral preparations should not be taken by patients with severe cardiac or renal disease due to the reputed diuretic effects.

Adverse effects: Allergic reactions (rash, urticaria, allergic rhinitis) and gastrointestinal complains (nausea, vomiting, diarrhoea) have been reported with orally administered birch leaf preparations (EMA 2008a; EMA 2008b).

Dosage: Adults only: birch leaf as a herbal tea 2–3 g; powdered herbal substance 650 mg; dry extract 0.25–1 g; other preparations as recommended by manufacturer, to a maximum daily dose of 12 g of herb in divided doses. Birch extracts are traditionally used over a 2- to 4-week period (EMA 2008b).

General plant information: Birches are the most common broad-leaved trees in northern Europe, and are very important in the biodiversity of coniferous forests, with a large number of fungal and animal species dependent on them (Hynynen et al. 2010). Birch sap, obtained by tapping the tree trunks of *B. pendula*, traditionally gathered in spring, is drunk as a tonic, or made into a wine or syrup. Birch oil, extracted by steam distillation of young buds, is used in the cosmetic and perfumery industries. Birch trees are known to be heavy metal accumulators.

References

Cîntă-Pînzaru S, Dehelean CA, Soica C, Culea M, Borcan F. (2012) Evaluation and differentiation of the *Betulaceae* birch bark species and their bioactive triterpene content using analytical FT-vibrational spectroscopy and GC-MS. *Chemistry Central Journal* 6(1): 67.

Dadáková E, Vrchotová N, Tříska J. (2010) Content of selected biologically active compounds in tea infusions of widely used European medicinal plants. *Journal of Agrobiology* 27(1): 27–34.

Duric K, Kovac-Besovic E, Niksic H, Sofic E. (2013) Antibacterial activity of methanolic extracts, decoction and isolated triterpene products from different parts of birch, *Betula pendula*, Roth. *Journal of Plant Studies* 2(2): 61–70.

EMA. (2008a) *Betula pendula* Roth.; *Betula pubescens* Ehrh., folium. Assessment report for the development of community monographs and for inclusion of herbal substance(s), preparation(s), or combinations thereof in the list. European Medicines Agency http://www.ema.europa.eu/docs/en_GB/document_library/Herbal_-_HMPC_assessment _report/2010/01/WC500069013.pdf (accessed August 2013).

EMA. (2008b) Community herbal monograph on *Betula pendula* Roth.;*Betula pubescens* Ehrh., folium. European Medicines Agency http://www.ema.europa.eu/docs/en_GB /document_library/Herbal_-_Community_herbal _monograph/2009/12/WC500018091.pdf (accessed August 2013).

Gründemann C, Gruber CW, Hertrampfe A, Zehl M, Kopp B, Huber R. (2011) An aqueous birch leaf extract of *Betula pendula* inhibits the growth and cell division of inflammatory lymphocytes. *Journal of Ethnopharmacology* 136(3): 444–451.

Huyke C, Laszczyk M, Scheffler A, Ernst R, Schempp CM. (2006) Treatment of actinic keratoses with birch bark extract: a pilot study. *Journal der Deutschen Dermatologischen Gesellschaft* 4(2): 132–136. [Article in German].

Hynynen J, Niemistö P, Viherä-Aarnio A, Brunner A, Hein S, Velling P. (2010) Silviculture of birch (*Betula pendula* Roth and *Betula pubescens* Ehrh.) in northern Europe. *Forestry* 83(1): 103–119.

Pisha, E; Chai, H, Lee IS; Chagwedera TE, Farnsworth NR, Cordell GA, Beecher CW, Fong HH, Kinghorn AD, Brown, DM, et al. (1995) Discovery of betulinic acid as a selective inhibitor of human melanoma that functions by induction of apoptosis. *Nature Medicine* 1(10): 1046–1051.

Shikov AN, Djachuk GI, Sergeev DV, Pzharitskaya ON, Esaulenko EV, Kosman VM, Makarov VG. (2011) Birch bark extract as therapy for chronic hepatitis C – a pilot study. *Phytomedicine* 18(10): 807–810.

Şoica C, Dehelean C, Danciu C, Wang HM, Wenz G, Ambrus R, Bojin F, Anghel M. (2012) Betulin complex in γ-Cyclodextrin derivatives: properties and antineoplasic activities in *in vitro* and *in vivo* tumor models. *International Journal of Molecular Sciences* 13(11): 14992–15011.

Wacker K, Gründemann C, Kern Y, Bredow L, Huber R, Reinhard T, Schwartzkopff J. (2012) Inhibition of corneal inflammation following keratoplasty by birch leaf extract. *Experimental Eye Research* 97(1): 24–30.

Wojnicz D, Kucharska AZ, Sokół-Łetowska A, Kicia M, Tichaczek-Goska D. (2012) Medicinal plants extracts affect virulence factors expression and biofilm formation by the uropathogenic *Escherichia coli*. *Urological Research* 40(6): 683–697.

Bitter Gourd   *Momordica charantia* L.

Synonyms: *Cucumis argyi* H.Lév; *C. intermedius* M.Roem.

Family: Cucurbitaceae

Other common names: Balsam pear; bitter melon; cerassie; karela; leprosy gourd; and numerous others

Drug name: Momordicae fructus

Botanical drug used: Fruit, less commonly, leaf

Indications/uses: The main use of the fruit is for the management of type 2 diabetes in most of the countries where the plant grows. A 'bush tea' made from the leaves is used in Jamaica and elsewhere (Cerassie tea) as a purgative and laxative, and is also known to reduce hyperglycaemia.

Evidence: Some clinical studies and a great deal of anecdotal and pharmacological evidence are available to support the hypoglycaemic effects, but robust clinical trial evidence is lacking.

Safety: The fruits are widely used as a vegetable with no serious adverse effects reported, but a range of toxic effects have been recorded from local and traditional medical uses, and most importantly, abortifacient effects. Only limited data is available, but the use of products containing *M. charantia* should be avoided during pregnancy and lactation.

Main chemical compounds: More than 200 compounds have been isolated from the fruit, seeds, leaves and roots (Pawar et al. 2013). The hypoglycaemic constituents are the proteins, the momorcharins (e.g. α- and β-momocharin) and the momordins (Choudhary et al. 2012). The triterpene glycosides (the cucurbitanes and cucurbitacins), such as the momordicosides, the goyaglycosides and charantin are found throughout the plant, including the fruit (e.g. Harinantenaina et al. 2006; Nguyen et al. 2010), sterols, saponins, other triterpenes and a range of phenolics including gallic acid, chlorogenic acid, catechin and epicatechin are also known (Chen et al. 2005; WHO 2009).

Clinical evidence: A 2012 Cochrane review concluded that there was no statistically significant difference in glycaemic control with *M. charantia* preparations compared to placebo. This review, however, found only four clinical studies suitable

---

Phytopharmacy: An evidence-based guide to herbal medicinal products, First Edition.
Sarah E. Edwards, Inês da Costa Rocha, Elizabeth M. Williamson and Michael Heinrich.
© 2015 John Wiley & Sons, Ltd. Published 2015 by John Wiley & Sons, Ltd.

for inclusion, and which were not considered to be of high quality. Overall, the data are inconclusive, but given the wide usage and abundant anecdotal evidence of hypoglycaemic effects, there clearly is a potential to use *M. charantia* preparations as a dietary food for special medical purposes (Chaturvedi 2012; Krawinkel and Keding 2006).

Pre-clinical evidence and mechanisms of action: *M. charantia* shows a wide range of beneficial effects relevant to managing diabetes (Chaturvedi 2012; Harinantenaina et al. 2006): it positively affects damaged β-cells, increases insulin levels and enhances the sensitivity of insulin. Glucose absorption is inhibited via interfering with glucosidases and disaccharidases. Energy homeostasis is modulated by enhancing the activity of AMP-activated protein kinases, stimulating the synthesis and release of thyroid hormones and adiponectin, the transport of glucose in the cells and of fatty acids in the mitochondria, by modulation of insulin secretion and by elevation of levels of uncoupling proteins in adipose and skeletal muscles (Chen et al. 2013). While there is a wide range of biochemical evidence, the use of bitter melon lacks an adequate clinical evidence base. Stimulating insulin release is probably less desirable than improving insulin sensitivity.

The momocharins have shown *in vitro* immunomodulatory, antiviral and antibacterial activity, linked to ribosome inactivating activity, but *in vivo* data are very limited (Grover and Yadav 2004; Puri et al. 2009).

Anticancer activities have been reported (Nerurkar and Ray 2010), which is unsurprising given the cytotoxicity of some of the constituents, but is outside the scope of this book.

Interactions with other drugs: No relevant data found, but may cause additive effects if taken with other hypoglycaemic agents.

Adverse effects: Abortifacient activity, both in traditional practice and in pharmacological experiments, is well documented. The sterol glycoside charantin, one of the hypoglycaemic constituents, is known to cause uterine haemorrhage and abortion in rabbits (Grover and Yadav 2004) and the cucurbitacins are well known for their *in vitro* cytotoxic effects (Chen et al. 2005). Other toxic effects have also been recorded, but in fact in the clinical studies carried out, no serious adverse effects were reported (Ooi et al. 2012). The WHO (2009) concludes '*Owing to potential abortifacient effects and possible teratogenicity ... the seeds of the crude drug should not be taken during pregnancy. Owing to reported adverse events such as severe hypoglycaemia and convulsions in children, the crude drug and its preparations should not be administered to children or taken during breastfeeding*'.

Dosage: Very limited reliable information is available. For adults, a daily oral dose of 10–15 ml of fresh juice or 2–15 g of dried herb (fruit) has been recommended (WHO 2009). Based on the Cochrane analysis, only a dose of 2000 mg of dried powder (unripe fruit) showed any relevant effectiveness (Ooi et al. 2012).

General plant information: The species is variable and a range of cultivars and varieties are known. It is a commonly used food item in Asian immigrant communities and can often be found in UK grocery stores. *M. balsamina* L. is a closely related species, often known under the same common names (like balsam pear) but it has been studied in far less detail. The fruit is one of the most bitter vegetables known, and is said to be an acquired taste.

References

Chaturvedi P. (2012) Antidiabetic potentials of *Momordica charantia*: multiple mechanisms behind the effects. *Journal of Medicinal Food* 15(2): 101–107.

Chen JC, Chiu MH, Nie RL, Cordell GA, Qiu SX. (2005) Cucurbitacins and cucurbitane glycosides: structures and biological activities. *Natural Product Reports* 22(3): 386–399.

Choudhary SK, Chhabra G, Sharma D, Vashishta A, Ohri S, Dixit A. (2012) Comprehensive evaluation of anti-hyperglycemic activity of fractionated *Momordica charantia* seed extract in alloxan-induced diabetic rats. *Evidence Based Complementary and Alternative Medicine* 2012: 293650.

Grover JK, Yadav SP. (2004) Pharmacological actions and potential uses of *Momordica charantia*: a review. *Journal of Ethnopharmacology* 93(1): 123–132.

Harinantenaina L, Tanaka M, Takaoka S, Oda M, Mogami O, Uchida M, Asakawa Y. (2006) *Momordica charantia* constituents and antidiabetic screening of the isolated major compounds. *Chemical and Pharmaceutical Bulletin (Tokyo)* 54(7): 1017–1021.

Krawinkel MB, Keding GB. (2006) Bitter gourd (*Momordica charantia*): A dietary approach to hyperglycemia. *Nutrition Reviews* 64(7 Pt 1): 331–337.

Leung L, Birtwhistle R, Kotecha J, Hannah S, Cuthbertson S. (2009) Anti-diabetic and hypoglycaemic effects of *Momordica charantia* (bitter melon): a mini review. *British Journal of Nutrition* 102(12): 1703–1708.

Nerurkar P, Ray RB. (2010) Bitter melon: antagonist to cancer. *Pharmaceutical Research* 27(6): 1049–1053.

Nguyen XN, Phan VK, Chau VM, Ninh KB, Nguyen XC, Le MH, Bui HT, Tran HQ, Nguyen HT, Kim YH. (2010) Cucurbitane-type triterpene glycosides from the fruits of *Momordica charantia*. *Magnetic Resonance in Chemistry* 48(5): 392–396.

Ooi CP, Yassin Z, Hamid TA. (2012) *Momordica charantia* for type 2 diabetes mellitus. *Cochrane Database of Systematic Reviews* 8: CD007845.

Pawar RS, Tamta H, Ma J, Krynitsky AJ, Grundel E, Wamer WG, Rader JI. (2013) Updates on chemical and biological research on botanical ingredients in dietary supplements. *Analytical and Bioanalytical Chemistry* 405(13): 4373–84.

Puri M, Kaur I, Kanwar RK, Gupta RC, Chauhan A, Kanwar JR. (2009) Ribosome inactivating proteins (RIPs) from Momordica charantia for anti viral therapy. *Current Molecular Medicine* 9(9): 1080–1094.

Shih CC, Shlau MT, Lin CH, Wu JB. (2014) *Momordica charantia* ameliorates insulin resistance and dyslipidemia with altered hepatic glucose production and fatty acid synthesis and AMPK phosphorylation in high-fat-fed mice. *Phytotherapy Research* 28(3): 363–371.

WHO. (2009) *WHO Monographs on Selected Medicinal Plants*. Vol. 4. WHO, Geneva, 456 pp.

Black Cohosh   *Actaea racemosa* L.

Synonyms: *A. monogyna* Walter; *Cimicifuga racemosa* (L.) Nutt.

Family: Ranunculaceae

Other common names: black snakeroot; bugbane; bugwort; Macrotys; rattle-weed; squawroot

Drug names: Cimicifugae racemosae rhizoma

Botanical drug used: Rhizomes, roots

Indications/uses: Black cohosh is used mainly to relieve menopausal symptoms, including hot flushes. It has also been used for symptomatic relief of backache, muscular and rheumatic aches and pains. Native American Indians used black cohosh for other indications including colds, cough, sore throat, constipation, kidney disorders, gynaecological disorders, and to induce lactation.

Evidence: Some clinical evidence supports the use of black cohosh to treat climacteric symptoms including hot flushes, sweating, sleep disorders and nervous irritability, although study results are inconsistent. Pharmacological studies have demonstrated anti-inflammatory properties, but other uses are not supported by clinical data.

Safety: In recent years, safety concerns have been raised due to a few serious case reports of liver toxicity associated with unlicensed products labelled as black cohosh. However, these appear to be due to adulteration or substitution with another plant species, either intentionally or as a result of confusion over nomenclature. Clinical trials have indicated that *A. racemosa* is relatively safe, with few side effects. Long term use of black cohosh (>6 months) has not been evaluated, and is not recommended. Current evidence does not support an association between black cohosh and increased risk of breast cancer, and a small trial of breast cancer survivors found evidence of efficacy with few adverse events. However, the safety in this group of patients cannot be guaranteed. Due to a lack of human data, black cohosh products should be avoided in patients on cancer chemotherapy or immunosuppressive treatment, as well as those on hormone therapies.

Phytopharmacy: An evidence-based guide to herbal medicinal products, First Edition.
Sarah E. Edwards, Inês da Costa Rocha, Elizabeth M. Williamson and Michael Heinrich.
© 2015 John Wiley & Sons, Ltd. Published 2015 by John Wiley & Sons, Ltd.

Main chemical compounds: The main active constituents are the triterpene glycosides, a great number of which have been reported and include actein, 27-deoxyactein, a series of cimicifugosides and the cimiracemosides. Polyphenolic compounds such as the cimiracemates, caffeic acid derivatives, ferulic and isoferulic acids, fukinolic acid and cimicifugic acids, quinolizidine alkaloids and other compounds are also present (Foster 2013; Williamson et al. 2013).

Most of the clinical studies have been carried out on standardised extracts (e.g. ZE450 and CR BNO 1055).

Clinical evidence:

A Cochrane review evaluated the clinical effectiveness and safety of black cohosh in treating menopausal symptoms in peri-menopausal and post-menopausal women. The review identified 16 randomised controlled trials, with 2027 women, which used a median daily dose of 40 mg of black cohosh, for a mean duration of 23 weeks. The results of the review showed that there was no significant statistical difference between black cohosh and placebo in the frequency of hot flushes, or menopausal symptom scores. However, while the authors conclude there is insufficient clinical evidence to support the use of black cohosh to treat menopausal symptoms, study results were mixed, with some studies suggesting that black cohosh may help in relieving menopausal symptoms. Variability in study design and levels of active constituents in phytopharmaceutical products may explain the inconsistencies found in the different studies, especially since only six of the studies were standardised to between 2.5% and 5.68% triterpene glycosides (Leach and Moore 2012).

A randomised, double-blind, placebo-controlled study in 80 menopausal women not included in Leach and Moore (2012) compared 8 mg/day of a black cohosh extract with placebo or conjugated oestrogens. At 12 weeks menopausal symptoms, including hot flushes, were significantly lower in the treated groups than the placebo group, with the black cohosh group scoring better than the oestrogen treatment (Warnecke 1985 summarised in NIH 2008).

Since the 2012 Cochrane review, an observational study has demonstrated that treatment with black cohosh extract Ze 450 in (unselected) patients with climacteric complaints resulted in a significant improvement of menopausal symptoms assessed by the total Kupperman Menopause Index scores (Drewe et al. 2013). A recent randomised, double-blind, placebo-controlled, 3-armed trial with 180 female outpatients with climacteric complaints has also been carried out. Women were treated for 12 weeks with black cohosh extract Ze 450 at doses of 6.5 mg or 13.0 mg, or placebo. Primary outcome assessed by the Kupperman Menopausal Index and secondary efficacy variables were patients' self-assessments of general quality of life (QoL), responder rates, and safety. Compared to placebo, patients receiving Ze 450 showed a significant reduction in the severity of menopausal symptoms in a dose-dependent manner from baseline to endpoint. Changes in symptoms and QoL were inversely correlated. Reported adverse events did not raise safety concerns and the authors concluded that extract Ze 450 is an effective and well-tolerated non-hormonal alternative to hormone treatment for symptom relief in menopausal women (Schellenberg et al. 2012).

Breast cancer patients: In a randomised, double-blind, placebo-controlled study in breast cancer survivors (69 women completed the trial) black cohosh treatment for 2 months was found to significantly reduce sweating compared to placebo, but intensity and frequency of hot flushes decreased in both treatment and control groups, and was not significantly different (Jacobson et al., 2001). Current evidence does not support an association between black cohosh and increased risk of breast cancer. Given conflicting but promising results, and apparent safety, further research is warranted (Fritz et al. 2014).

Pre-clinical evidence and mechanisms of action: Evidence for oestrogenic activity of black cohosh is contradictory, and its mechanism of action is unclear. Fukinolic acid has been demonstrated to have oestrogenic activity *in vitro*, but extracts of black cohosh were shown to not bind to oestrogen receptors and to be devoid of oestrogenic effects on mammary cancer cells *in vitro*, and on mammary gland and uterine histology in ovariectomised rats (Wuttke et al. 2014). Actein strongly inhibited the growth of human breast cancer cells *in vitro* and induced a dose dependent release of calcium into the cytoplasm. The ER IP3 receptor antagonist heparin blocked this release, indicating that the receptor is required for activity. Heparin partially blocked the growth inhibitory effect, while the MEK inhibitor U0126 enhanced it. Actein also preferentially inhibited the growth of 293T (NF-κB) cells (Einbond et al. 2013).

Black cohosh extracts act as an agonist and competitive ligand for the μ-opioid receptor, which may explain some of the ability of black cohosh to ameliorate symptoms associated with oestrogen withdrawal during the menopause and for its reputed analgesic properties (Johnson and Fahey 2012). Aqueous extracts inhibit pro-inflammatory cytokines in lipopolysaccharide-stimulated whole blood from healthy volunteers, with isoferulic acid being one of the main constituents responsible for the observed inhibition (Schmid et al. 2009).

Interactions: Although concern has been raised about possible herb-drug interactions, none have been verified in animals or humans except for possible augmentation of the anti-proliferative effect of tamoxifen. An *in vitro* study has suggested an interaction between black cohosh and docetaxol or cisplatin. An isolated report of a kidney transplant patient taking ciclosporin concomitantly with a combination herbal medicine containing alfalfa and black cohosh described acute rejection and vascultitis, although it is uncertain if black cohosh is to blame. In view of the lack of data, black cohosh should be avoided in combination with cancer and immunosuppressive treatments and also hormone therapies (EMA 2010; Williamson et al. 2013).

Contraindications: Black cohosh has a traditional use as a uterine stimulant, so its use during pregnancy is not recommended, especially during the first trimester. There is no safety data on lactation, so black cohosh should be avoided. Due to quality concerns, patients with hepatic impairment should avoid black cohosh products (see below). It is not appropriate for children (EMA 2010).

Adverse effects:

A review of 13 clinical trials indicated the relative safety of black cohosh, with 97% of all reported adverse effects minor, and with frequency observed similar for that of placebo (Borelli and Ernst 2008).The Cochrane review of 16 clinical trials also did not find any statistical differences of reported adverse events between women assigned black cohosh and those assigned placebo or hormone therapy. The adverse effects most frequently reported for black cohosh included breast pain/enlargement, infection, vaginal bleeding/spotting, musculoskeletal complaints and gastrointestinal upset (Leach and Moore 2012).

Liver toxicity reports: A meta-analysis of 5 randomised and controlled clinical trials for black cohosh extract found no evidence for hepatotoxicity, with no significant clinically relevant changes of hepatic enzymes observed (Naser et al. 2011).

Serious safety reports, however, have been associated with *unregistered* black cohosh products available in the UK (which have no guarantee of being manufactured using good manufacturing practices or quality controls). It is thought that these contained related species from China (genuine black cohosh only grows in North America), which have an uncertain safety profile. A number of related and unrelated species are traded in China under the name 'black cohosh'. It is thought that in some cases these may have been intentionally used to add bulk for economic reasons (Foster 2013). One case of fatal multi-organ failure and 42 cases of hepatoxicity have been reported, including acute hepatitis requiring liver transplantation. Analysis of the evidence has shown no confirmed causality for *A. racemosa* in these cases (Borrelli and Ernst 2008; Teschke and Schwarzenboeck 2009).

Dosage: Female adults in the menopause: for products, follow manufacturers' instructions.

General plant information: *A. racemosa* is native to the Appalachian mountain region of North America, where it is still predominantly wild harvested. It is currently threatened by habitat loss and overexploitation. The American Herbal Products Association estimated that between 1997 and 2005 more than 1 million kg of rhizomes of natural populations of *A. racemosa* were removed, with little effort to manage the plant as a natural resource (Chamberlain et al. 2013). There is thus a real danger of contaminated products being found on the market.

References

Borrelli F, Ernst E. (2008) Black cohosh (*Cimicifuga racemosa*) for menopausal symptoms: a systematic review of its efficacy. *Pharmacological Research* 58(1): 8–14.

Chamberlain JL, Ness G, Small CG, Bonner SJ, Hiebert EB. (2013) Modeling below-ground biomass to improve sustainable management of *Actaea racemosa*, a globally important medicinal forest product. *Forest Ecology and Management* 293(1): 1–8.

Drewe J, Zimmermann C, Zahner C. (2013) The effect of a *Cimicifuga racemosa* extracts Ze 450 in the treatment of climacteric complaints – an observational study. *Phytomedicine* 20(8–9): 659–66.

Einbond LS, Mighty J, Redenti S, Wu HA. (2013) Actein induces calcium release in human breast cancer cells. *Fitoterapia* 91: 28–38.

EMA. (2010) Community herbal monograph on *Cimicifugaracemosa* (L.)Nutt., rhizome. European Medicines Agency http://www.ema.europa.eu/docs/en_GB/document_library/Herbal_-_Community_herbal_monograph/2011/01/WC500100981.pdf (accessed August 2013).

Foster S. (2013) Exploring the peripatetic maze of black cohosh adulteration: a review of the nomenclature, distribution, chemistry, market status, analytical methods and safety. *HerbalGram* 98: 32–51.

Fritz H, Seely D, McGowan J, Skidmore B, Fernandes R, Kennedy DA, Cooley K, Wong R, Sagar S, Balneaves LG, Fergusson D. (2014) Black cohosh and breast cancer: a systematic review. *Integrative Cancer Therapy* 13(1): 12–29.

Jacobson JS, Troxel AB, Evans J, Klaus L, Vahdat L, Kinne D, Lo KM, Moore A, Rosenman PJ, Kaufman EL, Neugut AI, Grann VR. (2001) Randomized trial of black cohosh for the treatment of hot flashes among women with a history of breast cancer. *Journal of Clinical Oncology* 19(10): 2739–2745.

Johnson TL, Fahey JW. (2012) Black cohosh: coming full circle. *Journal of Ethnopharmacology* 141(3): 775–779.

Leach MJ, Moore MV. (2012) Black cohosh (*Cimicifuga spp.*) for menopausal symptoms (Review). *Cochrane Database of Systematic Reviews* (9): CD0000724.

Naser B, Schnitker J, Minkin MJ, de Arriba SG, Nolte KU, Osmers R. (2011) Suspected black cohosh hepatoxicity: no evidence by meta-analysis of randomized controlled clinical trials for isopropanolic black cohosh extract. *Menopause* 18(4): 366–375.

NIH. (2008) *Black Cohosh – Health Professional Factsheet.* Office of Dietary Supplements, National Institutes of Health (USA). http://ods.od.nih.gov/pdf/factsheets/BlackCohosh -HealthProfessional.pdf (accessed August 2013).

Schellenberg R, Saller R, Hess L, Melzer J, Zimmermann C, Drewe J, Zahner C. (2012) Dose-dependent effects of the *Cimicifuga racemosa* extract Ze 450 in the treatment of climacteric complaints: a randomized, placebo-controlled study. *Evidence Based Complementary and Alternative Medicine* 2012: 260301.

Schmid D, Woehs F, Svoboda M, Thalhammer T, Chiba P, Moeslinger T. (2009) Aqueous extracts of *Cimicifuga racemosa* and phenolcarboxylic constituents inhibit production of proinflammatory cytokines in LPS-stimulated human whole blood. *Canadian Journal of Physioliology and Pharmacology* 87(11): 963–972.

Teschke R, Schwarzenboeck A. (2009) Suspected hepatoxicity by *Cimicifugae racemosae rhizoma* (black cohosh, root): Critical analysis and structured causality assessment. *Phytomedicine* 16(1): 72–84.

Warnecke, G. (1985): Influencing of menopausal complaints with a phytodrug: successful therapy with *Cimicifuga* monoextract. *Medizinische Welt* 36: 871–874 [article in German].

Williamson EM, Driver S, Baxter K. (Eds.) (2013) *Stockley's Herbal Medicines Interactions.* 2nd Edition. Pharmaceutical Press, London, UK.

Wuttke W, Jarry H, Haunschild J, Stecher G, Schuh M, Seidlova-Wuttke D. (2014) The non-estrogenic alternative for the treatment of climacteric complaints: Black cohosh (*Cimicifuga* or *Actaea racemosa*). *Journal of Steroid Biochemistry and Molecular Biology* 139: 302–10.

Bladderwrack; Kelp   *Fucus vesiculosus* L., *F. serratus* L., *Ascophyllum nodosum* (L.) Le Jolis.

Family: Fucaceae

Other common names: Ocean kelp; sea kelp; *F. vesiculosus*: black tang; seawrack; *F. serratus*: serrated wrack; toothed wrack; *A. nodosum*: egg wrack; knotted kelp; knotted wrack; Norwegian kelp

Drug name: Fuci thallus

Botanical drug used: The thallus (whole plant) of *F. vesiculosus*, *F. serratus* L. and/or *A. nodosum* are all acceptable according to PhEur and BP monographs

---

Indications/uses: Kelp is used as an adjunct to weight loss and as a source of iodine. Traditionally, kelp was used to treat goitre, hypothyroidism, obesity, arthritis and rheumatism. Salts of alginic acid (alginates) are used in medicines for dyspepsia.

Evidence: Clinical studies to support the traditional uses of kelp are lacking. Pre-clinical studies have demonstrated a variety of different types of activity including effects on glucose metabolism (see the subsequent sections).

Safety: Safety and toxicity data for kelp preparations are currently unavailable. Hyperthyroidism has been associated with its use, due to its high iodine content. Seaweeds accumulate heavy metals and other toxic substances, including radioactive compounds, which may pose a risk to health. Kelp is not recommended during pregnancy and breastfeeding as safety has not been established. Kelp supplements should be avoided in combination with thyroid medications, lithium and amiodarone.

Main chemical compounds: Polysaccharides – predominantly alginic acid (algin), fucoidans and laminarum (sulphated polysaccharide esters); phenolic acids; iodine, and various vitamins and minerals, including ascorbic acid (vitamin C), tocopherol (vitamin E), calcium, magnesium, potassium (EMA 2013; Pharmaceutical Press Editorial Team 2013). The BP and PhEur specify a content of 0.03–0.2% of total iodine with reference to the dried drug.

Phytopharmacy: An evidence-based guide to herbal medicinal products, First Edition.
Sarah E. Edwards, Inês da Costa Rocha, Elizabeth M. Williamson and Michael Heinrich.
© 2015 John Wiley & Sons, Ltd. Published 2015 by John Wiley & Sons, Ltd.

Clinical evidence:

Weight loss: There is no clinical data available to support the traditional use of kelp for assisting in weight loss.

Effects on glucose and insulin responses: A study examining the impact of kelp on post-load plasma glucose and insulin concentrations in men and women in a double-blind, randomised, placebo-controlled crossover study using a commercially available blend of brown seaweed (*A. nodosum* and *F. vesiculosus*), found that compared with placebo, consumption of seaweed was associated with a reduction in the insulin incremental area under the curve and an increase in insulin sensitivity. This suggests that brown seaweed may alter insulin homeostasis in response to carbohydrate ingestion. Single doses of 500 mg of brown seaweed had no significant effect on the glucose response ($p = 0.24$, adjusted for baseline) and glucose and insulin responses were similar between men and women. No adverse events were reported (Paradis et al. 2011).

Immunological effects: Mekabu fucoidan (MF), a sulphated polysaccharide extracted from seaweed, which has previously been shown to have an immunomodulatory effect, was investigated for its clinical effects on antibody production after influenza vaccination in elderly Japanese men and women. A randomised, placebo-controlled, double-blind study conducted with 70 volunteers >60 years of age, found that MF (300 mg/day) produced higher antibody titres against all three strains in the seasonal influenza virus vaccine than the placebo group. The immune response against B antigen met the EU Licensure criteria regarding the geometric mean titre ratio in the MF group (2.4), but not in the placebo group (1.7). In the MF group, natural killer-cell activity tended to increase from baseline 9 weeks after MF intake, but in the placebo group no substantial increase was noted at 9 weeks, and in fact it decreased substantially from 9 to 24 weeks. In the immunocompromised elderly people, MF intake increased antibody production after vaccination, possibly preventing influenza epidemics (Negishi et al. 2013).

Effects on the menstrual cycle: A clinical case report investigated the effects of bladderwrack supplementation (0.5–1.25 g/day dry weight) on three pre-menopausal women suffering from abnormal menstrual cycle. No side effects were reported and results showed significant increase in menstrual cycle length, marked reduction in blood flow and average number of days of menstruation following treatment (5–7 cycles). Significant anti-oestrogenic and progestogenic effects were observed in one subject following administration. The author postulated that consumption of seaweed in Japanese populations may account for their reduced risk of oestrogen-related cancers, in addition to dietary intake of soy (Skibola 2004).

Pre-clinical evidence and mechanisms of action:

Anti-obesity effects: A study examining the anti-obesity effects of fucoidan in a mouse model of diet-induced obesity found that fucoidan supplementation significantly decreased body-weight gain, food efficiency ratio and relative liver and epididymal fat mass compared with a high fat diet (HFD) control group over 5 weeks. The mice supplemented with fucoidan showed significantly reduced triglyceride, total cholesterol and low-density lipoprotein levels in the plasma. Liver steatosis induced by the HFD also improved in the fucoidan group. Fucoidan affected the down-regulation expression patterns of genes including peroxisome proliferator-activated receptor γ, adipose-specific fatty acid binding protein and acetyl CoA carboxylase (Kim et al. 2014).

Effects on carbohydrate metabolism: Extracts of *A. nodosum* have a strong α-amylase inhibitory effect and extracts of *F. vesiculosus* are potent inhibitors of α-glucosidase (Lordan et al. 2013).

Anticancer effects: Fucoidan exhibits anti-metastatic effects on A549 lung cancer cells via the down-regulation of ERK1/2 and Akt-mTOR as well as NF-kB signaling pathways (Lee et al. 2012). Other studies show that fucoidan can induce cytotoxicity and apoptosis in various cancer cells, and also inhibit invasion, metastasis and angiogenesis (Senthilkumar et al. 2013).

Blood clotting: Fucoidan has recently been found to be a novel C-type lectin-like receptor 2 (CLEC-2) agonist in haemophilia and it has been suggested that decreased bleeding times could be achieved through activation of platelets in this way (Manne et al. 2013).

Antioxidant activity: Extracts of *F. vesiculosus* have high free radical scavenging activity, reducing power, inhibition of oxidation in liposomes and in fish oil, which correlated well with phenolic content. Pigments and tocopherols (in ethanolic extracts) and sulphated polysaccharides, proteins or peptides (in water extracts) may also contribute to the overall antioxidant activity (Farvin and Jacobsen 2013).

Thyroid effects: The high levels of iodine in kelp explain its traditional uses in treating thyroid diseases. Levels of iodine found in kelp dietary supplements have been found to vary considerably, even between different samples of the same product (Pharmaceutical Press Editorial Team 2013), although there is a standard available in the BP and PhEur (0.03–0.2% of total iodine in the dried herb).

Interactions:

Thyroid medication: Kelp should not be taken with other medicines or supplements containing iodine, as it may lead to overdose, and should not be used with thyroid medication, due to risk of interaction.

Amiodarone: An interaction between *F. vesiculosus* extract and amiodarone, which produced a considerable decrease on amiodarone bioavailability, has been observed in rats (Rodrigues et al. 2013). Although this has not been documented in humans, amiodarone has a narrow therapeutic window so the concurrent administration of supplements containing kelp should be avoided.

Lithium: A case was reported of a 60-year-old man with bipolar disorder taking lithium concomitantly with *F. vesiculosus* (as a laxative) who developed hyperthyroidism that resolved on discontinuing the kelp preparation (Arbaizar and Llorca 2011).

Contraindications: Safety of kelp use during pregnancy and lactation, in children and adolescents under 18 years of age, and in the elderly has not been established. Use of kelp is therefore not recommended for these groups. Patients with a thyroid disorder should not use kelp. Long-term use should be avoided.

Adverse effects: Reported side effects of kelp use include diarrhoea. Overdose (due to the iodine content) may lead to thyroid disorders, including hyperthyroidism, and such cases have been associated with the use of *F. vesiculosus* preparations, which resolved on cessation of use (EMA 2013; Pharmaceutical Press

Editorial Team 2013). A daily intake of 700 or 1,400 mg ($n = 3$) for several weeks produced an increase in menstrual cycle length (EMA 2013).

Dosage: Powdered herbal substance: a dose of 130 mg, twice daily with a glass of water, 2 hours before meals is used (EMA 2013). As an infusion: dried thallus – 5–10 g three times daily; liquid extract – 4–8 ml (1:1 in 25% alcohol) three time daily (Pharmaceutical Press Editorial Team 2013). Recommended doses vary according to the product but the major concern is the concentration of iodine contained in each preparation. The upper daily limit of 400 μg total iodine per day should not be exceeded. The use in children and adolescents under 18 years of age is not recommended. If patients taking kelp products have been unable to lose weight after 10 weeks, they should consult a doctor or a qualified health-care practitioner.

General plant information: All types of kelp are common on sea shores in temperate climates. Historically, the ash produced from burning seaweed was used as a source of soda ash and also potash used by the glass and soap industries. Kelp was also important as a crop fertiliser, notably in Jersey in the production of new potatoes and is still used in organic gardening. The gelling agent alginate, found in brown seaweeds including kelp, is extensively used in the dairy and baking industries.

References

Arbaizar B, Llorca J. (2011) *Fucus vesiculosus* induced hyperthyroidism in a patient undergoing concomitant treatment with lithium. *Actas Españolas de Psiquiatria* 39(6): 401–403.

EMA. (2013) Community herbal monograph on *Fucus vesiculosus* L., thallus. European Medicines Agency http://www.ema.europa.eu/docs/en_GB/document_library/Herbal_-_Community_herbal_monograph/2013/08/WC500148187.pdf.

Farvin KHS, Jacobsen C. (2013) Phenolic compounds and antioxidant activities of selected species of seaweeds from Danish coast. *Food Chemistry* 138(2-3): 1670–1681.

Kim MJ, Jeon J, Lee JS. (2014) Fucoidan Prevents High-Fat Diet-Induced Obesity in Animals by Suppression of Fat Accumulation. *Phytotheapy Research* 28(1): 137–143.

Lee H, Kim JS, Kim E. (2012) Fucoidan from seaweed *Fucus vesiculosus* inhibits migration and invasion of human lung cancer cell via PI3K-Akt-mTOR pathways. *PLoS One* 7(11): e50624.

Lordan S, Smyth TJ, Soler-Vila A, Stanton C, Ross RP. (2013) The α-amylase and α-glucosidase inhibitory effects of Irish seaweed extracts. *Food Chemistry* 141(3): 2170–2176.

Manne BK, Getz TM, Hughes CE, Alshehri O, Dangelmaier C, Naik UP, Watson SP, Kunapuli SP. (2013) Fucoidan is a novel platelet agonist for the C-type lectin-like receptor 2 (CLEC-2). *Journal of Biological Chemistry* 288(11): 7717–7726.

Negishi H, Mori M, Mori H, Yamori Y. (2013) Supplementation of elderly japanese men and women with fucoidan from seaweed increases immune responses to seasonal influenza vaccination. *Journal of Nutrition* 143(11): 1794–1798.

Paradis ME, Couture P, Lamarche B. (2011) A randomised crossover placebo-controlled trial investigating the effect of brown seaweed (*Ascophyllum nodosum* and *Fucus vesiculosus*) on postchallenge plasma glucose and insulin levels in men and women. *Applied Physiology, Nutrition, and Metabolism* 36(6): 913–919.

Pharmaceutical Press Editorial Team. (2013) *Herbal Medicines.* 4th Edition. Pharmaceutical Press, London, UK.

Rodrigues M, Alves G, Abrantes J, Falcão A. (2013) Herb-drug interaction of *Fucus vesiculosus* extract and amiodarone in rats: a potential risk for reduced bioavailability of amiodarone in clinical practice. *Food and Chemical Toxicology* 52: 121–128.

Senthilkumar K, Manivasagan P, Venkatesan J, Kim SK. (2013) Brown seaweed fucoidan: biological activity and apoptosis, growth signaling mechanism in cancer. *International Journal of Biological Macromolecules.* 60: 366–374.

Skibola CF. (2004) The effect of *Fucus vesiculosus*, an edible brown seaweed, upon menstrual cycle length and hormonal status in three pre-menopausal women: a case report. *BMC Complementary and Alternative Medicine* 4: 10.

Ulbricht C, Basch E, Boon H, Conquer J, Costa D, Culwell S, Dao J, Eisenstein C, Foppa IM, Hashmi S, Isaac R, Johnstone K, Kerbel BN, Leblanc YC, Mintzer M, Shkayeva M, Smith M, Sollars D, Woods J. (2013) Seaweed, kelp, bladderwrack (*Fucus vesiculosus*): an evidence-based systematic review by the natural standard research collaboration. *Alternative and Complementary Therapies* 19(4): 217–230.

Boldo   *Peumus boldo* Molina

Synonyms: *Boldea fragrans* (Pers.) Endl.; *Boldu boldus* (Molina) Lyons; *P. fragrans* Pers.

Family: Monimiaceae

Other common names: Boldus

Drug name: Boldi folium

Botanical drug used: Leaves

Indications/uses: Mild digestive disturbances, gallstones, cystitis, rheumatism; often used as an aid to slimming.

Evidence: There is well-documented support for traditional use but no clinical evidence available.

Safety: The leaf is considered safe but the essential oil contains a toxic component, ascaridole, so boldo should be used with caution and not in large doses or for extended periods of time. It is not suitable for use in children or in pregnancy and breastfeeding, and despite the traditional use for liver disorders, adverse effects have been reported with high doses and boldo should be avoided in patients with liver and kidney disease.

Main chemical compounds: The leaves contain alkaloids (mainly boldine, isoboldine and dehydroboldine, with isocorydine, laurotetanine, reticuline, pronuciferine, sinoacutine and other derivatives) and essential oil, the main components of which are *p*-cymene, ascaridole, 1,8–cineole, fenchone, linalool, α-terpineol and terpinen–4–ol (Latté 2014; Williamson et al. 2013).

Clinical evidence: No clinical trials have been recorded for boldo, for any indication.

Pre-clinical evidence and mechanisms of action: Boldine is a potent antioxidant and has anti-inflammatory effects. It has been suggested to protect against chemotherapy-induced lipoperoxidation (Fernández et al. 2009) and has been shown to reverse ROS formation and restore endothelial function in STZ-induced diabetes in animals (Lau et al. 2013). Boldine has been found to modulate the expression of regulators of adipogenesis and adiponectin levels, which may support the traditional use in obesity (Latté 2014; Yu et al. 2009).

Phytopharmacy: An evidence-based guide to herbal medicinal products, First Edition.
Sarah E. Edwards, Inês da Costa Rocha, Elizabeth M. Williamson and Michael Heinrich.
© 2015 John Wiley & Sons, Ltd. Published 2015 by John Wiley & Sons, Ltd.

Interactions: A single case report suggests that it may interact with warfarin, raising the INR slightly, although no other effects (bleeding, bruising, etc.) were observed, and the patient was also taking a fenugreek supplement (Lambert and Cornier 2001). No other reports have been found (Williamson et al. 2013).

Contraindications: Pregnancy, breastfeeding, obstruction of bile duct, cholangitis, liver disease, gallstones and any other biliary disorders that require medical supervision and advice (EMA 2008).

Adverse effects: Boldo has been widely used with few reports of adverse events. A case of hepatotoxicity was recorded in an elderly male with fatty liver who was taking a laxative herbal product containing boldo extract; transaminases returned to normal after withdrawal of the product (Piscaglia et al. 2005). A study in rats found that boldo extracts in high doses (800 mg/kg) caused biochemical and histological changes in the liver, in addition to teratogenic and abortive effects (Almeida et al. 2000). Ascaridole is also known to be toxic. These reports have led to recommendations that the duration of treatment of boldo should be limited and the labelling of products containing boldo must state '*If the symptoms persist more than 2 weeks during the use of the medicinal product, a doctor or a qualified health care practitioner should be consulted*' (EMA 2008).

Dosage: According to manufacturers' instructions; or dried leaf 60–200 mg three times daily; or extract equivalent, for up to 2 weeks (EMA 2008).

General plant information: Boldo is an evergreen shrub which grows in Peru, Brazil, Paraguay, and Argentina and is now occasionally naturalised in Europe and North America. It has leathery aromatic leaves and scented flowers, the female trees producing edible yellow fruits in the autumn.

References

Almeida ER, Melo AM, Xavier H. (2000) Toxicological evaluation of the hydro-alcohol extract of the dry leaves of *Peumus boldus* and boldine in rats. *Phytotherapy Research* 14(2): 99–102.

EMA. (2008) Community herbal monograph on *Peumus boldo* Molina, folium. European Medicines Agency http://www.emea.europa.eu/docs/en_GB/document_library/Herbal_-_Community_herbal_monograph/2009/12/WC500018097.pdf.

Fernández J, Lagos P, Rivera P, Zamorano-Ponce E. (2009) Effect of boldo (*Peumus boldus* Molina) infusion on lipoperoxidation induced by cisplatin in mice liver. *Phytotherapy Research* 23(7): 1024–1027.

Lambert JP, Cornier J. (2001) Potential interaction between warfarin and boldo-fenugreek. *Pharmacotherapy* 21 509–512.

Latté, KP. (2014) Boldoblätter. *Zeitschrift für Phytotherapie* 35: 40–46.

Lau YS, Tian XY, Huang Y, Murugan D, Achike FI, Mustafa MR. (2013) Boldine protects endothelial function in hyperglycemia-induced oxidative stress through an antioxidant mechanism. *Biochemical Pharmacology* 85(3): 367–375.

Piscaglia F, Leoni S, Venturi A, Graziella F, Donati G, Bolondi L. (2005) Caution in the use of boldo in herbal laxatives: a case of hepatotoxicity. *Scandinavian Journal of Gastroenterology* 40(2): 236–239.

Williamson EM, Driver S, Baxter K (Eds.) (2013) *Stockley's Herbal Medicines Interactions.* 2nd Edition. Pharmaceutical Press, London, UK.

Yu B, Cook C, Santanam N. (2009) The aporphine alkaloid boldine induces adiponectin expression and regulation in 3T3-L1 cells. *Journal of Medicinal Foods* 12(5): 1074–1083.

Brahmi   *Bacopa monnieri* (L.) Wettst.

Synonyms: '*B. monniera*', an erroneous name, which is not accepted by botanists but occurs frequently in the literature

Family: Plantaginaceae (formerly Scrophulariaceae)

Other common names: Bacopa; thyme-leafed gratiola; water hyssop

Botanical drug used: Aerial parts; dried whole plant

Indications/uses: Bacopa has a long history of use to enhance memory and cognition. It is a very important herb in Ayurveda where it is considered a 'Rasayana' – a general tonic that strengthens both body and mind and has adaptogenic properties.

Evidence: A systematic review of randomised controlled trials (dose of 300–450 mg extract per day) showed positive effects on some tests in the domain of memory free recall.

Safety: It is widely used but safety has not been assessed rigorously.

Main chemical compounds: The active compounds are the dammarane triterpenoid saponins based on the bacogenins and jujubogenins, most notably the bacosides, bacopasides and bacosaponins (Chakravarty et al. 2001; Chakravarty et al. 2002). Older publications refer 'bacosides A and B', but 'bacoside A' has now been identified as a mixture of four saponins, bacoside A3, bacopaside II, bacopasaponin C and the jujobogenin isomer of the latter. The identity of 'bacoside B' still needs to be clarified, as there is contradictory information in the scientific literature (Deepak and Amit 2013). The herb also contains phenylethanoid glycosides including monnierisides I-III, the alkaloids brahmine and herpestine and cucurbitacin derivatives known as bacobitacins (Engels and Brinckmann 2011; Williamson et al. 2013).

Clinical evidence: Research on *B. monnieri* continues to deliver new insights, and although mainly positive, the evidence is not clear-cut. A systematic review of randomised controlled trials of *B. monnieri* extract on memory (Pase et al. 2012) included six studies using three different Bacopa extracts at doses of 300–450 mg extract per day. Across the studies, Bacopa improved performance on 9 of 17 tests in the domain of memory free recall, but the authors highlighted that there 'were no cognitive tests in the areas of auditory perceptual abilities or idea production

---

Phytopharmacy: An evidence-based guide to herbal medicinal products, First Edition.
Sarah E. Edwards, Inês da Costa Rocha, Elizabeth M. Williamson and Michael Heinrich.
© 2015 John Wiley & Sons, Ltd. Published 2015 by John Wiley & Sons, Ltd.

and only a paucity of research in the domains of reasoning, number facility and language behaviour'. Another meta-analysis (Kongkeaw et al. 2014) included nine studies with a chronic >12 weeks dosing of standardised extracts of *B. monnieri* without any co-medication and the authors concluded that *B. monnieri* improves cognition, particularly speed of attention.

Several clinical trials demonstrate improvements in the following: speed of visual information processing, learning rate, memory consolidation and state anxiety (Stough et al. 2001), retention of new information (Roodenrys et al. 2002), performance in delayed-recall memory tasks and task reaction times (Calabrese et al. 2008), spatial working memory accuracy (Stough et al. 2008) and improved memory acquisition and retention (Morgan and Stevens 2010). For example, in a double-blind, randomised placebo-controlled independent group design study over three months with two groups, a Bacopa group ($n = 37$) and a placebo group ($n = 39$), a significant improvement was observed only in one out of four tests (the delayed recall of word pairs task) in the Bacopa group. Two recent studies using acute administration of Bacopa extract CDRI 08 on sustained cognitive performance, multitasking stress reactivity and mood, found positive effects on some of the parameters measured, but not others (Benson et al. 2014; Downey et al. 2013).

Pre-clinical evidence and mechanisms of action: While a large and diverse range of effects have been studied *in vivo*, *in vitro* and *in silico*, the current picture is far from clear and the quality of the studies varies. Often high doses were needed to demonstrate an effect. Bacopa is mainly used to enhance brain function and a large body of studies point to central nervous system effects, including *in vivo* models such as the Morris water maze test, which is used as a model for early Alzheimer's disease (Uabundit et al. 2010). In an acute stress (AS) and chronic unpredictable stress (CUS) model in rats, the extract normalised stress-induced alterations in plasma corticosterone and other cerebral stress indicators (Sheikh et al. 2007).

The neuroprotective effects of Bacopa extract have been shown in primary cortical neurone culture: it was able to protect neurons from beta-amyloid-induced cell death, but not glutamate-induced excitotoxicity (Limpeanchob et al. 2008).

Interactions with other drugs: No robust data are available (Williamson et al. 2013).

Adverse effects: No major adverse effects have been observed in the clinical studies, but again there is a lack of data. It is not recommended during pregnancy and lactation, as safety has not been established.

Dosage: In most clinical studies, a dose of 300–450 mg extract per day was generally used, with the composition of the extracts varying. Bacopa is used in a variety of preparations: herbal teas, encapsulated dried powder and fresh leaf.

General plant information: The name Brahmi is derived from the Sanskrit term 'Brahman', referring to life, and taken from the name of the Hindu god Brahma, the creator. *B. monnieri* is native to large regions of Southeast Asia including India, Bangladesh, Sri Lanka, China, Taiwan, and Thailand, and is now widely cultivated. NB: the name 'Brahmi' is also (less commonly) applied to *Centella asiatica* (see p. 91), so care must be taken if the species is not named scientifically.

References

Benson S, Downey LA, Stough C, Wetherell M, Zangara A, Scholey A. (2014) An acute, double-blind, placebo-controlled cross-over study of 320 mg and 640 mg doses of *Bacopa monnieri* (CDRI 08) on multitasking stress reactivity and mood. *Phytotherapy Research* 28(4):551–559.

Calabrese, C., W. L. Gregory, et al. (2008) Effects of a standardized *Bacopa monnieri* extract on cognitive performance, anxiety, and depression in the elderly: a randomized, double-blind, placebo-controlled trial. *Journal of Alternative and Complementary Medicine* 14(6): 707–713.

Chakravarty AK, Sarkar T, Masuda K, Shiojima K, Nakane T, Kawahara N. (2001) Bacopaside I and II: two pseudojujubogenin glycosides from *Bacopa monniera*. *Phytochemistry* 58: 553–6.

Chakravarty AK, Sarkar T, Nakane T, Kawahara N, Masuda K. (2002) New phenylethanoid glycosides from *Bacopa monniera*. *Chemical and Pharmaceutical Bulletin (Tokyo)* 50(12): 1616–1618.

Deepak M, Amit A. (2013) 'Bacoside B' – the need remains for establishing identity. *Fitoterapia* 87: 7–10.

Downey LA, Kean J, Nemeh F, Lau A, Poll A, Gregory R, Murray M, Rourke J, Patak B, Pase MP, Zangara A, Lomas J, Scholey A, Stough C. (2013) An acute, double-blind, placebo-controlled crossover study of 320 mg and 640 mg doses of a special extract of *Bacopa monnieri* (CDRI 08) on sustained cognitive performance. *Phytotherapy Research* 27(9): 1407–1413.

Engels G, Brinckmann J. (2011) Bacopa. *HerbalGram* 91: 1–4.

Kongkeaw Ch, Dilokthornsakul B, Thanarangsarit P, Limpeanchob N, Scholfield CN. (2014) Meta-analysis of randomized controlled trials on cognitive effects of *Bacopa monnieri* extract. *Journal of Ethnopharmacology* 151(1): 528–535.

Limpeanchob N, Jaipan S, Rattanakaruna S, Phrompittayarat W, Ingkaninan K. (2008) Neuroprotective effect of *Bacopa monnieri* on beta-amyloid-induced cell death in primary cortical culture. *Journal of Ethnopharmacology* 120(1): 112–117.

Morgan A, Stevens J. (2010) Does *Bacopa monnieri* improve memory performance in older persons? Results of a randomized, placebo-controlled, double-blind trial. *Journal of Alternative and Complementary Medicine* 16(7): 753–759.

Pase MP, Kean J, Sarris J, Neale C, Scholey AB, Stough C. (2012) The cognitive-enhancing effects of *Bacopa monnieri*: a systematic review of randomized, controlled human clinical trials. *The Journal of Alternative and Complementary Medicine* 18(7): 647–652.

Roodenrys, S, Booth D, et al. (2002) Chronic effects of Brahmi (*Bacopa monnieri*) on human memory. *Neuropsychopharmacology* 27(2): 279–281.

Sheikh N, Ahmad A, Siripurapu KB, Kuchibhotla VK, Singh S, Palit G. (2007) Effect of *Bacopa monnieri* on stress induced changes in plasma corticosterone and brain monoamines in rats. *Journal of Ethnopharmacology* 111(3): 671–676.

Stough, C, Lloyd J, Clarke J, Downey LA, Hutchison CW, Rodgers T, Nathan PJ et al. (2001) The chronic effects of an extract of *Bacopa monnieri* (Brahmi) on cognitive function in healthy human subjects. *Psychopharmacology* 156(4): 481–484.

Stough C, Downey Lloyd J LA, Silber B, Redman S, Hutchison C, Wesnes K, Nathan PJ. (2008) Examining the nootropic effects of a special extract of *Bacopa monniera* on human cognitive functioning: 90 day double-blind placebo-controlled randomized trial. *Phytotherapy Research* 22(12): 1629–1634.

Uabundit N, Wattanathorn J, Mucimapura S, Ingkaninan K. (2010) Cognitive enhancement and neuroprotective effects of *Bacopa monnieri* in Alzheimer's disease model. *Journal of Ethnopharmacology* 127(1): 26–31.

Williamson EM, Driver S, Baxter K. (Eds.) (2013) *Stockley's Herbal Medicines Interactions*. 2nd Edition. Pharmaceutical Press, London, UK.

Burdock   *Arctium lappa* L.

Synonyms: *A. majus* (Gaertn.) Bernh.; *Bardana arctium* Hill; *B. lappa* Hill; and many others

Family: Asteraceae (Compositae)

Other common names: bardana; beggar's buttons; cockle buttons; gobo; greater burdock; lappa; thorny burr

Drug names: Arctii radix; Arctii fructus

Botanical drug used: primarily roots; dried, ripe fruits

Indications/uses: Traditionally used as a diuretic (to increase urine output to 'flush' the urinary tract); as a 'blood purifier' to clear the body of toxins; and as an appetite stimulant. Burdock is also used topically to treat skin complaints such as eczema, acne and psoriasis. In TCM, it is often used with other herbs to treat sore throat and colds.

Evidence: There is a lack of clinical evidence to support the traditional uses of burdock. *In vitro* and animal studies have demonstrated various pharmacological actions of burdock and its constituents including anti-inflammatory, diuretic, hypoglycaemic, antimicrobial, anti-allergic, anti-tumour, hepatoprotective and protective against mutagens (see below).

Safety: There is a lack of safety data, and so use in children and during pregnancy and lactation should be avoided.

Main chemical compounds: All parts of the plant contain lignans, including arctigenin, arctiin, matairesinol and polyphenolic derivatives including caffeic acid derivatives. The leaves also contain sesquiterpenes, mainly fukinanolide and eudesmol derivatives such as arctiol, and others. The roots contain carbohydrates (including inulin up to 45–50%, mucilage, pectin and sugars), arctic acid and polyacetylenes, and the fruits contain a series of lignans known as lappaols (Chan et al. 2011; Pharmaceutical Press Editorial Team 2013).

Clinical evidence: Clinical data to support the traditional uses of burdock is lacking.

Pre-clinical evidence and mechanisms of action: Various pharmacological properties have been described for burdock, many of which support the traditional

Phytopharmacy: An evidence-based guide to herbal medicinal products, First Edition.
Sarah E. Edwards, Inês da Costa Rocha, Elizabeth M. Williamson and Michael Heinrich.
© 2015 John Wiley & Sons, Ltd. Published 2015 by John Wiley & Sons, Ltd.

uses, and most of which appear to be due to the lignan content. Topically applied burdock extract inhibited acute ear swelling to some extent in an *in vivo* allergic model in mice (Knipping et al. 2008) and arctigenin has been shown to have anti-inflammatory properties *in vitro*, probably via the regulation of inducible nitric oxide synthase (iNOS; Zhou et al. 2009).

Lignans extracted from burdock have been shown to be potent platelet activating factor (PAF) antagonists, indicating their anti-allergic potential, as well as calcium antagonists (Chan et al. 2011). They also inhibit aldose reductase (Xu et al. 2010) and lower blood sugar concentrations, together with an increase in carbohydrate tolerance and a reduction in toxicity (Xu et al. 2008).

In vivo studies in rats have illustrated the gastroprotective activity of burdock extracts and the regeneration of damaged gastric mucosa using models of colitis and ulceration, thought to be through anti-secretory and antioxidative mechanisms (de Almeida et al. 2013; da Silva et al. 2013).

Other pharmacological evidence indicates that burdock protects against mutagenic activity, hepatotoxicity and has anti-tumour properties (Pharmaceutical Press Editorial Team 2013). Arctiin also significantly decreased levels of urinary albumin, prevented sclerosis of glomeruli and restored glomerular filtration barrier damage in diabetic nephropathy in rats, thought to be via regulating the expression of nephrin and podocin (Ma et al. 2013). The polyacetylenes from burdock root possess potent antifungal and antibacterial activities. Other constituents, such as caffeic acid and arctigenin, also have demonstrated antiviral activity (including against herpes virus, adenovirus and HIV) *in vitro* and *in vivo* (Chan et al. 2011).

Interactions: None recorded (Williamson et al. 2013).

Contraindications: Uterine stimulant action has been reported in animal studies, so, together with a lack of safety data, burdock should be avoided during pregnancy and lactation. Allergic reactions may occur in individuals hypersensitive to the herb or plants from the Asteraceae.

Adverse effects: The most common side effect reported is contact dermatitis after topical use of the 'root oil' of burdock. There has been a case of anaphylaxis 'due to burdock consumption', which turned out to be a case of misidentification, and also a case report of atropine-like poisoning was found to be caused by contamination of a burdock root product, probably with deadly nightshade (*Atropa belladonna* L.) (Pharmaceutical Press Editorial Team 2013). These cases again highlight the potential risk posed by unlicensed products where quality and identification of the herb cannot be guaranteed.

Dosage: For oral administration in adults, dried root: 2–6 g by infusion three times daily or equivalent as extract. For proprietary products, the manufacturers' instructions should be followed.

General plant information: Burdock is native to Europe and northern Asia, and is naturalised in North America. Burdock root is commonly eaten as a cooked vegetable in Asia (da Silva et al. 2013) and the leaves have been used to prepare beverages.

References

de Almeida ABA, Sánchez-Hildago M, Martin AR, Luiz-Ferreira A, Trigo JR, Vilegas W, dos Santos LC, Souza-Brito AR, de la Lastra CA. (2013) Anti-inflammatory intestinal activity of *Arctium lappa* L. (Asteraceae) in TNBS colitis model. *Journal of Ethnopharmacology* 146(1): 300–310.

Chan YS, Cheng LN, Wu JH, Chan E, Kwan YW, Lee SM, Leung GP, Yu PH, Chan SW. (2011) A review of the pharmacological effects of *Arctium lappa* (burdock). *Inflammopharmacology* 19(5): 245–254.

Knipping K, van Esch E, Wijering SC, van der Heide S, Dubois AE, Garssen J. (2008) *In vitro* and *in vivo* anti-allergic effects of *Arctium lappa* L. *Experimental Biology and Medicine (Maywood)* 233(11): 1469–1477.

Ma ST, Liu DL, Deng JJ, Niu R, Liu RB. (2013) Effect of arctiin on glomerular filtration barrier damage in STZ-induced diabetic nephropathy rats. *Phytotherapy Research* 27(10): 1474–1480.

Pharmaceutical Press Editorial Team. (2013) *Herbal Medicines*. 4th Edition. Pharmaceutical Press, London, UK.

da Silva LM, Allemand A, Mendes DA, dos Santos AC, André E, de Souza LM, Cipriani TR, Dartora N, Margues MC, Baggio CH, Werner MF. (2013) Ethanolic extracts of roots of Arctium lappa L. accelerates the healing of acetic-acid induced gastric ulcer in rats: Involvement of the antioxidant system. *Food and Chemical Toxicology* 51: 179–187.

Williamson EM, Driver S, Baxter K. (Eds.) (2013) *Stockley's Herbal Medicines Interactions*. 2nd Edition. Pharmaceutical Press, London, UK.

Xu Y, Wang X, Zhou M, Ma L, Deng Y, Zhang H, Zhao A, Zhang Y, Jia W. (2008) The antidiabetic activity of total lignan from *Fructus Arctii* against alloxan-induced diabetes in mice and rats. *Phytotherapy Research* 22(1): 97–101.

Xu Z, Yang H, Zhou M, Feng Y, Jia W. (2010) Inhibitory effect of total lignan from *Fructus Arctii* on aldose reductase. *Phytotherapy Research* 24(3): 472–473.

Zhou F, Wang L, Liu K. (2009) *In vitro* anti-inflammatory effects of arctigenin, a lignan from *Arctium lappa* L., through inhibition of iNOS pathway. *Journal of Ethnopharmacology* 122(3): 457–462.

Butcher's Broom   *Ruscus aculeatus L.*

Family: Asparagaceae (previously Liliaceae)

Other common names: Knee holly; kneeholm; sweet broom

Drug name: Rusci rhizoma

Botanical drug used: Rhizome

Indications/uses: To relieve symptoms of venous insufficiency such as discomfort and heaviness of the legs, including in premenstrual oedema, to relieve symptoms of itching and burning associated with haemorrhoids, and for diabetic retinopathy. Preparations are available for both internal and external use, and cosmetic products for cellulite and stretch marks of pregnancy often contain butcher's broom extracts.

Evidence: Only few qualitative studies and a one-off meta-analysis study showed that butcher's broom preparations seem to be efficacious for the treatment of chronic venous insufficiency (CVI). No clinical studies to support the cosmetic uses are available.

Safety: Overall considered to be safe, although allergic reactions have been reported.

Main chemical compounds: The main constituents are steroidal saponins based on the aglycones ruscogenin and neoruscogenin, and include ruscin, neoruscin, deglucoruscin, ruscoside, with their various sulphated and acetylated derivatives, ruscozepines A and B, aculeosides A and B and others. Coumarins such as esculetin, triterpenes and flavonoids are also present (Anon. 2001; Barbič et al. 2013; EMA 2008).

Clinical evidence: A meta-analysis (Boyle et al. 2003), and three other qualitative studies–two observational, multi-centre, open clinical studies (Aguilar Peralta et al. 2007; Guex et al. 2009), and one observational, single-arm, multi-centre prospective trial (Guex et al. 2010)) – focused on patients suffering from chronic CVI and taking a formulation product containing butcher's broom (150 mg) with hesperidin methyl chalcone (150 mg) and ascorbic acid (100 g). The preparation significantly reduced the severity of the symptoms and improved quality of life compared to placebo.

Phytopharmacy: An evidence-based guide to herbal medicinal products, First Edition.
Sarah E. Edwards, Inês da Costa Rocha, Elizabeth M. Williamson and Michael Heinrich.
© 2015 John Wiley & Sons, Ltd. Published 2015 by John Wiley & Sons, Ltd.

Similar effects were observed in an open-label clinical trial where functional signs of venous insufficiency were correlated with an improvement of venous refilling time in patients suffering from CVI (Allaert et al. 2011).

In one multi-centre, double-blind, randomised, placebo-controlled trial, women with CVI received capsules of butcher's broom rhizome extract (36.0–37.5 mg), twice daily for 12 weeks. The study showed that the extract was a safe and effective treatment for patients suffering from CVI. Further clinical trials which include men should be considered (Vanscheidt et al. 2002).

Pre-clinical evidence and mechanisms of action: The mechanism of action is unclear, but pre-clinical studies suggest that butcher's broom extracts reduce vascular permeability and have a vasoconstrictive effect. The mechanism is thought to be mediated via alpha-1 adrenergic activity by the steroidal saponins (Anon. 2001). Anti-inflammatory and anti-thrombotic activities are other potential mechanisms (Huang et al. 2008) and recently, several of the steroidal saponins and also esculin were found to reduce thrombin-induced hyperpermeability of endothelial cells *in vitro*, with the highest activities observed for deglucoruscin, ruscin and esculin (Barbič et al. 2013).

Interactions: None reported (Williamson et al. 2013).

Contraindications: Not recommended during pregnancy and lactation, nor for children or adolescents under 18 years old, as safety has not been established. Hypersensitivity has been reported. Internal use should be avoided in patients with hypertension (Anon. 2001).

Adverse effects: Nausea, gastrointestinal complaints, diarrhoea and lymphocytic colitis may occur with internal administration. A doctor should be consulted if there is inflammation of the skin or subcutaneous induration, ulcers, sudden swelling of one or both legs, cardiac or renal insufficiency (EMA 2008).

Dosage: According to manufacturers' instructions for products, but up to 350 mg, three times daily, for dried powdered root extracts. Most clinical trials used a combination product containing butcher's broom (150 mg), twice a day, for 12 weeks.

General plant information: The herb is native to Mediterranean Europe and Africa. The young shoots have been eaten like those of asparagus, to which it is closely related. The flat shoots, known as cladodes, give the appearance of stiff, spine-tipped leaves, and the tough mature branches used to be bound into bundles and sold to butchers for sweeping their blocks, hence the name 'butcher's broom'.

References

Aguilar Peralta GR, Arevalo Gardoqui J, Llamas Macias FJ, Navarro Ceja VH, Mendoza Cisneros SA, Martinez Macias CG. (2007) Clinical and capillaroscopic evaluation in the treatment of chronic venous insufficiency with *Ruscus aculeatus*, hesperidin methylchalcone and ascorbic acid in venous insufficiency treatment of ambulatory patients. *International Angiology* 26(4): 378–384.

Allaert FA, Hugue C, Cazaubon M, Renaudin JM, Clavel T, Escourrou P. (2011) Correlation between improvement in functional signs and plethysmographic parameters during venoactive treatment (Cyclo 3 Fort). *International Angiology* 30(3): 272–277.

Anon. (2001) *Ruscus aculeatus* (Butcher's Broom). Monograph. *Alternative Medicine Review* 6(6): 608–612.

Barbič M, Willer EA, Rothenhöfer M, Heilmann J, Fürst R, Jürgenliemk G. (2013) Spirostanol saponins and esculin from Rusci rhizoma reduce the thrombin-induced hyperpermeability of endothelial cells. *Phytochemistry* 90: 106–113.

Boyle P, Diehm C, Robertson C. (2003) Meta-analysis of clinical trials of Cyclo 3 Fort in the treatment of chronic venous insufficiency. *International Angiology* 22(3): 250–262.

EMA. (2008) Community herbal monograph on *Ruscus aculeatus* L., rhizoma. European Medicines Agency http://www.ema.europa.eu/docs/en_GB/document_library/Herbal_-_Community_herbal_monograph/2009/12/WC500018286.pdf.

Guex JJ, Avril L, Enrici E, Enriquez E, Lis C, Taieb C. (2010) Quality of life improvement in Latin American patients suffering from chronic venous disorder using a combination of *Ruscus aculeatus* and hesperidin methyl-chalcone and ascorbic acid (quality study). *International Angiology* 29(6): 525–532.

Guex JJ, Enriquez Vega DM, Avril L, Boussetta S Taieb C. (2009) Assessment of quality of life in Mexican patients suffering from chronic venous disorder – impact of oral *Ruscus aculeatus*-hesperidin-methyl-chalcone-ascorbic acid treatment – 'QUALITY Study'. *Phlebology* 24(4): 157–165.

Huang YL, Kou JP, Ma L, Song JX, Yu BY. (2008) Possible mechanism of the anti-inflammatory activity of ruscogenin: role of intercellular adhesion molecule-1 and nuclear factor-kappaB. *Journal of Pharmacological Sciences* 108(2): 198–205.

Vanscheidt W, Jost V, Wolna P, Lucker PW, Muller A, Theurer C, Patz B, Grutzner KI. (2002) Efficacy and safety of a Butcher's broom preparation (*Ruscus aculeatus* L. extract) compared to placebo in patients suffering from chronic venous insufficiency. *Arzneimittelforschung* 52(4): 243–250.

Williamson EM, Driver S, Baxter K. (Eds.) (2013) *Stockley's Herbal Medicines Interactions.* 2nd Edition. Pharmaceutical Press, London, UK.

Butterbur

Petasites hybridus (L.) G.Gaertn., B.Mey. & Schreb.

Synonyms: *P. officinalis* Moench; *P. vulgaris* Desf.; *Tussilago hybrida* L.; *T. petasites* L.; and others

Family: Asteraceae (Compositae)

Other common names: Petasites; purple butterbur; sweet coltsfoot

Drug name: Petasitidis folium

Botanical drug used: Leaf (rhizome and radix are also used occasionally)

Indications/uses: At present, butterbur cannot be recommended until the reasons for the reported toxicity are resolved. Usually, it is used for the prophylactic treatment of migraines and as an antispasmodic agent for chronic cough or asthma.

Evidence: There is moderate evidence for its use for migraine prophylaxis and treatment of seasonal allergic rhinitis.

Safety: If the pyrrolizidine alkaloids have been removed from the extract, it is safer, but recently, commercially available products in which the alkaloids have been removed have also shown toxic effects. It would therefore be wise to avoid butterbur until this problem is resolved.

Main chemical compounds: The main constituents are the sesquiterpenes petasin and isopetasin, volatile oils, flavonoids and tannins (Aydin et al. 2013; Prieto 2014). Toxic pyrrolizidine alkaloids are present in the fresh plant, but preparations for internal use need to have these removed.

Clinical evidence: For migraine prophylaxis, two clinical trials have shown that 25 mg of standardised Petasites extract (in capsule form) twice daily for 12 weeks can significantly reduce the frequency and duration of migraine attacks compared to placebo, with no adverse reactions (Grossman and Schmidramsl 2001; Diener et al. 2004). However, a three-arm, parallel group, randomised trial comparing Petasites extract (75 mg and 50 mg twice daily) with placebo over 4 months found that only the higher dose of extract was more effective than placebo (Lipton et al. 2004).

Phytopharmacy: An evidence-based guide to herbal medicinal products, First Edition.
Sarah E. Edwards, Inês da Costa Rocha, Elizabeth M. Williamson and Michael Heinrich.
© 2015 John Wiley & Sons, Ltd. Published 2015 by John Wiley & Sons, Ltd.

Overall, the clinical evidence is promising (Prieto 2014), but systematic reviews suggest that there is only moderate evidence for its effectiveness in migraine prophylaxis. Further studies on dose/day and the products' efficacy are needed (Agosti et al. 2006; Sun-Edelstein and Mauskop 2009).

Butterbur extracts have also shown encouraging results for the treatment of seasonal allergic rhinitis (Schapowal 2005), and a similar effect was observed in a post-marketing surveillance study for a specific extract, which was found to be safe and efficacious at a dose of two tablets daily for 2 weeks (8 g petasins per tablet; Kaufeler et al. 2006; Guo et al. 2007: review).

A study using a combination of butterbur *root* with valerian root, lemon balm leaf and passionflower herb, was found to be efficacious and safe in a short-term study of patients with somatoform disorders (Melzer et al. 2009).

Pre-clinical evidence and mechanisms of action: The anti-inflammatory properties are likely to be due to inhibition of lipoxygenase activity and down-regulation of leukotriene synthesis. Petasin has antispasmodic properties including in cerebral blood vessels (Aydin et al. 2013) and butterbur also has a long tradition of use as an antispasmodic for gastrointestinal conditions. A pre-clinical study showed that ethanolic extracts of butterbur blocked ethanol-induced gastric damage and reduced ulcerations of the small intestine caused by indometacin in rats. The possible mechanism of action was thought to be due to inhibition of leukotriene (rather than prostaglandin) biosynthesis (Brune et al. 1993).

Interactions: None known.

Contraindications: Not recommended, especially during pregnancy and lactation, as safety has not been established, and in fact is being questioned.

Adverse effects: Liver toxicity of the pyrollizidine alkaloids is well documented and based on the carcinogenicity in animals and documented cases in humans (MHRA 2012; Prieto 2014; Williamson et al. 2013). Unlicensed butterbur products in the United Kingdom have been associated with 40 cases of liver toxicity, including two cases of liver transplant, and the MHRA issued a safety warning in January 2012 stating that butterbur can result in serious liver damage and organ failure. Butterbur is not registered or licensed by the MHRA as an HMP, but is found in some food supplements (MHRA 2012). Rashes (Danesch and Rittinghausen 2003) and gastrointestinal problems have been reported as side effects (Agosti et al. 2006) with specific standardised and patented extracts.

Dosage: At present, this herb cannot be recommended. If the reasons for the toxicity are resolved, manufacturers' instructions should be followed.

General plant information: Butterbur is a very common plant throughout Europe and in parts of Asia and North America. Cases of accidental confusion of *P. hybridus* with coltsfoot, *Tussilago farfara* L., have resulted in isolated cases of serious and occasionally lethal liver toxicity.

References

Agosti R, Duke RK, Chrubasik JE, Chrubasik S. (2006) Effectiveness of *Petasites hybridus* preparations in the prophylaxis of migraine: a systematic review. *Phytomedicine* 13(9–10): 743–746.

Aydin AA, Zerbes V, Parlar H, Letzel T. (2013) The medical plant butterbur (Petasites): analytical and physiological (re)view. *Journal of Pharmaceutical and Biomedical Analysis* 75: 220–229.

Brattström A, Schapowal A, Maillet I, Schnyder B, Ryffel B, Moser R. (2010) Petasites extract Ze 339 (PET) inhibits allergen-induced Th2 responses, airway inflammation and airway hyperreactivity in mice. *Phytotherapy Research* 24(5): 680–685.

Brune K, Bickel D, Peskar BA. (1993) Gastro-protective effects by extracts of *Petasites hybridus*: the role of inhibition of peptido-leukotriene synthesis. *Planta Medica* 59(6): 494–496.

Danesch U, Rittinghausen R. (2003) Safety of a patented special butterbur root extract for migraine prevention. *Headache* 43(1): 76–78.

Diener HC, Rahlfs VW, Danesch U. (2004) The first placebo-controlled trial of a special butterbur root extract for the prevention of migraine: reanalysis of efficacy criteria. *European Neurology* 51(2): 89–97.

Grossman W, Schmidramsl H. (2001) An extract of *Petasites hybridus* is effective in the prophylaxis of migraine. *Alternative Medicine Review* 6(3): 303–310.

Guo R, Pittler MH, Ernst E. (2007) Herbal medicines for the treatment of allergic rhinitis: a systematic review. *Annals of Allergy, Asthma & Immunology* 99(6): 483–495.

Kaufeler R, Polasek W, Brattstrom A, Koetter U. (2006) Efficacy and safety of butterbur herbal extract Ze 339 in seasonal allergic rhinitis: postmarketing surveillance study. *Advances in Therapy* 23(2): 373–384.

Lipton RB, Gobel H, Einhaupl KM, Wilks K, Mauskop A. (2004) *Petasites hybridus* root (butterbur) is an effective preventive treatment for migraine. *Neurology* 63(12): 2240–2244.

Melzer J, Schrader E, Brattström A, Schellenberg R, Saller R. (2009) Fixed Herbal Drug Combination With and Without Butterbur (Ze 185) for the Treatment of Patients with Somatoform Disorders: Randomized, Placebo-controlled Pharmacoclinical Trial. *Phytotherapy Research* 23(9) 1303–1308.

MHRA. (2012). Consumers are advised not to take unlicensed Butterbur (*Petasites hybridus*) herbal remedies http://www.mhra.gov.uk/Safetyinformation/Generalsafetyinformation andadvice/Herbalmedicines/Herbalsafetyupdates/Allherbalsafetyupdates/CON140849 (accessed July 2012).

Prieto J. (2014) Update on the efficacy and safety of Petadolex®, a butterbur extract for migraine prophylaxis. *Botanics: Targets and Therapy* 4: 1–9.

Schapowal A. (2005) Treating Intermittent Allergic Rhinitis: A Prospective, Randomized, Placebo and Antihistamine-controlled Study of Butterbur Extract Ze 339. *Phytotherapy Research* 19(6): 530–537.

Sun-Edelstein C, Mauskop A. (2009) Foods and supplements in the management of migraine headaches. *Clinical Journal of Pain* 25(5): 446–452.

Williamson EM, Driver S, Baxter K. (Eds.) (2013) *Stockley's Herbal Medicines Interactions*. 2nd Edition. Pharmaceutical Press, London, UK.

Calendula
Calendula officinalis L.

Synonyms: *Caltha officinalis* (L.) Moench; and others

Family: Asteraceae (Compositae)

Other common names: Gold bloom; marigold; pot marigold

Drug name: Calendulae flos

Botanical drug used: Flowers (dried ray florets)

Indications/uses: Calendula extract-containing creams and gels are used to treat skin irritation, inflammation and burns, especially after radiotherapy, and for nipple soreness when breastfeeding, for nappy rash (diaper dermatitis) and to aid wound healing. Internally, it has been used for stomach disorders and inflammation, but this is much less frequent.

Evidence: Calendula is a popular and well-known topical anti-inflammatory agent with a very long history of use. There is evidence from *in vitro* and animal studies to support its use as a treatment for burns, including radiation-induced, and as a wound-healing agent, but there is little clinical trial evidence.

Safety: *In vivo* studies show that calendula has low toxicity when applied topically, but may cause allergic reactions. When taken internally, the extract is of low toxicity in animals, but no data is available from human studies. There is a lack of safety data to recommend its use during pregnancy and lactation.

Main chemical compounds: The major actives are the triterpene saponins, known as calendulosides and calendasaponins, the triterpene alcohols such as calenduladiol and the heliantriols; sesquiterpene glycosides known as officinosides; carotenoids including lycopene and beta-carotene; flavonoids including quercetin, isoquercitrin and rutin); essential oil; polysaccharides (Basch et al. 2006; Della Loggia et al. 2004; Williamson 2003).

Clinical evidence: Clinical studies evaluating calendula are sparse and often of poor quality. Calendula cream applied topically has been reported to reduce the incidence of skin reactions (grade 2 or 3) to radiation treatment in women with breast cancer (McQuestion 2011), but other studies have found no difference

Phytopharmacy: An evidence-based guide to herbal medicinal products, First Edition.
Sarah E. Edwards, Inês da Costa Rocha, Elizabeth M. Williamson and Michael Heinrich.
© 2015 John Wiley & Sons, Ltd. Published 2015 by John Wiley & Sons, Ltd.

between calendula formulations and placebo (e.g. Sharp et al. 2013). Even though the evidence based on clinical studies is somewhat limited, its use for skin irritation, inflammation and burns based on long-standing medical experience provides evidence for its use.

Pre-clinical evidence and mechanisms of action: Calendula extract in a gel formulation protected against UV-irradiation-induced oxidative stress in the skin of hairless mice. It was thought to be due to an increase in collagen synthesis in the sub-epidermal connective tissue and maintenance of glutathione levels (e.g. Fonseca et al. 2011). Wound-healing effects have been demonstrated using an *in vitro* scratch assay in 3T3 fibroblasts (Fronza et al. 2009) and an *in vivo* excision wound model in rats (Preethi and Kuttan 2009).

In vitro studies have indicated that calendula extract has antiviral, anti-HIV and antifungal properties (e.g. Efstratiou et al. 2012; Kalvatchev et al. 1997), and constituents from calendula (oleanolic acid and its glycosides) have been demonstrated to be antibacterial and antiparasitic (Szakiel et al. 2008). Both *in vitro and in vivo* studies have demonstrated topical anti-inflammatory activity (Della Loggia et al. 1994; Preethi et al. 2009). The mechanisms of action are not well understood and are probably multiple, and conferred by several different constituents. Triterpenes, notably faradiol, have potent anti-inflammatory activity (Della Loggia et al. 1994) and the presence of carotenoids such as lycopene (which reduces proinflammatory cytokines), may also contribute to the anti-inflammatory action (Preethi et al. 2009).

Anti-angiogenic and anticancer effects have been reported for calendula extracts, but these indications are outside the scope of this book and there are no clinical studies to support such uses.

Interactions: No reports of interactions are available for calendula administered either internally or externally. Calendula is reputed to cause drowsiness, although there is no evidence for this, so it may be wise to avoid internal use in combination with central nervous system (CNS) sedatives (Williamson et al. 2013).

Contraindications: Contraindicated in cases of allergy to plants from the Asteraceae (daisy family). *In vitro* studies in isolated rabbit and guinea pig uterine tissues have demonstrated that calendula has a weak 'uterotonic' effect. Anecdotal reports indicate spermatocidal and abortifacient effects of calendula. There is a lack of safety data to recommend its use during pregnancy and lactation, and systemic effects from topical use are unclear (Basch et al. 2006).

Adverse effects: Toxicity studies in rodents indicate that calendula is relatively non-toxic (Lagarto et al. 2011). However, anaphylactic shock has been reported in one case after gargling with calendula infusion (Basch et al. 2006).

Dosage: For adults, a topical application of 2–5% ointment should be applied three to four times daily as needed.

General plant information: The name 'marigold', most likely derives from 'Mary's Gold', as medieval Christians dedicated the plant to the Virgin Mary and the flowers were placed around statues of Mary as an offering in place of coins. Calendula is also often called 'pot marigold', since historically it was used as a condiment to flavour stews, or as a substitute for the expensive saffron. Calendula should not be confused with *Tagetes* species, commonly known as French or African marigolds.

References

Basch E, Bent E, Foppa I, et al. (2006) Marigold (*Calendula officinalis* L.): an evidence-based systematic review by the Natural Standard Research Collaboration. *Journal of Herbal Pharmacotherapy* 6(3–4): 135–159.

Della Loggia R, Tubaro A, Sosa S, et al. (1994) The role of triterpenoids in the topical anti-inflammatory activity of *Calendula officinalis* flowers. *Planta Medica* 60(6): 516–520.

Efstratiou E, Hussain AI, Nigam PS, et al. (2012) Antimicrobial activity of *Calendula officinalis* petal extracts against fungi, as well as Gram-negative and Gram-positive clinical pathogens. *Complementary Therapies in Clinical Practice* 18(3): 173–176.

Fonseca YM, Catini CD, Vicentini FT, Cardoso JC, Cavalcanti De Albuquerque Jr RL, Viera Fonseca MJ. (2011) Efficacy of marigold extract-loaded formulations against UV-induced oxidative stress. *Journal of Pharmaceutical Sciences* 100(6): 2182–2193.

Fronza M, Heinzmann B, Hamburger M, Laufer S, Merfort I. (2009) Determination of the wound healing effect of *Calendula* extracts using the scratch assay with 3T3 fibroblasts. *Journal of Ethnopharmacology* 126(3): 463–467.

Kalvatchev Z, Walder R, Garzaro D. (1997) Anti-HIV activity of extracts from *Calendula officinalis* flowers. *Biomedicine & Pharmacotherapy* 51(4): 176–180.

Lagarto A, Bueno V, Guerra I, et al. (2011) Acute and subchronic oral toxicities of *Calendula officinalis* extract in Wistar rats. *Experimental and Toxicologic Pathology* 63(4): 387–391.

McQuestion M. (2011) Evidence-based skin care management in radiation therapy: clinical update. *Seminars in Oncology Nursing* 27(2): e1–17.

Preethi KC, Kuttan G, Kuttan R. (2009) Anti-inflammatory activity of flower extract of *Calendula officinalis* Linn. and its possible mechanism of action. *Indian Journal of Experimental Biology* 47(2): 113–120.

Preethi KC, Kuttan R. (2009) Wound healing activity of flower extract of *Calendula officinalis*. *Journal of Basic and Clinical Physiology and Pharmacology* 20(1): 73–79.

Sharp L, Finnilä K, Johansson H, Abrahamsson M, Hatschek T, Bergenmar M. (2013) No differences between Calendula cream and aqueous cream in the prevention of acute radiation skin reactions – results from a randomised blinded trial. *European Journal of Oncology Nursing* 17(4): 429–435.

Szakiel A, Ruszkowski D, Grudniak A, et al. (2008) Antibacterial and antiparasitic activity of oleanolic acid and its glycosides isolated from marigold (*Calendula officinalis*). *Planta Medica* 74(14): 1709–1715.

Williamson EM. (2003) *Potter's Cyclopedia of Herbal Medicines.* C W Daniels, UK.

Williamson EM, Driver S, Baxter K. (Eds.) (2013) *Stockley's Herbal Medicines Interactions.* 2nd Edition. Pharmaceutical Press, London, UK.

Cannabis   *Cannabis sativa* L.

Synonym: *C. indica* Lam; *C. ruderalis* Janisch

Family: Cannabaceae

Other common names: Ganja; hashish; Indian hemp; marihuana; marijuana

Drug Name: *Cannabis sativa* L., folium cum flore

Botanical drug used: Extract of leaves and flowers

Indications/uses: It is not appropriate for pharmacists or other health care professionals to recommend cannabis for therapeutic use, although the licensed product may be *prescribed* for the treatment of pain and intractable spasms in multiple sclerosis. There is anecdotal and pre-clinical evidence for its use in epilepsy, chronic pain, depression, appetite regulation and many other conditions, where it has been taken illicitly as a form of self-medication, as well as the recreational use to produce euphoria.

Evidence: The commercial product, Sativex, has sufficient clinical evidence to be licensed for use for symptoms of multiple sclerosis, and there is a considerable body of data to support (but not prove unequivocally) other medicinal uses of cannabis, many of which are currently under investigation or in clinical trial, including epilepsy and cancer.

Safety: The registered product is very safe in therapeutic doses, but illicit cannabis use is associated with an increase in mental illness and concerns over competence in driving, especially if taken in conjunction with alcohol or other psychoactive drugs.

Main chemical compounds: Cannabis contains a wide range of cannabinoids, which are the major active compounds and unique to cannabis. Delta-9-tetrahydrocannabinol (THC) is the main psychoactive constituent and the cause of many of the central nervous system (CNS) effects of cannabis. However, other cannabinoids, which are non-cannabimimetic (such as cannabidiol, cannabinol, cannabigerol, cannabichromene, tetrahydrocannabivarin and cannabidivarin) are increasingly being investigated for their pharmacological and therapeutic properties. Cannabinoids are often found in the plant as their acid metabolites, but

---

Phytopharmacy: An evidence-based guide to herbal medicinal products, First Edition.
Sarah E. Edwards, Inês da Costa Rocha, Elizabeth M. Williamson and Michael Heinrich.
© 2015 John Wiley & Sons, Ltd. Published 2015 by John Wiley & Sons, Ltd.

are decarboxylated at high temperatures (i.e. when smoked), whereas medicinal cannabis products are heat-treated to ensure they are present in the non-acid form. There are many other, non-cannabinoid, components including terpenes, carotenoids and flavonoids (Russo 2011). Pharmaceutically, a product made from a standardised extract of cannabis herb grown from genetically characterised strains of the plant is becoming more widely available as a sublingual spray. It contains specified amounts of THC and cannabidiol. Delta-9-THC is also made synthetically (known as dronabinol) and is marketed in the form of capsules for oral administration.

Clinical evidence: This brief information is provided so that pharmacists can assess the risk of drug interactions through the illicit use of the drug, as well as in the event of prescription use. Much of the clinical evidence is from small trials (Amar 2006), with the exception of the licensed use in multiple sclerosis (Pryce and Baker 2012) and appetite stimulation in HIV (Farrimond et al. 2011). There is anecdotal and pre-clinical evidence that cannabis and/or cannabinoids may be useful in the treatment of epilepsy, depression, intractable pain and cancer, and many trials are under way to assess the potential benefit of cannabis in these conditions. Cannabis has many other properties which may be exploited therapeutically in the future for CNS disorders (Hill et al. 2012), appetite regulation (Farrimond et al. 2011), cardiovascular disease (Montecucco and Di Marzo 2012), inflammatory bowel disorders (Esposito et al. 2013) and other non-CNS disorders (Izzo et al. 2009).

Pre-clinical evidence and mechanisms of action: The mechanisms of actions are wide and varied, and some remain to be elucidated. THC (the psychoactive component) interacts with the cannabinoid (CBR)-1 receptor. Other cannabinoids act by different mechanisms, including CBR-2 and other receptors and enzymes. Some cannabinoids and extracts (depending on composition) have been shown to inhibit neurotransmission, confer neuroprotection in certain types of trauma, induce apoptosis in cancer cells and modulate immune responses (Izzo et al. 2009; Hill et al. 2012). Other non-cannabinoid compounds are thought to contribute to the therapeutic effects of cannabis (Russo 2011).

Interactions with other drugs: The evidence for the interactions of cannabis is weak at present, since it has only recently been legally authorised for therapeutic use. Although cannabis can modulate CYP3A4 activity, clinical studies show that it does not affect the pharmacokinetics of drugs such as docetaxel or indinavir. However, it may affect ciclosporin metabolism, causing increased ciclosporin blood levels and thus a possible increase in toxic side effects. Full details of current knowledge of cannabis pharmacokinetics and drug interactions can be obtained from *Stockley's Herbal Medicines Interactions* (Williamson et al. 2013). The most likely and relevant combinations are explained briefly below.

THC may decrease phenytoin levels to a degree which may affect efficacy, although only experimental evidence is available to support this. Patients with epilepsy (one of the main indications for phenytoin) sometimes take cannabis for therapeutic reasons, albeit illicitly, as well as recreationally, so if phenytoin suddenly becomes less effective, this possible combination should be borne in mind. Cannabis has complex effects in epilepsy, even if taken alone.

Patients who are taking clozapine and who give up smoking cannabis may develop higher blood levels of clozapine and be at risk of adverse reactions, since plasma levels of clozapine are lower in smokers than in non-smokers. Clozapine has been shown to act at cannabinoid receptors, which may help reduce cannabis consumption, but any therapeutic effect (beneficial or otherwise) has yet to be proved. Chlorpromazine blood levels are also lowered by cannabis smoking, but significance of the clinical effect is not known.

A case of sildenafil toxicity has been reported in a patient who smoked cannabis at the same time, but the causality was not proved. This possible combination should be borne in mind if unexpected sildenafil toxicity occurs.

Patients taking cannabis frequently take alcohol or other illicit drugs at the same time. It has long been accepted that alcohol potentiates some of the effects of cannabis, but a study has shown that experienced cannabis users may be able to overcome some of the negative cognitive effects. There is also a pharmacological interaction between cannabis and cocaine, and with ecstasy (MDMA, 3,4-methylenedioxy-N-methylamphetamine) and phencyclidine. The interactions between these agents are in addition to the expected psychoactive additive effects.

Low doses of cannabis have been shown to enhance the analgesic effects of opioids and it has been suggested that this could be used to advantage in patients.

Contraindications: Cannabis use is normally illegal, but medicinal products based on cannabis are licensed for certain indications. It is not appropriate for pharmacists to recommend that patients should take cannabis, although they could suggest a discussion with their medical specialist.

Adverse effects: Sedation, anxiety, impairment of cognition. These have been shown to be rare when the licensed product is used (Wade 2012). Recreational cannabis use is associated with an increased risk of psychiatric disorders such as schizophrenia and bipolar disorder, but causality is disputed.

Dosage: According to manufacturer's instructions.

General plant information: Cannabis grows all over the world and has many economic uses unrelated to its medicinal or recreational applications. The stems, usually of low-THC-containing varieties, are a source of fibre for cloth and sail making, and the seeds are used to produce hemp oil, which is rich in omega-3 fatty acids and gamma-linolenic acid. It is one of the oldest economically important plants and its use has been recorded for at least 2500 years (Jiang et al. 2006).

References

The primary literature for cannabis is extensive; this is a selection of recent reviews containing very many primary reference sources.

Amar MB. (2006) Cannabinoids in medicine: A review of their therapeutic potential. *Journal of Ethnopharmacology* 105(1–2): 1–25.

Esposito G, Filippis DD, Cirillo C, Iuvone T, Capoccia E, Scuderi C, Steardo A, Cuomo R, Steardo L. (2013) Cannabidiol in inflammatory bowel diseases: a brief overview. *Phytotherapy Research* 27(5): 633–636.

Farrimond JA, Mercier MS, Whalley BJ, Williams CM. (2011) *Cannabis sativa* and the endogenous cannabinoid system: therapeutic potential for appetite regulation. *Phytotherapy Research* 25(2): 170–188.

Hill AJ, Williams CM, Whalley BJ, Stephens GJ. (2012) Phytocannabinoids as novel therapeutic agents in CNS disorders. *Pharmacology & Therapeutics* 133(1): 79–97.

Izzo AA, Borrelli F, Capasso R, Di Marzo V, Mechoulam R. (2009) Non-psychotropic plant cannabinoids: new therapeutic opportunities from an ancient herb. *Trends in Pharmacological Sciences* 30(10): 515–27.

Jiang HE, Li X, Zhao YX, Ferguson DK, Hueber F, Bera S, Wang YF, Zhao LC, Liu CJ, Li CS. (2006) A new insight into *Cannabis sativa* (Cannabaceae) utilization from 2500-year-old Yanghai Tombs, Xinjiang, China. *Journal of Ethnopharmacology* 108(3): 414–422.

Montecucco F, Di Marzo V. (2012) At the heart of the matter: the endocannabinoid system in cardiovascular function and dysfunction. *Trends in Pharmacological Sciences* 33(6): 331–340.

Pryce G, Baker D. (2012) Potential control of multiple sclerosis by cannabis and the endocannabinoid system. *CNS & Neurological Disorders – Drug Targets* 11(5): 624–641.

Russo EB. (2011) Taming THC: potential cannabis synergy and phytocannabinoid-terpenoid entourage effects. *British Journal of Pharmacology* 163(7): 1344–1364.

Wade D. (2012) Evaluation of the safety and tolerability profile of Sativex: is it reassuring enough? *Expert Review of Neurotherapeutics* 12(4 Suppl): 9–14.

Williamson EM, Driver S, Baxter K. (Eds.) (2013) *Stockley's Herbal Medicines Interactions*. 2nd Edition. Pharmaceutical Press, London, UK.

Centaury   *Centaurium erythraea* Rafn

Synonyms: *C. majus* (Boiss.) Druce; *C. minus* Moench; *C. umbellatum* Gilib. ex Beck; *Erythraea centaurium* (L.) Pers.; *Gentiana centaurium* L.; and others

Family: Gentianaceae

Other common names: Bitter-herb; common centaury; European centaury; feverwort

Drug name: Centaurii herba

Botanical drug used: Dried flowering aerial parts

Indications/uses: Traditionally, centaury has been used to treat dyspepsia, abdominal colic, flatulence, bloating and other gastrointestinal complaints. It is also used to treat temporary loss of appetite. Traditionally, it was used as a liver and kidney tonic. Under the THR scheme, it is part of a combination product used to 'help flushing of the urinary tract and to assist in minor urinary complaints associated with cystitis in women only' (in combination with rosemary leaf, *Rosmarinus officinalis* L. and lovage root, *Levisticum officinale* W.D.J. Koch). In North African traditional medicine it is used to treat diabetes mellitus.

Evidence: Clinical data to support the traditional uses of centaury are lacking. The intensely bitter chemical constituents (iridoids) make its use as an appetite stimulant plausible.

Safety: Limited toxicological data from a study in rodents indicates that centaury is relatively safe if taken orally (Tahraoui et al. 2010) and the constituents are found in many other bitter herbs. However, there is a lack of safety data in humans. The use of centaury during pregnancy and breastfeeding, and in patients under 18 years of age should thus be avoided. Long-term use in adults is also not recommended.

Main chemical compounds: Iridoids, including approximately 2% gentiopicroside, with secoridioid glycosides including centapicrin, centauroside and swertiamarin (swertiamaroside); xanthones such as eustomin and methylswertianin; triterpenoids (e.g. amyrin, erythrodiol, oleanolic acid, sitosterol); phenolic acids (e.g. protocatechuic, hydrobenzoic, ρ-coumaric, vanillic, caffeic) and alkaloids (e.g. traces of gentianine, gentioflavine, gentianidine) (Aberham et al. 2011; Pharmaceutical Press Editorial Team 2013).

Phytopharmacy: An evidence-based guide to herbal medicinal products, First Edition.
Sarah E. Edwards, Inês da Costa Rocha, Elizabeth M. Williamson and Michael Heinrich.
© 2015 John Wiley & Sons, Ltd. Published 2015 by John Wiley & Sons, Ltd.

Clinical evidence: There are no systematic clinical studies available on the medicinal uses of centaury. A combination product also containing *Levisticum officinale* W.D.J. Koch and *Rosmarinus officinalis* L. was shown to be effective in the treatment and prophylaxis of urinary tract infection (UTI) compared with standard therapy, both in adults and children, and there was a reduced number of relapses with a very good safety profile (Naber 2013).

Pre-clinical evidence and mechanisms of action: The herb is very bitter, which is the basis for the digestive use. Centapicrin has a bitterness value of about 4,000,000 and is one of the most intensely bitter compounds known, and sweroside has a bitterness value of about 12,000. A study in rats showed that a 50% aqueous-ethanolic centaury extract exhibited protective effects against aspirin-induced gastric ulcer and it was suggested that this was a result of its antioxidant activity (Tuluce et al. 2011; Valentão et al. 2003).

An aqueous extract of centaury lowered blood glucose *in vivo* in streptozotocin-induced diabetic rats. Degenerative changes in pancreatic β-cells were also minimised in the centaury treated diabetic rats, indicating the protective nature of the extract. The authors concluded that the therapeutic effect of centaury in diabetes was attributable to its antioxidant potential (Sefi et al. 2011). However, more recently, the antidiabetic effects of swertiamarin have been shown to be due to an active metabolite of swertiamarin, gentianine (also present in centaury), which increased adipogenesis, and was associated with a significant increase in the mRNA expression of PPAR-γ, GLUT-4 and adiponectin, in rats (Vaidya et al. 2013).

Animal studies have shown anti-inflammatory and antipyretic (but not analgesic) actions of aqueous centaury extract (Berkan et al. 1991; Pharmaceutical Press Editorial Team 2013). The methanol extract has been shown to exhibit hepatoprotective activity against paracetamol (acetaminophen)-induced liver toxicity in rats (Mroueh et al. 2004) and the ethanolic extract has been shown to possess antimutagenic effects (Valentão et al. 2003).

Interactions: None known.

Contraindications: Centaury should not be used by those suffering from active peptic ulcers, or if hypersensitive to the active substance (EMA 2009).

Adverse effects: None reported.

Dosage: In adults and elderly (oral administration): (a) dried herb for tea preparation: 1–4 g up to four times daily; (b) powdered herb: 0.25–2 g, up to three times daily (EMA 2009). For other products and formulations, follow the manufacturers' instructions.

General plant information: Centaury is native to much of Europe, the Mediterranean region and western Asia (including Afghanistan, Pakistan and Iran). It is widely naturalised in other parts of the world, including North America and Australia. The name 'centaury' comes from its association in Greek mythology with the centaur Chiron, who was skilled in medicinal herbs and used the plant to cure himself of a poisoned arrow wound (Grieve 1931). The flowers are particularly rich in centapicrin, so if the herb has a high content of flowers, it will have a much higher bitterness value.

References

Aberham A, Pieri V, Croom EM Jr, Ellmerer E, Stuppner H. (2011) Analysis of iridoids, secoiridoids and xanthones in *Centaurium erythraea, Frasera caroliniensis* and *Gentiana lutea* using LC-MS and RP-HPLC. *Journal of Pharmaceutical and Biomedical Analysis* 54(3): 517–525.

Berkan T, Üstünes L, Lermioglu F, Ozer A. (1991) Antiinflammatory, analgesic, and antipyretic effects of an aqueous extract of *Erythraea centaurium*. *Planta Medica* 57(1): 34–37.

EMA. (2009) Community Herbal Monograph on *Centaurium erythraea* Rafn, herba. European Medicines Agency http://www.ema.europa.eu/docs/en_GB/document_library /Herbal_-_Community_herbal_monograph/2009/12/WC500018164.pdf (accessed July 2013).

Grieve M. (1931) *A Modern Herbal*. Harcourt, Brace & Company / (1971) Dover Publications, USA / (1995) online version Ed Greenwood, USA http://www.botanical.com /botanical/mgmh/c/centau46.html (accessed July 2013).

Mroueh M, Saab Y, Rizkallah R. (2004) Hepatoprotective activity of *Centaurium erythraea* on acetaminophen-induced hepatotoxicity in rats. *Phytotherapy Research* 18(5): 431–433.

Naber KG. (2013) Efficacy and safety of the phytotherapeutic drug Canephron® N in prevention and treatment of urogenital and gestational disease: review of clinical experience in Eastern Europe and Central Asia. *Research and Reports in Urology* 5: 39–46.

Pharmaceutical Press Editorial Team. (2013) *Herbal Medicines*. 4th Edition. Pharmaceutical Press, London, UK.

Sefi M, Fetoui H, Lachkar N, Tahraoui A, Lyoussi B, Boudawara T, Zeghal N. (2011) *Centaurium erythrea* [sic] (Gentianaceae) leaf extract alleviates streptozotocin-induced oxidative stress and β-cell damage in rat pancreas. *Journal of Ethnopharmacology* 135(2): 243–250.

Tahraoui A, Israili ZH, Lyoussi B. (2010) Acute and sub- chronic toxicity of a lyophilised aqueous extract of *Centaurium erythraea* in rodents. *Journal of Ethnopharmacology* 132(1): 48–55.

Tuluce Y, Ozkol H, Koyuncu I, Ine H. (2011) Gastroprotective effect of small centaury (Centaurium erythraea L) on aspirin-induced gastric damage in rats. *Toxicology and Industrial Health* 27(8): 760–768.

Valentão P, Fernandes E, Carvalho F, Andrade PB, Seabra RM, Bastos ML. (2003) Hydroxyl radical and hypochlorous acid scavenging activity of small Centaury (*Centaurium erythraea*) infusion. A comparative study with green tea (*Camellia sinensis*). *Phytomedicine* 10(6-7): 517–522.

Vaidya H, Goyal RK, Cheema SK. (2013) Anti-diabetic activity of swertiamarin is due to an active metabolite, gentianine, that upregulates PPAR-γ gene expression in 3T3-L1 cells. *Phytotherapy Research* 27(4): 624–627.

Centella   *Centella asiatica* (L.) Urb.

Synonyms: *Hydrocotyle asiatica* L.; *Trisanthus cochinchinensis* Lour.; and others

Family: Apiaceae (Umbelliferae)

Other common names: Asiatic pennywort; gotu kola; hydrocotyle; Indian pennywort; Indian water navelwort

Drug name: Centellae herba

Botanical drug used: Dried, fragmented aerial parts

Indications/uses: Centella is now mainly used for symptoms of chronic venous insufficiency (CVI), including varicose veins, varicose ulcers, diabetic and airline flight microangiopathy. The herb is also used to aid wound healing and treat skin conditions such as burns, scarring, eczema and psoriasis, as a topical application and oral tablets. It is used in cosmetic preparations to reduce the signs of skin ageing. Centella is also taken orally to relieve anxiety and improve cognition. It has traditionally been used for many other conditions including diarrhoea, fever and amenorrhoea.

Evidence: The clinical evidence for use in CVI and some skin conditions is good, and several licensed preparations containing standardised Centella extract are available. These include creams and powders for cutaneous use, and tablets containing the total triterpenic fraction of the extract. For other indications, the evidence is weak.

Safety: Generally safe at recommended doses, but avoid in pregnancy and breastfeeding. Allergic reactions have been reported after topical application.

Main chemical compounds: The dried aerial parts should contain a minimum of 6% of total triterpenoid derivatives, based on Asiatic acid and madecassic acid and their derivatives, and include asiaticoside B, asiaticoside A (=madecassoside), braminoside, brahmoside, brahminoside, thankuniside isothankuniside. In addition to about 0.1% essential oil composed of β-caryophyllene, trans-p-farnesene and germacrene, *C. asiatica* contains a wide range of other substances including carotenoids and flavonoids (EMA 2012; Orhan 2012, Ghedira and Goetz 2013).

Phytopharmacy: An evidence-based guide to herbal medicinal products, First Edition.
Sarah E. Edwards, Inês da Costa Rocha, Elizabeth M. Williamson and Michael Heinrich.
© 2015 John Wiley & Sons, Ltd. Published 2015 by John Wiley & Sons, Ltd.

Clinical Evidence: A systematic review of the efficacy of *C. asiatica* for improvement of the signs and symptoms of CVI, including varicose veins, found eight studies which met the inclusion criteria. It concluded that the herb significantly improved microcirculatory parameters, such as rate of ankle swelling and veno-arteriolar response (Chong and Aziz 2013; EMA 2010). Small clinical studies have also found a potential use in scar management and prevention of diabetic and airline flight microangiopathy (Anon. 2007; EMA 2010).

Human studies have shown that oral administration of Centella extract can produce an enhancement of mood and a reduction in anxiety, and also an improvement in measures of cognition (Anon. 2007; EMA 2010). For other indications, the evidence is very weak.

Pre-clinical evidence and mechanisms of action: The properties of Centella have been investigated widely in animal and cell-based studies. Most of the experimental work has been carried out using the triterpene fraction or asiaticoside, or extracts standardised to the triterpene content. The mechanisms behind the wound-healing effects include enhancing microcirculation and collagen synthesis, and stimulation of epithelisation and angiogenesis. Raised levels of monocyte chemoattractant protein-1 (MCP-1) in keratinocytes have been found following Centella administration, and increased vascular endothelial growth factor (VEGF) and interleukin (IL)-1β levels have been found in burn wound exudates. This suggests that the enhancement of burn wound healing might be due to the promotion of angiogenesis during wound repair as a result of the stimulation of VEGF production. These mechanisms are also relevant in the treatment of chronic venous sufficiency.

The extract has anticholinesterase, neuroprotective and antioxidant effects, supporting its use in cognitive and depressive disorders. A GABA-stimulating effect has been shown, and after orally administered leaf extract, an improvement in recovery with increased axonal regeneration following nerve damage and neuronal dendritic growth stimulation have been observed in brain areas involved in memory and learning processes. It has been suggested that the ERK/RSK signalling pathway mediates central effects. Centella extract decreases β-amyloid levels, reduces lipid peroxidation and protects against DNA damage in an animal model of Alzheimer's disease (EMA 2010; Orhan 2012).

Interactions with other drugs: No clinically significant interactions have been recorded (Williamson et al. 2013). Caution should be advised in patients taking other sedative preparations in case of additive effects, but no clinical cases have been reported (EMA 2010).

Contraindications: Allergy to plants of the Apiaceae. Centella should be avoided during pregnancy, due to its reported emmenagogue action (Anon. 2007).

Adverse effects: The tolerability of oral *C. asiatica* preparations is generally good and Centella has no reported toxicity in recommended doses. Contact dermatitis has been reported on a few occasions using topical preparations, and gastric complaints and nausea have occasionally been seen following oral administration. High doses are said to have sedative properties (EMA 2010).

Dosage: 0.6 g dried herb as an infusion, tincture or extract, up to four times daily. For products, use according to manufacturers' instructions. Tablets containing a total triterpenic fraction of the extract (oral dose of 60–120 mg daily), and for external use, creams containing 1% extract and powders with 2% extract, are available (EMA 2010).

General plant information: Centella is found in most tropical and subtropical countries growing in swampy areas, including parts of India, Pakistan, Sri Lanka, Madagascar, South Africa, the South Pacific and Eastern Europe. It has a long history of use, dating back to ancient Chinese and Ayurvedic literature. In traditional Chinese medicine it is known as Leigonggen, and in Ayurveda, as Mandukparni, where it is one of the main herbs for revitalising the nervous system and has been used to treat emotional disorders such as depression. The leaves are edible and widely used in beverages, salads and curries throughout South and Southeast Asia. It is sometimes also called 'Brahmi' (see *Bacopa monnieri* (L.) Wettst.), p. 69.

References

The primary literature for Centella is extensive; this is a selection of good recent reviews, which include many referenced individual studies. The EMA assessment is particularly comprehensive.

Anonymous. (2007) *Centella asiatica*. Monograph. *Alternative Medicine Review* 12(1): 69–72.

Chong NJ, Aziz Z. (2013) A Systematic Review of the Efficacy of *Centella asiatica* for Improvement of the signs and symptoms of chronic venous insufficiency. *Evidence Based Complementary and Alternative Medicine* 2013: 627182.

Ghedira K, Goetz P. (2013) Hydrocotyle: *Centella asiatica* (L.) Urban (Apiaceae). *Phytotherapie* 11(5): 310–315.

EMA. (2010) Assessment report on *Centella asiatica* (L.) Urban, herba. European Medicines Agency. http://www.ema.europa.eu/ema/index.jsp?curl=pages/medicines/herbal/medicines /herbal_med_000046.jsp&mid=WC0b01ac058001fa1d.

Orhan IE. (2012) *Centella asiatica* (L.) Urban: from traditional medicine to modern medicine with neuroprotective potential. *Evidence Based Complementary and Alternative Medicine* 2012: 946259.

Williamson EM, Driver S, Baxter K. (Eds.) (2013) *Stockley's Herbal Medicines Interactions*. 2nd Edition. Pharmaceutical Press, London, UK.

Chamomile, German   *Matricaria chamomilla* L.

Synonyms: *Chamomilla recutita* (L.) Rausch.; *C. officinalis* K.Koch; *M. recutita* L.; and others

Family: Asteraceae (Compositae)

Other common names: Blue chamomile; hungarian chamomile; matricaria; sweet false chamomile; wild chamomile

Drug names: Matricariae Flos

Botanical drug used: Flowerhead, flowering top

Indications/uses: Chamomile is widely used both internally and externally to treat inflammatory conditions. Topically, chamomile cream, gel and ointment are applied to treat nappy rash, sore and cracked skin (especially sore nipples during breastfeeding), haemorrhoids, bruises, insect bites and burns, including after radiotherapy. Chamomile products are also used to soothe inflammation of the mouth and gums, and for infant teething. Internally, usually in the form of a tea, chamomile is used to treat digestive ailments, including bloating and flatulence, and as an anxiolytic, for restlessness and mild insomnia due to nervous disorders.

Evidence: Chamomile is very popular and there is much anecdotal evidence for its anti-inflammatory effects both topically and internally. Apart from some small studies on skin inflammation and itching, there is little clinical evidence to prove efficacy.

Safety: Generally regarded as safe but allergic reactions may occur in some individuals.

Main chemical compounds: *M. chamomilla* produces an essential oil (0.4–1.5%) which is very variable in content depending on the chemotype and origin. It has an intense blue colour due to its chamazulene content. The major active constituents are considered to be α-bisabolol and its oxides, with farnesene, azulene and spiroethers. Flavonoid glycosides including apigenin, quercetin, patuletin, luteolin and their glucosides constitute up to 8% (dry weight) together with sesquiterpene lactones and coumarins (Mckay and Blumberg 2006; Tschiggerl and Bucar 2012; Tisserand and Young 2014).

Phytopharmacy: An evidence-based guide to herbal medicinal products, First Edition.
Sarah E. Edwards, Inês da Costa Rocha, Elizabeth M. Williamson and Michael Heinrich.
© 2015 John Wiley & Sons, Ltd. Published 2015 by John Wiley & Sons, Ltd.

Clinical evidence:

There is plenty of supporting pharmacological evidence, but few good clinical studies for the use of chamomile (McKay and Blumberg 2006).

Even though the evidence based on clinical studies is somewhat limited, for some of the key indications, especially for the management of gastrointestinal complaints and for a range of inflammatory conditions, use of chamomile is well-established based on medical experience.

Anxiolytic effects: A recent exploratory study suggested that a standardised extract may be useful in depression induced by anxiety (Amsterdam et al. 2012) and a previous study showed modest effects in generalised anxiety disorder (Amsterdam et al. 2009).

Topical effects on skin inflammation: Topical application of a chamomile extract solution was found to be more effective for relieving itching and inflammation of peristomal skin lesions than hydrocortisone ointment (1%) in colostomy patients, although further studies are needed to confirm this result (Charousaei et al. 2011). A topical lukewarm compress of chamomile extract (2.5–5%) was also found to be useful in the topical treatment of phlebitis due to chemotherapy (Reis et al. 2011).

Pre-clinical evidence and mechanisms of action:

Anti-inflammatory and anti-allergic effects: Anti-inflammatory and anti-allergic activities of German chamomile have been demonstrated in studies many times both *in vitro* and *in vivo* (e.g. Chandrashekhar et al. 2011; Kobayashi et al. 2005; Petronilho et al. 2012). Data suggests that chamomile has a mechanism of action involving selective COX-2 inhibition (Srivastava et al. 2009).

Miscellaneous CNS and metabolic effects: The oil has been shown to have a stimulant effect in mice (Can et al. 2012). Antimutagenic and cholesterol-lowering activities, as well as antispasmodic and anxiolytic effects, have also been shown (McKay and Blumberg 2006).

The traditional use for healing stomach ulcers is supported by a demonstrated anti-*Helicobacter pylori* effect (Shikov et al. 2008).

Interactions: There is a possible risk of interaction of German chamomile when taken internally with warfarin, following one case report. Chamomile tea reduces absorption of iron, but to a lesser extent than black tea (Williamson et al. 2013).

Contraindications: Chamomile is contraindicated in patients who are sensitive or allergic to plants from the Asteraceae (e.g. ragweed, aster, chrysanthemum). An Argentinean study found that loose chamomile flowers were contaminated with *Clostridium botulinum* spores and recommended that use in under 1-year-old infants should be avoided (Bianco et al. 2008).

Adverse effects: None known.

Dosage: Powdered flowers 2–8 g, as infusion; liquid extract 1:1, 2–4 ml. Ointment containing 10% of an extract standardised to bisabolol (Rankin-Box and Williamson 2006).

General plant information: *M. chamomilla* originates from southern and eastern Europe and is today distributed throughout temperate regions. The species is a common wild plant found in gardens and wasteland. If plants are collected from the wild, care needs to be taken to authenticate accurately any material intended for medicinal use.

References

Amsterdam JD, Li Y, Soeller I, Rockwell K, Mao JJ, Shults J. (2009) A randomized, double-blind, placebo-controlled trial of oral *Matricaria recutita* (chamomile) extract therapy for generalized anxiety disorder. *Journal of Clinical Psychopharmacology* 29(4): 378–382.

Amsterdam JD, Shults J, Soeller I, Mao JJ, Rockwell K, Newberg AB. (2012) Chamomile (*Matricaria recutita*) may provide antidepressant activity in anxious, depressed humans: an exploratory study. *Alternative Therapies, Health and Medicine* 18(5): 44–49.

Bianco MI, Lúquez C, de Jong LIT, et al. (2008) Presence of *Clostridium botulinum* spores in *Matricaria chamomilla* (chamomile) and its relationship with infant botulism. *International Journal of Food Microbiology* 121(3): 357–360.

Can OD, Demir Özkay U, Kıyan HT, Demirci B. (2012) Psychopharmacological profile of Chamomile (*Matricaria recutita* L.) essential oil in mice. *Phytomedicine* 19(3–4): 306–310.

Chandrashekhar VM, Halagali KS, Nidavani RB, Shalavadi MH, Biradar BS, Biswas D, Muchchandi IS. (2011) Anti-allergic activity of German chamomile (*Matricaria recutita* L.) in mast cell mediated allergy model. *Journal of Ethnopharmacology* 137(1): 336–340.

Charousaei F, Dabirian A, Mojab F. (2011) Using chamomile solution or a 1% topical hydrocortisone ointment in the management of peristomal skin lesions in colostomy patients: results of a controlled clinical study. *Ostomy Wound Management* 57(5): 28–36.

Kobayashi, Y, Takashi, R, Ogino, F. (2005) Antipruritic effect of the single oral administration of German chamomile flower extract and its combined effect with antiallergic agents in ddY mice. *Journal of Ethnopharmacology* 101(1–3): 308–312.

McKay DL, Blumberg JB. (2006) A review of the bioactivity and potential health benefits of chamomile tea (*Matricaria recutita* L.). *Phytotherapy Research* 20(7): 519–530.

Petronilho S, Maraschin M, Coimbra MA, et al. (2012) *In vitro* and *in vivo* studies of natural products: A challenge for their valuation. A case study of chamomile (*Matricaria recutita* L.). *Industrial crops and Products* 40: 1–12.

Rankin-Box D, Williamson EM. (2006) *Complementary medicine: a guide for pharmacists.* Churchill Livingstone Elsevier, Edinburgh, London, New York, etc., 292 pp.

Reis PE, Carvalho EC, Bueno PC, Bastos JK. (2011) Clinical application of *Chamomilla recutita* in phlebitis: dose response curve study. *Revista Latino-Americana de Enfermagem* 19(1): 3–10.

Shikov AN, Pozharitskaya ON, Makarov VG, Kvetnaya AS. (2008) Antibacterial activity of *Chamomilla recutita* oil extract against *Helicobacter pylori*. *Phytotherapy Research* 22(2): 252–253.

Srivastava JK, Pandey M, Gupta S. (2009) Chamomile, a novel and selective COX-2 inhibitor with anti-inflammatory activity. *Life Sciences* 85(19–20): 663–669.

Tisserand R, Young R. (2014) *Essential Oil Safety.* 2nd Edition. Churchill Livingstone Elsevier, UK, pp 242–243.

Tschiggerl C, Bucar F. (2012) Guaianolides and volatile compounds in chamomile tea. *Plant Foods for Human Nutrition* 67(2): 129–135.

Williamson EM, Driver S, Baxter K. (Eds.) (2013) *Stockley's Herbal Medicines Interactions.* 2nd Edition. Pharmaceutical Press, London, UK.

Chamomile, Roman *Chamaemelum nobile* (L.) All.

Synonyms: *Anthemis nobilis* L.; *Chamomilla nobilis* (L.) Godr.; and others

Family: Asteraceae (Compositae)

Other common names: English chamomile

Drug names: Chamomillae romanae flos

Botanical drug used: Dried flowerheads

Indications/uses: Despite the chemical differences, this species is used in a way similar to the more widely used *Matricaria recutita* (German chamomile, see page 94). Internally, Roman chamomile is used to treat digestive ailments, including bloating and flatulence; and as a sedative for restlessness and mild insomnia due to nervous disorders. Under the THR scheme it is registered only 'for the relief of flatulence, bloating and mild upset stomach'. Externally, chamomile extracts are used to treat skin inflammation and irritation, infections of the mouth and gums, and are widely used in cosmetics, especially hair care products. The essential oil is popular in aromatherapy.

Evidence: Although chamomile is very popular and there is much anecdotal evidence, there is little clinical evidence available for either internal or external use.

Safety: Generally regarded as safe, although there may be some quality issues with the herb and some products, due to contamination or misidentification of species.

Main chemical compounds: *C. nobile* contains an essential oil (up to about 1.75%) composed mainly of esters of angelic and butyric acids, with some chamazulene, farnesene, terpinen-4-ol and others; sesquiterpene lactones of the germacranolide type, mainly nobilin and 3-epinobilin. Other major constituents of the flowers include spiro-ethers (cis-en-yn-dicycloether and trans-en-yn-dicycloether) and many phenolic compounds, primarily the flavonoids apigenin, its derivative chamaemeloside, quercetin and patuletin as glucosides, caffeoylquinic acid and other phenolic acids (Carnat et al. 2004; Guimarães et al. 2013; Ma et al. 2007; Srivastava et al. 2010; Tisserand and Young 2014).

Phytopharmacy: An evidence-based guide to herbal medicinal products, First Edition.
Sarah E. Edwards, Inês da Costa Rocha, Elizabeth M. Williamson and Michael Heinrich.
© 2015 John Wiley & Sons, Ltd. Published 2015 by John Wiley & Sons, Ltd.

Clinical evidence: Clinical studies for the use of Roman chamomile are lacking, despite the wide usage of the herb, most commonly as a mild sedative and carminative, and in cosmetic preparations.

Pre-clinical evidence and mechanisms of action: Roman chamomile has antioxidant and antimicrobial effects (Srivastava et al. 2010). Extracts have been shown to be hypoglycaemic (Eddouks et al. 2005) and anti-hypertensive in animals (Zeggwagh et al. 2009).

Interactions: No details available.

Contraindications: Chamomile is contraindicated in patients who are sensitive or allergic to plants from the Asteraceae (daisy family).

Adverse effects: Allergic reactions have been reported, including when extracts are taken orally. The essential oil is stated to be safe when used in aromatherapy (Tisserand and Young 2014).

Dosage: Powdered flowers 2–8 g, as infusion (herbal tea). For manufactured products and other formulations, see manufacturers' instructions.

General plant information: Roman chamomile is used to create scented lawns in the United Kingdom, such as the famous chamomile lawn at Buckingham Palace.

References

Carnat A, Carnat AP, Fraisse D, Ricoux L, Lamaison JL. (2004) The aromatic and poly-phenolic composition of Roman camomile tea. *Fitoterapia* 75(1): 32–38.

Eddouks M, Lemhadri A, Zeggwagh NA, Michel JB. (2005) Potent hypoglycaemic activity of the aqueous extract of *Chamaemelum nobile* in normal and streptozotocin-induced diabetic rats. *Diabetes Research and Clinical Practice* 67(3): 189–195.

Guimarães R, Barros L, Dueñas M, et al. (2013) Nutrients, phytochemicals and bioactivity of wild Roman chamomile: A comparison between the herb and its preparations. *Food Chemistry* 136: 718–725.

Ma CM, Winsor L, Daneshtalab M. (2007) Quantification of spiroether isomers and herniarin of different parts of *Matricaria matricarioides* and flowers of *Chamaemelum nobile*. *Phytochemical Analysis* 18(1): 42–49.

Srivastava JK, Shankar E, Gupta S. (2010) Chamomile: A herbal medicine of the past with bright future. *Molecular Medicine Reports* 3(6): 895–901.

Tisserand R, Young R. (2014) *Essential Oil Safety*. 2nd Edition. Churchill Livingstone Elsevier, UK, pp. 244–245.

Zeggwagh NA, Moufid A, Michel JB, Eddouks M. (2009) Hypotensive effect of *Chamaemelum nobile* aqueous extract in spontaneously hypertensive rats. *Clinical and Experimental Hypertension* 31(5): 440–450.

Chasteberry	*Vitex agnus-castus* L.

Synonyms: *Agnus-castus robusta* (Lebas) Carrière; *A. vulgaris* Carrière; *V. agnus* Stokes; and others

Family: Lamiaceae (Labiatae)

Other common names: Agnus castus; chaste tree; monk's pepper

Drug name: Agni casti fructus

Botanical drug used: Whole dried ripe fruit (berry)

Indications/uses: In women, *V. agnus-castus* berries are traditionally used to alleviate symptoms of premenstrual syndrome, including mastalgia, migraine and depression, as well as menstrual disorders such as amenorrhoea and dysmenorrhoea. Chasteberry has also been used to treat female infertility, as a galactagogue, and to alleviate menopausal symptoms. In men, the fruit has been used to suppress libido and treat acne.

Evidence: Clinical evidence suggests some benefits for use of chasteberry extracts in the treatment of premenstrual syndrome and latent hyperprolactinaemia. Although traditionally it has been used to prevent miscarriages and to promote breast milk production, clinical evidence to support this is lacking – and it may actually reduce lactation due to inhibition of prolactin secretion.

Safety: The berries have been widely used for centuries without ill effects, but safety, including during pregnancy and lactation, has not been established.

Main chemical compounds: Essential oil (up to 2%), with bornyl acetate, 1,8-cineol, limonene, α- and β-pinene as major constituents; diterpenes including viteagnusin, vitexilactone, rotundifuran, vitexilactam A, viteagnusides A-C, vite-trifolin D and others; flavonoids including casticin, chrysoplenetin, chrysosplenol D, cynaroside, 6-hydroxykaempferol, apigenin, isorhamnetin, luteolin and their derivatives; iridoids including agnuside and *p*-hydroxybenzoic acid derivatives have been identified. *V. agnus-castus* is often standardised to the content of the flavonoid casticin and sometimes also to the iridoid glycoside agnuside (Chen et al. 2011; Ono et al. 2011; WHO 2009; Williamson et al. 2013).

Phytopharmacy: An evidence-based guide to herbal medicinal products, First Edition.
Sarah E. Edwards, Inês da Costa Rocha, Elizabeth M. Williamson and Michael Heinrich.
© 2015 John Wiley & Sons, Ltd. Published 2015 by John Wiley & Sons, Ltd.

Clinical evidence:

Premenstrual syndrome: A systematic review assessed 12 randomised controlled trials of *V. agnus-castus*, six of which involved over 100 participants (range 110 to 217), in alleviating symptoms of premenstrual syndrome (PMS). It was found superior to placebo (five out of six studies), pyridoxine (vitamin B6, 200 mg/day; 1 study) and magnesium oxide (1 study). The one exception found that soya-based 'placebo' (1800 mg/day) significantly outperformed chasteberry.

Two studies investigated the use of *V. agnus-castus* in premenstrual dysphoric disorder: one reported that it was equivalent to fluoxetine, while the other found that fluoxetine outperformed chasteberry. In a double-blind, placebo-controlled parallel-group study involving 162 women aged 18–45 years with PMS, the proprietary extract 'Ze 440' (standardised to casticin) was administered over three menstrual cycles. A daily dose of 20 mg of Ze 440 was effective in relieving symptoms of PMS (Schellenberg et al. 2012).

A recent open-label clinical observational study reported that chasteberry was able to reduce the frequency of migraine attacks in women with premenstrual syndrome (Ambrosini et al. 2013).

Hyperprolactinaemia: One trial reported *V. agnus-castus* extract to be superior to placebo in reducing prolactin secretion, normalising a shortened luteal phase and increasing mid-luteal progesterone and 17β-oestradial levels in latent hyperprolactinaemia. Another found it comparable to bromocriptine for reducing serum prolactin levels and ameliorating cyclic mastalgia (van Die et al. 2013).

Menopausal symptoms: A questionnaire survey of German gynaecologists on complementary and alternative medicine (CAM) in the treatment of climacteric symptoms found that chasteberry was assessed as 'effective' by 17.4% ($n = 421$) of respondents, 59.9% ($n = 1448$) considered it 'sometimes effective', while 20.8% ($n = 504$) rated it 'unimportant' (von Studnitz et al. 2013).

Pre-clinical evidence and mechanisms of action:

Hormonal effects: *In vitro, V. agnus-castus* extracts have been shown to bind to oestrogen receptor β, but not α. They also bind to the μ-opiate receptor in Chinese hamster ovary. Agonistic effects on pituitary dopamine (D2) receptors have been demonstrated with high concentrations, and inhibition of basal as well as thyrotropin-releasing hormone-stimulated prolactin secretion has been demonstrated in cultured rat pituitary cells (Gardner and McGuffin 2013). Chasteberry extract administered to rats resulted in a reduction in prolactin secretion. It has been suggested that it improves luteal function resulting in an increase in progesterone levels, which may explain its action for miscarriage prevention (Gardner and McGuffin). In a study in orchidectomised male rats, oral administration of chasteberry for 12 weeks gave protection from osteoporosis, preserving both cortical and trabecular bone (Sehmisch et al. 2009).

Anti-inflammatory and immunomodulatory effects: Antioxidant activity of aqueous and ethanol extracts has been reported (Sağlam et al. 2007). The *p*-hydroxybenzoic acid derivatives were found to exhibit significant anti-inflammatory activity, while casticin and artemetin were reported to be potent inhibitors of lipoxygenase *in vitro* (Choudhary et al. 2009). Casticin was also found to be a potent immunomodulatory compound *in vitro*, comparable to prednisolone, with a suppressive effect on PHA-stimulated T-cell activation.

Cytotoxicity: Cytotoxic effects of casticin (against MDBK cell line) have been reported at low concentrations during prolonged incubation (Mesaik et al. 2009). Strong anticancer activity of chasteberry powder was found in another *in vitro* study using a murine neuroblastoma cell line (Mazzio and Soliman 2009), but this activity is out of scope of this book.

Anxiogenic-like effects: Rats orally administered chasteberry exhibited anxiogenic effects, which was reported to be via involvement of the $5HT_{1A}$ receptor (Yaghmaei et al. 2012).

Interactions: No drug interactions were identified. However, chasteberry has dopamine agonist properties and therefore may interact with drugs with dopamine agonist or dopamine antagonist action. Active constituents of chasteberry may also have additive effects with opiods due to similar pharmacological activity. Presence of oestrogenic compounds in chasteberry may result in additive or opposing effects with oestrogens or oestrogen antagonists (e.g. tamoxifen). Chasteberry is not recommended for use concurrently with hormonal contraceptives, HRT or with IVF treatment (Williamson et al. 2013).

Contraindications: Use of *V. agnus-castus* fruit in children and adolescents under 18 years is not recommended due to lack of safety data. It should not be used during pregnancy and lactation. Patients who suffer or suffered from an oestrogen-sensitive cancer, or those with a history of a pituitary disorder should consult a medical practitioner before use. Patients using dopamine agonists or antagonists, oestrogens and anti-oestrogens should also consult a medical practitioner before use. Intake of chasteberry may mask symptoms of prolactin-secreting tumours of the pituitary gland (EMA 2010).

Adverse effects: From reviews of clinical trials of chasteberry, it appears that adverse events reported were mild and transient, and that it is generally well tolerated. However, severe allergic reactions with face swelling, dyspnoea and swallowing difficulties have been reported. The most common adverse events include gastrointestinal disturbance (mainly nausea) and skin conditions, with other occasional side effects such as headache, fatigue and hormone-related symptoms (EMA 2010; Gardner and McGuffin 2013).

Dosage: For products, see manufacturers' instructions. Powdered berries: (adults): 400 mg twice daily; tincture (ratio of herb to aqueous ethanol 1:5), 40 drops once daily, corresponding to approximately 33 mg herbal substance (EMA 2010).

General plant information: *V. agnus-castus* grows in humid regions of central Europe and central Asia and has been used as a medicine since antiquity. The medieval name for the plant, monk's pepper and the name chasteberry, derive from the use of the seeds as a spice in monastic kitchens where they were thought to repress the libido, but were also a cheap alternative to the more expensive black pepper (Odenthal 1998).

References

Ambrosini A, Di Lorenzo C, Coppola G, Pierelli F. (2013) Use of *Vitex agnus-castus* in migrainous women with premenstrual syndrome: an open-label clinical observation. *Acta Neurologica Belgica* 113(1): 25–29.

Chen SN, Friesen JB, Webster D, Nikolic D, van Breemen RB, Wang ZJ, Fong HHS, Farnsworth NR, Pauli GF. (2011) Phytoconstituents from *Vitex agnus-castus* fruits. *Fitoterapia* 82(4): 528–533.

Choudhary MI, Azizuddin, JS, Nawaz SA, Khan KM, Tareen RB, Atta-ur-Rahmann. (2009) Antiinflammatory and lipoxygenase inhibitory compounds from *Vitex agnus-castus*. *Phytotherapy Research* 23(9): 1336–1339.

van Die MD, Burger HG, Teede HJ, Bone KM. (2013) *Vitex agnus-castus* extracts for female reproductive disorders: a systematic review of clinical trials. *Planta Medica* 79(7): 562–575.

EMA. (2010) Community herbal monograph on *Vitex agnus-castus* L. European Medicines Agency http://www.ema.europa.eu/docs/en_GB/document _library/Herbal_-_Community_herbal_monograph/2011/01/WC500101541.pdf (accessed February 2014).

Gardner Z, McGuffin M. (Eds.) (2013) *American Herbal Product Association's Botanical Safety Handbook*. 2nd Edition. CRC Press, USA, 1072 pp.

Mazzio EA, Soliman KFA. (2009) *In vitro* screening for the tumoricidal properties of international medicinal herbs. *Phytotherapy Research* 23(3): 385–398.

Mesaik MA, Azizuddin, MS, Khan KM, Tareen RB, Ahmed A, Atta-ur-Rahmann, Choudhary MI. (2009) Isolation and immunomodulatory properties of a flavonoid, casticin from *Vitex agnus-castus*. *Phytotherapy Research* 23(11): 1516–1520.

Odenthal KP. (1998) *Vitex agnus-castus* L. – traditional drug and actual indications. *Phytotherapy Research* 12(S1): S160–S161.

Ono M, Eguchi K, Konoshita M, Furusawa C, Sakamoto J, Yasuda S, Ikeda T, Okawa M, Kinjo J, Yoshimitsu H, Nohara T. (2011) A new diterpenoid glucoside and two new diterpenoids from the fruit of *Vitex agnus-castus*. *Chemical and Pharmaceutical Bulletin (Tokyo)*. 59(3): 392–396.

Sağlam H, Pabuçcuoğlu A, Kivçak B. (2007) Antioxidant activity of *Vitex agnus-castus* L. extracts. *Phytotherapy Research* 21(11): 1059–1060.

Schellenberg R, Zimmermann C, Drewe J, Hoexter G, Zahner C. (2012) Dose-dependent efficacy of the *Vitex agnus castus* extract Ze 440 in patients suffering from premenstrual syndrome. *Phytomedicine* 19(14): 1325–1331.

Sehmisch S, Boeckhoff J, Willie J, Seidlova-Wuttke D, Rack T, Tezval M, Wuttke W, Stuermer KM, Stuermer EK. (2009) *Vitex agnus castus* as prophylaxis for osteopenia after orchidectomy in rats compared with estradiol and testosterone supplementation. *Phytotherapy Research* 23(6): 851–858.

von Studnitz FSG, Eulenburg C, Mueck AO, Buhling KJ. (2013) The value of complementary and alternative medicine in the treatment of climacteric symptoms: results of a survey among German gynecologists. *Complementary Therapies in Medicine* 21(5): 492–495.

WHO. (2009) *WHO Monographs on Selected Medicinal Plants*. Vol. 4. WHO, Geneva, 456 pp.

Williamson EM, Driver S, Baxter K. (Eds.) (2013) *Stockley's Herbal Medicines Interactions*. 2nd Edition. Pharmaceutical Press, London, UK.

Yaghmaei P, Oryan S, Gharehlar LF, Salari AA, Solati J. (2012) Possible modulation of the anxiogenic effects of *Vitex agnus-castus* by the serotonergic system. *Iranian Journal of Basic Medical Sciences* 15(2): 768–776.

Chilli/Capsicum   *Capsicum annuum* L., *C. frutescens* L., *C. pubescens* Ruiz & Pav. and other *Capsicum spp.*

Family: Solanaceae

Other common names: Cayenne; chilli pepper; hot pepper; tabasco pepper

Drug names: Capsici fructus; Capsici oleoresina

Botanical drug used: Fruit, oleoresin (capsaicin, the ethereal extract)

Indications/uses: Post-herpetic neuralgia and rheumatic pain, and to increase blood flow in the peripheral circulation (external use); to relieve colic, flatulent dyspepsia, as an expectorant and more recently as a weight-loss supplement (internal use).

Evidence: Overall, externally-applied capsaicin is effective in treating certain types of pain. Internally, for weight loss, a recent systematic review has concluded that there is the evidence that it could play a role as part of a weight management programme.

Safety: Considered to be safe if used according to manufacturers' instructions and as a very widely-consumed food item in moderate doses.

Main chemical compounds: The pungent principles are the capsaicinoids, present in concentrations up to 1.5%, but more usually around 0.1%. The major capsaicinoids are capsaicin, dihydrocapsaicin, nordihydrocapsaicin, homodihydrocapsaicin and homocapsaicin. The capsinoids, which share similar structural features with the capsaicinoids and consist of capsiate, dihydrocapsiate, and nordihydrocapsiate, and so on, are less toxic and less pungent, and are being investigated for their anti-tumour, anti-obesity and other effects (Luo et al. 2011). Other constituents include the carotenoid pigments (capsanthin, capsorubin, carotene, lutein), vitamins including A and C and a small amount of volatile oil (Williamson et al. 2013).

Clinical evidence: There is evidence for efficacy for the use of externally applied capsaicin products in treating pain, many of which carry full marketing authorisations. For weight loss, a systematic review of 90 clinical trials, of which 20 were

---

Phytopharmacy: An evidence-based guide to herbal medicinal products, First Edition.
Sarah E. Edwards, Inês da Costa Rocha, Elizabeth M. Williamson and Michael Heinrich.
© 2015 John Wiley & Sons, Ltd. Published 2015 by John Wiley & Sons, Ltd.

selected for inclusion, involving 563 participants, found three areas of *potential* benefit: increased energy expenditure; increased lipid oxidation; and reduced appetite. Capsaicinoid consumption modestly increased energy expenditure (by approximately 50 kcal/day), and so would only produce significant weight loss after 1–2 years. Regular consumption reduced abdominal adipose tissue levels and reduced appetite and energy intake (Whiting et al. 2012). They may therefore play a role as part of a weight management program.

Pre-clinical evidence and mechanisms of action: Capsaicin is a known activator of the TRPV1 receptor (vanilloid receptor 1) and many of its effects are caused by this mechanism (Luo et al. 2011), including the analgesic effects (O'Neill et al. 2012) and the respiratory effects (Banner et al. 2011). It also stimulates prostaglandin biosynthesis and inhibits protein synthesis *in vitro*, and inhibits growth of various pathogenic bacteria such as *Escherichia coli, Bacillus subtilis* (Anon. 2007).

Contrary to public perception, capsaicin does not stimulate but inhibits acid secretion and stimulates mucus secretions and gastric mucosal blood flow which help in the prevention and healing of ulcers (Satyanarayana 2006).

Interactions with other drugs: Capsicum may moderately reduce the absorption of dietary iron when taken internally. Several potential interactions have been reported using *in vitro* studies, for example, aspirin, digoxin, ciprofloxacin, cefalexin and theophylline, but none have been assessed as clinically significant (Williamson et al. 2013).

Contraindications: Capsicum and capsaicin products are strongly irritant and should not be applied to mucous membranes, broken or sensitive skin. Internally, doses should be restricted to levels normally found in foods.

Adverse effects: Skin irritation (topical application), alveolitis (by inhalation), gastric pain (high internal doses) (Anon. 2007). These are mainly due to the capsaicinoids; the more recently discovered capsinoids are less irritant and less toxic (Watanabe et al. 2011).

Dosage: According to manufacturers' instructions. Chillies are very variable in potency depending on the variety and its capsaicin content.

General plant information: Capsicum is widely used as a spice, and sweeter, less pungent, varieties of peppers are eaten all over the world.

References

The primary literature for capsicum is extensive; this is a selection of recent and relevant reviews.

Anon. (2007) Final report on the safety assessment of *Capsicum annuum* extract, *Capsicum annuum* fruit extract, *Capsicum annuum* resin, *Capsicum annuum* fruit powder, *Capsicum frutescens* fruit, *Capsicum frutescens* fruit extract, *Capsicum frutescens* resin, and capsaicin. *International Journal of Toxicology* 26(Suppl 1): 3–106.

Banner KH, Igney F, Poll C. (2011) TRP channels: emerging targets for respiratory disease. *Pharmacology & Therapeutics* 130(3): 371–384.

Luo XJ, Peng J, Li YJ. (2011) Recent advances in the study of capsaicinoids and capsinoids. *European Journal of Pharmacology* 650(1): 1–7.

O'Neill J, Brock C, Olesen AE, Andresen T, Nilsson M, Dickenson AH. (2012) Unravelling the mystery of capsaicin: a tool to understand and treat pain. *Pharmacological Reviews* 64(4): 939–971.

Satyanarayana MN. (2006) Capsaicin and gastric ulcers. *Critical Reviews in Food Science and Nutrition* 46(4): 275–328.

Watanabe T, Ohnuki K, Kobata K. (2011) Studies on the metabolism and toxicology of emerging capsinoids. *Expert Opinion on Drug Metabolism & Toxicology* 7(5): 533–542.

Whiting S, Derbyshire E, Tiwari BK. (2012) Capsaicinoids and capsinoids. A potential role for weight management? A systematic review of the evidence. *Appetite* 59(2): 341–348.

Williamson EM, Driver S, Baxter K. (Eds.) (2013) *Stockley's Herbal Medicines Interactions*. 2nd Edition. Pharmaceutical Press, London, UK.

Cinnamon;   *Cinnamomum verum* J. Presl;
Chinese Cinnamon/Cassia   *C. cassia* (L.) J. Presl

Synonyms: *C. verum: C. aromaticum* J. Graham; *C. zeylanicum* Blume; and others; *C. cassia: C. aromaticum* Nees; *C. longifolium* Lukman; and others

Family: Lauraceae

Other common names: *C. verum:* Ceylon cinnamon; true cinnamon; *C. cassia:* cassia bark; Chinese cassia
   Several other species of *Cinnamomum* are also commonly traded using the name 'cinnamon'.

Drug name: *C. verum:* Cinnamomi cortex; Cinnamomi corticis aetheroleum (Cinnamon oil). *C. cassia:* Cinnamomi cassaie cortex; Cinnamomi cassiae aetheroleum (Cassia oil).

Botanical drug used: Dried inner bark from shoots of *C. verum* or stripped trunk bark from *C. cassia*; bark oil (obtained by steam distillation from the cortex)

Indications/uses: Cinnamon bark and oil (both types) are traditionally used for the symptomatic relief of mild, spasmodic gastrointestinal complaints including bloating and flatulence, and in oral preparations such as pastilles and cough mixtures for sore throats and coughs, and also dentifrices. The bark is also used to lower blood glucose levels in diabetes, and for symptomatic treatment of diarrhoea. Cassia is used in traditional Chinese medicine for circulatory disorders.

Evidence: Some limited clinical data indicates that short-term use of cinnamon lowers blood pressure in individuals with type 2 diabetes and pre-diabetes. While one meta-analysis found that cinnamon also showed a beneficial effect on glycaemic control, a Cochrane review found that cassia was no more effective than placebo on blood sugar levels and other outcomes in patients with diabetes mellitus.

Safety: Cinnamon used in foodstuffs is not of concern. However, larger doses should not be used during pregnancy, as animal studies have indicated that the constituent cinnamaldehyde increases risk of foetal abnormalities. Safety has not been established during lactation. Coumarin, a constituent of Chinese cinnamon (cassia), is known to cause liver and kidney damage at high concentrations.

Phytopharmacy: An evidence-based guide to herbal medicinal products, First Edition.
Sarah E. Edwards, Inês da Costa Rocha, Elizabeth M. Williamson and Michael Heinrich.
© 2015 John Wiley & Sons, Ltd. Published 2015 by John Wiley & Sons, Ltd.

Main chemical compounds: The major constituent of both *C. verum* and *C. cassia* is cinnamaldehyde at 65–80% and 90% of the volatile oil, respectively. *C. verum* also contains *o*-methoxycinnamaldehyde. Its volatile oil contains 10% eugenol and terpenoids including linalool. In *C. cassia,* only a trace of eugenol is found, and *C. cassia* is also lacking in monoterpenoids and sesquiterpenoids. Coumarin is present in *C. cassia*, but not in *C. verum* (Barceloux 2009; Pharmaceutical Press Editorial Team 2013; WHO 1999).

Clinical evidence:

A number of the clinical and pre-clinical studies do not clearly distinguish between the various species of cinnamon, which may result in slightly conflicting data, since the species differ slightly in phytochemical composition. Overall, the evidence for blood-sugar-lowering effects is positive but conflicting.

Effects associated with blood sugar levels and diabetes control: A Cochrane review identified 10 randomised controlled trials, involving 577 participants with diabetes mellitus, who were administered a mean daily dose of 2 g 'cinnamon' (predominantly *C. cassia*), for a period of 4–16 weeks. The effect on fasting blood glucose was inconclusive, and no significant difference between cinnamon and control groups was found in levels of glycosylated haemoglobin A1c (HbA1c), serum insulin or postprandial glucose (Leach and Kumar 2012).

In contrast to the Cochrane findings, another systematic review and meta-analysis (of six randomised controlled trials lasting 40 days–4 months, $n = 435$) found that cinnamon (daily dose 1–6 g) exerted a beneficial effect on glycaemic control, both HbA1c and fasting plasma glucose (FPG) (Akilen et al. 2012).

A recent systematic review of randomised controlled trials of 'cinnamon' (both *C. cassia* and *C. verum*) concluded that its consumption is associated with a statistically significant decrease in levels of fasting plasma glucose, total cholesterol, LDL-C, and triglyceride levels, and an increase in HDL-C levels; however, no significant effect on hemoglobin A1c was found. The authors suggested that the high degree of heterogeneity in the studies may limit their applicability to patient care, because the dose and duration of therapy were unclear (Allen et al. 2013).

A meta-analysis of three randomised controlled clinical trials (from a possible 93 studies found), two of which involved individuals with type 2 diabetes and one with pre-diabetic syndrome, found that short-term administration of cinnamon was associated with a notable reduction in both systolic and diastolic blood pressure. However, the findings were based on a low number of patients, and further longer term randomised controlled trials involving a large number of patients are required (Akilen et al. 2013).

In a cross-over study, 30 healthy young adults (aged 18–30), both normal weight and obese, were fed 50 g available carbohydrate (in instant farina cereal) either plain or with 6 g ground cinnamon. Blood glucose levels were measured at intervals from 0–120 minutes following the meal. Addition of cinnamon to the cereal resulted in significant reduction in 120-minute glucose area under the curve, and blood glucose at 15, 30, 45 and 60 minutes. The results suggested that cinnamon may be effective in moderating postprandial glucose response in normal weight and obese healthy adults (Magistrelli 2012).

Pre-clinical evidence and mechanisms of action:

Cinnamon has been shown to exhibit a large number of pharmacological properties, some of which are highlighted here.

Antidiabetic effects: A dose-dependent decrease in plasma glucose levels was observed in diabetic rats orally administered 5, 10 or 20 mg/kg daily of the constituent cinnamaldehyde for 45 days (Subash Babu et al. 2007). In another *in vivo* study, a dose-dependent reduction in glucose levels was observed in diabetic mice orally administered 50, 100, 150 or 200 mg/kg daily of a *C. cassia* extract for 6 weeks (Kim et al. 2006). Procyanidin oligomers from *C. cassia* bark extract have also been shown to exhibit hypoglycaemic activity (Lu et al. 2011). The antidiabetic activity of an aqueous cinnamon extract and cinnamon-polyphenol-enriched defatted soya flour (CDSF) were demonstrated *in vivo*, by acutely lowering fasting blood glucose in diet-induced obese hyperglycaemic mice at 300 mg/kg and 600 mg/kg, respectively. The preparations inhibited hepatic glucose production in rat hepatoma cells and the cinnamon extract decreased gene expression of the regulators of hepatic gluconeogenesis, phosphoenolpyruvate carboxykinase and glucose-6-phosphatase (Cheng et al. 2012).

Antimicrobial properties: Strong antibacterial activity of cinnamon oil (*C. verum*) has been demonstrated *in vitro* against bacteria associated with respiratory infections (Fabio et al. 2007). Another study showed that the essential oil of *C. cassia* has strong antifungal activity against *Candida albicans* (MIC 80% = 0.169 μl/ml), and potentiates amphotericin B *in vitro* (Giordani et al. 2006), substantiating its use in respiratory infections.

Anti-inflammatory effects: The anti-inflammatory activity of *C. cassia* and its constituents, notably cinnamaldehyde, has been demonstrated *in vitro* and *in vivo*. Cinnamaldehyde was shown to inhibit inflammatory mediators including nitrous oxide (NO), tumour necrosis factor-alpha (TNF-α) and prostaglandin E_2 (PGE_2), and decreased inducible nitric oxide synthase (iNOS), cyclooxygenase-2 (COX-2) and nuclear transcription factor-kappa B (NF-κB) expression (Liao et al. 2012). Another *in vitro* study demonstrated that ethanolic extract of *C. cassia* bark powder exhibited anti-neuroinflammatory activity, with cinnamaldeyhde being the most active constituent (Ho et al. 2013). An aqueous extract of *C. cassia* was also shown to protect against radiation-induced oxidative and inflammatory damage in rats (Azab et al. 2011).

Anti-tumour effects: *C. cassia* bark aqueous extract inhibited pro-angiogenic factors and regulators of tumour progression in melanoma cell lines and in an *in vivo* melanoma model (Kwon et al. 2009).

Anti-oestrogenic properties: An ethanolic extract of *C. cassia* was reported to have anti-oestrogenic activity in a recombinant yeast system (Kim et al. 2008).

Interactions: No case reports of interactions were identified and it was considered unlikely that cinnamon would markedly affect the control of diabetes with conventional drugs (Williamson et al. 2013).

Contraindications: Hypersensitivity to cinnamon or its constituents. People with diabetes are advised to use cinnamon with caution, as animal studies indicate that it may modify glucose levels. Therapeutic doses of cinnamon during pregnancy should be avoided (Gardner and McGuffin 2013). Internal ingestion of the essential oil, except when diluted in manufactured products, is not recommended.

Adverse effects: At standard therapeutic doses (2–4 g daily bark), no side effects from cinnamon are expected. Overdose causes increases in heart rate, peristalsis, respiration and perspiration, which may be followed by sedation and sleepiness.

Cinnamaldehyde, the main compound in the essential oil, may cause skin irritation in sensitive individuals. Oral inflammation or lesions have been reported following prolonged oral exposure to cinnamon-containing products such as toothpaste. Cases of topical dermatitis from cinnamon, its essential oil and constituents have been documented, and occupational asthma, skin and eye irritation and hair loss have been observed in workers regularly exposed to cinnamon dust. Vomiting, diarrhoea and loss of consciousness were reported in a child following ingestion of 60 ml of the essential oil (Gardner and McGuffin 2013).

Dosage: Cinnamon bark: adults – herbal tea: 0.5–1 g comminuted herbal substance as an infusion up to four times daily; liquid extract: 0.5–1 ml three times daily; tincture: 2–4 ml daily (EMA 2011a). Cinnamon bark oil: adults – 50–200 mg daily in 2–3 doses. Undiluted oil may cause irritation and is not recommended (EMA 2011b).

General plant information: *C. verum* is native to Sri Lanka, from where the majority of the world's cinnamon bark is produced, while *C. cassia* is native to China. Cinnamon has been traded as a spice for millennia: there are several references to it in the Bible, and it was used in ancient Egypt in embalming fluid (Barceloux 2009). In the United States, cassia bark is often referred to simply as 'cinnamon'.

References

Akilen R, Pimlott Z, Tsiami A, Robinson N. (2013) Effect of short-term administration of cinnamon on blood pressure in patients with prediabetes and type 2 diabetes. *Nutrition* 29(10): 1192–1196.

Akilen R, Tsiami A, Devendra D, Robinson N. (2012) Cinnamon in glycaemic control: Systematic review and meta analysis. *Clinical Nutrition* 31: 609–615.

Allen RW, Schwartzman E, Baker WL, Coleman CI, Phung OJ. (2013) Cinnamon use in type 2 diabetes: an updated systematic review and meta-analysis. *Annals of Family Medicine* 11(5): 452–459.

Azab KS, Mostafa AHA, Ali EMM, Abdul-Aziz MAS. (2011) Cinnamon extract ameliorates ionizing radiation-induced cellular injury in rats. *Ecotoxicology and Environmental Safety* 74(8): 2324–2329.

Barceloux DG. (2009) Cinnamon (*Cinnamomum* species). *Disease-a-Month* 55(6): 327–335.

Cheng DM, Kuhn P, Poulev A, Rojo LE, Lila MA, Raskin I. (2012) *In vivo* and *in vitro* antidiabetic effects of aqueous cinnamon extract and cinnamon polyphenol-enhanced food matrix. *Food Chemistry* 135(4): 2994–3002.

EMA. (2011a) Community herbal monograph on *Cinnamomum verum* J.S. Presl, cortex. European Medicines Agency http://www.ema.europa.eu/docs/en_GB/document_library /Herbal_-_Community_herbal_monograph/2011/08/WC500110095.pdf (accessed August 2013).

EMA. (2011b) Community herbal monograph on *Cinnamomum verum* J.S. Presl, corticis aetheroleum. European Medicines Agency http://www.ema.europa.eu/docs/en_GB /document_library/Herbal_-_Community_herbal_monograph/2011/08 /WC500110091.pdf (accessed February 2014).

Fabio A, Cermelli C, Fabio G, Nicoletti P, Quaglio P. (2007) Screening of the antibacterial effects of a variety of essential oils on microorganisms responsible for respiratory infections. *Phytotherapy Research* 21(4): 374–377.

Gardner Z, McGuffin M. (Eds.) (2013) *American Herbal Product Association's Botanical Safety Handbook*. 2nd Edition. CRC Press, USA, 1072 pp.

Giordani R, Regli P, Kaloustian J, Portugal H. (2006) Potentiation of antifungal activity of amphotericin B by essential oil from *Cinnamomum cassia*. *Phytotherapy Research* 20(1): 58–61.

Ho SC, Chang KS, Chang PW. (2013) Inhibition of neuroinflammation by cinnamon and its main components. *Food Chemistry* 138(4): 2275–2282.

Kim SH, Hyun SH, Choung SY. (2006) Anti-diabetic effect of cinnamon extract on blood glucose in db/db mice. *Journal of Ethnopharmacology* 104(1–2): 119–123.

Kim IG, Kang SC, Kim KC, Choung ES, Zee OP. (2008) Screening of estrogenic and antiestrogenic activities from medicinal plants. *Environmental Toxicology and Pharmacology* 25(1): 75–82.

Kwon HK, Jeon WK, Hwang JS, Lee CG, So JS, Park JA, Ko BS, Im SH. (2009) Cinnamon extract suppresses tumor progression by modulating angiogenesis and the effector function of CD8+T cells. *Cancer Letters* 278(2): 174–182.

Leach MJ, Kumar S. (2012) Cinnamon for diabetes mellitus. *Cochrane Database of Systematic Reviews* (9): CD007170.

Liao JC, Deng JS, Chiu CS, Hou WC, Huang SS, Shie PH, Huang GJ. (2012) Anti-inflammatory activities of *Cinnamomum cassia* constituents *in vitro* and *in vivo*. *Evidence-Based Complementary and Alternative Medicine* 2012: 429320.

Lu Z, Jia Q, Wang R, Wu X, Wu Y, Huang C, Li Y. (2011) Hypoglycemic activities of A- and B-type procyanidin oligomer-rich extracts from different Cinnamon barks. *Phytomedicine* 18(4): 298–302.

Magistrelli A. (2012) Effect of ground cinnamon on postprandial blood glucose concentration in normal-weight and obese adults. *Journal of the Academy of Nutrition and Dietetics* 112(11): 1806–1809.

Pharmaceutical Press Editorial Team. (2013) *Herbal Medicines*. 4th Edition. Pharmaceutical Press, London, UK.

Subash Babu P, Prabuseenivasan S, Ignacimuthu S. (2007) Cinnamaldehyde – a potential antidiabetic agent. *Phytomedicine* 14(1): 15–22.

WHO. (1999) *WHO Monographs on Selected Medicinal Plants*. Vol. 1. WHO, Geneva, 295 pp.

Williamson EM, Driver S, Baxter K. (Eds.) (2013) *Stockley's Herbal Medicines Interactions*. 2nd Edition. Pharmaceutical Press, London, UK.

Cola	*Cola nitida* (Vent.) Schott & Endl., *C. acuminata* (P.Beauv.) Schott & Endl.

Synonyms: *Sterculia nitida* Vent. (= *C. nitida*); *C. grandiflora* Schott & Endl. (= *C. acuminata*)

Family: Sterculiaceae (Malvaceae)

Other common names: Cola nut; kola

Drug name: Colae semen

Botanical drug used: Whole or fragmented dried seeds, freed from the testa

---

Indications/uses: Cola is used as a stimulant in depression and fatigue, to enhance the appetite and as a diuretic.

Evidence: Although there is no clinical evidence, cola contains significant amounts of caffeine and is widely used for symptoms of temporary fatigue.

Safety: Considered to be safe, despite the limited safety information available, due to its long history of use as a flavouring agent in the food and beverage industry.

Main chemical compounds: The key bioactive compound is caffeine, with traces of other xanthine derivatives such as theophylline and theobromine also found. Other compounds present are flavonoids (catechin and epicatechin), anthocyanins and tannins including colatin, colatein and colanin (Burdock et al. 2009; EMAa 2011; Williamson et al. 2013).

Clinical Evidence: No clinical data is available for cola and its preparations. However, the action of caffeine as a proven stimulant has been extensively studied for its mental and physical effects (Glade 2010; Heckman et al. 2010).

Pre-clinical evidence and mechanisms of action: Only few pharmacological studies have been reported for cola seeds. A behavioural study in rats, using a locomotion grid with number of squares crossed and resistance to capture as outcome measures, compared the effects of fresh cola seed extract (*C. nitida*; standardised to 6.2% caffeine, 0.9% theobromine and 15% catechin) with pure

---

Phytopharmacy: An evidence-based guide to herbal medicinal products, First Edition.
Sarah E. Edwards, Inês da Costa Rocha, Elizabeth M. Williamson and Michael Heinrich.
© 2015 John Wiley & Sons, Ltd. Published 2015 by John Wiley & Sons, Ltd.

caffeine. The extract had an effect on behaviour similar to that of pure caffeine but the onset was more gradual for cola than caffeine (Scotto et al. 1987). In another study, the effect of an extract of cola seeds in rats (320 mg/kg/day, equivalent to 20 mg/kg/day of caffeine) was assessed using electroencephalogram readings. The extract showed results on cortical activity similar to that of caffeine, supporting its corticostimulatory effects (Vaille et al. 1993).

Interactions: Interactions are similar to other caffeine-containing stimulants. Cola can reduce the sedative action of drugs. Caution is advised in patients taking monoamine oxidase-inhibiting drugs including furazolidone, procarbazine and selegiline. It may cause serious additive effects if combined with other central nervous system (CNS) stimulants and sympathomimetic drugs such as ephedra. Cola may reduce the bioavailability of the antimalarial drug, halofantrine (EMAb 2011; Williamson et al. 2013).

Contraindications: Not recommended during pregnancy and lactation or in children and adolescents under 18 years of age due to lack of safety data. Not recommended in patients with gastric and duodenal ulcers, cardiovascular disorders (hypertension and arrhythmia) or hyperthyroidism due to the CNS-stimulant effect of caffeine. Caution is advised in patients with psychological disorders as it may exacerbate depression symptoms or induce anxiety (EMAb 2011).

Adverse effects: Cola seeds contain significant amounts of caffeine; thus, it may cause headaches, drowsiness, anxiety and nausea, depending on the individual's sensitivity to caffeine (Heckman et al. 2010). It may increase the risk of developing hypertension. Not recommended before bedtime as it may cause insomnia. Cola may cause gastrointestinal tract (GI) irritation (Burdock et al. 2009) and bright yellow pigmentation of gums if chewed regularly (Ashri and Gazi 1990).

Dosage: For preparations, as recommended by the manufacturer. Products are usually standardised to a minimum of 1.5–2.5% of caffeine or theobromine. As a herbal tea, 1–3 g of powder three times a day (Burdock et al. 2009; EMAa 2011; EMAb 2011).

General plant information: The tree is native to the West African rainforest (indigenous to Togo, Sierra Leone and Angola) but nowadays cultivated in other tropical climates where the seeds are commonly chewed for their stimulant effects. Cola extract is widely used in popular soft drinks.

References

Ashri N, Gazi M. (1990) More unusual pigmentations of the gingiva. *Oral Surgery, Oral Medicine, Oral Pathology* 70(4): 445–449.

Burdock GA, Carabin IG, Crincoli CM. (2009) Safety assessment of kola nut extract as a food ingredient. *Food and Chemical Toxicology* 47(8): 1725–1732.

EMAa. (2011) Assessment report on *Cola nitida* (Vent.) Schott et Endl. and its varieties and *Cola acuminata* (P. Beauv.) Schott et Endl., semen. European Medicines Agency http://www.ema.europa.eu/docs/en_GB/document_library/Herbal_-_HMPC_assessment_report/2012/02/WC500122486.pdf.

EMAb. (2011) Community herbal monograph on *Cola nitida* (Vent.) Schott et Endl. and its varieties and *Cola acuminata* (P. Beauv.) Schott et Endl., semen. European Medicines Agency http://www.ema.europa.eu/docs/en_GB/document_library/Herbal_-_Community_herbal_monograph/2011/04/WC500105363.pdf.

Glade MJ. (2010) Caffeine-Not just a stimulant. *Nutrition* 26(10): 932–938.

Heckman MA, Weil J, Gonzalez de Mejia E. (2010) Caffeine (1, 3, 7-trimethylxanthine) in foods: a comprehensive review on consumption, functionality, safety, and regulatory matters. *Journal of Food Science* 75(3): R77–87.

Scotto G, Maillard C, Vion-Dury J, Balansard G, Jadot G. (1987) Behavioral effects resulting from sub-chronic treatment of rats with extract of fresh stabilized cola seeds. *Pharmacology Biochemistry and Behavior* 26(4): 841–845.

Vaille A, Balansard G, Jadot G. (1993) Effects of a subacute treatment in rats by a fresh cola extract on EEG and pharmacokinetics. *Pharmacology Biochemistry and Behavior* 45(4): 791–796.

Williamson EM, Driver S, Baxter K. (Eds.) (2013) *Stockley's Herbal Medicines Interactions.* 2nd Edition. Pharmaceutical Press, London, UK.

Comfrey   *Symphytum officinale* L.

Related species include rough comfrey, *S. asperum* Lepech, and Russian comfrey, *Symphytum × uplandicum* Nyman (a hybrid of *S. officinale × S. asperum*).

Family: Boraginaceae

Other common names: Boneset; knitbone

Drug name: Symphyti radix; Symphyti herba

Botanical drug used: Root, aerial parts

Indications/uses: Traditionally comfrey is used externally for the symptomatic relief of bruises and minor sprains. In European popular medicine comfrey was applied as a poultice to treat broken bones (hence comfrey's other common names 'knitbone' and 'boneset'), to reduce joint inflammation and to promote wound healing. THR products, mainly creams, are now available which contain pyrrolizidine alkaloid (PA)-free extracts.

Evidence: There is some clinical evidence to support short-term external use of comfrey as an anti-inflammatory and analgesic, for the relief of swelling of muscles and joints and bruises and sprains.

Safety: Comfrey should not be used internally due to the presence of the hepatotoxic PAs. Long-term use or application on broken skin should be avoided unless the special PA-free extracts are used. Safety has not been established for use of comfrey during pregnancy and lactation and is not recommended.

Main chemical compounds: Mucilage polysaccharides (29%) composed of fructose and glucose units, especially in the root; allantoin, phenolic acids, including rosmarinic, chlorogenic, caffeic and α-hydroxy caffeic acids; glycopeptides and amino acids; triterpene saponins, as monodesmosidic and bidesmosidic glycosides based on hederagenin, oleanolic acid and lithospermic acid. Comfrey also contains PAs, the content of which depends on the source of plant material and storage conditions (the alkaloids are labile). The PAs include intermedine, lycopsamine, symphytine, 7-acetylintermedine and 7-acetyllycopsamine (Staiger 2012). Symphytum species vary in alkaloid content, with Russian comfrey containing a higher proportion of the more toxic retronecine diester form than *S. officinale* (Rode 2002).

Phytopharmacy: An evidence-based guide to herbal medicinal products, First Edition.
Sarah E. Edwards, Inês da Costa Rocha, Elizabeth M. Williamson and Michael Heinrich.
© 2015 John Wiley & Sons, Ltd. Published 2015 by John Wiley & Sons, Ltd.

Clinical evidence:

There is increasing clinical evidence to support the effectiveness of comfrey root extract ointment in the treatment of various musculoskeletal and joint complaints, as highlighted in the subsequent section.

Back pain: In a placebo-controlled, double-blind, randomised clinical trial of 120 patients with acute upper or lower back pain, a proprietary comfrey cream (Kytta-Salbe®) containing comfrey root fluid extract (1:2, 35.0 g, extraction solvent ethanol 60% v/v, <0.35 ppm pyrrolizidine content) was applied three times per day, 4 g per application, and compared to placebo over 5 days. Results showed a significant ($p < 0.001$) reduction in pain in the group treated with comfrey extract compared to placebo. The comfrey ointment was also found to be fast acting, with pain intensity reduced by about 33% after 1 hour (Gianetti et al. 2010).

A double-blind, placebo-controlled, three-arm randomised trial ($n = 379$) assessed a combination of 35% comfrey root extract plus 1.2% methyl nicotinate, versus methyl nicotinate alone or placebo, applied topically three times daily for 5 days, for relief of acute upper or low back pain. Pain scores were found to be significantly lower ($p < 0.0001$) in the combination treatment group than in the methyl nicotinate alone or placebo groups, demonstrating superiority of the combination treatment (Pabst et al. 2013).

Osteoarthritis: The proprietary ointment Kytta-Salbe was assessed in the treatment of osteoarthritis in a randomised, double-blind, placebo-controlled trial of 220 patients suffering from painful osteoarthritis of the knee. Patients were administered 2 g of either the comfrey cream or placebo three times daily for 21 days. The comfrey treated group had significant ($p < 0.001$) reduction in pain and improvement in quality of life, mobility of the knee, and clinical global assessment measured by both physicians and patients (Grube et al. 2007).

Another study compared two concentrations of topically applied comfrey-based creams containing a blend of tannic acid and eucalyptus to a eucalyptus reference cream, assessing pain, stiffness and physical functioning in 43 patients with osteoarthritis of the knee. Patients were randomly assigned to a treatment group (receiving cream containing either 10 or 20% *S. officinale* root extract) or a placebo cream, three times per day for six weeks, and were evaluated every two weeks during treatment. Significant ($p < 0.01$) differences were found in all categories measured (pain, stiffness and daily function), confirming that both comfrey creams were more effective than the reference cream (Smith and Jacobsen 2011).

A study assessing the effect of a combination of standardised comfrey extract (200 mg/g), tannic acid, plus other ingredients including aloe vera gel (300 mg/g), eucalyptus oil and frankincense oil, applied topically three times daily for 12 weeks in the treatment of 133 patients with osteoarthritic knee pain found that the cream reduced pain and increased muscle strength, but had no effect on systemic inflammation or cartilage breakdown during the 12 week treatment (Staiger 2013).

Sprains: In a double-blind, placebo-controlled, randomised study, 142 patients with unilateral acute ankle sprains were treated with four daily applications of comfrey extract ointment, or placebo, for eight days. The active treatment was notably superior, with a significant reduction in pain ($p < 0.0001$) and ankle swelling ($p = 0.0001$) and an improvement in ankle mobility (Koll et al. 2004).

Another study of patients with acute unilateral ankle sprains ($n = 164$) compared treatment with comfrey root extract cream to topically applied diclofenac gel (containing 1.16 g diclofenac diethylamine salt). The comfrey treatment was found to be superior in global efficacy as evaluated by physicians and patients, especially with greater reduction of pain on pressure and movement (Predel et al. 2005).

Rheumatism: In a 4-week pilot study, 41 patients with musculoskeletal rheumatism were treated topically with a cream containing a combination of 35% comfrey root extract plus 1.2% methyl nicotinate or placebo. Assessment of the pain parameter 'tenderness when pressure applied' found the ointment superior to placebo in patients with epicondylitis and tendovaginitis, but not in patients with periarthritis (Staiger 2013).

Inflammation: In a controlled study, the effects of dermatological preparations containing 5 or 10% of a comfrey root extract (2:7, 50% ethanol) on the process of healing of experimentally induced UV-B erythema in 29 volunteers were evaluated. The extract was found to have an anti-inflammatory potency equal to or greater than diclofenac, with a correlation between efficacy and the concentration of α-hydroxy caffeic acid in the extract (Staiger 2013).

Other studies: In non-interventional post-marketing surveillance studies summarised in a review, topical preparations containing comfrey root extract were used to treat a number of conditions including bruises, strains, sprains and painful joint and muscle complaints in both children and adults, with notable clinical improvement, reducing the need for NSAIDS in a number of cases (Staiger 2013).

A topically applied proprietary product (Traumaplant®), made from aerial parts of Russian comfrey (*S.* × *uplandicum*), has also been investigated in randomised double-blind clinical trials. The preparation, containing 10% active ingredient, was found to significantly improve wound healing compared to an identical low-dose (1%) preparation in a study with 278 patients. In another study of 215 patients with pain in the lower or upper back, a significant reduction in pain was found following treatment with the same preparation, compared to the 1% reference preparation (Staiger 2012).

Pre-clinical evidence and mechanisms of action: The main active constituents responsible for pharmacological activity and mechanisms of action of comfrey have not been completely elucidated. It is thought that allantoin and rosmarinic acid are responsible for the main effects (Horinouchi and Otuki 2013; Staiger 2012). Rosmarinic acid has been shown to demonstrate anti-inflammatory activity in various tests, although significant absorption through the skin has not been shown. The anti-inflammatory activity of comfrey root extracts has also been demonstrated in animal studies. A glycopeptide isolated from comfrey root dose-dependently inhibited release of prostaglandins and arachidonic acid in rat stomach preparations. An *in vitro* study found that a 60% ethanolic comfrey root extract exerted an immunomodulatory effect on elements of the human immune system. Wound healing effects have also been demonstrated in a test model of fibroblasts in a collagen matrix: 40% ethanolic comfrey root extract and its high molecular weight fraction both inhibited shrinkage of the collagen matrix (Staiger 2012).

Interactions: None reported.

Contraindications: Comfrey should not be taken internally, or used on broken skin due to the potential presence of PAs. The fresh plant should be avoided as the alkaloid content is higher. Comfrey products should not be used during pregnancy and lactation and is not recommended for use in children and adolescents under 18 (EMA 2011; Pharmaceutical Press Editorial Team 2013).

Adverse effects: Several cases of hepatic veno-occlusive disease have been associated with oral ingestion of comfrey, particularly the fresh herb used as a tea, and attributed to the PAs, which are well known to cause liver damage (Pharmaceutical Press Editorial Team 2013). The PA-free extract used in several products has been shown to be absent of mutagenic effects *in vitro* (Benedek et al. 2010; Frost et al. 2013).

Dosage: Adults and elderly – liquid root extract (DER 2:1, extraction solvent ethanol 65% v/v) in an ointment base (100 g ointment contains 10 g extract): topical application twice daily (EMA 2011).

General plant information: *S. officinale* is native to Europe and temperate Asia, and has naturalised through much of the United States of America (Staiger 2013). A mulch of the plants is used by organic gardeners as a 'green fertiliser'.

References

Benedek B, Ziegler A, Ottersbach P. (2010) Absence of mutagenic effects of a particular *Symphytum officinale* L. liquid extract in the bacterial reverse mutation assay. *Phytotherapy Research* 24(3): 466–8.
EMA. (2011) Community herbal monograph on *Symphytum officinale* L., radix. Draft. European Medicines Agency http://www.ema.europa.eu/docs/en_GB/document_library /Herbal_-_Community_herbal_monograph/2011/08/WC500110650.pdf (accessed August 2013).
Frost R, MacPherson H, O'Meara S. (2013) A critical scoping review of external uses of comfrey (*Symphytum spp.*). *Complementary Therapies in Medicine* 21(6): 724–745.
Gianetti BM, Staiger C, Bulitta M, Predel HG. (2010) Efficacy and safety of a comfrey root extract ointment in the treatment of acute upper or lower back pain: results of a double-blind, randomised, placebo controlled, multicentre trial. *British Journal of Sports Medicine* 44(9): 637–641.
Grube B, Grünwald J, Krug L, Staiger C. (2007) Efficacy of a comfrey root (Symphyti offic. radix) extract ointment in the treatment of patients with painful osteoarthritis of the knee: results of a double-blind, randomised, bicenter, placebo-controlled trial. *Phytomedicine* 14(1): 2–10.
Horinouchi CDS, Otuki MF. (2013) Botanical Briefs: Comfrey (*Symphytum officinale*). *Cutis* 91(5): 225–228.
Koll R, Buhr M, Dieter R, Pabst H, Predel H-G, Petrowicz O, Giannetti B, Klingenburg S, Staiger C. (2004) Efficacy and tolerance of a comfrey root extract (Extr. Rad. Symphyti) in the treatment of ankle distorsions: results of a multicenter, randomized, placebo-controlled, double-blind study. *Phytomedicine* 11(6): 470–477.
Pabst H, Schaefer A, Staieger C, Junker-Samek M, Predel HG. (2013) Combination of comfrey root extract plus methyl nicotinate in patients with conditions of acute upper or low back pain: a multicentre randomised controlled trial. *Phytotherapy Research* 27(6): 811–817.
Pharmaceutical Press Editorial Team. (2013) *Herbal Medicines*. 4th Edition. Pharmaceutical Press, London, UK.
Predel H-G, Giannetti B, Koll R, Bulitta M, Staiger C. (2005) Efficacy of a Comfrey root extract ointment in comparison to Diclofenac gel in the treatment of ankle distortions: results of an observer-blind, randomized, multicenter study. *Phytomedicine* 12(1): 707–714.
Rode D. (2002) Comfrey toxicity revisited. *Trends in Pharmacological Sciences* 23(11): 497–499.
Smith DB, Jacobsen BH. (2011) Effect of a blend of comfrey root extract (*Symphytum officinale* L.) and tannic acid creams in the treatment of osteoarthritis of the knee: randomized, placebo-controlled, double-blind, multiclinical trials. *Journal of Chiropractic Medicine* 10(3): 147–156.
Staiger C. (2012) Comfrey: a clinical overview. *Phytotherapy Research* 26(10): 1441–1448.
Staiger C. (2013) Comfrey root: from tradition to modern clinical trials. *Wiener Medizinische Wochenschrift* 163(3–4): 58–64.

Cramp Bark   *Viburnum opulus* L.

Synonyms: *Opulus edulis* J.Presl; *V. americanum* Mill.; *V. opulus* var. *roseum* L.; and others

Family: Adoxaceae

Other common names: Guelder rose; European cranberrybush; high-bush cranberry

Drug name: Viburni opulus cortex

Botanical drug used: Stem bark

Indications/uses: Cramp bark, as the name suggests, is mainly used as an antispasmodic for all types of smooth and skeletal muscle, including colic, but especially for female complaints such as dysmenorrhoea, excessive menstrual bleeding, to prevent miscarriage, and improve uterine tone and facilitate childbirth.

Evidence: There is no clinical evidence for any indication but a long history of use.

Safety: No toxicity has been reported and the constituents do not suggest any safety concerns; however, there is no data confirming safety either.

Main chemical compounds: The bark contains many phenolic derivatives, in particular catechins and epicatechin; ellagic, caffeic, chlorogenic, neochlorogenic, *p*-coumaric, ferulic, gallic, protocatechuic, homogentisic and syringic acids; flavonoids including astragalin, paeonoside; the hydroquinone arbutin and the coumarins esculetin and scopoletin. There are also triterpene derivatives including α- and β-amyrin, ursolic and oleanolic acids, fatty acids and a small amount of volatile oil (Turek and Cisowski 2007; Upton 2000).

Clinical evidence: There is no clinical evidence for the use of cramp bark for any indication.

Pre-clinical evidence and mechanisms of action: Muscle relaxant effects have been demonstrated for the extract in several studies, and the activity attributed to the coumarin and volatile oil fractions (Upton 2000). However, the evidence is generally weak.

Phytopharmacy: An evidence-based guide to herbal medicinal products, First Edition.
Sarah E. Edwards, Inês da Costa Rocha, Elizabeth M. Williamson and Michael Heinrich.
© 2015 John Wiley & Sons, Ltd. Published 2015 by John Wiley & Sons, Ltd.

Interactions: Cramp bark extract has been tested for inhibition of CYP1A1, CYP1A2, CYP2C9, CYP2C19, CYP2D6 and CYP3A4, using cDNA-expressed CYP450 isoforms. It inhibited some of the enzymes at around 1 μg/ml, but the authors concluded that it is unlikely to lead to any clinically significant interactions. However, confirmation is required using *in vivo* studies (Ho et al. 2011).

Contraindications: None known.

Adverse effects: No data available.

Dosage: Powdered bark 1–4 g or equivalent as extract (decoction or tincture).

General plant information: The guelder rose is widely cultivated as an ornamental for its beautiful flowers. The berries are a rich source of anthocyanins with antimicrobial activity (Česonienė et al. 2012). The herbal drug is often adulterated with the bark of other species of *Viburnum,* especially black haw, *V. prunifolium* L., and the toxic *V. alnifolium* Marshall.

References

Česonienė L, Daubaras R, Viškelis P, Sarkinas A. (2012) Determination of the total phenolic and anthocyanin contents and antimicrobial activity of *Viburnum opulus* fruit juice. *Plant Foods for Human Nutrition* 67(3): 256–261.

Ho SH, Singh M, Holloway AC, Crankshaw DJ. (2011) The effects of commercial preparations of herbal supplements commonly used by women on the biotransformation of fluorogenic substrates by human cytochromes P450. *Phytotherapy Research* 25(7): 983–989.

Turek S, Cisowski W. (2007) Free and chemically bonded phenolic acids in barks of *Viburnum opulus* L. and *Sambucus nigra* L. *Acta Poloniae Pharmaceutica* 64(4): 377–383.

Upton R (Ed). (2000) Cramp Bark. *American Herbal Pharmacopoeia and Therapeutic Compendium.* AHP, Santa Cruz, CA, USA.

Cranberry

Vaccinium macrocarpon Aiton, *V. oxycoccos* L.

Family: Ericaceae

Other common names: Large or American cranberry (*V. macrocarpon*); small, European, common or northern cranberry (*V. oxycoccos*)

Drug name: Vacciniae fructus

Botanical drug used: Fresh or dried ripe fruit, juice derived from the fruit; cranberry liquid preparation

---

Indications/uses: The main use of cranberry juice and cranberry products is for the prevention and treatment of urinary tract infections (UTIs). Traditionally, cranberries have also been used for blood and digestive disorders, asthma, fever, loss of appetite, gallbladder and liver disease.

Evidence: Despite a large number of clinical trials on the use of cranberry for UTIs, limited evidence of a beneficial effect has been demonstrated.

Safety: Overall considered to be safe. Cranberries are widely used in food and beverages, and have been sometimes described as a 'superfood'.

Main chemical compounds: The main active components are thought to be the A-type proanthocyanidins (PACs), and other polyphenolics including chlorogenic and *p*-coumaric acids, flavonoids and organic acids such as malic, citric, quinic and benzoic acids (WHO 2009; Williamson et al. 2013). Cranberry varieties and products vary widely in PAC content and recent research shows that it is important to characterise these in order to ensure therapeutic efficacy (Chrubasik-Hausmann and Vlachojannis 2014).

Clinical evidence:

Despite many clinical studies, the clinical benefits of cranberry are not conclusively proven.

Prevention and treatment of urinary tract infection: Despite beneficial outcomes from many clinical studies and positive assessments from other systematic reviews, the latest Cochrane review on the use of cranberries for the prevention of UTIs concludes that there is still limited evidence of efficacy compared to placebo (Jepson et al. 2012). The equivocal results are exacerbated by differences in the products

---

Phytopharmacy: An evidence-based guide to herbal medicinal products, First Edition.
Sarah E. Edwards, Inês da Costa Rocha, Elizabeth M. Williamson and Michael Heinrich.
© 2015 John Wiley & Sons, Ltd. Published 2015 by John Wiley & Sons, Ltd.

tested, and it is now recognised that preparations standardised to clinically relevant active compounds (e.g. A-type PACs) are needed to clarify the clinical recommendations (Chrubasik-Hausmann and Vlachojannis 2014).

Other infections: A randomised, placebo-controlled parallel intervention study showed that consumption of cranberry polyphenols enhanced human T cell proliferation and reduced the number of symptoms associated with colds and influenza. Subjects drank a low calorie cranberry beverage (450 ml) made with a juice-derived, powdered cranberry fraction ($n = 22$) or a placebo beverage ($n = 23$), daily, for 10 weeks. Although incidence of illness was not reduced, significantly fewer symptoms of illness were reported in the cranberry treated group (Nantz et al. 2013).

Cardiovascular effects: Several small studies have investigated the effect of cranberry juice on vascular function, blood cholesterol levels and other indicators of cardiovascular health. The results, whilst generally positive, did not conclusively demonstrate therapeutic benefits (e.g. Dohadwala et al. 2011; Ruel et al. 2006).

Pre-clinical evidence and mechanisms of action:

Antimicrobial effects: The mechanism of action in preventing and treating UTIs and other infections is thought to be mainly due to the antimicrobial effect of the proanthocyanidins, which prevent *Escherichia coli* from adhering to the endothelium (in the urinary tract, the urothelium walls) (Foo et al. 2000; Howell et al. 2005; Ruz et al. 2009). An *in vitro* study demonstrated that a constituent of cranberry juice inhibited the adhesion of *Helicobacter pylori* to human gastric cells and mucus, as well as human erythrocytes – thus it may also prove useful for the prevention of stomach ulcers (Burger et al. 2002).

General health benefits: Proanthocyanidins and flavonoids are acclaimed to be beneficial in helping to prevent many diseases, from cardiovascular conditions to cancer, but the evidence is limited. For mechanisms and examples see Xu et al. (2012) and Romano et al. (2013).

Interactions: Despite the fact that in clinical trials no interactions have been found to date, there are case reports of increases in the international normalisation ratio (INR) in patients taking warfarin, so it may be advisable to avoid excessive cranberry consumption in those patients (Williamson et al. 2013). A study evaluating the effects of anthocyanidins and anthocyanins, including those found in cranberry, on the expression and catalytic activity of CYP2C9, CYP2A6, CYP2B6, and CYP3A4 in primary cultures of human hepatocytes and human liver microsomes, concluded that significant induction or inhibition of these enzymes would not be expected with cranberry ingestion in the normal ranges of food or dietary supplement (Srovnalova et al. 2014).

Contraindications: A health care professional should be consulted prior to the use of cranberry juice to rule out more serious conditions (e.g. pyelonephritis) and in patients with impaired kidney function or kidney stones (WHO 2009). Diabetic patients should use sugar-free preparations due to the high content of sugars in the juice. A cohort study in Norway showed that regular cranberry ingestion in pregnancy did not have any adverse effects for mother or baby (Heitmann et al. 2013).

Adverse effects: None known.

Dosage: According to manufacturers' instructions. The daily recommended dose for adults should be of at least 300 ml and not exceed 960 ml or equivalent of cranberry juice (Ruz et al. 2009; WHO 2009).

General plant information: Cranberries grow in acidic bogs in cooler regions of the northern hemisphere, especially Canada and the USA, and are widely cultivated in shallow beds which can be regularly irrigated.

References

Burger O, Weiss E, Sharon N, Tabak M, Neeman I, Ofek I. (2002) Inhibition of *Helicobacter pylori* adhesion to human gastric mucus by a high-molecular-weight constituent of cranberry juice. *Critical Reviews in Food Science and Nutrition* 42(3): 279–284.

Chrubasik-Hausmann S, Vlachojannis C. (2014) Proanthocyanin content in Cranberry CE medicinal products. *Phytotherapy Research* 28(11): 1612–1614.

Dohadwala MM, Holbrook M, Hamburg NM, Shenouda SM, Chung WB, Titas M, Kluge MA, Wang N, Palmisano J, Milbury PE, Blumberg JB, Vita JA. (2011) Effects of cranberry juice consumption on vascular function in patients with coronary artery disease. *American Journal of Clinical Nutrition* 93(5): 934–940.

Foo LY, Lu Y, Howell AB, Vorsa N. (2000) The structure of cranberry proanthocyanidins which inhibit adherence of uropathogenic P-fimbriated *Escherichia coli in vitro*. *Phytochemistry* 54(2): 173–181.

Heitmann K, Nordeng H, Holst L. (2013) Pregnancy outcome after use of cranberry in pregnancy – the Norwegian Mother and Child Cohort Study. *BMC Complementary and Alternative Medicine* 13: 345.

Howell AB, Reed JD, Krueger CG, Winterbottom R, Cunningham DG, Leahy M. (2005) A-type cranberry proanthocyanidins and uropathogenic bacterial anti-adhesion activity. *Phytochemistry* 66(18): 2281–2291.

Jepson RG, Williams G and Craig JC. (2012) Cranberries for preventing urinary tract infections. *Cochrane Database of Systematic Reviews* 10: CD001321.

Nantz MP, Rowe CA, Muller C, Creasy R, Colee J, Khoo C, Percival SS. (2013) Consumption of cranberry polyphenols enhances human γδ-T cell proliferation and reduces the number of symptoms associated with colds and influenza: a randomized, placebo-controlled intervention study. *Nutrition Journal* 12: 161.

Ruel G, Pomerleau S, Couture P, Lemieux S, Lamarche B, Couillard C. (2006) Favourable impact of low-calorie cranberry juice consumption on plasma HDL-cholesterol concentrations in men. *British Journal of Nutrition* 96(2): 357–364.

Romano B, Pagano E, Montanaro V, Fortunato AL, Milic N, Borrelli F. (2013) Novel insights into the pharmacology of flavonoids. *Phytotherapy Research* 27(11): 1588–1596.

Ruz EN, Gonzalez CC, Jaen Sde L, Escoto PG, Urquiza EK, Rosenfield LO, Ortiz CS and Castellanos PV. (2009) Cranberry juice and its role in urinary infections. *Ginecologia y obstetricia de Mexico* 77(11): 512–517.

Srovnalova A, Svecarova M, Kopecna Zapletalova M, Anzenbacher P, Bachleda P, Anzenbacherova E, Dvorak Z. (2014) Effects of Anthocyanidins and Anthocyanins on the Expression and Catalytic Activities of CYP2A6, CYP2B6, CYP2C9, and CYP3A4 in Primary Human Hepatocytes and Human Liver Microsomes. *Journal of Agricultural and Food Chemistry*. 62(3): 789–797.

WHO. (2009) *WHO Monographs on Selected Medicinal Plants*. Vol. 4. WHO, Geneva, 456 pp.

Williamson EM, Driver S, Baxter K. (Eds.) (2013) *Stockley's Herbal Medicines Interactions*. 2nd Edition. Pharmaceutical Press, London, UK.

Xu Z, Du P, Meiser P, Jacob C. (2012) Proanthocyanidins: oligomeric structures with unique biochemical properties and great therapeutic promise. *Natural Product Communications* 7(3): 381–388.

Damiana   *Turnera diffusa* Willd. ex Schult.

Synonyms: *T. aphrodisiaca* Ward; *T. diffusa* var. *aphrodisiaca* (Ward) Urb.; *T. diffusa* var. *diffusa*; and others

Family: Passifloraceae (formerly Turneraceae)

Other common names: Mexican holly; old woman's broom

Drug name: Turnerae diffusae folium

Botanical drug used: Leaf

Indications/uses: Damiana is used to relieve fatigue and nervous exhaustion and has been included in formulations with other medicinal plants to treat male dysfunction and impotence. It is regarded as a tonic and sexual stimulant. Traditionally, damiana has a number of other uses, and is taken to alleviate menopausal symptoms and for its reputed digestive, diuretic, antimalarial, anti-ulcer and hypoglycaemic properties. It is found mainly in THR products in combination with other medicinal plants (including cola and saw palmetto). Damiana is listed as a flavour ingredient by the Council of Europe and is permitted to be added in small quantities in foodstuffs.

Evidence: There is limited pharmacological evidence to support the traditional uses of damiana leaf. Its reputed aphrodisiac properties have not been fully substantiated in humans, although studies have shown that it improves sexual function in male rats. There is some evidence that preparations containing damiana may help induce weight loss when taken regularly for 45 days.

Safety: Safety and toxicological data are limited for damiana leaf preparations, although results from an *in vivo* study in mice indicate that it is relatively non-toxic, having an $L_{D50} > 5,000$ mg/kg with a single i.p. dose. However, due to the reported presence of toxic cyanogenic glycosides and the constituent arbutin, long-term use or high doses are not recommended. No information on the safety of damiana during pregnancy and lactation is available; but doses in excess of those found in food should be avoided.

Main chemical compounds: Chemical studies on damiana leaf are limited. Damiana leaf has been reported to contain traces of cyanogenic glycosides (tetraphyllin B) and the phenolic glycoside arbutin, an essential oil (0.2–0.9%) containing α- and

---

β-pinene, thymol, α-copaene, δ-cadinene calamenene and others, flavonoid glycosides and traces of β-sitosterol (EMEA 1999; Pharmaceutical Press Editorial Team 2013; Zhao et al. 2007).

Clinical evidence: Clinical data are limited. One small double-blind placebo-controlled study of 34 women given a nutritional supplement ('Argin Max') containing damiana extract for 4 weeks, showed improvement in sexual desire, reduction in vaginal dryness, increase in sexual intercourse and orgasm and improvement in clitoral sensation (Kumar et al. 2005). Small clinical studies have shown significant weight loss effects with damiana extracts given for 45 days (Kumar et al. 2005), as did a double-blind placebo-controlled study in overweight patients given a herbal preparation containing damiana in combination with maté tea (leaves of *Ilex paraguariensis* A.St.-Hil.) and guarana (seeds of *Paullinia cupana* Kunth), which also produced a significant delay in gastric emptying.

Even though some clinical data are available (but in one case of a complex mixture), overall, the evidence is very limited.

Pre-clinical evidence and mechanisms of action:

Damiana has demonstrated *in vitro* oestrogenic activity using a yeast screen and a tritiated-water release assay. A methanol extract of damiana leaf was shown to dose-dependently inhibit aromatase (oestrogen synthase) (Zhao et al. 2008). The antioxidant activity of the methanolic extracts of damiana leaves was shown to be similar to that of quercetin. Extracts from damiana roots and stems had even greater antioxidant activity, thought to be due to other phenolic constituents (Salazar et al. 2008).

Gastric ulcers: Acute administration of damiana in rats did not protect against gastric ulcers, in contrast to an earlier study in mice where damiana extracts were shown to inhibit ulceration. This may be attributable either to the model or to the type of extracts used (Bezerro et al. 2011).

Sexual function: A damiana leaf hydroalcoholic extract administered to sexually potent and sexually sluggish/impotent rats was found to improve the copulatory performance of the sexually sluggish/impotent rats, although it had no effect on sexually potent rats (Arletti et al. 1999). Damiana produced a faster recovery following sexual exhaustion in another study in male rats (Estrada-Reyes et al. 2009). Pro-sexual effects of damiana aqueous extract in rats have been shown to involve participation of the nitric oxide (NO) pathway, in common with sildenafil. In the same study, damiana was also shown to exhibit anxiolytic-like effects (Estrada-Reyes et al. 2013).

Diabetes: Studies have demonstrated conflicting results on the effects of damiana on glucose levels in healthy and diabetic animals. A decoction of damiana leaves exhibited a hypoglycaemic effect in temporarily hyperglycaemic rabbits in one study, whereas a water ethanolic extract of damiana leaf was shown to have no hypoglaemic activity in normal or alloxan-diabetic mice (Alarcon-Aguilera et al. 1998; Alarcon-Aguilera et al. 2002). This contradictory evidence of hypoglycaemic effects may be due to improper selection of plant material, geographical or seasonal variation or discrepancies in time of collection (Kumar et al. 2005).

Interactions: None known.

Contraindications: None known. Damiana leaf is not recommended for use during pregnancy and lactation or in children or adolescents under 18 years of age due to a lack of safety data. Alcoholic extracts of damiana roots have been reported to exhibit oxytocic activity and should be avoided during pregnancy (Kumar et al. 2005).

Adverse effects: There has been one reported cases of possible cyanide poisoning (symptoms similar to those of rabies or strychnine poisoning were described) following ingestion of 8 oz (226 g) of damiana extract, in a person with a history of alcohol abuse. Internal use of damiana leaf is also reported to interfere with iron absorption (Gardner and McGuffin 2013; Kumar et al. 2005). Damiana leaves have a slight laxative effect and may cause loosening of the stools if taken at higher than recommended doses (Kumar et al. 2005).

Dosage: For products, manufacturers' instructions should be followed. According to standard herbal reference texts, recommended oral dosages of damiana leaf in adults are as follows: dried leaf – 2–4 g as an infusion (tea) three times daily (Pharmaceutical Press Editorial Team 2013).

General plant information: *T. diffusa* grows in tropical and sub-tropical areas of Central and South America and Africa. It is reported that smoking of damiana leaves induces a euphoric effect, characterised by relaxation and increased imagination (Kumar et al. 2005). Damiana is an ingredient in some so-called 'legal highs' (Schäffer et al. 2013).

References

Alarcon-Aguilera FJ, Roman-Romas R, Perez-Gutierrez S, Aguilar-Contreras A, Contreras-Weber CC, Flores-Saenz JL. (1998) Study of the anti-hyperglycemic effect of plants used as antidiabetics. *Journal of Ethnopharmacology* 61(2): 101–110.

Alarcon-Aguilera FJ, Roman-Romas R, Flores-Saenz JL, Aguirre-Garcia F. (2002) Investigation on the hypoglycaemic effects of four Mexican medicinal plants in normal and alloxan-diabetic mice. *Phytotherapy Research* 16(4): 383–386.

Arletti R, Benelli A, Cavazzuti E, Scarpetta G, Bertolini A. (1999) Stimulating property of *Turnera diffusa* and *Pfaffia paniculata* extracts on the sexual behavior of male rats. *Psychopharmacology* 143(1): 15–19.

Bezerro AG, Mendes FR, Tabach R, Carlini EA. (2011) Effects of a hydroalcoholic extract of *Turnera diffusa* in tests for adaptogenic activity. *Revista Brasileira de Farmacognosia (Brazilian Journal of Pharmacognosy)* 21(1): 121–127.

EMEA. (1999) Committee for Veterinary Medicinal Products: *Turnera diffusa* Summary report. European Medicines Agency http://www.ema.europa.eu/docs/en_GB/document_library/Maximum_Residue_Limits_-_Report/2009/11/WC500015756.pdf (accessed August 2013).

Estrada-Reyes R, Ortiz-López P, Gutiérrez-Ortíz JM, Martínez-Mota L. (2009) *Turnera diffusa* Wild (Turneraceae) recovers sexual behavior in sexually exhausted males. *Journal of Ethnopharmacology* 123(3): 423–429.

Estrada-Reyes R, Carro-Juárez M, Martínez-Mota L. (2013) Pro-sexual effects of *Turnera diffusa* Wild (Turneraceae) in male rats involves the nitric oxide pathway. *Journal of Ethnopharmacology* 146(1): 164–172.

Gardner Z, McGuffin M. (Eds.) (2013) *American Herbal Product Association's Botanical Safety Handbook*. 2nd Edition. CRC Press, USA, 1072 pp.

Kumar S., Taneja R., Sharma A. (2005) The genus *Turnera*: a review update. *Pharmaceutical Biology* 43(5): 383–391.

Pharmaceutical Press Editorial Team. (2013) *Herbal Medicines.* 4th Edition. Pharmaceutical Press, London, UK.

Salazar R, Pozos ME, Cordero P, Perez J, Salinas MC, Waksman N. (2008) Determination of the antioxidant activity of plants from Northeast Mexico. *Pharmaceutical Biology* 46(3): 166–170.

Schäffer M, Gröger T, Pütz M, Zimmerman R. (2013) Assessment of the presence of damiana in herbal blends of forensic interest based on comprehensive two-dimensional gas chromatography. *Forensic Toxicology* 31(2): 251–262.

Zhao J, Dasmahapatra AK, Khan SI, Khan IA. (2008) Anti-aromatase activity of the constituents from damiana (*Turnera diffusa*). *Journal of Ethnopharmacology* 120(3): 387–393.

Zhao J, Pawar RS, Ali Z, Khan IA. (2007) Phytochemical investigation of *Turnera diffusa.* *Journal of Natural Products* 70(2): 289–292.

Dandelion   *Taraxacum officinale aggr.* F.H.Wigg., *T. mongolicum* Hand.-Mazz.

Synonyms: *Leontodon taraxacum* L.; *T. campylodes* G.E.Haglund; *T. vulgare* (Lam.) Schrank

Family: Asteraceae (Compositae)

Other common names: Lion's tooth; piss-a-bed

Drug name: Taraxaci radix cum herba; Taraxaci radix
  T. mongolicum is used in Chinese medicine.

Botanical drug used: Root and herb

Indications/uses: As part of multi-component THR products, dandelion is used as a diuretic and adjuvant in minor urinary complaints, and also to relieve the symptoms of mild digestive and liver disorders. It is mainly used in herbal combination products for indigestion, flatulence associated with over-indulgence, for the short-term relief of occasional constipation, and to relieve bloating associated with premenstrual water retention. It has anti-inflammatory properties and in Europe it has been used traditionally for conditions such as rheumatism, urinary tract infections, fever, sore throat, jaundice and as a galactogogue, laxative and tonic.

Evidence: There is a lack of clinical evidence, but its uses are supported by traditional folk medicine and pharmacological data.

Safety: Overall considered to be safe. Young dandelion leaves are eaten in salads and the roasted root has been used as an addition or substitute for coffee.

Main chemical compounds: The root and leaf contain sesquiterpene lactones, the eudesmanolides taraxinic acid, dihydrotaraxinic acid, taraxacoside and taraxacolide; sterols including taraxasterol and β-taraxasterol; flavonoids such as luteolin and quercetin, the caffeic acid derivatives chlorogenic and cichoric acids, and coumarins including cichoriin and aesculin. Carotenoids and vitamin A are also present, and in the root, oligosaccharides (EMA 2009a; Schütz et al. 2006; Williamson et al. 2013).

Clinical evidence:

Clinical studies for the use of dandelion are lacking.

Phytopharmacy: An evidence-based guide to herbal medicinal products, First Edition.
Sarah E. Edwards, Inês da Costa Rocha, Elizabeth M. Williamson and Michael Heinrich.
© 2015 John Wiley & Sons, Ltd. Published 2015 by John Wiley & Sons, Ltd.

Diuretic effects: A pilot study assessing the diuretic effect of an ethanolic extract of the leaf in 17 healthy volunteers was carried out over a single day. The extract increased the frequency of urination and excretion ratio of fluids and showed promise, but further studies are needed to support this indication (Clare et al. 2009).

Sore throat: A Cochrane review on the efficacy and safety of Chinese medicines including *T. mongolicum* for sore throat concluded that a meta-analysis was not possible due to the low quality of the studies. One non-blinded study (50 participants) found that the combination of 'compound dandelion soup' with sodium penicillin was more effective than sodium penicillin alone for acute purulent tonsillitis (Huang et al. 2012).

Pre-clinical evidence and mechanisms of action:

Anti-inflammatory effects: Dandelion extracts and compounds show antioxidant and anti-inflammatory activities which translate into diverse biological effects. The sesquiterpene lactones and the phenylpropanoids exert anti-inflammatory effects, and the sesquiterpene lactones also exhibit anticancer and antimicrobial activities. Signalling pathways and molecules such as NF-κB, Akt, MEK, ERK, TNF and IL, among others, are involved in mediating the effects (reviewed by Gonzalez-Castejon et al. 2012).

Liver protective effects: Dandelion root extract decreased hepatic superoxide dismutase activity, reduced hepatic fibrinous deposits and restored histological architecture in CCl(4)-induced liver fibrosis in mice, by inactivating hepatic stellate cells and enhancing hepatic regenerative capabilities (Domitrović et al. 2010). A dandelion leaf extract significantly suppressed lipid accumulation in the liver and reduced insulin resistance in high-fat diet-induced fatty liver disease in C57BL/6 mice, via the AMPK pathway (Davaatseren et al. 2013). *T. mongolicum* whole herb extract demonstrated a potent antiviral effect against hepatitis virus B and a protective effect on hepatocytes *in vitro*. It was suggested that the antiviral properties may be attributed in part to blocking protein synthesis and DNA replication, and the protective effect to its ability to ameliorate oxidative stress (Jia et al. 2014).

Inhibition of adipocyte differentiation and lipogenesis: Three dandelion extracts (leaf, root and a commercial root powder) containing caffeic and chlorogenic acids as the main phenolic constituents were tested in 3T3-L1 preadipocytes. All three decreased lipid and triglyceride accumulation, with no cytotoxicity as assessed with the MTT assay at the concentrations tested. DNA microarray analysis showed that the extracts regulated the expression of a number of genes and long non-coding RNAs that play a major role in the control of adipogenesis (González-Castejón et al. 2014).

Immunological and anti-fatigue effects: An extract of *T. officinale* (TO), orally administered to mice, was examined for anti-fatigue and immune-enhancing effects *in vivo* by performing a forced swimming test (FST) and *in vitro* by using peritoneal macrophages. FST immobility time was significantly decreased in the TO-treated group (100 mg/kg) and the level of lactic dehydrogenase, an indicator of muscle damage, tended to decline after TO administration. Blood urea nitrogen levels also decreased significantly. The production of cytokines and nitric oxide (NO) in mouse peritoneal macrophages was examined, and when TO was used in combination with recombinant interferon-gamma, a noticeable cooperative induction of tumour necrosis factor (TNF)-α, interleukin (IL)-12, and IL-10 production was observed.

These results suggest that TO improves fatigue-related indicators and immunological parameters in mice (Lee et al. 2012).

Interactions: None known. There is some evidence that *T. mongolicum* might alter the absorption of ciprofloxacin, based on an animal study (Williamson et al. 2013).

Contraindications: Not recommended during pregnancy and lactation or in children under 12 years of age due to lack of safety data. Dandelion extracts should not be used in cases of active peptic ulcer or biliary diseases except under expert professional supervision. Not recommended in patients with renal failure or heart failure, due to potential hyperkalaemia (EMA 2009b, ESCOP 2003).

Adverse effects: Epigastric pain, hyperacidity, and allergic reactions may occur (EMA 2009b). *T. officinale* extracts have been shown to decrease fertility in male rats, although no human studies have confirmed this, and despite a traditional use in Jordan as a male fertility enhancer (Tahtamouni et al. 2011).

Dosage: Commercial products: as recommended by the manufacturer. As dried root with herb, 3 to 4 g as a decoction or 4 to 10 g as an infusion up to 3 times a day; dry extract, up to 600 mg a day (EMA 2009b).

General plant information: Endemic to Europe but widely distributed in the warmer temperate zones of the Northern Hemisphere as an invasive species. *Taraxacum* species have also been used for over 2000 years as a diuretic in both traditional Chinese Medicine and Ayurveda medicine (Clare et al. 2009). Dandelion is used in Chinese medicine, both internally and externally, for abscesses and eye inflammations.

References

Clare BA, Conroy RS, Spelman K. (2009) The diuretic effect in human subjects of an extract of *Taraxacum officinale* folium over a single day. *Journal of Alternative and Complementary Medicine* 15(8): 929–934.

Davaatseren M, Hur HJ, Yang HJ, Hwang JT, Park JH, Kim HJ, Kim MJ, Kwon DY, Sung MJ. (2013) *Taraxacum official* [sic] (dandelion) leaf extract alleviates high-fat diet-induced nonalcoholic fatty liver. *Food and Chemical Toxicology* 58: 30–36.

Domitrović R, Jakovac H, Romić Z, Rahelić D, Tadić Z. (2010) Antifibrotic activity of *Taraxacum officinale* root in carbon tetrachloride-induced liver damage in mice. *Journal of Ethnopharmacology* 130(3): 569–577.

EMA. (2009a) Assessment report on *Taraxacum officinale* Weber ex Wigg., radix cum herba. European Medicines Agency http://www.ema.europa.eu/docs/en_GB/document _library/Herbal_-_HMPC_assessment_report/2011/03/WC500102972.pdf.

EMA. (2009b) Community herbal monograph on *Taraxacum officinale* Weber ex Wigg., radix cum herba. European Medicines Agency http://www.ema.europa.eu/docs/en_GB /document_library/Herbal_-_Community_herbal_monograph/2011/01/WC500101484.pdf.

ESCOP. (2003) *ESCOP Monographs: The Scientific Foundation for Herbal Medicinal Products.* 2nd Edition. Thieme, Exeter and London, UK.

Gonzalez-Castejon M, Visioli F, Rodriguez-Casado A. (2012) Diverse biological activities of dandelion. *Nutrition Reviews* 70(9): 534–547.

González-Castejón M, García-Carrasco B, Fernández-Dacosta R, Dávalos A, Rodriguez-Casado A. (2014) Reduction of Adipogenesis and Lipid Accumulation by *Taraxacum officinale* (Dandelion) Extracts in 3T3L1 Adipocytes: An in vitro Study. *Phytotherapy Research* 28(5): 745–752.

Jia YY, Guan RF, Wu YH, Yu XP, Lin WY, Zhang YY, Liu T, Zhao J, Shi SY, Zhao Y. (2014) *Taraxacum mongolicum* extract exhibits a protective effect on hepatocytes and an antiviral effect against hepatitis B virus in animal and human cells. *Molecular Medicine Reports* 9(4): 1381–1387.

Huang Y, Wu T, Zeng L, Li S. (2012) Chinese medicinal herbs for sore throat. *Cochrane Database of Systematic Reviews* 3: CD004877.

Lee BR, Lee JH, An HJ. (2012) Effects of *Taraxacum officinale* on fatigue and immunological parameters in mice. *Molecules.* 17(11): 13253–13265.

Schütz K, Carle R, Schieber A. (2006) Taraxacum – a review on its phytochemical and pharmacological profile. *Journal of Ethnopharmacology* 107(3): 313–323.

Tahtamouni LH, Alqurna NM, Al-Hudhud MY, Al-Hajj HA. (2011) Dandelion (*Taraxacum officinale*) decreases male rat fertility *in vivo*. *Journal of Ethnopharmacology* 135(1): 102–109.

Williamson EM, Driver S, Baxter K. (Eds.) (2013) *Stockley's Herbal Medicines Interactions.* 2nd Edition. Pharmaceutical Press, London, UK.

Devil's Claw	*Harpagophytum procumbens* (Burch.) DC. ex Meissner, *H. zeyheri* Decne.

Family: Pedaliaceae

Other common names: Grapple plant

Drug name: Harpagophyti Radix

Botanical drug used: Tuberous secondary roots

Indications/uses: In THR products, devil's claw extracts are used to relieve rheumatic or muscular pain, general aches and pains in the muscles and joints and backache. In its region of origin, the Kalahari Desert of southern Africa, preparations are consumed for the same purposes and also as a general health tonic and for treating general pain, fevers, ulcers and boils.

Evidence: Overall there is – compared to other traditional herbal medicinal products – good clinical evidence for some of devil's claw products in terms of their clinical effectiveness for treating pain, especially lower back pain. There is also pre-clinical evidence for its anti-inflammatory effects.

Safety: Despite a lack of studies on genotoxicity, carcinogenicity or reproduction, the available data do not highlight any potential risks of registered products.

Main chemical compounds: Iridoids are the best known compounds from the genus and include harpagoside, often considered to be a key active constituent, harpagide and procumbide. Other compounds include phenolic glycosides such as acteoside and isoacteoside, harpagoquinones, flavonoids, phytosterols and carbohydrates (EMA 2009; Mncwangi et al. 2012). Different preparations vary widely in their composition and efficacy, so it is advisable to use a specific, standardised product which has been tested clinically (Ouitas and Heard 2010).

Clinical evidence:

Anti-inflammatory activity: Considerable clinical evidence for the use of devil's claw as an anti-inflammatory and analgesic (commonly for lower back pain), and as an anti-rheumatic agent, has now accumulated. While many of the published trials

Phytopharmacy: An evidence-based guide to herbal medicinal products, First Edition.
Sarah E. Edwards, Inês da Costa Rocha, Elizabeth M. Williamson and Michael Heinrich.
© 2015 John Wiley & Sons, Ltd. Published 2015 by John Wiley & Sons, Ltd.

do not conform to the highest methodological quality criteria, the data from good quality studies indicated that devil's claw appeared effective in the reduction of the main clinical symptom of pain (Brien et al. 2006). Most importantly, pain reduction was also found in a Cochrane meta-analysis using an extract equivalent to 50 mg or 100 mg harpagoside per daily dose; a daily dose of 60 mg reduced pain about the same as a daily dose of 12.5 mg of Vioxx (Gagnier et al. 2006).

In an open, single-group study of 8 weeks' duration, 259 patients suffering from arthritis and other rheumatic conditions (AORC) were included. A total of 222 patients were included in the effectiveness assessments (Intention to Treat [ITT] population). The patients self-administered a daily dose of 960 mg, one tablet (480 mg) each morning and evening. There were statistically significant ($p < 0.0001$) improvements in patient assessment of global pain, stiffness and function and in mean pain scores for hand, wrist, elbow, shoulder, hip, knee and back pain. Quality of life measurements (SF-12) were significantly increased from baseline, and 60% of patients either reduced or stopped concomitant pain medication (Warnock et al. 2007).

Pre-clinical evidence and mechanisms of action: A wide range of activities for *H. procumbens* have been identified, including antidiabetic, antimicrobial, antioxidant and other activities, but the most important are in terms of pain modulation and anti-inflammatory effects. Considerable evidence points to inhibitory effects on TNF-α and on arachidonic acid metabolism by acting on the COX-2 mRNA expression (Fiebich et al. 2001, 2010). A range of *in vivo* data point to analgesic effects of extracts (Mncwangi et al. 2012).

Interactions: No data suggesting an interaction between *Harpagophytum* and oral anticoagulants, or sulfonylureas or other drugs have been reported (EMA 2009). Based on an *in vitro* evaluation, Modarai et al. (2011) concluded that 'Devil's claw preparations are unlikely to interact with conventional medications via the CYP P450 enzyme system, which is in agreement with the pharmacovigilance data, i.e. the lack of reported interactions.'

Contraindications: Due to lack of data, the drug is contraindicated in children and during pregnancy and lactation. Based on some data on gastric or duodenal side effects, devil's claw is contraindicated in cases of gastric or duodenal ulcers.

Adverse effects: While there are no data on genotoxicity, carcinogenicity or reproductive toxicity, the available data on single and repeated dose toxicity point to a very low risk of registered products.

Dosage: A range of products containing 200–480 mg of extract with a defined drug-extract ratio are commonly used, and several of these preparations have been studied clinically and/or pharmacologically. Several special extracts are on the market.

General plant information: The common English name of the plant refers to the fruit's shape and specifically its long protrusions with sharp, grapple-like hooks, and straight thorns on the upper surface, which can cause severe injuries in humans and cattle.

While the two closely related species *H. procumbens* and *H. zeyheri* are currently accepted as source plants for the drug, further research will need to establish the equivalence between these two species, and thus the drugs derived from them.

References

Brien S, Lewith GT, McGregor G. (2006) Devil's Claw (*Harpagophytum procumbens*) as a treatment for osteoarthritis: a review of efficacy and safety. *Journal of Alternative and Complementary Medicine* 12(10): 981–993.

Chrubasik S, Junck H, Breitschwerdt H, Conradt C, Zappe H. (1999) Effectiveness of Harpagophytum extract WS 1531 in the treatment of exacerbation of low back pain: a randomized, placebo-controlled, double-blind study. *European Journal of Anesthesiology* 16(2): 118–129.

EMA. (2009) Assessment Report on *Harpagophytum procumbens* DC. and/or *Harpagophytum zeyheri* Decne, Radix. European Medicines Agency. Available from: http://www.ema.europa.eu/docs/en_GB/document_library/Herbal_HMPC_assessment_report/2010/01/WC500059019.pdf (accessed June 2013).

Fiebich BL, Heinrich M, Hiller KO, Kammerer N. (2001) Inhibition of TNF-alpha synthesis in LPS-stimulated primary human monocytes by *Harpagophytum* extract STEIHAP 69. *Phytomedicine* 8(1): 28–30.

Fiebich BL, Muñoz E, Rose T, Weiss G, McGregor GP. (2012) Molecular Targets of the Anti-inflammatory *Harpagophytum procumbens* (Devil's claw): Inhibition of TNFα and COX-2 Gene Expression by Preventing Activation of AP-1. *Phytotherapy Research* 26(6): 806–811.

Gagnier JJ, van Tulder MW, Berman BM, Bombardier C. (2006) Herbal medicine for low back pain. *Cochrane Database of Systematic Reviews* 2: CD004504. (with additional edits 2010).

Mncwangi N, Chen W, Vermaak I, Viljoen AM, Gericke N. (2012) Devil's Claw-a review of the ethnobotany, phytochemistry and biological activity of *Harpagophytum procumbens*. *Journal of Ethnopharmacology* 143(3): 755–71.

Modarai M, Suter A, Kortenkamp A, Heinrich M. (2011) A luminescence-based screening platform for assessing the interaction potential of herbal medicinal products and its use in the evaluation of Devils Claw (Harpagophyti Radix). *Journal of Pharmacy and Pharmacology* 63: 429–438.

Ouitas NA, Heard C. (2010) Estimation of the Relative Antiinflammatory Efficacies of Six Commercial Preparations of *Harpagophytum procumbens* (Devil's Claw). *Phytotherapy Research* 24(3): 333–338.

Warnock M, McBean D, Suter A, Tan J, Whittaker P. (2007) Effectiveness and safety of Devil's Claw tablets in patients with general rheumatic disorders. *Phytotherapy Research* 21(12): 1228–1233.

Echinacea

Echinacea purpurea (L.) Moench., *E. angustifolia* DC., *E. pallida* (Nutt.) Nutt.

Family: Asteraceae (Compositae)

Other common names: Black Sampson; Coneflower; *E. purpurea, E. angustifolia*: purple coneflower; *E. angustifolia*: Kansas snakeroot; narrow-leaf purple cone-flower; *E. pallida*: pale coneflower; pale purple coneflower; sometimes it is also called *Rudbeckia*, which is, however, another genus of the Asteraceae.

Drug name: Echinaceae herba; Echinaceae radix

Botanical drugs used: *E. purpurea*: rhizome and root/fresh flowering aerial parts/dried herb/fresh expressed juice; *E. pallida*: root; *E. angustifolia*: herb

---

Indications/uses: Echinacea is used to treat common cold and influenza type infections and other upper respiratory tract infections; inflammation of the mouth and throat. It is also used for minor skin conditions (*E. purpurea* and *E. angustifolia* in combination with *Baptisia tinctoria* (L.) R.Br. and *Fumaria officinalis* L.) based on traditional use.

Evidence: Overall, a rather large number of clinical studies and a series of meta-analyses provide evidence that Echinacea preparations appear to be efficacious both in the treatment (reducing symptoms and duration) and prevention of the common cold.

Safety: Overall, it is considered to be safe, although it should be avoided in patients allergic to plants from the Asteraceae (daisy family).

Main chemical compounds: The phytochemical composition of the various species is slightly different and this leads to confusion as to the potential for drug interactions. *E. purpurea*: The root contains alkamides, including dodecatetraenoic acid isobutylamide, caffeic acid derivatives and saturated (non-hepatotoxic) pyrrolizidine alkaloids. The herb contains similar alkamides to the root, caffeic acid derivatives and polysaccharides. The pressed juice (from the aerial parts) contains heterogeneous polysaccharides, inulin-type compounds, arabinogalactan polysaccharides and glycoproteins. *E. pallida*: The root contains similar caffeic acid

---

Phytopharmacy: An evidence-based guide to herbal medicinal products, First Edition.
Sarah E. Edwards, Inês da Costa Rocha, Elizabeth M. Williamson and Michael Heinrich.
© 2015 John Wiley & Sons, Ltd. Published 2015 by John Wiley & Sons, Ltd.

esters and glycosides to *E. purpurea*. Polyenes and polyacetylenes, including a range of ketoalkenes and ketopolyacetylenes, have been reported and polysaccharides and glycoproteins are also present. *E. angustifolia:* The root contains alkamides and similar caffeic acid esters and glycosides to *E. purpurea*. Alkylketones and the saturated pyrrolizidine alkaloids are also present (Barnes et al. 2005; Williamson et al. 2013).

Clinical evidence: Clinical evidence for its use as an immunomodulator is available for some of the standardised extracts. Overall, Echinacea preparations appear to be efficacious both in the treatment (reducing symptoms and duration) and prevention of the common cold. However, Echinacea preparations tested in clinical trials differ greatly. Better evidence is shown with preparations based on the aerial parts of *E. purpurea* as they might be effective for the early treatment of colds in adults, but the results are not fully consistent (Barnes et al. 2005; Linde et al. 2006; Nahas and Balla 2011). A randomised, double-blind, placebo-controlled trial assessed the safety and efficacy of a specific *E. purpurea* product over a period of 4 months. Participants (755 healthy subjects) were given the herbal preparation (standardised to contain 5 mg/100 g of dodecatetraenoic acid isobutylamide; 0.9 ml 3 times a day corresponding to 2400 mg of extract) for illness prevention or placebo. The study concluded that this herbal preparation was safe and effective compared to placebo and should be recommended as a prophylactic treatment (Jawad et al. 2012). However, in a separate randomised clinical trial with four parallel groups (no pills; placebo pills, blinded; echinacea pills, blinded; echinacea pills, open-label), another Echinacea preparation did not show statistical significance with respect to shorter illness duration when compared to the other subgroups (Barrett et al. 2010). For the subgroup who believed in Echinacea and received pills, however, illnesses were substantively shorter and less severe, regardless of whether the pills contained Echinacea or not (Barrett et al. 2011).

A multicentre, randomised, double-blind, double-dummy controlled trial showed that an Echinacea and sage preparation was as efficacious and well tolerated as a chlorhexidine/lidocaine spray in the treatment of acute sore throats (Schapowal et al. 2009).

As far as the traditional use in skin conditions is concerned, the evidence is weaker. A standardised preparation of *E. purpurea* inhibited the proliferation of *Propionibacterium acnes* bacteria (one of the causative organisms for acne) *in vitro*. The herbal extract also reversed the bacterially induced pro-inflammatory cytokines IL-6 and IL-8, thus potentially providing a benefit to individuals with acne, not only by inhibiting proliferation of the organism, but also by reversing the associated inflammation (Sharma et al. 2011).

Pre-clinical evidence and mechanisms of action: The exact mechanism of action for the immunomodulating effects of Echinacea preparations is unclear. Several components in Echinacea appear to be responsible for its activity. *In vitro* and *in vivo* studies suggest that Echinacea stimulates phagocytosis, enhances mobility of leucocytes, stimulates TNF and interleukin (IL)-1 secretion from macrophages and lymphocytes, and improves respiratory activity (Groom et al. 2007; Melchart et al. 2002).

Interactions: There are no clinical reports of an interaction, but possibly a theoretical interaction with immunosuppressants (such as ciclosporin and methotrexate) due to an antagonistic effect. Echinacea is unlikely to raise caffeine levels in most patients but it may affect it. *In vitro* data point to a very low risk of interactions for chemically well-defined preparations (Modarai et al. 2010). If side effects are observed, advise the patient to either stop taking Echinacea and/or reduce their caffeine intake. There is one report that clearance of intravenous midazolam may be modestly increased in patients taking Echinacea (Williamson et al. 2013).

Contraindications: Echinacea is not recommended during pregnancy and lactation as safety has not been established. The MHRA has recommended that Echinacea should not be used in children under 12 years old to avoid allergic reactions (MHRA 2009), although numerous unlicensed paediatric preparations are available on the market. Contraindications for Echinacea include hypersensitivity to any of the compounds present in the herb, or to composite plants in general. Due to its immunostimulating activity, Echinacea should not be used in cases of progressive systemic disorders (e.g. tuberculosis and sarcoidosis), autoimmune disorders (e.g. collagenosas and multiple sclerosis), immunodeficiences (e.g. HIV infection; AIDS), immunosuppression (e.g. oncological cytostatic therapy; history of organ or bone marrow transplant), diseases of white blood cells system (agranulocytosis, leukemia) and allergic diathesis (urticaria, atopic dermatitis and asthma) based on its possible mode of action. It should not be used for more than 8 weeks (MHRA 2009).

Adverse effects: Echinacea may trigger allergic reactions. Skin rashes and itching have been observed in isolated cases. Facial swelling, difficulty in breathing, dizziness and reduction of blood pressure are rare side effects. Hypersensitivity reaction (rash, urticaria, angioedema of the skin, Quincke oedema, bronchospasm with obstruction, asthma and anaphylactic shock) may occur. Echinacea may trigger allergic reaction in atopic patients. Association with autoimmune diseases (encephalitic disseminate, erythema nodosum, immunothrombocytopenia, Evans Syndrome and Sjögren Syndrome with renal tubular dysfunction) has been reported. Leucopenia may occur in long-term use (more than 8 weeks) (MHRA 2010).

Dosage: Dosage varies according to type of extract and plant part used, e.g. 40.68 mg/daily (crude extract) of *E. purpurea* herb and root; 176 mg/daily (dried pressed juice) of *E. purpurea* herb. Do not use for longer than 10 days or if symptoms persist consult a health-care professional.

General plant information: Echinacea is endemic to eastern and central North America.

References

The literature on Echinacea is extensive; these are reviews and papers selected to support the information provided in the monograph:

Barnes J, Anderson LA, Gibbons S, Phillipson JD. (2005) *Echinacea* species (*Echinacea angustifolia* (DC.) Hell., *Echinacea pallida* (Nutt.) Nutt., *Echinacea purpurea* (L.) Moench): a review of their chemistry, pharmacology and clinical properties. *Journal of Pharmacy and Pharmacology* 57(8): 929–954.

Barrett B, Brown R, Rakel D, Mundt M, Bone K, Barlow S, Ewers T. (2010) *Echinacea* for treating the common cold: a randomized trial. *Annals of the Internal Medicine* 153(12): 769–777.

Barrett B, Brown R, Rakel D, Rabago D, Marchand L, Scheder J, Mundt M, Thomas G, Barlow S. (2011) Placebo effects and the common cold: a randomized controlled trial. *Annals of the Family Medicine* 9(4): 312–322.

Groom SN, Johns T, Oldfield PR. (2007) The potency of immunomodulatory herbs may be primarily dependent upon macrophage activation. *Journal of Medicinal Food* 10(1): 73–79.

Jawad M, Schoop R, Suter A, Klein P, Eccles R. (2012) Safety and efficacy profile of *Echinacea purpurea* to prevent common cold episodes: a randomized, double-Blind, placebo-controlled trial. *Evidence Based Complementary and Alternative Medicine* 2012: 841315.

Linde K, Barrett B, Wolkart K, Bauer R and Melchart D. (2006) *Echinacea* for preventing and treating the common cold. *Cochrane Database of Systematic Reviews* (1): CD000530.

Melchart D, Clemm C, Weber B, Draczynski T, Worku F, Linde K, Weidenhammer W, Wagner H and Saller R. (2002) Polysaccharides isolated from *Echinacea purpurea* herba cell cultures to counteract undesired effects of chemotherapy – a pilot study. *Phytotherapy Research* 16(2): 138–142.

MHRA. (2009) Echinaforce forte cold and flu tablets. Medicines and Healthcare Product Regulatory Agency, http://www.mhra.gov.uk/home/groups/par/documents/websiteresources/con052092.pdf (accessed May, 2012).

MHRA. (2010) Echinaforce Junior cold and flu tablets. Medicines and Healthcare Product Regulatory Agency, http://www.mhra.gov.uk/home/groups/par/documents/website resources/con076219.pdf.

Modarai, M, Suter A, Kortenkamp, A. Heinrich M. (2010) Metabolomic profiling of liquid Echinacea medicinal products inhibiting Cytochrome P450 3A4 (CYP3A4). *Planta Medica* 76(4): 378–385.

Nahas R, Balla A. (2011) Complementary and alternative medicine for prevention and treatment of the common cold. *Canadian Family Physician* 57(1): 31–36.

Schapowal A, Berger D, Klein P and Suter A. (2009) Echinacea/sage or chlorhexidine/lidocaine for treating acute sore throats: a randomized double-blind trial. *European Journal of Medical Research* 14(9): 406–412.

Sharma M, Schoop R, Suter A and Hudson JB. (2011) The potential use of Echinacea in acne: control of *Propionibacterium acnes* growth and inflammation. *Phytotherapy Research* 25(4): 517–521.

Williamson EM, Driver S, Baxter K. (Eds.) (2013) *Stockley's Herbal Medicines Interactions.* 2nd Edition. Pharmaceutical Press, London, UK.

Elderberry, Elderflower   *Sambucus nigra* L.

Family: Adoxaceae (formerly Caprifoliaceae)

Other common names: Black elder; European elder; Sambucus

Drug name: Sambuci fructus; Sambuci flores

Botanical drug used: Fruit (berries), dried flower

Indications/uses: Elderberry and elderflower extracts have a long history of use for the relief of the symptoms of the common cold and flu, chills and sore throats. There is also a traditional folk medicine use of the flowers in the treatment of diabetes.

Evidence: There is limited clinical evidence to show that the berries may help alleviate and shorten the duration of cold and flu symptoms. No studies have been carried out on the flower extracts.

Safety: Generally considered to be safe, but the berries should not be eaten raw. Heat treatment destroys the toxic constituents and these are not present in commercial products.

Main chemical compounds:

Berries: Anthocyanins including cyanidin-3-sambubioside and cyanidin-3-glucoside, flavonoids such as quercetin and rutin, the hemagglutinin protein Sambucus nigra agglutinin III (SNA-III), cyanogenic glycosides, for example, sambunigrin, viburnic acid and vitamins A and C (Anon. 2005; Williamson et al. 2013).

Flowers: Flavonoids including rutin, quercetin, hyperoside, nicotiflorin, kaempferol, naringenin; triterpenes based on oleanolic and ursolic acids. More recently, a number of acyl spermidines have been identified, with N,N-diferuloylspermidine and N-acetyl-N,N-diferuloylspermidine being most abundant (Kite et al. 2013), as well as α-linolenic and linoleic acids (Christensen et al. 2010).

Clinical evidence: Two randomised, double-blind, placebo-controlled studies investigated the efficacy of a commercially available preparation of elderberries for the treatment of symptoms associated with influenza A and B. The extract was found to reduce the duration of symptoms in patients with laboratory-confirmed

Phytopharmacy: An evidence-based guide to herbal medicinal products, First Edition.
Sarah E. Edwards, Inês da Costa Rocha, Elizabeth M. Williamson and Michael Heinrich.
© 2015 John Wiley & Sons, Ltd. Published 2015 by John Wiley & Sons, Ltd.

influenza virus infection, but further clinical studies are required (reviewed by Guo et al. 2007; Ulbricht et al. 2014; Vlachojannis et al. 2010).

Pre-clinical evidence and mechanisms of action:

Antimicrobial effects (berries): Several mechanisms are thought to be responsible for the antimicrobial effects of the extracts. Elderberry extracts have antimicrobial and antiviral activity against both gram-positive and negative bacteria, as well as on influenza viruses (Kinoshita et al. 2012; Krawitz et al. 2011). Its constituents neutralise the activity of the haemagglutinin spikes found on the surface of several viruses, which are vital for virus replication (Zakay-Rones et al. 1995; Zakay-Rones et al. 2004). It has antioxidant properties due to anthocyanin flavonoid content (Wu et al. 2002). Elderberry extracts also have immune-modulating activity by increasing the production of cytokines (Barak et al. 2001; Barak et al. 2002).

Antidiabetic and related effects (flowers): Extracts of elderflowers were found to activate PPARγ and to stimulate insulin-dependent glucose uptake, suggesting that they have a potential use in the prevention and/or treatment of insulin resistance. α-Linolenic acid and linoleic acids, as well as naringenin, activated PPARγ without stimulating adipocyte differentiation. These compounds were not able to fully account for the observed PPARγ activation of the crude elderflower extracts, so synergistic or other mechanisms may be involved. Flavonoid glycosides were not able to activate PPARγ, whereas some of their aglycones may be potential agonists of PPARγ (Christensen et al. 2010).

Antidiabetic and related effects (berries): A *S. nigra* berry extract was found to improve bone mineral density in experimental diabetes in rats, an effect thought to be related to its antioxidant activity (Badescu et al. 2012).

Interactions: None known. Very weak experimental evidence suggest that elder extracts may have additive effects with antidiabetic drugs and phenobarbital, and may antagonise the effects of morphine. However, this is unlikely to be of much clinical relevance. Patients with diabetes should be advised to monitor blood sugar closely when using flower extracts (Williamson et al. 2013).

Contraindications: Not recommended during pregnancy and lactation, as safety has not been established. Not recommended for children under 12 years of age due to lack of adequate data (EMEA 2008).

Adverse effects: None known.

Dosage: For herbal products, follow manufacturers' instructions. Liquid extract: 3–5 ml three times daily; tincture: 10–25 ml three times daily (EMA 2008). Also used in combination products for treatment of common cold. Do not use for longer than 7 days, or if symptoms persist consult a health care professional (EMA 2008).

General plant information: A deciduous shrub or small tree, native to Europe, Asia and North Africa. In Britain, the flowers mark the beginning of the summer, while the berries mark the end of it. The flowers are commonly used to make refreshing drinks. The unripe berries of elder contain toxic constituents which are lost during the drying and heating process and are not present in the medicinal product, or jams or syrups prepared from the fruit (Williamson et al. 2013).

References

Anon. (2005) *Sambucus nigra* (elderberry). Monograph. *Alternative Medicine Review* 10(1): 51–54.

Badescu L, Badulescu O, Badescu M, Ciocoiu M. (2012) Mechanism by *Sambucus nigra* Extract Improves Bone Mineral Density in Experimental Diabetes. *Evidence Based Complementary and Alternative Medicine* 2012: 848269.

Barak V, Birkenfeld S, Halperin T. Kalickman I. (2002) The effect of herbal remedies on the production of human inflammatory and anti-inflammatory cytokines. *Israel Medical Association Journal* 4(11 Suppl): 919–922.

Barak V, Halperin T, Kalickman I. (2001) The effect of Sambucol, a black elderberry-based, natural product, on the production of human cytokines: I. Inflammatory cytokines. *European Cytokine Network* 12(2): 290–296.

Christensen KB, Petersen RK, Kristiansen K, Christensen LP. (2010) Identification of bioactive compounds from flowers of black elder (*Sambucus nigra* L.) that activate the human peroxisome proliferator-activated receptor (PPAR) gamma. *Phytotherapy Research* 24(Suppl 2): S129–S132.

EMEA. (2008) Assessment report for the development of Community monographs and for inclusion of herbal substance(s), preparation(s) or combinations thereof in the list – *Sambucus nigra* L., flos. European Medicines Agency http://www.ema.europa.eu/docs/en _GB/document_library/Herbal_-_HMPC_assessment_report/2009/12/WC500018238.pdf.

EMA. (2008) Community Herbal Monograph on *Sambucus nigra* L., flos. European Medicines Agency http://www.ema.europa.eu/docs/en_GB/document_library/Herbal _-_Community_herbal_monograph/2009/12/WC500018227.pdf.

Guo R, Pittler MH, Ernst E. (2007) Complementary medicine for treating or preventing influenza or influenza-like illness. *American Journal of Medicine* 120(11): 923–929.

Kinoshita E, Hayashi K, Katayama H, Hayashi T, Obata A. (2012) Anti-influenza virus effects of elderberry juice and its fractions. *Bioscience, Biotechnology, and Biochemistry* 76(9): 1633–1638.

Kite GC, Larsson S, Veitch NC, Porter EA, Ding N, Simmonds MS. (2013) Acyl spermidines in inflorescence extracts of elder (*Sambucus nigra* L., Adoxaceae) and elderflower drinks. *Journal of Agricultural and Food Chemistry* 61(14): 3501–3508.

Krawitz C, Mraheil M A, Stein M, Imirzalioglu C, Domann E, Pleschka S, Hain T. (2011) Inhibitory activity of a standardized elderberry liquid extract against clinically-relevant human respiratory bacterial pathogens and influenza A and B viruses. *BMC Complementary and Alternative Medicine* 11: 16.

Ulbricht C, Basch E, Cheung L, Goldberg H, Hammerness P, Isaac R, Khalsa KP, Romm A, Mills E, Rychlik I, Varghese M, Weissner W, Windsor RC, Wortley J. (2014) An Evidence-Based Systematic Review of Elderberry and Elderflower (*Sambucus nigra*) by the Natural Standard Research Collaboration. *Journal of Dietary Supplements* 11(1): 80–120.

Vlachojannis JE, Cameron M, Chrubasik S. (2010) A systematic review on the Sambuci fructus effect and efficacy profiles. *Phytotherapy Research* 24(1): 1–8.

Williamson E, Driver S, Baxter K. (Eds.) (2013) *Stockley's Herbal Medicines Interactions*. 2nd Edition. Pharmaceutical Press, London UK.

Wu X, Cao G, Prior RL. (2002) Absorption and metabolism of anthocyanins in elderly women after consumption of elderberry or blueberry. *Journal of Nutrition* 132(7): 1865–1871.

Zakay-Rones Z, Thom E, Wollan T, Wadstein J. (2004) Randomized study of the efficacy and safety of oral elderberry extract in the treatment of influenza A and B virus infections. *Journal of International Medical Research* 32(2): 132–140.

Zakay-Rones Z, Varsano N, Zlotnik M, Manor O, Regev L, Schlesinger M, Mumcuoglu M. (1995) Inhibition of several strains of influenza virus *in vitro* and reduction of symptoms by an elderberry extract (*Sambucus nigra* L.) during an outbreak of influenza B Panama. *Journal of Alternative and Complementary Medicine* 1(4): 361–369.

Eucalyptus — *Eucalyptus globulus* Labill., *E. polybractea* F.Muell. ex R.T.Baker, *E. smithii* F.Muell. ex R.T.Baker, and other *Eucalyptus spp.*

Family: Myrtaceae

Other common names: Blue gum

Drug name: Eucalypti aetheroleum (Eucalyptus oil); Eucalypti folium

Botanical drug used: Essential oil distilled from leaves; more rarely, leaves

Indications/uses: The oil is used widely for the relief of symptoms of colds, cough, catarrh and sore throat, and as a decongestant, in oral and pastille preparations. It may be administered as an inhalant in hot water, in skin vapour rubs or via room vaporisers. Eucalyptus oil is an effective insect repellent and insecticide, and is a component of liniments applied for the relief of muscular-skeletal pain and inflammation. It is used for its antiseptic and flavouring properties in dentifrices and cosmetics.

Evidence: Due to the widespread and long-standing use, eucalyptus oil has a good empirical evidence base for use as a decongestant in the symptomatic relief of cold and coughs, and as an insect repellant.

Safety: Considered safe for external use and as an inhalation, but should only be taken internally in very small doses. Less than 5 ml or a few drops of the oil (in children) can result in lethal poisoning.

Main chemical compounds: The oil contains 1,8-cineole (eucalyptol) as the major component (>70%) with cymene, α-pinene, β-pinene, terpineol, pinocarveol, ledol, aromadendrene and others. The leaf also contains the flavonoids rutin, hyperoside and quercitrin, with tannins and other polyphenolics including tellimagrandin, catechins, and a series of euglobals, macracarpals and cypellogins (Boulekbache-Makhlouf et al. 2013; EMA 2012a; Williamson 2003).

Phytopharmacy: An evidence-based guide to herbal medicinal products, First Edition.
Sarah E. Edwards, Inês da Costa Rocha, Elizabeth M. Williamson and Michael Heinrich.
© 2015 John Wiley & Sons, Ltd. Published 2015 by John Wiley & Sons, Ltd.

Clinical evidence: Few clinical studies on eucalyptus oil and preparations have been carried out for the treatment of upper respiratory tract infections such as cold and cough, although there is a very long history of use for these purposes (Sadlon and Lamson 2010). Several clinical studies have been published on its major constituent (1,8-cineole) (EMA 2012b; ESCOP 2003). Cineole showed some beneficial effects in rhinosinusitis (Kehrl et al. 2004) and obstructive pulmonary diseases (Worth et al. 2009) and has anti-inflammatory activity (Juergens et al. 2003). The oil has been shown to have insect-repellent as well as insecticidal and larvicidal properties (Williamson 2003).

Pre-clinical evidence and mechanisms of action: Eucalyptus leaves have antimicrobial, antifungal and antiviral properties. The oil has antiseptic effects against a variety of bacteria and yeasts (Mulyaningsih et al. 2010). It has weak hyperaemic effects when applied topically, and has expectorant, decongestant, antitussive, immunomodulating and anti-inflammatory effects (EMA 2012a; Sadlon and Lamson 2010; Williamson 2003). Insecticidal effects, including against human lice and dust mites, have been described (Williamson et al. 2007).

Interactions: *E. globulus* extract (6 mg/kg and 3.25 mg/kg), when administered concurrently with diazepam (2 mg/kg), had an inhibitory effect on muscle relaxation and spontaneous motor activity produced by diazepam (Quílez et al. 2012).

Contraindications: Not recommended during pregnancy and lactation, as safety has not been established. The use in children and adolescents under 18 years of age has not been established due to lack of adequate data. There is a risk that cineole containing preparations, like other essential oils, can induce laryngospasm, bronchospasm, or asthma-like attacks, and/or respiratory arrest in children under 2 years. It should not be used internally by patients with gastrointestinal or biliary tract inflammation or severe hepatic disorders. If dyspnoea, fever or purulent sputum are present, a qualified health care practitioner should be consulted (EMA 2012a).

Adverse effects: Rare side effects include nausea, vomiting and diarrhoea. Care should be taken if used internally as overdoses have been reported. Only a few drops in children, and 4–5 ml in adults, can lead to life-threatening poisoning. Signs of poisoning include epigastric burning, nausea and vomiting, CNS depression, cyanosis, ataxia, miosis, pulmonary damage, and delirium. Convulsions or coma may occur and deaths have been reported. The use of eucalyptus oil in nasal oily preparations is now considered unsuitable, as the vehicle inhibits ciliary movements (Sweetman 2011).

Dosage: Oral use: 1.5–3 g leaf as herbal tea (in 150 ml of boiling water) up to four times daily; inhalation: 2–3 g of the leaf in 150 ml of boiling water, or a few drops of oil, up to three times daily. If the symptoms persist for more than 1 week during the use of the medicinal product, consult a health care professional (EMA 2012a).

General plant information: The genus is native to Australia but cultivated throughout the tropics and sub-tropics. Aboriginal Australians use eucalyptus as a decongestant by rubbing the leaves together (releasing the volatile oil), and inhaling. The essential oil is used in the form of Menthol and Eucalyptus Inhalation BP for steam inhalation as a decongestant. The oil is used as a solvent to remove grease or oily stains.

References

Boulekbache-Makhlouf L, Meudec E, Mazauric JP, Madani K, Cheynier V. (2013) Qualitative and semi-quantitative analysis of phenolics in *Eucalyptus globulus* leaves by high-performance liquid chromatography coupled with diode array detection and electrospray ionisation mass spectrometry. *Phytochemical Analysis* 24(2): 162–170.

EMA. (2012a) Community herbal monograph on *Eucalyptus globulus* Labill., folium. European Medicines Agency http://www.ema.europa.eu/docs/en_GB/document_library/Herbal_-_Community_herbal_monograph/2012/06/WC500129043.pdf.

EMA. (2012b) Assessment report on *Eucalyptus globulus* Labill., folium. European Medicines Agency http://www.ema.europa.eu/docs/en_GB/document_library/Herbal_-_HMPC_assessment_report/2012/06/WC500129041.pdf.

ESCOP. (2003) *ESCOP Monographs: The Scientific Foundation for Herbal Medicinal Products.* 2nd Edition. Thieme, Exeter and London, UK.

Juergens UR, Dethlefsen U, Steinkamp G, Gillissen A, Repges R and Vetter H. (2003) Anti-inflammatory activity of 1.8-cineol (eucalyptol) in bronchial asthma: a double-blind placebo-controlled trial. *Respiratory Medicine* 97(3): 250–256.

Kehrl W, Sonnemann U, Dethlefsen U. (2004) Therapy for acute nonpurulent rhinosinusitis with cineole: results of a double-blind, randomized, placebo-controlled trial. *Laryngoscope* 114(4): 738–742.

Mulyaningsih S, Sporer F, Zimmermann S, Reichling J, Wink M. (2010) Synergistic properties of the terpenoids aromadendrene and 1,8-cineole from the essential oil of *Eucalyptus globulus* against antibiotic-susceptible and antibiotic-resistant pathogens. *Phytomedicine* 17(13): 1061–1066.

Quílez AM, Saenz MT, García MD. (2012) *Uncaria tomentosa* (Willd. ex. Roem. & Schult.) DC. and *Eucalyptus globulus* Labill. interactions when administered with diazepam. *Phytotherapy Research* 26(3): 458–461.

Sadlon AE, Lamson DW. (2010) Immune-modifying and antimicrobial effects of Eucalyptus oil and simple inhalation devices. *Alternative Medicine Review* 15(1): 33–47.

Sweetman S. (Ed.) (2011) *Martindale: The Complete Drug Reference.* 37th Edition. Pharmaceutical Press, London.

Williamson EM. (2003) *Potter's Cyclopedia of Herbal Medicines.* C W Daniels, UK.

Williamson EM, Priestley CM, Burgess IF. (2007) An investigation and comparison of the bioactivity of selected essential oils on human lice and house dust mites. *Fitoterapia* 78(7–8): 521–525.

Worth H, Schacher C and Dethlefsen U. (2009) Concomitant therapy with Cineole (Eucalyptole) reduces exacerbations in COPD: a placebo-controlled double-blind trial. *Respiratroy Research* 10: 69.

Evening Primrose (Oil) *Oenothera biennis L.*

Synonyms: *O. grandiflora* L'Hér.; and others

Family: Onagraceae

Other common names: Evening star; King's cure-all; suncup; sundrop

Drug name: Oenotherae biennis oleum

Botanical drug used: Seed oil

Indications/uses: Evening primrose oil is taken to provide essential fatty acids, and for the symptomatic relief of itching in acute and chronic dry skin conditions. It is also used for mastalgia, premenstrual syndrome, menopausal symptoms, rheumatoid arthritis, diabetic neuropathy and attention deficit hyperactivity disorder (ADHD). Evening primrose oil is used topically to treat dry skin conditions.

Evidence: A Cochrane review of clinical trials found that oral ingestion of evening primrose oil lacked an effect on eczema, although a recent study found it helped alleviate symptoms of atopic dermatitis in children and adolescents. In the United Kingdom, evening primrose oil prescription medicines for the treatment of eczema and breast pain had their licences withdrawn in 2002 due to lack of evidence of effectiveness. Another systematic review found evening primrose no better than placebo in alleviating symptoms of premenstrual syndrome. Limited evidence suggests benefit for reducing pain in rheumatoid arthritis and diabetic neuropathy, and pharmacological studies have confirmed anti-inflammatory effects.

Safety: Short-term oral use of evening primrose oil is generally well tolerated. However, long-term use (>1 year) is not recommended due to increased risk of inflammation, thrombosis and immunosuppression. Use of evening primrose oil during pregnancy and lactation is not recommended due to lack of safety data. Safety of use in children under 12 years of age has also not been established.

Main chemical compounds: Essential fatty acids of the omega-6 series including linoleic acid (about 65–85%) and gamma-linolenic (gamolenic) acid (GLA) (about 7–14%) are the major constituents of evening primrose oil. Other fatty acids include

Phytopharmacy: An evidence-based guide to herbal medicinal products, First Edition.
Sarah E. Edwards, Inês da Costa Rocha, Elizabeth M. Williamson and Michael Heinrich.
© 2015 John Wiley & Sons, Ltd. Published 2015 by John Wiley & Sons, Ltd.

oleic acid, α-linolenic acid, palmitic acid and stearic acid. The sterols β-sitosterol and campesterol and triterpene alcohols are also present (Montserrat-de la Paz et al. 2012; Pharmaceutical Press Editorial Team 2013; WHO 2004; Williamson et al. 2013).

Clinical evidence:

Many clinical studies have been carried out on evening primrose oil, with conflicting results. It may be helpful in some patients and when supplementation with essential fatty acids is needed.

Atopic dermatitis and eczema: A Cochrane review assessed 19 studies that investigated oral evening primrose oil in the treatment of eczema, but found that it failed to significantly improve overall eczema symptoms compared to placebo (Bamford et al. 2013). An earlier review based on 10 studies of the use of evening primrose oil in atopic dermatitis also found no convincing evidence to support its use (Williams 2003).

A more recent (uncontrolled) study investigated the dose-dependent effects on clinical symptoms and serum fatty acids of evening primrose oil in children and adolescents (aged 2–15 years) with atopic dermatitis. Patients were randomly divided into two groups: those receiving 160 mg evening primrose oil daily ($n = 20$) and those receiving 320 mg daily ($n = 20$) for 8 weeks. Serum fatty acid levels were significantly higher in the 320 mg group that the 160 mg group, and while both groups showed a reduction in Eczema Area Severity Index (EASI) scores after 8 weeks, the improvement was greater in the 320 mg group (Chung et al. 2013).

Inflammation: A preliminary study found that topically applied evening primrose oil reduced inflammation at 5-azacitine injection sites of patients with myelodysplastic syndrome (Platzbecker et al. 2010).

Premenstrual syndrome: In a systematic review of randomised controlled trials assessing herbal treatments for alleviating symptoms of premenstrual syndrome, evening primrose oil was found to be no more effective than placebo (Dante and Facchinetti 2011).

Menopausal symptoms: A six-week randomised, placebo-controlled clinical trial evaluated oral evening primrose oil (1000 mg daily) in reducing frequency, severity and duration of menopausal hot flushes in women aged 45–59 years. The results showed improvement in hot flushes for both intervention and control groups, with evening primrose outperforming placebo, although statistical significance was found only in comparison of severity of hot flushes (Farzaneh et al. 2013).

Mastalgia: In a review evaluating treatments for severe persistent mastalgia in 291 women, patients were orally administered either evening primrose oil (3 g), bromocriptine (5 mg) or danazol (200 mg) daily for 3–6 months. In cases of cyclical mastalgia, good responses were obtained in 45% of patients treated with evening primrose oil, in 47% in those treated with bromocriptine and in 70% treated with danazol. In patients with non-cyclical mastalgia, the response rate was 27%, 20% and 31%, respectively. Adverse events were far greater in the bromocriptine and danazol groups than the evening primrose oil group (WHO 2004).

In a randomised, double-blind controlled study in 120 women investigating oral use of evening primrose oil or fish oil in treatment of severe chronic mastalgia for 6 months, all groups showed a decrease in pain, but neither demonstrated clear benefit over the control oils (corn and wheatgerm) (Blommers et al. 2002).

Another randomised, double-blind, placebo-controlled trial in 41 patients found that daily doses of evening primrose oil of 3000 mg, or in combination with 1200 IU vitamin E, taken for 6 months reduced the severity of cyclical mastalgia (Pruthi et al. 2010).

Rheumatoid arthritis: A systematic review of 11 clinical trials investigating the use of GLA (from borage seed oil, blackcurrent seed oil as well as evening primrose oil) found that it significantly reduced pain compared with placebo (Stonemetz 2008). In a meta-analysis of herbal medicinal products used to treat rheumatoid arthritis GLA doses equal or greater than 1400 mg/day showed benefit in the alleviation of rheumatic complaints. However, lower doses (500 mg) were ineffective (Cameron et al. 2009).

Diabetic neuropathy: Limited and inconclusive evidence exists for the use of evening primrose oil in treatment of diabetic neuropathy. Two of three studies evaluated in a review demonstrated beneficial results, and the authors conceded that evening primrose oil may be useful in cases of mild diabetic neuropathy, or as an adjunct therapy for patients with mild to moderate neuropathy (Stonemetz 2008).

Attention deficit hyperactivity disorder (ADHD): In a randomised, placebo-controlled, crossover Australian study, 104 children with ADHD (aged 7–12 years) were either administered polyunsaturated fatty acids (PUFAs), PUFAs with micronutrients, or placebo. The PUFAs were administered daily in the form of six capsules each containing 400 mg fish oil, 100 mg evening primrose oil, 29 mg docosahexaenoic acid, 10 mg GLA and 1.8 mg vitamin E. The results showed that treatment with PUFAs significantly improved parental-evaluated ratings of attention, impulsivity and hyperactivity (Stonemetz 2008).

Pre-clinical evidence and mechanisms of action:

Anti-allergic effects: *In vivo* evening primrose oil added to the diet (1 g/kg body weight for 5 days) was found to have anti-allergic properties, reducing the severity of bronchial reactions following allergen challenge in sensitised guinea-pigs (WHO 2004).

Anti-inflammatory effects: As a rich source of essential fatty acids, the mechanism of action of evening primrose oil is thought to involve modulation of prostaglandins. GLA in the diet is metabolised to dihomo-gamma-linolenic acid (DGLA), which can inhibit synthesis of arachidonic acid metabolites, exerting an anti-inflammatory effect (Pruthi et al. 2010). Pharmacological studies have also confirmed anti-inflammatory effects of evening primrose oil sterols, which inhibited the release of proinflammatory mediators by murine peritoneal macrophages stimulated with lipopolysaccharide. Nitric oxide (NO) reduction was a consequence of the inhibition of inducible nitric oxide synthetase (iNOS) expression. The sterols also reduced tumour necrosis factor (TNF)-α, interleukine (IL)-1β and thromboxane B_2 (TxB_2), but did not reduce prostaglandin E_2 (PGE$_2$) (Montserrat-de la Paz et al. 2012).

Other effects: Animal studies have demonstrated that evening primrose oil reduced total serum cholesterol and triglyceride levels and increased high-density lipoprotein levels. Other activities include antihypertensive, antiulcer and anti-arthritic activities (WHO 2004).

Interactions: Evening primrose oil may potentially interact with antiplatelet drugs such as warfarin due to a potential additive effect, as it can inhibit platelet aggregation and increase bleeding time. An isolated case report described an HIV-positive man who took evening primrose oil with a product containing aloes, rhubarb and liquorice, in addition to lopanivir boosted with ritonavir, tenofovir and lamuvidine and who developed raised levels of lopinavir with persistent diarrhoea. *In vitro cis*-linoleic acid was found to be a modest inhibitor of cytochrome P450 isoenzyme CYP2C9 and modest to minor inhibitor of CYP1A2, CYP2C19, CYP3A4 and CYP2D6, although clinical relevance of these, and therefore an effect on lopanivir, is unlikely. As a precursor of prostaglandin E_1 (PGE_1), GLA from evening primrose oil theoretically could interact with NSAIDs, although this appears to be of little clinical importance. Seizures have occurred in a few schizophrenic patients taking evening primrose oil with phenothiazine. However, no adverse effects were observed in others and there is no conclusive evidence that evening primrose oil was the cause (Williamson et al. 2013).

Contraindications: Hypersensitivity to the active substance. Oral evening primrose oil is not recommended during pregnancy as it has been associated with an increased need for obstetric intervention during labour and delivery. Oral evening primrose oil should be used with caution by patients with a bleeding disorder or who are using anticoagulants. Individuals undergoing surgery under general anaesthetic or those with seizure disorders should also avoid evening primrose oil (EMA 2011; Gardner and McGuffin 2013; Stonemetz 2008; Williamson et al. 2013).

Adverse effects: Evening primrose oil is generally well tolerated, with no significant adverse effects reported in a review involving approximately 500,000 users in the United Kingdom. Reported adverse events from clinical trials of evening primrose oil have included headaches, abdominal pain, nausea and loose stools (Gardner and McGuffin 2013).

Dosage: Adolescents (over 12 years), adults and elderly: single oral dose – 2 g; daily oral dose – 4–6 g (EMA 2011).

General plant information: *O. biennis* is native to Europe and has naturalised in North America. It has yellow fragrant flowers that open in the evening and wilt after one night (WHO 2004).

References

Bamford JTM, Ray S, Musekiwa A, van Gool C, Humphreys R, Ernst E. (2013) Oral evening primrose oil and borage oil for eczema. *Cochrane Database of Systematic Reviews* 2013(4): CD004416.

Blommers J, de Lange-de Klerk ESM, Kuik DJ, Bezemer PD, Meijer S. (2002) Evening primrose oil and fish oil for severe chronic mastalgia: a randomised, double-blind controlled, trial. *American Journal of Obstetrics and Gynecology* 187(5): 1389–1394.

Cameron M, Gagnier JJ, Little CV, Parsons TJ, Blümle A, Chrubasik S. (2009) Evidence of effectiveness of herbal medicinal products in the treatment of arthritis. *Phytotherapy Research* 23(12): 1647–1662.

Chung BY, Kim JH, Cho SI, Ahn IS, Kim HO, Park CW, Lee CH. (2013) Dose-dependent effects of evening primrose oil in children and adolescents with atopic dermatitis. *Annals of Dermatology* 25(3): 285–291.

Dante G, Facchinetti F. (2011) Herbal treatments for alleviating symptoms of premenstrual symptoms: a systematic review. *Journal of Psychosomatic Obstetrics and Gynecology* 32(1): 42–51.

EMA. (2011) Community herbal monograph on *Oenothera biennis* L.; *Oenothera lamarckiana* L., oleum. European Medicines Agency. http://www.ema.europa.eu/docs/en_GB/document _library/Herbal_-_Community_herbal_monograph/2012/WC500124923.pdf.

Farzaneh F, Fatehi S, Sohrabi MR, Alizadeh K. (2013) The effect of oral evening primrose oil on menopausal hot flashes: a randomized clinical trial. *Archives of Gynecology and Obstetrics* 288(5): 1075–1079.

Gardner Z, McGuffin M. (Eds.) (2013) *American Herbal Product Association's Botanical Safety Handbook*. 2nd Edition. CRC Press, USA, 1072 pp.

Montserrat-de la Paz S, Fernández-Arche A, Angel-Martín M, García-Giménez MD. (2012) The sterols isolated from Evening Primrose oil modulate the release of proinflammatory mediators. *Phytomedicine* 19(12): 1072–1076.

Pharmaceutical Press Editorial Team. (2013) *Herbal Medicines*. 4th Edition. Pharmaceutical Press, London, UK.

Platzbecker U, Aul C, Ehninger G, Giagounidis A. (2010) Reduction of 5-azacitidine induced skin reactions in MDS patients with evening primrose oil. *Annals of Hematology* 89(4): 427–428.

Pruthi S, Wahner-Roedler DL, Torkelson CJ, Cha SS, Thicke LA, Hazleton JH, Bauer BA. (2010) Vitamin E and evening primrose oil for management of cyclical mastalgia: a randomised pilot study. *Alternative Medicine Review* 15(1): 59–67.

Stonemetz D. (2008) A review of the clinical efficacy of evening primrose oil. *Holistic Nursing Practice* 22(3): 171–174.

WHO. (2004) *WHO Monographs on Selected Medicinal Plants*. Vol. 2. WHO, Geneva, 358 pp.

Williams HC. (2003) Evening primrose oil for atopic dermatitis. *BMJ* 27(7428): 1358–1359.

Williamson EM, Driver S, Baxter K. (Eds.) (2013) *Stockley's Herbal Medicines Interactions*. 2nd Edition. Pharmaceutical Press, London, UK.

Fennel	*Foeniculum vulgare* Mill.

Synonyms: *F. dulce* Mill.; *F. officinale* All.; *F. officinale* var. *dulce* (Mill.) Alef.; *F. vulgare* var. *dulce* (Mill.) Batt. & Trab.; *F. vulgare* var. *vulgare*; and others

Family: Apiaceae (Umbelliferae)

Other common names: Bitter fennel; sweet fennel

Drug name: Foeniculi amari fructus; Foeniculi dulcis fructus

Botanical drug used: Dried ripe fruits (sometimes erroneously referred to as 'seeds')
Note – different chemotypes of fennel are recognised. In medical preparations fennel is usually described as either "bitter" or "sweet" fennel type.

Indications/uses: Bitter fennel is the most important type for medicinal purposes. It is used for dyspepsia and mild spasmodic ailments of the GI tract, including for infants and babies where it is sometimes given in the form of a tea (infusion) to treat colic. Fennel is also used for respiratory problems such as catarrh, and is reputed to promote lactation. It is also used for menstrual cramps.

Evidence: The evidence is largely empirical and relates to the essential oil content. The preclinical data support the use as an antispasmodic and bronchodilator.

Safety: In recent years, controversy has developed over the safety of fennel, despite its long history of both food and medicinal use. It has always been considered as very safe and the German Commission E lists no risks. These safety concerns are based on the content of estragole (methylchavicol), which is known for its potential carcinogenicity, although no clinical reports of toxicity for fennel have been documented and the relevance of some studies using pure estragole at high doses has been disputed.

Main chemical compounds:

Bitter fennel: Anethole and fenchone are the major components of the essential oil, which should contain no more than 5% estragole (methylchavicol). α-Pinene, limonene, camphene, *p*-cymene, β-pinene, β-myrcene, α-phellandrene, sabinene, γ-terpinene and terpinolene are other constituents (ESCOP 2009; Mimica-Dukić et al. 2003; Williamson 2003).

Sweet fennel: Anethole is the major component of the oil, with not more than 10% estragole and not more than 7.5% fenchone. Other components include α-pinene, limonene, β-pinene, β-myrcene and *p*-cymene.

Other constituents of the fruits of both bitter and sweet fennel include water-soluble glycosides of the same monoterpenoids found in the oil, and other phenolic compounds (ESCOP 2009).

NB: Estragole levels are higher in fennel oils after long distillation times. In one study, after 6 hours the content was 3.78%, whereas after 12 hours, it was 5.29% for the same sample (Mimica-Dukić et al. 2003).

Clinical evidence: The evidence for the use of fennel as an antispasmodic for colic and in dysmenorrhea is largely based on empirical use, but clinical data for fennel preparations show encouraging results. A meta-analysis identified three studies relevant for effects in colic, although the composition of the preparations and the interventions varied. A well-conducted double-blind, placebo controlled, two-arm, two centre-study in a group of 125 infants with colic (2–12 weeks of age), reported a significant improvement in symptoms in infants given fennel extract compared with placebo (Perry et al. 2011).

Several small studies have shown that fennel can be effective in reducing the severity of uterine spasms in dysmenorrhea (e.g. Bokaie et al. 2013).

Pre-clinical evidence and mechanisms of action: A wide range of studies have been conducted on fennel and its essential oil. Early pharmacological *in vivo* research demonstrated spasmolytic, bronchodilatory, secretolytic and expectorant effects in rabbits. Antimicrobial effects have been widely reported, as well as antioxidant, antiulcerogenic and hypotensive effects (Pharmaceutical Press Editorial Team 2013; Williamson 2003).

Interactions: No clinical data are available and drug interactions are not thought to be a significant problem with fennel. The methanolic extract showed inactivation of erythromycin N-demethylation mediated by human liver microsomal enzymes in a study which identified 5-methoxypsoralen (5-MOP; bergaptene) as the strongest inhibitor of CYP3A4 (Subehan Zaidi et al. 2007). However, this is a very minor component of the oil and is not always found in samples (Mimica-Dukić et al. 2003).

Adverse effects: Based on the available clinical studies and the very few reported side effects for fennel, the herb is generally considered to be very safe and has been widely used in infants and by breastfeeding mothers for centuries. As a result of animal data, concerns were raised recently about the potential carcinogenic effects of estragole, a component of the essential oil, which also has weak genotoxic effects (Bristol 2011). However, these studies used an extremely high dose (about 100–1000 higher than the anticipated human dose), and the validity of these data has been questioned (see discussion in Gori et al. 2012). Fennel oil is also known to have oestrogenic effects in high doses (Williamson 2003); and fennel preparations are not recommended for small children for long periods of time. However, if used in the form of a herbal tea in normal dose ranges there seem to be no risk to most patient groups, and while there also is no evidence for toxicological risks in case of infants and breastfeeding mothers, further research is needed.

Dosage: Fennel is generally used as a herbal tea (infusion), with a dose range of 1.5–2.5 g of dried herb per serving. 'Instant' teas, syrups or honeys containing fennel and other herbal ingredients are also available.

General plant information: Fennel is an important mythological plant in many parts of the world. The Greek cultural hero Prometheus is said to have used it to steal fire from the Gods. In Asian cultures, fennel was ingested to speed the elimination of poisons and as one of the ancient Saxon nine sacred herbs, it was credited with the power to cure. In the Middle Ages, it was draped over doorways on Midsummer's Eve to protect the household from evil spirits.

Fennel has yielded a wide range of useful cultivars and products. Today Florence fennel is an important aromatic vegetable. The (preferably green) seeds are widely used in cookery including as an after-meal digestive.

References

Alexandrovich I, Rakovitskaya O, Kolmo E, Sidorova T, Shushunov S. (2003) The effect of Fennel (*Foeniculum vulgare*) seed oil emulsion in infantile colic: a randomized, placebo-controlled trial. *Alternative Therapies in Health and Medicine* 9(4): 58–61.

Bokaie M, Farajkhoda T, Enjezab B, Khoshbin A, Zarchi Mojgan K. (2013) Oral fennel (*Foeniculum vulgare*) drop effect on primary dysmenorrhea: Effectiveness of herbal drug. *Iranian Journal of Nursing and Midwifery Research* 18(2): 128–32.

Bristol DW. (2011) NTP 3-month toxicity studies of estragole (CAS No. 140-67-0) administered by gavage to F344/N rats and B6C3F1 mice. *Toxicity Report Series.* 82: 1–111.

ESCOP. (2009) *ESCOP Monographs: The Scientific Foundation for Herbal Medical Products.* 2nd Edition Supplement 2009. Thieme, Stuttgart, New York.

Gori L, Gallo E, Mascherini V, Mugelli A, Vannacci A, Firenzuoli F. (2012) Can estragole in fennel seed decoctions really be considered a danger for human health? A fennel safety update. *Evidence-Based Complementary and Alternative Medicine* 2012: 860542.

Martins C, Cação R, Cole KJ, Phillips DH, Laires A, Rueff J, Rodrigues AS. (2012) Estragole: a weak direct-acting food-borne genotoxin and potential carcinogen. *Mutation Research* 747(1): 86–92.

Mimica-Dukić N, Kujundžić S, Soković M, Couladis M. (2003) Essential oil composition and antifungal activity of *Foeniculum vulgare* Mill. obtained by different distillation conditions. *Phytotherapy Research.* 17(4): 368–371.

Perry R, Hunt K, Ernst E. (2011) Nutritional supplements and other complementary medicines for infantile colic: a systematic review. *Pediatrics* 127(4): 720–733. doi: 10.1542 /peds.2010-2098.

Pharmaceutical Press Editorial Team. (2013) *Herbal medicines.* 4th Edition. Pharmaceutical Press, London, UK.

Subehan Zaidi SF, Kadota S, Tezuka Y. (2007) Inhibition of human liver cytochrome P450 3A4 by constituents of fennel (*Foeniculum vulgare*): identification and characterization of a mechanism-based inactivator. *Journal of Agricultural and Food Chemistry* 55(25): 10162–10167.

Williamson EM. (2003) *Potter's Cyclopedia of Herbal Medicines.* C W Daniels, UK.

Feverfew   *Tanacetum parthenium* (L.) Sch.Bip.

Synonyms: *Chrysanthemum parthenium* (L.) Bernh.; *Leucanthemum parthenium* (L.) Godr. & Gren.; *Pyrethrum parthenium* (L.) Sm.; and others

Family: Asteraceae (Compositae)

Other common names: Altamisa; featherfew; featherfoil

Drug name: Tanaceti parthenii herba

Botanical drug used: Herb

> **Indications/uses:** Registered products are available for the prevention of migraine headaches. Feverfew has traditionally been used for fever, as the name suggests, as well as rheumatic conditions, coughs and colds.
>
> **Evidence:** There is some evidence of efficacy for the prophylaxis of migraines but – as often – the quality and the size of the studies makes an assessment problematic.
>
> **Safety:** Overall considered to be safe, but side effects are common, especially with higher doses. Should be avoided in pregnancy and breast-feeding.

Main chemical compounds: The main active constituents are sesquiterpene lactones, mainly parthenolide and its esters, with santamarin, reynosin, artemorin, partholide and chrysanthemonin. Other compounds present are flavonoids such as kaempferol, tanetin and quercetagetin, and volatile oil consisting of α-pinene, bornyl acetate, bornyl angelate, costic acid, camphor and spirotekal ethers (EMA 2010a; Pareek et al. 2011; Williamson et al. 2013).

Clinical evidence:

Migraine prophylaxis: A Cochrane review on the efficacy and safety of feverfew mono-preparations for preventing migraine included five randomised, placebo-controlled, double-blind trials. The authors concluded that there was not enough evidence to support its efficacy in migraine prophylaxis beyond placebo (Pittler and Ernst 2004). Many other clinical studies on feverfew alone or in combination for the treatment of migraines have been conducted, but the EMA has concluded that there is still not enough evidence to support a full product licence based on well-established use and more robust clinical trials assessing larger patient samples are required (EMA 2010b).

Phytopharmacy: An evidence-based guide to herbal medicinal products, First Edition.
Sarah E. Edwards, Inês da Costa Rocha, Elizabeth M. Williamson and Michael Heinrich.
© 2015 John Wiley & Sons, Ltd. Published 2015 by John Wiley & Sons, Ltd.

Rheumatoid arthritis: A double-blind, placebo-controlled study of a dried powdered feverfew leaf extract (70–86 mg, equivalent to 2–3 µmol parthenolide, daily for 6 weeks) in women (41 participants) with symptomatic rheumatoid arthritis showed no apparent benefit from feverfew compared to placebo (Pattrick et al. 1989).

Pre-clinical evidence and mechanisms of action:

Anti-migraine effects: The mechanisms of action are not fully clear but the anti-inflammatory and antiplatelet activities are likely to be associated with the inhibition of NF-κB and IκB kinases (which play an important role in pro-inflammatory cytokine-mediated signalling) by parthenolide and derivatives. Parthenolide and the flavonoid tanetin inhibit prostaglandins biosynthesis, while a whole feverfew extract has been shown to decrease vascular smooth muscle spasm (Bork et al. 1997; Pareek et al. 2011).

Anticancer effects: The potential anticancer effects of parthenolide and its mechanisms of anti-tumour and anti-inflammatory effects have been reviewed (Ghantous et al. 2013; Mathema et al. 2012).

Antimicrobial effects: Feverfew has antimicrobial properties, due mainly to the essential oil constituents (Mohsenzadeh et al. 2011).

Interactions: May interact with antiplatelet (e.g. aspirin, clopidogrel) and anticoagulant drugs (e.g. warfarin). Patients should avoid feverfew with these drugs, and should also discuss any bleeding or bruising episodes that occur with a health care professional (Williamson et al. 2013).

Contraindications: Not recommended during pregnancy and lactation due to lack of safety data, and because parthenolide is cytotoxic. Not recommended in children and adolescents under 18 years of age or in patients with hypersensitivity to plants from the Asteraceae (EMA 2010a).

Adverse effects: Feverfew can cause allergic reactions and cytotoxicity symptoms such as mouth ulcers, due to the presence of sesquiterpene lactones (especially parthenolide) (Williamson et al. 2013). Gastrointestinal disturbances have also been reported (EMA 2010a).

Dosage: As recommended by the manufacturer. An average dose is 100 mg of powdered dried feverfew herb a day (EMA 2010a), but folklore use also recommends 2–3 fresh leaves daily taken with food.

General plant information: Feverfew is native to the Balkan Peninsula but now is a widespread weed. It was recorded in the first century as an anti-inflammatory by the Greek physician Dioscorides.

References

Bork PM, Schmitz ML, Kuhnt M, Escher C, Heinrich M. (1997) Sesquiterpene lactone containing Mexican Indian medicinal plants and pure sesquiterpene lactones as potent inhibitors of transcription factor κB (NF-κB). *FEBS-Letters* 402(1): 85–90.
EMA. (2010a) Community herbal monograph on *Tanacetum parthenium* (L.) Schulz Bip., herba. European Medicines Agency. http://www.ema.europa.eu/docs/en_GB/document _library/Herbal_-_Community_herbal_monograph/2011/01/WC500100983.pdf.
EMA. (2010b) Assessment report on *Tanacetum parthenium* (L.) Schulz Bip., herba. European Medicines Agency. http://www.ema.europa.eu/docs/en_GB/document_library /Herbal_-_HMPC_assessment_report/2011/06/WC500107719.pdf.

Ghantous A, Sinjab A, Herceg Z, Darwiche N. (2013) Parthenolide: from plant shoots to cancer roots. *Drug Discovery Today* 18(17–18): 894–905.

Mathema VB, Koh YS, Thakuri BC, Sillanpää M. (2012) Parthenolide, a sesquiterpene lactone, expresses multiple anti-cancer and anti-inflammatory activities. *Inflammation* 35(2): 560–565.

Mohsenzadeh F, Chehregani A, Amiri H. (2011) Chemical composition, antibacterial activity and cytotoxicity of essential oils of *Tanacetum parthenium* in different developmental stages. *Pharmaceutical Biology* 49(9): 920–926.

Pareek A, Suthar M, Rathore GS, Bansal V. (2011) Feverfew (*Tanacetum parthenium* L.): A systematic review. *Pharmacognosy Reviews* 5(9): 103–110.

Pattrick M, Heptinstall S, Doherty M. (1989) Feverfew in rheumatoid arthritis: a double blind, placebo controlled study. *Annals of the Rheumatic Diseases* 48(7): 547–549.

Pittler MH, Ernst E. (2004) Feverfew for preventing migraine. *Cochrane Database of Systematic Reviews* (1): CD002286.

Williamson EM, Driver S, Baxter K. (Eds.) (2013) *Stockley's Herbal Medicines Interactions*. 2nd Edition. Pharmaceutical Press, London, UK.

Ganoderma *Ganoderma lucidum* (Curtis) P. Karst.; or G. lucidum (Leyss. ex Fr.) Karst.

Family: Ganodermataceae

Other common names: Ling Zhi; Reishi

Botanical drug used: Fruiting body, spores

Indications/uses: Adaptogen and general tonic; for immunotherapy and as an adjunct to cancer treatments including chemotherapy and radiotherapy. It is also used traditionally in China and Japan as a sedative, liver protectant and cholesterol-lowering agent.

Evidence: There are few clinical studies available to support the uses and they are of variable quality.

Safety: Considered safe but as an immune modulator, it would probably be wise to avoid use in immune disorders or concurrent immunosuppressant therapy.

Main chemical compounds: The mature fruiting body contains triterpenoids, mainly lanostanes (such as ganoderic acids, ganoderals ganoderiols, ganoderenic acids, lucidenic acids, ganolucidic acids, applanoxidic acids; lucidone and others (Boh et al. 2007). *G. sinense* J.D. Zhao, L.W. Hsu & X.Q. Zhang is also used medicinally, but there are chemical differences in the triterpene profile (Da et al. 2012). Both species contain polysaccharides, mainly glucans and peptidoglycans (known as ganoderans).

Clinical evidence: A recent Cochrane review (Jin et al. 2012) did not find sufficient evidence to justify the use of *G. lucidum* as a first-line treatment for cancer and concluded that it also remains uncertain whether it helps prolong long-term cancer survival. However, it was suggested that it could be administered as an adjunct to conventional treatment due to its properties of enhancing tumour response and stimulating host immunity.

Pre-clinical evidence and mechanisms of action: Studies on *G. lucidum* polysaccharide (GLPS) fractions have suggested that the anti-tumour activities are mediated by immunomodulatory effects, and recent data also suggest that GPLS may suppress tumourigenesis or inhibit tumour growth through direct cytotoxic effect

---

Phytopharmacy: An evidence-based guide to herbal medicinal products, First Edition.
Sarah E. Edwards, Inês da Costa Rocha, Elizabeth M. Williamson and Michael Heinrich.
© 2015 John Wiley & Sons, Ltd. Published 2015 by John Wiley & Sons, Ltd.

and anti-angiogenic actions. *In vitro* and *in vivo* studies have shown that GLPS affects cells including B lymphocytes, T lymphocytes, dendritic cells, macrophages, and natural killer cells (Xu et al. 2011).

Evidence that *G. lucidum* extracts may have a potential inhibitory effect on cancer metastasis through activation of kinases including ERK1/2, PI 3-kinase and protein kinase B has been reported. Activation of these kinases inhibits activator protein-1 (AP-1) and nuclear factor-kappa B (NF-κB), resulting in the down-regulation of many other urokinase plasminogen activators, matrix metalloproteinase-9, VEGF, GF-β1, interleukin (IL)-8, inducible NO and β1-integrin in various cell lines or animal models. *G. lucidum* may therefore be useful in the prevention of cancer metastasis (Weng and Yen 2010).

The triterpenoids have been reported to possess hepatoprotective, anti-hypertensive, hypocholesterolemic and anti-histaminic effects, anti-tumour and anti-angiogenic activity, effects on platelet aggregation and complement inhibition. Polysaccharides, especially β-D-glucans, have been known to possess anti-tumour effects through immunomodulation and anti-angiogenesis and have a protective effect against free radicals, reducing cell damage caused by mutagens (Boh et al. 2007).

The different parts of the fruiting body (whole fruiting body, pileus and stipe) as well as ganoderma spores have been compared for their anti-tumour and immunomodulatory activities in S-180 sarcoma-bearing mice. The GL whole fruiting body, stipe and sporoderm-broken spore possessed stronger inhibitory activities on sarcoma growth when compared with the pileus extract (Yue et al. 2008).

Interactions: Ganoderma polysaccharides have been found *in vitro* to enhance the sensitivity of several tumour cells lines to a number of anticancer drugs such as cisplatin, arsenic trioxide (Huang et al. 2010), etoposide and doxorubicin (Sadava et al. 2009).

Contraindications: None known, but as an immune modulator, it would probably be wise to avoid use in immune disorders or with concurrent immunosuppressant therapy.

Adverse effects: The Cochrane review (Jin et al. 2012) stated that *G. lucidum* was generally well tolerated by most participants, with only a few minor adverse events and no major toxicity observed across the studies.

Dosage: According to manufacturers' instructions.

General plant information: Ganoderma is a basidiomycete fungus. It is rarely found in the wild but is widely cultivated, and in S.E. Asia it is used to prepare medicinal soups.

References

The primary literature for ganoderma is extensive; this is a selection of recent and relevant reviews.

Boh B, Berovic M, Zhang J, Zhi-Bin L. (2007) *Ganoderma lucidum* and its pharmaceutically active compounds. *Biotechnology Annual Review* 13: 265–301.

Da J, Wu WY, Hou JJ, Long HL, Yao S, Yang Z, Cai LY, Yang M, Jiang BH, Liu X, Cheng CR, Li YF, Guo DA. (2012) Comparison of two officinal Chinese pharmacopoeia species of *Ganoderma* based on chemical research with multiple technologies and chemometrics analysis. *Journal of Chromatography A* 1222: 59–70.

Huang CY, Chen JY, Wu JE, Pu YS, Liu GY, Pan MH, Huang YT, Huang AM, Hwang CC, Chung SJ, Hour TC. (2010) Ling-Zhi polysaccharides potentiate cytotoxic effects of anticancer drugs against drug-resistant urothelial carcinoma cells. *Journal of Agricultural and Food Chemistry* 58(15): 8798–8805.

Jin X, Ruiz Beguerie J, Sze DM, Chan GC. (2012) *Ganoderma lucidum* (Reishi mushroom) for cancer treatment. *Cochrane Database of Systematic Reviews* 6: CD007731.

Paterson RR. (2006) Ganoderma – a therapeutic fungal biofactory. *Phytochemistry* 67(18): 1985–2001.

Sadava D, Still DW, Mudry RR, Kane SE. (2009) Effect of Ganoderma on drug-sensitive and multidrug-resistant small-cell lung carcinoma cells. *Cancer Letters* 277(2): 182–189.

Sanodiya BS, Thakur GS, Baghel RK, Prasad GB, Bisen PS. (2009) *Ganoderma lucidum*: a potent pharmacological macrofungus. *Current Pharmaceutical Biotechnology* 10(8): 717–742.

Weng CJ, Yen GC. (2010) The *in vitro* and *in vivo* experimental evidences disclose the chemopreventive effects of *Ganoderma lucidum* on cancer invasion and metastasis. *Clinical and Experimental Metastasis* 27(5): 361–369.

Xu Z, Chen X, Zhong Z, Chen L, Wang Y. (2011) *Ganoderma lucidum* polysaccharides: immunomodulation and potential anti-tumor activities. *American Journal of Chinese Medicine* 39(1): 15–27.

Yue GG, Fung KP, Leung PC, Lau CB. (2008) Comparative studies on the immunomodulatory and antitumor activities of the different parts of fruiting body of *Ganoderma lucidum* and *Ganoderma* spores. *Phytotherapy Research* 22(10): 1282–1291.

Garlic   *Allium sativum* L.

Family: Amaryllidaceae (Alliaceae)

Drug name: Allii sativi bulbus

Botanical drug used: Bulb

Indications/uses: It is claimed to possess anti-hypertensive, anti-thrombotic, fibrinolytic, antimicrobial, anticancer, antidiabetic and lipid-lowering properties. It is also used to relieve the symptoms of catarrh, rhinitis and nasal congestion.

Evidence: There is some limited evidence of beneficial effects to support its main uses, but further clinical studies are needed.

Safety: Overall considered to be safe although the usual precautions are required (see below). Avoid garlic supplements (approximately 1 week) before and after major surgery.

Main chemical compounds: The main constituents are the sulphur-containing amino acid alliin, which in the presence of the enzyme alliinase will be converted to allicin, ajoene, and others. Other constituents are terpenes (citral, geraniol, linalool), proteins, amino acids, minerals, vitamins, trace elements, lipids and prostaglandins (ESCOP 2003; Pharmaceutical Press Editorial Team 2013). A series of saponins, such as the sativosides and erubosides, have been isolated, and garlic also contains organo-selenium compounds (Amagase 2006).

Clinical evidence: A Cochrane review on the use of garlic as monotherapy in patients diagnosed with hypertension found insufficient evidence to support its use in lowering the risk of cardiovascular morbidity and mortality, but it may have some blood pressure lowering effect (Stabler et al. 2012). The evidence for its use for reducing some of the risk factors associated with atherosclerosis, such as peripheral arterial occlusive disease, is also not statistically significant (Jepson et al. 2013). A review of the use of garlic for the prevention of pre-eclampsia and its complications for pregnant women and their babies also did not provide sufficient evidence to recommend an increase of garlic intake (Meher and Duley 2010). A meta-analysis of the effect of garlic on hypercholesterolemia showed that prolonged use (longer than 2 months) was effective in reducing total

---

Phytopharmacy: An evidence-based guide to herbal medicinal products, First Edition.
Sarah E. Edwards, Inês da Costa Rocha, Elizabeth M. Williamson and Michael Heinrich.
© 2015 John Wiley & Sons, Ltd. Published 2015 by John Wiley & Sons, Ltd.

serum cholesterol (TC) and low-density lipoprotein (LDL-c) in patients with high cholesterol, reducing the risk of coronary problems and with minor side effects. Garlic might be a safer alternative therapeutic for patients with slightly elevated cholesterol, but further studies are required (Ried et al. 2013).

Randomised controlled trials on the effectiveness of garlic either for the prevention or treatment of the common cold suggested that there is still limited clinical evidence to support this therapeutic effect. However, there is one study suggesting it may prevent occurrences of the common cold (Lissiman et al. 2012). Nevertheless, more studies are required to validate this finding. The use of garlic as an antibiotic adjuvant therapy also showed no significant evidence to support its use for lung infection (Hurley et al. 2013).

Pre-clinical evidence and mechanisms of action: The exact mechanism of action is unknown and many of the different types of constituents have biological activity (Amagase 2006). Pre-clinical studies have shown that garlic extracts inhibited platelet aggregation (Hiyasat et al. 2009), activated NO synthase (Pedraza-Chaverri et al. 1998) and inhibited angiotensin I converting enzyme (Rietz et al. 1993). They also showed an effect on reducing cholesterol biosynthesis by affecting the activity of HMG-CoA reductase (Qureshi et al. 1983), but other targets might also be affected, such as the activity of acylCoA:cholesterol acyltransferase (ACAT), among others (Zeng et al. 2013). There are several studies supporting its antibacterial activity (Casella et al. 2013; Dini et al. 2011). Most of the cardiovascular effects are due to the content of allicin and its subsequent decomposition products. The antimicrobial effects are due mainly to the sulphides and the presence of the allyl group is fundamental for this activity (Casella et al. 2013).

Interactions: Garlic supplements may interact with antiplatelet drugs, warfarin and related drugs, increasing the risk of bleeding. They may also have a beneficial effect on blood lipids when used concomitantly with fish oil. Although limited information is available, there was a report on the reduction of plasma levels of an HIV-protease inhibitor (saquinavir) when used concomitantly with a garlic supplement. Garlic supplements should probably be avoided when taking HIV-protease inhibitors. It may interact with ACE inhibitors (lisinopril) but this is based on one report only (Williamson et al. 2013). Garlic phytochemicals are able to modulate CYP3A4 and P-glycoprotein activity, but no detrimental effects have yet been observed (Berginc and Kristl 2013) and there is no reason to avoid garlic in the diet.

Contraindications: None known.

Adverse effects: May cause gastrointestinal irritation or allergic reaction but it is rare (ESCOP 2003). Garlic supplementation should be stopped at least 1 week before and after major surgery as a precaution, to decrease the chance of bleeding.

Dosage: As recommended by the manufacturer. A daily dose of 4 g of fresh garlic or the equivalent of a commercial preparation (Pharmaceutical Press Editorial Team 2013). Products are usually standardised according to the content of the sulphur-containing compounds, alliin, allicin and/or γ-glutamyl-(S)-allyl-L-cysteine (Williamson et al. 2013).

General plant information: Garlic is a widely used herb with many culinary and medicinal uses dating back to antiquity. A case–control study on southern European populations investigated the relationship between the frequency of onion/garlic use and different types of cancer. The study concluded that the consumption of *Allium* vegetables might have a protective effect for certain types of cancers (Galeone et al. 2006).

References

Amagase H. (2006) Clarifying the real bioactive constituents of garlic. *Journal of Nutrition* 136(Suppl 3): 716S–725S.

Berginc K, Kristl A. (2013) The mechanisms responsible for garlic – drug interactions and their *in vivo* relevance. *Curr Drug Metabolism* 14(1): 90–101.

Casella S, Leonardi M, Melai B, Fratini F, Pistelli L. (2013) The role of diallyl sulfides and dipropyl sulfides in the in vitro antimicrobial activity of the essential oil of garlic, *Allium sativum* L., and leek, *Allium porrum* L. *Phytotherapy Research* 27(3): 380–383.

Chan JY, Yuen AC, Chan RY, Chan SW. (2013) A review of the cardiovascular benefits and antioxidant properties of allicin. *Phytotherapy Research* 27(5): 637–646.

Dini C, Fabbri A, Geraci A. (2011) The potential role of garlic (*Allium sativum*) against the multi-drug resistant tuberculosis pandemic: a review. *Annali dell'Istituto Superiore di Sanita* 47(4): 465–473.

ESCOP. (2003) *ESCOP Monographs: The Scientific Foundation for Herbal Medicinal Products.* 2nd Edition. Thieme, Exeter and London, UK.

Galeone C, Pelucchi C, Levi F, Negri E, Franceschi S, Talamini R, Giacosa A, La Vecchia C. (2006) Onion and garlic use and human cancer. *American Journal of Clinical Nutrition* 84(5): 1027–1032.

Hiyasat B, Sabha D, Grotzinger K, Kempfert J, Rauwald JW, Mohr FW, Dhein S. (2009) Antiplatelet activity of *Allium ursinum* and *Allium sativum*. *Pharmacology* 83(4): 197–204.

Hurley MN, Forrester DL, Smyth AR. (2013) Antibiotic adjuvant therapy for pulmonary infection in cystic fibrosis. *Cochrane Database of Systematic Reviews* 6: CD008037.

Jepson RG, Kleijnen J, Leng GC. (2013) Garlic for peripheral arterial occlusive disease. *Cochrane Database of Systematic Reviews* 4: CD000095.

Lissiman E, Bhasale AL, Cohen M. (2012) Garlic for the common cold. *Cochrane Database of Systematic Reviews* 3: CD006206.

Meher S, Duley L. (2010) Garlic for preventing pre-eclampsia and its complications. *Cochrane Database of Systematic Reviews* (3): CD006065.

Pedraza-Chaverri J, Tapia E, Medina-Campos ON, de los Angeles Granados M, Franco M. (1998) Garlic prevents hypertension induced by chronic inhibition of nitric oxide synthesis. *Life Sciences* 62(6): PL 71–77.

Pharmaceutical Press Editorial Team. (2013) *Herbal Medicines.* 4th Edition. Pharmaceutical Press, London, UK.

Qureshi AA, Din ZZ, Abuirmeileh N, Burger WC, Ahmad Y, Elson CE. (1983) Suppression of avian hepatic lipid metabolism by solvent extracts of garlic: impact on serum lipids. *Journal of Nutrition* 113(9): 1746–1755.

Ried K, Toben C, Fakler P. (2013) Effect of garlic on serum lipids: an updated meta-analysis. *Nutrition Reviews* 71(5): 282–299.

Rietz B, Isensee H, Strobach H, Makdessi S, Jacob R. (1993) Cardioprotective actions of wild garlic (*Allium ursinum*) in ischemia and reperfusion. *Molecular and Cellular Biochemistry* 119(1–2): 143–150.

Stabler SN, Tejani AM, Huynh F, Fowkes C. (2012) Garlic for the prevention of cardiovascular morbidity and mortality in hypertensive patients. *Cochrane Database of Systematic Reviews* 8: CD007653.

Williamson EM, Driver S, Baxter K. (Eds.) (2013) *Stockley's Herbal Medicines Interactions.* 2nd Edition. Pharmaceutical Press, London, UK.

Zeng T, Zhang CL, Zhao XL, Xie KQ. (2013) The roles of garlic on the lipid parameters: a systematic review of the literature. *Critical Reviews in Food Science and Nutrition* 53(3): 215–230.

Gentian   *Gentiana lutea* L.

Synonyms: *G. major* Bubani; *Swertia lutea* Vest ex Rchb.; and others

Family: Gentianaceae

Other common names: Bitterwort; bitter root; yellow gentian

Drug name: Gentianae luteae radix

Botanical drug used: Rhizomes and roots

Indications/uses: Traditionally, gentian is used to treat loss of appetite following convalescence, and gastrointestinal complaints such as dyspepsia, bloating and flatulence. Historically, it was used as an emmenagogue and to eradicate intestinal worms. It is used by the food industry in the manufacture of bitter beverages and aperitifs. A multi-component preparation containing the ground drug of *G. lutea* is registered under the THR scheme to relieve nasal congestion and sinusitis.

Evidence: Clinical data to support the traditional uses of gentian are lacking, although the highly bitter iridoids stimulate digestive secretions, making its use as an appetite stimulant plausible.

Safety: Limited safety data are available. Genotoxicity studies suggest that gentian may potentiate genotoxic agents, and mutagenic activity has also been reported, although gentian has been used medicinally and in beverages for many years. It should be avoided during pregnancy, lactation and in children. Long-term use or excessive doses in adults is not recommended.

Main chemical compounds: Bitter terpenes, including the secoiridoids gentiopricroside (gentiomarin/gentiopicrin), amarogentin and swertiamarine; xanthones including gentisin; alkaloids including gentianine (0.6–0.8%). It also contains small amounts of tannins, sugars and essential oil (Aberham et al. 2011; Pharmaceutical Press Editorial Team 2013; WHO 2007).

Clinical evidence: There are no robust clinical data available for the medicinal uses of gentian. Uncontrolled studies suggest that gentian stimulates secretion of gastric and gall bladder secretions and may be effective at treating dyspepsia, but controlled studies are lacking (Pharmaceutical Press Editorial Team 2013; WHO 2007).

Phytopharmacy: An evidence-based guide to herbal medicinal products, First Edition.
Sarah E. Edwards, Inês da Costa Rocha, Elizabeth M. Williamson and Michael Heinrich.
© 2015 John Wiley & Sons, Ltd. Published 2015 by John Wiley & Sons, Ltd.

Pre-clinical evidence and mechanisms of action: The gastrointestinal effects are closely linked to the very high bitterness of gentian. The 'minimum bitterness value' (=10,000) is defined in the European Pharmacopoeia. Amarogentin has a bitterness value of 58,000,000 and is therefore, despite being present in very low concentrations, considered to be the key active constituent. *In vitro* and animal studies have demonstrated that gentian extracts exhibit a range of pharmacological properties including antimicrobial, antispasmodic, anti-inflammatory, antioxidant, choleretic and secretory activities (Nastasijević et al. 2012; WHO 2007).

Antioxidant activity has been attributed to the phenolic content of gentian extracts, especially gentiopicroside (Nastasijević et al. 2012), which also exhibited analgesic properties, and has been shown to inhibit the expression of GluN2B-containing N-methyl-D-aspartate (NMDA) receptors in animal studies (Liu et al. 2012).

Extracts of gentian have been shown to dose-dependently inhibit sorbitol accumulation in human erythrocytes under high glucose conditions. Amarogentin was shown to be a potential inhibitor of aldose reductase, indicating that gentian or its constituents may be useful in the prevention or treatment of diabetes complications (Akileshwari et al. 2012).

Aqueous gentian root extract has been reported to inhibit the proliferation of rat aortic smooth muscle cells induced by platelet-derived growth factor (PDGF-BB). The results indicated that the extract exerted its effects through the blockade of extracellular signal-regulated protein kinases (ERK1/2) activation, and consequent inducible nitric oxide synthase (iNOS) expression. The authors suggested that gentian extracts may therefore provide novel candidates for the prevention and treatment of atherosclerosis (Kesavan et al. 2013).

Interactions: None known.

Contraindications: Due to its reported mutagenic activity and historical use as an emmenagogue, gentian should not be used during pregnancy and lactation or in children. It is contraindicated in gastric or duodenal ulcer, high blood pressure, hyperacidity, or if hypersensitive to the active substance (EMA 2009; Pharmaceutical Press Editorial Team 2013; WHO 2007).

Adverse effects: Headaches have been reported (ESCOP 2003). Gentian is also reported to exert an effect on the menstrual cycle (Pharmaceutical Press Editorial Team 2013). Results of a study investigating genotoxicity of gentian suggest that gentian itself is not genotoxic in somatic cells of *Drosophila melanogaster* (fruit fly), but that it interfered with genotoxic agents *in vivo*, potentiating their effects (Patenković et al. 2013). Mutagenic activity has also been reported for gentian using the Ames Assay, with 100 g of gentian root yielding approximately 100 mg total mutagenic components, predominantly gentisin and isogentisin (Pharmaceutical Press Editorial Team 2013).

Dosage: In adults and elderly (oral administration): (a) cut herb for tea preparation: single dose 1–4 g, up to four times daily; (b) powdered herbal substance: single dose 0.25–2 g up to three times daily; (c) liquid extract: single dose 2–4 ml, up to three times daily; (d) tincture: single dose 1.5–5 g, up to three times daily; (e) soft extract: single dose 0.2 g; daily dose 1–2 g (EMA 2009).

General plant information: Gentian is a perennial herb endemic to mountainous regions of central and southern Europe and western Asia, and is found growing on uncultivated ground in France, Spain and the Balkan mountains (Nastasijević et al. 2012). Gentian should not be confused with gentian violet – the latter is a synthetic triarylmethane dye, formerly important as a topical antiseptic.

References

Aberham A, Pieri V, Croom EM Jr, Ellmerer E, Stuppner H. (2011) Analysis of iridoids, sec-oiridoids and xanthones in *Centaurium erythraea, Frasera caroliniensis* and *Gentiana lutea* using LC-MS and RP-HPLC. *Journal of Pharmaceutical and Biomedical Analysis* 54(3): 517–525.

Akileshwari C, Muthenna P, Nastasijević B, Goksić J, Petrash JM, Reddy GP. (2012) Inhibition of aldose reductase by *Gentiana lutea* extracts. *Experimental Diabetes Research* 2012: 147965, 8 pages.

EMA. (2009) Community herbal monograph on *Gentiana lutea* L., radix. European Medicines Agency http://www.ema.europa.eu/docs/en_GB/document_library/Herbal_-_Community _herbal_monograph/2010/03/WC500075295.pdf (accessed 7 July 2013).

ESCOP. (2003) *ESCOP Monographs: The Scientific Foundation for Herbal Medicinal Products.* 2nd Edition. Thieme, Exeter and London, UK.

Kesavan R, Potunuru UR, Nastasijecić B, Avaneesh T, Joksić G, Dixit M. (2013) Inhibition of vascular smooth muscle cell proliferation by *Gentiana lutea* root extracts. *PLoS ONE* 8(4): e61393.

Liu SB, Ma L, Guo HJ, Feng B, Guo YY, Li XQ, Sun WJ, Zheng LH, Zhao MG. (2012) Gentiopicroside attenuates morphine rewarding effect through downregulation of GluN2B receptors in nucleus accumbens. *CNS Neuroscience & Therapeutics* 18(8): 652–658.

Nastasijević B, Lazarević-Pašti T, Dimitrijević-Branković S, Pašti I, Vujačić A, Joksić G, Vasić V. (2012) Inhibition of myeloperoxidase and antioxidative activity of *Gentiana lutea* extracts. *Journal of Pharmaceutical and Biomedical Analysis* 66: 191–196.

Patenković A, Stamenković-Radak M, Nikolić D, Marković T, Andelković M. (2013) Synergistic effect of *Gentiana lutea* L. on methyl methanesulfonate genotoxicity in the *Drosophila* wing spot test. *Journal of Ethnopharmacology* 146: 632–636.

Pharmaceutical Press Editorial Team. (2013) *Herbal Medicines.* 4th Edition. Pharmaceutical Press, London, UK.

WHO. (2007) *WHO Monographs on Selected Medicinal Plants.* Vol. 3. WHO, Geneva, 390 pp.

Ginger   *Zingiber officinale* Roscoe

Family: Zingiberaceae

Other common names: Ginger root; stem ginger

Drug name: Zingiberis rhizoma

Botanical drug used: Rhizome

Indications/uses: Ginger is one of the most widely used herbal medicines in the world and has a history of traditional use in most countries. Under the THR scheme, it is registered for use for the relief of minor digestive complaints such as indigestion, dyspepsia, flatulence and temporary loss of appetite, and to relieve symptoms of travel sickness. It is also an important ingredient in Asian Indian and Chinese herbal medicines for catarrh, rheumatism, constipation, vomiting and other digestive disorders. It is popularly used to relieve symptoms of morning sickness in pregnancy.

Evidence: There is plausible scientific evidence to support its use as an antiemetic and for digestive complaints.

Safety: Overall, considered to be safe, including during pregnancy, as supported by its food use. Ideally, a robust safety assessment should be conducted to confirm this (see below). Special attention should be taken in pregnancy.

Main chemical compounds: The main constituents are the pungent principles, the gingerols and their derivatives the gingerdiols, gingerdiones and dihydrogingerdiones. The gingerols convert to the more pungent shogaols on drying. The other major actives are the terpenoids, mainly the sesquiterpenoids zingiberene and bisabolene, with zingerone, zingiberol, zingiberenol, curcumene, camphene and linalool (EMA 2012; Palatty et al. 2013; Williamson et al. 2013).

Clinical evidence:

Motion sickness: In a small cross-over design, double-blind, randomised placebo-controlled study, 18 healthy volunteers with a history of motion sickness were included. Motion sickness was induced by movement and 13 out of the 18 participants developed nausea. The effect of ginger (1000 and 2000 mg ginger capsules or placebo with water 1 hour before circular vection) on nausea and gastric motion

---

Phytopharmacy: An evidence-based guide to herbal medicinal products, First Edition.
Sarah E. Edwards, Inês da Costa Rocha, Elizabeth M. Williamson and Michael Heinrich.
© 2015 John Wiley & Sons, Ltd. Published 2015 by John Wiley & Sons, Ltd.

was evaluated. Pre-treatment with 1000 mg ginger capsules effectively reduced the nausea, gastric activity and plasma vasopressin release caused by circular vection (Lien et al. 2003) but further studies are needed, as this is the only clinical trial reported for motion sickness.

Pregnancy-induced nausea and vomiting (PNV): A systematic review of the effectiveness of ginger, which included four randomised clinical trials, found its use to be safe and effective in the treatment of PNV compared to placebo and vitamin B_6. Nevertheless, the authors highlighted that the use of ginger in this population is not without risk (Ding et al. 2013). A previous assessment on its use during pregnancy advised women experiencing nausea and vomiting in pregnancy to be cautious (Tiran 2012). A Cochrane review found that ginger may be helpful to women with PNV, but the evidence was limited and inconsistent (Matthews et al. 2010). No beneficial effects were observed when ginger oil was used as aromatherapy for pain management during childbirth (Smith et al. 2011).

Chemotherapy-induced nausea and vomiting (CINV): A systematic literature review found mixed results with two out of seven studies showing no benefits, three with some benefit on acute nausea and/or delayed nausea (when combined with standard anti-CINV treatment), and two reported an equal effect to metoclopramide (Marx et al. 2013). The lack of consistency in the trials' methodology makes it difficult to draw any clinical recommendation for its use in CINV as yet.

Prevention and treatment of postoperative nausea and vomiting (PONV): A meta-analysis which included 363 patients on a fixed dose of ginger showed that a dose of at least 1 g of ginger was more effective than placebo on 24-hour PONV (Chaiyakunapruk et al. 2006). A previous systematic review found no clinical relevance for its use in PONV (Morin et al. 2004) but this review included various dose regimens which were not comparable.

Irritable bowel syndrome (IBS): A small clinical study found no beneficial effect for ginger, although it was considered well tolerated and safe (van Tilburg et al. 2014).

Pain and inflammation: A review of the use of ginger for pain found that in five out of eight clinical trials, the use of ginger for the treatment of osteoarthritic pain and experimentally induced muscle pain improved pain ratings compared to placebo, while two other reports found no effect (Terry et al. 2011). More robust studies are needed to draw recommendations since previous studies vary in methodological detail and quality.

Dysmenorrhoea: A recent study comparing ginger with zinc sulphate and placebo on the improvement of primary dysmenorrhoea in young women found that participants receiving ginger or zinc sulphate reported better relief of pain during the intervention ($p < 0.05$) than placebo. Ginger and zinc sulphate had similar effects (Kashefi et al. 2014).

Pre-clinical evidence and mechanisms of action:

Antiemetic and other digestive effects: The mechanism of antiemetic action has been attributed mainly to the gingerol, shogaol and zingiberene content, and is at least partly due to serotonin (5-HT) antagonist effects, which suppress vasopressin and reduce gastric activity. Ginger is weakly cholinergic, based on pre-clinical studies (Tiran 2012), and cholagogic, carminative, antispasmodic and appetite stimulant effects have been recorded (Palatty et al. 2013).

Anti-inflammatory effects: Anti-inflammatory activity of ginger has also been reported *in vivo* and *in vitro* (Palatty et al. 2013; WHO 1999). The mechanism is complex but gingerols are known vanilloid receptor (VR1) agonists, as well as NF-kappa B and cyclooxygenase inhibitors (Dedov et al. 2002; Li et al. 2012a).

Other metabolic effects: A wide range of activities, from antidiabetic and anticancer to anti-obesity effects, have been described for ginger; see Li et al. (2012b); Kubra and Rao (2012) for reviews.

Interactions: Due to case reports on the increased response to anticoagulant drugs (e.g. warfarin) when used concomitantly with ginger, patients should be advised to consider increased monitoring. However, this effect has been disputed as no interaction has been identified in controlled clinical trials (Jiang et al. 2005). Possible synergistic effect with antiplatelet drugs (nifedipine) may occur, but this report was based on only one study (Williamson et al. 2013).

Contraindications: None known. Several sources do not recommend its use during pregnancy due to lack of safety data (Tiran 2012), and warnings have been issued in several European countries (including Finland and Denmark). However, many experts do not support this recommendation and two recent reviews have concluded that ginger is safe during pregnancy (Ding et al. 2013; Viljoen et al. 2014). It has been suggested that ginger supplements should be avoided in cases of gallstones as there is suggestion of a cholagogic action, stimulating the secretion of bile (Tiran 2012). Not recommended in children below 6 years of age.

Adverse effects: Mainly gastrointestinal related (stomach upset, eructation, heartburn and nausea) but these are moderate to mild and occur with low frequency (EMA 2012).

Dosage: Powdered, dried rhizome: in cases of prophylaxis for motion sickness, 0.5–2 g daily (single or divided doses) 30 minutes before travel; for pregnancy-induced nausea and vomiting, 500 mg three times daily, while for PONV 1000 mg 1 hour before induction of anaesthesia has been suggested (EMA 2012).

General plant information: Ginger is extremely popular and has been used as a food and medicine for more than 5000 years. In traditional Chinese Medicine and Ayurveda, it forms part of many medicinal formulations. Ginger constituents vary according to origin and are different in the fresh or dried rhizomes, although both are used for similar purposes.

References

Chaiyakunapruk N, Kitikannakorn N, Nathisuwan S, Leeprakobboon K, Leelasettagool C. (2006) The efficacy of ginger for the prevention of postoperative nausea and vomiting: a meta-analysis. *American Journal of Obstetrics and Gynecology* 194(1): 95–99.

Dedov VN, Tran VH, Duke CC, Connor M, Christie MJ, Mandadi S, Roufogalis BD. (2002) Gingerols: a novel class of vanilloid receptor (VR1) agonists. *British Journal of Pharmacology* 137(6): 793–8.

Ding M, Leach M, Bradley H. (2013) The effectiveness and safety of ginger for pregnancy-induced nausea and vomiting: a systematic review. *Women Birth* 26(1): e26–e30.

EMA. (2012) Assessment report on *Zingiber officinale* Roscoe, rhizoma. European Medicines Agency http://www.ema.europa.eu/docs/en_GB/document_library/Herbal_-_HMPC _assessment_report/2012/06/WC500128140.pdf.

Kashefi F, Khajehei M, Cher MT, Alavinia M, Asili J. (2014) Comparison of the Effect of Ginger and Zinc Sulfate on Primary Dysmenorrhea: A Placebo-Controlled Randomized Trial. *Pain Management Nursing.* 52(4): 826–833.

Kubra IR, Rao LJ. (2012) An impression on current developments in the technology, chemistry, and biological activities of ginger (*Zingiber officinale* Roscoe). *Critical Reviews in Food Science and Nutrition* 52(8): 651–88.

Li XH, McGrath KC, Nammi S, Heather AK, Roufogalis BD. (2012a) Attenuation of liver pro-inflammatory responses by *Zingiber officinale* via inhibition of NF-kappa B activation in high-fat diet-fed rats. *Basic and Clinical Pharmacology and Toxicology* 110(3): 238–44.

Li Y, Tran VH, Duke CC, Roufogalis BD. (2012b) Preventive and protective properties of *Zingiber officinale* (Ginger) in diabetes mellitus, diabetic complications, and associated lipid and other metabolic disorders: a brief review. *Evidence-Based Complementary and Alternative Medicine* 2012: 516870.

Lien HC, Sun WM, Chen YH, Kim H, Hasler W, Owyang C. (2003) Effects of ginger on motion sickness and gastric slow-wave dysrhythmias induced by circular vection. *American Journal of Physiology Gastrointestinal and Liver Physiology* 284(3): G481–G489.

Marx WM, Teleni L, McCarthy AL, Vitetta L, McKavanagh D, Thomson D, Isenring E. (2013) Ginger (*Zingiber officinale*) and chemotherapy-induced nausea and vomiting: a systematic literature review. *Nutrition Reviews* 71(4): 245–254.

Matthews A, Dowswell T, Haas DM, Doyle M, O'Mathuna DP. (2010) Interventions for nausea and vomiting in early pregnancy. *Cochrane Database of Systematic Reviews* (9): CD007575.

Morin AM, Betz O, Kranke P, Geldner G, Wulf H, Eberhart LH. (2004) Is ginger a relevant antiemetic for postoperative nausea and vomiting? *Anasthesiol Intensivmed Notfallmed Schmerzther* 39(5): 281–285.

Palatty PL, Haniadka R, Valder B, Arora R, Baliga MS. (2013) Ginger in the prevention of nausea and vomiting: a review. *Critical Reviews in Food Science and Nutrition* 53(7): 659–669.

Smith CA, Collins CT, Crowther CA. (2011) Aromatherapy for pain management in labour. *Cochrane Database of Systematic Reviews* (7): CD009215.

Terry R, Posadzki P, Watson LK, Ernst E. (2011) The use of ginger (*Zingiber officinale*) for the treatment of pain: a systematic review of clinical trials. *Pain Medicine* 12(12): 1808–1818.

Tiran D. (2012) Ginger to reduce nausea and vomiting during pregnancy: evidence of effectiveness is not the same as proof of safety. *Complementary Therapies in Clinical Practice* 18(1): 22–25.

van Tilburg MA, Palsson OS, Ringel Y, Whitehead WE. (2014) Is ginger effective for the treatment of irritable bowel syndrome? A double blind randomized controlled pilot trial. *Complementary Therapies in Medicine* 22(1): 17–20.

Viljoen E, Visser J, Koen N, Musekiwa A. (2014) A systematic review and meta-analysis of the effect and safety of ginger in the treatment of pregnancy-associated nausea and vomiting. *Nutrition Journal* 13(1): 20.

WHO. (1999) *WHO Monographs on Selected Medicinal Plants* Vol. 1. WHO, Geneva, 295 pp.

Williamson EM, Driver S, Baxter K. (Eds.) (2013) *Stockley's Herbal Medicines Interactions.* 2nd Edition. Pharmaceutical Press, London, UK.

Ginkgo   *Ginkgo biloba* L.

Family: Ginkgoaceae

Other common names: Fossil tree; Kew tree; maidenhair tree

Drug name: Ginkgo folium

Botanical drug used: Leaf

Indications/uses: Gingko is used mainly to enhance cognition, and to prevent or reduce memory deterioration during old age, and in the milder forms of dementia including the early stages of Alzheimer's disease. It is also used for the prevention of acute mountain sickness (AMS) and is registered under the THMPD for the relief of symptoms of Raynaud's syndrome and tinnitus. In 2014, the EMA issued a daft monograph proposing that 240 mg/day of a dry extract (DER 35-67:1, extraction solvent: acetone 60% m/m) could be used for the improvement of (age-associated) cognitive impairment and of quality of life in mild dementia (EMA 2014).

Evidence: The evidence from clinical studies is extensive and mainly positive, but results are inconsistent. The evidence is best (5) for cognitive impairment and quality of life in mild dementia if chemically well characterised extracts are used (see below). Based on recent systematic reviews there is more limited evidence at a clinical level, but trials are very varied as to baseline and outcome measures used, so comparisons are difficult.

Safety: Overall considered to be safe, even in conjunction with most conventional drugs, but as a precaution should not be taken with antiplatelet or anticoagulant drugs. Caution is advised in combination with anti-epileptic drugs (valproate and/or phenytoin) as there have been case reports of seizures, although causality was not proved, and also with aminoglycosides, ciclosporin, tacrolimus, calcium channel blockers (diltiazem, nicardipine, nifedipine), haloperidol, proton pump inhibitors and anti-HIV drugs such as efavirenz, which all have a high potential for interaction.

Main chemical compounds: The main active constituents are considered to be the biflavone glycosides, such as ginkgetin, isoginkgetin and bilobetin; and the terpene lactones, the ginkgolides A, B, C, and so on, and bilobalide. Ginkgolic acids

---

Phytopharmacy: An evidence-based guide to herbal medicinal products, First Edition.
Sarah E. Edwards, Inês da Costa Rocha, Elizabeth M. Williamson and Michael Heinrich.
© 2015 John Wiley & Sons, Ltd. Published 2015 by John Wiley & Sons, Ltd.

are found in the seeds (which are eaten as a food in Asia) and are present in trace amounts in the leaves (ESCOP 2003; Kraft and Hobbs 2004; Williamson et al. 2013).

Much of the research, and most of the clinical studies, have been conducted using special extracts such as EGb 761® (or Li 1370®) which are standardised to approximately 24% (25%) ginkgo flavone glycosides and 6% terpene lactones. These contain limited amounts of ginkgolic acids (usually <5 ppm) which are potentially toxic, although no clinical cases of harm have been reported from products with greater amounts.

Note: The main threat to the safety of a ginkgo product is a lack of quality control. Analysis of commercial products in the Netherlands showed a different chemical composition to that stated on the label in 25 out of 29 sampled products (Fransen et al. 2010). A reproducible, standardised herbal extract of good quality, free from adulteration or contamination, can only be guaranteed in a THR product.

Clinical evidence:

Memory and cognitive decline: Numerous clinical trials have been performed using *G. biloba* extracts (mostly EGb761, Li 1370, and chemically similar products) and there is considerable experience of the clinical use of these extracts, mostly in Continental Europe. When assessed for cognitive impairment and dementia, the most recent Cochrane review showed that the evidence available for ginkgo is inconsistent and unreliable (Birks and Grimley Evans 2009). It has been concluded that future research with EGb761 targeting cognitive impairment in old age is warranted as there is promising data from both pre-clinical and clinical studies (Lautenschlager et al. 2012). A study assessing the association between intake of EGb761 and cognitive function in 3612 *non-demented* elderly adults over a 20-year period found that cognitive decline in this population was lower in subjects who reported using EGb761 than in those who did not (Amieva et al. 2013). Gavrilova et al. (2014) showed beneficial effects of EGb 761® in geriatric patients with mild cognitive disjunction.

Ginkgo preparations were included in a Cochrane review (as were donepezil, memantine and rivastigmine) on treatments for memory disorder in adult patients with multiple sclerosis (MS). No convincing evidence was found to support any of the agents as effective treatments for memory disorder in MS patients, although all were found to be safe and well tolerated (He et al. 2011).

A systematic review and meta-analysis on the use of ginkgo in neuropsychiatric disorders (three randomised controlled trials in patients with schizophrenia, and eight in patients with dementia), concluded that there is sufficient available evidence to support its use in patients with dementia and as an adjunctive therapy in schizophrenic patients, but larger, multicentre studies should be carried out (Brondino et al. 2013).

Cardiovascular and related disorders: The evidence for use of gingko to promote recovery after stroke (Zeng et al. 2005) and relieve intermittent claudication (Nicolai et al. 2013) are inconclusive, with larger clinical studies needed.

Tinnitus: A Cochrane review on the use of ginkgo in the management of tinnitus included four randomised clinical trials (1543 participants) and showed that there is limited evidence to demonstrate its effectiveness on tinnitus when this is the primary indication. However, one study on patients with mild-to-moderate dementia, some of whom had tinnitus, found a small but significant reduction of the symptoms of tinnitus in those patients taking *G. biloba* (Hilton et al. 2013). The effect of ginkgo in different subgroups of tinnitus patients should be considered, since ginkgo has been shown to affect vascular permeability and neuronal metabolism, so there is a rationale for this.

Macular degeneration: Two randomised clinical trials (119 participants) on the use of ginkgo for age-related macular degeneration suggested possible benefit on vision, but further studies are needed (Evans 2013).

Prevention of altitude cerebral oedema: Gingko (240 mg extract bd) has been used for the prevention of acute mountain sickness (AMS). However, a clinical study comparing two different products showed one extract to be effective, but not the other. This was attributed to differences in the chemical composition of the products (Leadbetter et al. 2009). Ginkgo supplementation was found not to reduce pulmonary artery systolic pressure in AMS, unlike acetozolamide (Ke et al. 2013).

Pre-clinical evidence and mechanisms of action: The literature is extensive and the multiple mechanisms of action have not been completely resolved. The effects on memory seem to be via a combination of increased blood flow in the brain, neuronal protection, inhibition of amyloid formation and inhibition of caspase activity, all shown in animals (Luo et al. 2002). The ginkgolides and the extract EGb 761 are also potent PAF-antagonists with anti-inflammatory effects contributing to an improvement in cerebral insufficiency and stimulation of endothelium-derived relaxing factor, and may also be responsible for their anecdotal positive effects in asthma (Chu et al. 2011). The antioxidant, anti-inflammatory and antiplatelet effects and protective actions against brain degeneration are thought to be mainly due to the bilobalide and ginkgolide content (Smith and Luo 2004). Ginkgolide B has recently been shown to reduce neuronal cell apoptosis in the hemorrhagic rat brain, possibly via the involvement of Toll-like receptor 4 and the nuclear factor-kappa B pathway (Yu et al. 2012). It also has a protective effect on cerebral oedema induced by high altitude in rats, attributed to its antioxidant properties and suppression of the caspase-dependent apotosis pathway (Botao et al. 2013), suggesting it is an important constituent for efficacy in AMS as well as other cerebral indications.

Interactions: It was concluded in 2013 that 'the intake of the standardised GLE [Ginkgo leaf extract], EGb 761, together with synthetic drugs appears to be safe as long as daily doses up to 240 mg are consumed. If this applies to other extracts prepared according to the European Pharmacopoeia remains uncertain' (Unger 2013). Ginkgo may interact with anticoagulant, antiplatelet and non-steroidal anti-inflammatory drugs (NSAIDs) (aspirin, clopidogrel and ticlopidine), increasing the risk of bleeding. Caution is advised when ginkgo is taken concomitantly with aminoglycosides or ciclosporin, although this is only based on pre-clinical studies, and when in combination with anti-epileptics drugs (valproate, or valproate and phenytoin) as there have been case reports describing seizures in patients (causality not established). Care should also be taken with calcium channel blockers (diltiazem, nicardipine, nifedipine), haloperidol, proton pump inhibitors and anti-HIV drugs, all of which are very important therapeutically and have a high potential for interaction (Williamson et al. 2013).

Contraindications: Not recommended during pregnancy and lactation due to lack of safety data. Use should be stopped prior to surgery due to a potential risk of increased bleeding or interaction with peri-operative drug treatment.

Adverse effects: Ginkgo leaf extract is generally considered to be safe (e.g. Unger 2013), although as with any drug, hypersensitivity may occur. Gastrointestinal complains, headaches and allergies have been reported (ESCOP 2003) along with skin reactions, nausea, dizziness, restlessness, heart palpitation and weakness. Other adverse effects reported in case studies were related to blood clotting disorders (Fransen et al. 2010).

In 2013, in the United States, concerns were raised about potential adulteration of products. Due to the high drug extract ratio (35–67 g of drug for 1 g of extract), production costs are high and this can result in a high rate of adulteration (Daniels 2013).

Long-term use of ginkgo extract (2-year study) caused an increase in liver cancer in mice, and in cancer of the thyroid gland in rats and male mice (National Toxicology Program 2013), although concerns were raised about the preparation (and thus the composition) of the extract, as well as other aspects of the study (Smith 2013). Anticancer properties have been reported for ginkgo extracts (e.g. You et al. 2013), including inhibition of caspase activity (e.g. Botao et al., 2013) and enhancement of the effect of antineoplastic agents and hormone antagonists (e.g. Dias et al. 2013).

Dosage: As recommended by the manufacturer. A daily dose of 120–240 mg of standardised ginkgo extract, two to three times a day is recommended (ESCOP 2003).

General plant information: Concerns were raised about adulterated herbal medicinal products found on the Australian and Danish markets containing free flavonol aglycones and also containing the isoflavone genistein, which is not native to ginkgo (Wohlmuth et al. 2014).

Ginkgo is renowned as one of the oldest living species and it is widely distributed in eastern Asia, especially China, Japan and Korea. The tree is now also widely cultivated in Europe and North America. The seeds have been used as a food and in traditional Chinese medicine, have been used for the treatment of asthma, cough and enuresis for over 5000 years (Smith et al. 2004).

References

Amieva H, Meillon C, Helmer C, Barberger-Gateau P, Dartigues JF. (2013) *Ginkgo biloba* extract and long-term cognitive decline: a 20-year follow-up population-based study. *PLoS One*. 2013;8(1): e52755.

Birks J, Grimley Evans J. (2009) *Ginkgo biloba* for cognitive impairment and dementia. *Cochrane Database of Systematic Reviews* (1): CD003120.

Botao Y, Ma J, Xiao W, Xiang Q, Fan K, Hou J, Wu J, Jing W. (2013) Protective effect of ginkgolide B on high altitude cerebral edema of rats. *High Altitude Medicine and Biology* 14(1): 61–4.

Brondino N, De Silvestri A, Re S, Lanati N, Thiemann P, Verna A, Emanuele E, Politi P. (2013) A Systematic Review and Meta-Analysis of *Ginkgo biloba* in Neuropsychiatric Disorders: From Ancient Tradition to Modern-Day Medicine. *Evidence-Based Complementary and Alternative Medicine* 2013: 915691.

Chu X, Ci X, He J, Wei M, Yang X, Cao Q, Li H, Guan S, Deng Y, Pang D, Deng X. (2011) A novel anti-inflammatory role for ginkgolide B in asthma via inhibition of the ERK/MAPK signaling pathway. *Molecules* 16(9): 7634–7648.

Daniels, St. (2013) Gingko adulteration very widespread. http://www.nutraingredients -usa.com (accessed 05 August 2013).

Dias MC, Furtado KS, Rodrigues MA, Barbisan LF. (2013) Effects of *Ginkgo biloba* on chemically-induced mammary tumors in rats receiving tamoxifen. *BMC Complementary and Alternative Medicine* 13: 93.

EMA. (2014) Community herbal monograph on *Ginkgo biloba* L., folium. Draft. European Medicines Agency http://www.ema.europa.eu/docs/en_GB/document_library/Herbal_-_Community_herbal_monograph/2014/02/WC500161210.pdf.

ESCOP. (2003) *ESCOP Monographs: The Scientific Foundation for Herbal Medicinal Products*. 2nd Edition. Thieme, Exeter and London, UK.

Evans JR. (2013) *Ginkgo biloba* extract for age-related macular degeneration. *Cochrane Database of Systematic Reviews* 1: CD001775.

Fransen H P, Pelgrom S M, Stewart-Knox B, de Kaste D and Verhagen H. (2010) Assessment of health claims, content, and safety of herbal supplements containing *Ginkgo biloba*. *Food and Nutrition Research* 54: 5221.

Gavrilova SI, Preuss UW, Wong JWM, Hoerr R, Kaschel R, Bachinskaya N. (2014) Efficacy and safety of *Ginkgo biloba* extract EGb 761® in mild cognitive impairment with neuropsychiatric symptoms: a randomized, placebo-controlled, double-blind, multi-center trial *International Journal of Geriatric Psychiatry* 29(10): 1087–1095

He D, Zhou H, Guo D, Hao Z, Wu B. (2011) Pharmacologic treatment for memory disorder in multiple sclerosis. *Cochrane Database of Systematic Reviews* (10): CD008876.

Hilton MP, Zimmermann EF, Hunt WT. (2013) *Ginkgo biloba* for tinnitus. *Cochrane Database of Systematic Reviews* 3: CD003852.

Ke T, Wang J, Swenson ER, Zhang X, Hu Y, Chen Y, Liu M, Zhang W, Zhao F, Shen X, Yang Q, Chen J, Luo W. (2013) Effect of acetazolamide and *gingko biloba* on the human pulmonary vascular response to an acute altitude ascent. *High Altitude Medicine and Biology* 14(2): 162–7.

Kraft K, Hobbs C. (2004) *Pocket Guide to Herbal Medicine*. Thieme, Stuttgart.

Lautenschlager N T, Ihl R and Muller WE. (2012) *Ginkgo biloba* extract EGb 761(R) in the context of current developments in the diagnosis and treatment of age-related cognitive decline and Alzheimer's disease: a research perspective. *International Psychogeriatrics* 24(Suppl 1): S46–S50.

Leadbetter G, Keyes LE, Maakestad KM, Olson S, Tissot van Patot MC, Hackett PH. (2009) *Ginkgo biloba* does – and does not – prevent acute mountain sickness. *Wilderness & Environmental Medicine* 20(1): 66–71.

Luo Y, Smith JV, Paramasivam V, Burdick A, Curry KJ, Buford JP, Khan I, Netzer WJ, Xu H, Butko P. (2002) Inhibition of amyloid-beta aggregation and caspase-3 activation by the *Ginkgo biloba* extract EGb761. *Proceedings of the National Academy of Sciences of the United States of America (PNAS)* 99(19): 12197–12202.

National Toxicology Program. (2013) Toxicology and carcinogenesis studies of *Ginkgo biloba* extract (CAS No. 90045-36-6) in F344/N rats and B6C3F1/N mice (Gavage studies). *National Toxicology Program technical report series* (578): 1–183.

Nicolai SP, Kruidenier LM, Bendermacher BL, Prins MH, Stokmans RA, Broos PP, Teijink JA. (2013) *Ginkgo biloba* for intermittent claudication. *Cochrane Database of Systematic Reviews* 6: CD006888.

Smith JV, Luo Y. (2004) Studies on molecular mechanisms of *Ginkgo biloba* extract. *Applied Microbiology and Biotechnology* 64(4): 465–472.

Smith T. (2013) Experts question relevance of ginkgo toxicology report. *HerbalEGram* 10(5).

Unger M. (2013) Pharmacokinetic drug interactions involving *Ginkgo biloba*. *Drug Metabolism Review* 45(3): 353–385.

Williamson EM, Driver S, Baxter K. (Eds.) (2013) *Stockley's Herbal Medicines Interactions*. 2nd Edition. Pharmaceutical Press, London, UK.

Wohlmuth H, Savage, K, Dowell A., Mouatt, P. (2014) Adulteration of *Ginkgo biloba* products and a simple method to improve its detection. *Phytomedicine* 21(6): 912–918.

You OH, Kim SH, Kim B, Sohn EJ, Lee HJ, Shim BS, Yun M, Kwon BM, Kim SH. (2013) Ginkgetin induces apoptosis via activation of caspase and inhibition of survival genes in PC-3 prostate cancer cells. *Bioorganic and Medicinal Chemistry Letters* 23(9): 2692–2695.

Yu WH, Dong XQ, Hu YY, Huang M, Zhang ZY. (2012) Ginkgolide B reduces neuronal cell apoptosis in the traumatic rat brain: possible involvement of Toll-like receptor 4/nuclear factor-kappa B pathway. *Phytotherapy Research* 26(12): 1838–1844.

Zeng X, Liu M, Yang Y, Li Y and Asplund K. (2005) *Ginkgo biloba* for acute ischaemic stroke. *Cochrane Database of Systematic Reviews* (4): CD003691.

Ginseng

Panax ginseng C.A.Mey., *P. quinquefolius* L.

Synonyms: *P. ginseng*: *P. schin-seng* var. *coraiensis* T.Nees, *P. verus* Oken; *P. quinquefolius*: *P. americanus* (Raf.) Raf., *P. quinquefolius* var. *americanus* Raf.

Family: Araliaceae

Other common names: *P. ginseng*: Asian ginseng, Chinese ginseng, Korean ginseng, Oriental ginseng; *P. quinquefolius*: American ginseng, Wisconsin ginseng (when collected and grown in Wisconsin, the source of most American ginseng; it is highly prized, even in Asia)

Drug names: *P. ginseng*: Ginseng radix; Ginseng radix rubra (used to describe 'red ginseng', which is the species' steamed root); *P. quinquefolius*: Panacis quinquefolii radix

Botanical drug used: Secondary storage root

Indications/uses: Ginseng is used as a tonic and 'adaptogen' (i.e. to increase the body's ability to adapt to stress), to overcome fatigue, improve cognitive function and combat ageing. It is marketed as an immune-enhancer and as a 'mental and physical booster'. *P. ginseng* is an important component of many traditional Chinese medicines, where it has been used for centuries to increase well-being, manage diabetes and treat a number of diseases, including erectile dysfunction. Both species of *Panax* are used for broadly similar purposes, although in TCM terms, American ginseng is said to be more cooling and soothing, and Asian ginseng to be 'hot' and more stimulating.

Evidence: Ginseng (*Panax spp.*) appears to have beneficial effects on some aspects of cognitive function, well-being and quality of life in healthy study participants, but evidence is limited. There is no high-quality evidence of its efficacy in dementia patients. There is a huge amount of pharmacological evidence supporting its uses – but these do not always translate into clinical effects.

Safety: *Panax* seems safe, with no acute or chronic toxicity or serious adverse events reported. However, adulterated or substituted ginseng products are found widely.

Phytopharmacy: An evidence-based guide to herbal medicinal products, First Edition.
Sarah E. Edwards, Inês da Costa Rocha, Elizabeth M. Williamson and Michael Heinrich.
© 2015 John Wiley & Sons, Ltd. Published 2015 by John Wiley & Sons, Ltd.

Main chemical compounds: The closely related species *P. quinquefolius* and *P. ginseng* both contain similar triterpene saponins, which include those commonly referred to as ginsenosides; an essential oil with polyacetylenes (e.g. falcarinol, panaxydol) and sesquiterpenes, polysaccharides (e.g. panaxans A-F), peptidoglycans, fatty acids, carbohydrates and phenolic compounds (Li et al. 2012; Williamson et al. 2013). Over 150 ginsenosides have been identified so far (ginsenosides series Ra to Rs), but the content of the active ginsenosides and their metabolites varies depending on the species, plant age, preservation method, season of harvest and extraction method. Variability of ginsenoside content may also partly be attributed to natural variations in environmental growing conditions (soil, weather, geographical location) in addition to production procedures, including fermentation (Jia and Zhao 2009; Li et al. 2012; Smith et al. 2014).

Clinical evidence:

Cognitive function: A Cochrane review (Geng et al. 2010) investigating the efficacy of the genus *Panax* on cognitive performance suggested that ginseng improved some aspects of cognitive function, well-being and quality of life in healthy participants, but there was a lack of convincing evidence. No high-quality evidence was found for efficacy in dementia. A further meta-analysis of 65 studies investigating other therapeutic areas produced the following key conclusions (Shergis et al. 2013).

Psychomotor performance (17 studies): The psychomotor performance seemed to improve in people given *P. ginseng*, but the significance of these results is difficult to interpret.

Physical performance (10 studies): No significant effect on physical performance was seen.

Circulatory system (8 studies): A varied set of study groups, protocols and dose regimens used and evaluation parameters makes an assessment difficult.

Effects on glucose metabolism (6 studies): There is some evidence that ginseng can help decrease circulating glucose levels.

Respiratory system (5 studies): Ginseng may reduce respiratory inflammation and infection and improve lung function and respiratory muscle strength.

Erectile dysfunction (4 studies): Of three studies carried out on red ginseng, two showed no beneficial effect, one produced a significant improvement. The other positive study used a cultured ginseng product.

Quality of life and mood (4 studies): Limited effects in some areas were seen with questionnaire studies.

Menopausal symptoms: (4 studies): No indication of beneficial effects. A further review stated that 'firm conclusions cannot be drawn', but some data point to the possibility of an effective treatment for managing menopausal symptoms (Kim et al. 2013).

Cancer-related fatigue: A preliminary study indicated that *P. quinquefolius* improved cancer-related fatigue at doses of 1000–2000 mg/day (Barton et al. 2010).

NB: In most studies the risk of bias was unclear, and despite the number of research reports, very limited conclusions can be made. In general terms, these assessments are similar to those reported earlier by Ernst (2010) and more recently by Choi et al. (2013), who reviewed the literature in the Korean language.

Pre-clinical evidence and mechanisms of action:
The overall effect of the herb is the result of the complex and diverse actions of the many ginsenosides. It is not possible to ascribe particular effects to each one.

Cognitive effects: Mechanisms and actions of ginseng and ginsenosides have recently been reviewed (Smith et al. 2014), and, in general, support the reported cognitive enhancement effects of the herb.

Other effects: Anti-apoptotic, anti-inflammatory, radioprotective, neuroprotective, vasodilatory and antioxidative effects have been described, as well as modulation of the hypothalamus-pituitary-adrenal axis, angiogenesis, autoimmune transmitter release, glucose metabolism ion channels and the immune response; increase in neural plasticity; increase in nitric oxide production; normalisation of lipid profiles; preservation of mitochondrial membrane integrity; inhibition of platelet adhesion; stimulation of interferon production, natural killer cell activity and phagocytosis (Ernst 2010).

Interactions: There have been reports of interactions of unspecified ginseng products with phenelzine, a monoamine oxidase inhibitor (MAOI), and suspected potentiation of nifedipine (via CYP3A4) in an animal study. Depending on the nature of the extract, ginseng exerts varied enzyme-selective effects on other CYPs. At high concentrations, ginsenosides are known to moderately inhibit P-glycoprotein. Ginseng should be avoided if taking MAOIs, nifedipine and in cancer chemotherapy. American ginseng (*P. quinquefolius*) was shown to reduce the effect of warfarin in healthy volunteers, although Asian ginseng (*P. ginseng*) had no effect. Ginseng extracts contain antiplatelet agents, so changes in bleeding times may occur but are not predictable. Ginseng lowers blood glucose, so caution should be observed in patients taking oral antidiabetic agents (Williamson et al. 2013).

Contraindications: None reported, although there is a lack of safety data for its use in children or during pregnancy/lactation. Diabetics should be cautious using ginseng, as it may reduce blood glucose levels.

Adverse effects: If taken at the recommended dosage, ginseng is not associated with serious adverse effects. In an open study of patients ingesting excessive quantities (as much as 15 g) of ginseng (type unspecified) per day, it induced hypertension, nervousness, irritability, diarrhoea, skin eruptions and insomnia. One case of cerebral arteritis (inflammation of small and medium-sized arteries in the brain) associated with high dose (6 g) ethanol extract of ginseng root was reported. Oestrogenic-like side effects have been reported in both pre- and post-menopausal women following the use of ginseng. However, clinical studies have shown that standardised ginseng extract does not cause a change in hormonal levels in either men or women (WHO 1999).

In a four week, long placebo-controlled, three-armed study including 170 healthy male and female volunteers (placebo, 1000 mg twice a day, 500 mg twice a day), a 20% ethanol extract obtained from 4-year-old *P. ginseng* root was shown to be safe, tolerable, and free of any serious unwanted effects. Some mild side effects were observed, but were similar to placebo, with the most common reported being headache, and sleep and gastrointestinal complaints (Ernst 2010).

Dosage: Daily dose: dried root 0.5–2 g by decoction; doses of other preparations should be calculated accordingly (WHO 1999). A wide range of preparations are sold, which recommend doses not based on adequate dose-finding studies.

General plant information: Ginseng is one of the most widely used herbal medicines globally. *P. ginseng* is practically extinct in the forests of China and the Korean peninsula. Wild ginseng is considered superior to cultivated ginseng, and as demand in China has soared, so has the pressure on wild populations. Most of the American ginseng produced is exported to Asia. *P. quinquefolius* is listed in the Convention on International Trade in Endangered Species of Wild Flora and Fauna (CITES) Appendix II, which regulates international trade of the species. Illegal poaching of the species from the wild is known to occur due to the high price it commands. Adulteration is known to occur widely as the herb is expensive (Chandler 1988).

Siberian ginseng (*Eleutherococcus senticosus*) while belonging to the same plant family, is a different, unrelated species. It contains different saponins, so although used for similar purposes, it has been discussed in a separate monograph (see pp. 177–179).

References

A huge body of literature is available on ginseng of all types, and these are a selection of reviews and more recent studies where the botanical drug has been properly authenticated and usually characterised.

Barton DL, Soori GS, Bauer BA, Sloan JA, Johnson PA, Figueras C, Duane S, Mattar B, Liu H, Atherton PJ, Christensen B, Loprinzi CL. (2010) Pilot study of *Panax quinquefolius* (American ginseng) to improve cancer-related fatigue: a randomized, double-blind, dose-finding evaluation: NCCTG trial N03CA. *Supportive Care in Cancer* 18(2): 179–187.

Chandler RF. (1988) Ginseng – an aphrodisiac. *Canadian Pharmaceutical Journal* 122: 36–38.

Choi J, Kim TH, Choi TY, Lee MS. (2013) Ginseng for health care: a systematic review of randomized controlled trials in Korean literature. *PLoS One* 8(4): e59978.

Ernst E. (2010) *Panax ginseng*: An overview of the clinical evidence. *Journal of Ginseng Research* 34(4): 259–263.

Geng J, Dong J, Ni H, Lee MS, Wu T, Jiang K, Wang G, Zhou AL, Malouf R. (2010) Ginseng for cognition. *Cochrane Database of Systematic Reviews* (12): CD007769.

Jia L, Zhao Y. (2009) Current evaluation of the millennium phytomedicine – Ginseng (1): etymology, pharmacognosy, phytochemistry, market and regulations. *Current Medicinal Chemistry* 16(19): 2475–2484.

Kim MS, Lim HJ, Yang HJ, Lee MS, Shin BC, Ernst E. (2013) Ginseng for managing menopause symptoms: a systematic review of randomized clinical trials. *Journal of Ginseng Research* 37(1): 30–36.

Lee NH, Yoo SR, Kim HG, Cho JH, Son CG. (2012) Safety and tolerability of *Panax ginseng* root extract: a randomized, placebo-controlled, clinical trial in healthy Korean volunteers. *The Journal of Alternative and Complementary Medicine* 18(11): 1061–1069.

Li X, Yan YZ, Kim YK, Uddin MR, Bae H, Kim HH, Park SU. (2012) Ginsenoside content in the leaves and roots of *Panax ginseng* at different ages. *Life Science Journal* 9(4): 679–683.

Shergis JL, Zhang AL, Zhou W, Xue CC. (2013) *Panax ginseng* in randomised controlled trials: a systematic review. *Phytotherapy Research* 27(7): 949–965.

WHO. (1999) *WHO Monographs on Selected Medicinal Plants.* Vol. 1. WHO, Geneva, 295 pp.

Smith I, Williamson EM, Putnam S, Farrimond J, Whalley BJ. (2014) Effects and mechanisms of ginseng and ginsenosides on cognition. *Nutrition Reviews* 72(5): 319–333.

Williamson EM, Driver S and Baxter K. (Eds.) (2013) *Stockley's Herbal Medicines Interactions.* 2nd Edition. Pharmaceutical Press, London, UK.

♠ ♠ ♠ ♠ ♠

Ginseng, Siberian *Eleutherococcus senticosus (Rupr. & Maxim.) Maxim.*

Synonyms: *Acanthopanax asperatus* Franch. & Sav.; *A. senticosus* (Rupr. & Maxim) Harms; and others

Family: Araliaceae

Other common names: Eleuthero; taigo; thorny ginseng (NB: It is illegal to market products containing this herb as 'Siberian ginseng' in the United States, as the name 'ginseng' is reserved for *Panax* species; see p. 173)

Drug name: Eleutherococci radix

Botanical drug used: Root

Indications/uses: The uses of *E. senticosus* are similar to those of Asian and American ginseng, notably to improve the immune function. In this context it is used as an 'adaptogen', increasing the ability to adapt to stress. Siberian ginseng is also used to overcome fatigue, enhance cognitive function and combat ageing.

Evidence: Clinical data for the efficacy of *E. senticosus* is limited, although *in vitro* and *in vivo* studies support its use as an adaptogen, demonstrating that it possesses anti-stress, anti-ulcer, anti-irradiation, anticancer, anti-inflammatory, and hepatoprotective properties.

Safety: There are limited clinical data on the safety and toxicity of *E. senticosus*. Avoid in uncontrolled hypertension.

Main chemical compounds: Siberian ginseng contains triterpene saponins, which are chemically distinct from American and Asian ginseng, and known as eleutherosides; lignans such as syringaresinol; coumarins; and polysaccharides; all of which may contribute to the effects (Huang et al. 2011a; Wichtl 2008).

Clinical evidence: *E. senticosus* root extract (120 mg/day for two days) was assessed for its effect when used as part of stress management training. While the training showed benefits, the addition of Siberian ginseng did not show additional benefits (Schaffler et al. 2013), indicating that it is not useful for short-term interventions.

Phytopharmacy: An evidence-based guide to herbal medicinal products, First Edition.
Sarah E. Edwards, Inês da Costa Rocha, Elizabeth M. Williamson and Michael Heinrich.
© 2015 John Wiley & Sons, Ltd. Published 2015 by John Wiley & Sons, Ltd.

When chronic fatigue was treated with *E. senticosus* root extract (2 g/day, containing approximately 9 mg of eleutherosides, 1 month duration) the Rand Vitality Index scores improved significantly compared to placebo. A 2-month treatment was significantly effective in a subset of subjects with mild-to-moderate fatigue, but not in the entire group (Hartz et al. 2004; Panossian 2013).

Eleutherococcus root reduced total and LDL cholesterol, triglycerides and glucose in healthy human volunteers. An ergospirometric test revealed a higher oxygen plateau, suggesting that active components in the root affect cellular defence and physical fitness, as well as lipid metabolism (Szołomicki et al. 2000).

Some evidence exists for potential benefits in cases of bipolar disorder and neurosis (Panossian 2013), but not only are the studies of limited quality but these are also conditions not suitable for self-medication. For example, a randomised, double-blind clinical trial compared *E. senticosus* extract and lithium, with fluoxetine and lithium, as treatment for adolescents with bipolar disorder. The results demonstrated no significant difference in efficacy between the two treatments, which were generally well tolerated (Weng et al. 2007).

Pre-clinical evidence and mechanisms of action: Numerous *in vitro* and *in vivo* studies have been performed assessing the various properties of *E. senticosus* and its extracts. Eleutherosides, the active constituents isolated from *E. senticosus*, have been shown to alleviate both mental and physical fatigue in mice (Huang et al. 2011). It has been postulated that adaptogens such as Siberian ginseng and its constituents affect numerous signalling pathways, eventually regulating NF-κB-mediated defence responses, and overall by having an influence on transcriptional control of metabolic regulation, both at a cellular level, and at the level of the whole organism (Huang et al. 2011a; Panossian and Wagner 2005; Panossian et al. 2013). *E. senticosus* exhibits hypoglycaemic effects and has been proposed as a potential agent for preventing type II diabetes (Watanabe et al. 2010).

Interactions: Izzo (2013) concluded that *E. senticosus*, at generally recommended over-the-counter doses, is unlikely to alter the disposition of co-administered medications primarily metabolised by CYP2D6 or CYP3A4. Increased levels of digoxin following co-administration of Siberian ginseng and digoxin were reported, but may be based on the chemical similarity of the compounds resulting in interferences with the assay.

Contraindications: None reported, although there is a lack of safety data for its use in children or during pregnancy and lactation. High doses should be avoided in patients with uncontrolled hypertension (Rasmussen et al. 2012).

Adverse effects: Few reported, but may cause a mild increase in blood pressure. See also ginseng (*P. quinquefolius* and *P. ginseng*), p. 173–175.

Dosage: Daily dose: dried root 0.5–2 g by decoction (WHO 2004); for products, follow the manufacturers' instructions.

General plant information: Siberian ginseng is native to East Asia, including Siberia, China, Japan and Korea, and is still collected from the wild, but also from China and Korea. Its superficial similarity to 'ginseng' has given rise to similar uses; however, it is in fact not justified based on the existing evidence. Confusion about the species' chemical composition and imprecise definition of the materials used in clinical studies has also hampered research.

References

Hartz AJ, Bentler S, Noyes R, Hoehns J, Logemann C, Sinift S, Butani Y, Wang W, Brake K, Ernst M, Kautzman H. (2004) Randomized controlled trial of Siberian ginseng for chronic fatigue. *Psychological Medicine* 34(1): 51–61.

Huang L, Zhao H, Huang B, Zheng C, Peng W, Qin L. (2011a) *Acanthopanax senticosus*: review of botany, chemistry and pharmacology. *Pharmazie* 66(2): 83–97.

Huang LZ, Huang BK, Ye Q, Qin LP. (2011b) Bioactivity-guided fractionation for anti-fatigue property of *Acanthopanax senticosus*. *Journal of Ethnopharmacology* 133(1): 213–219.

Izzo AA. (2013) Interactions between herbs and conventional drugs: overview of the clinical data. *Medical Principles and Practice* 21(5): 404–428.

Panossian, AG. (2013) Adaptogens in Mental and Behavioral Disorders, *Psychiatric Clinics of North America* 36(1): 49–64.

Panossian AG, Wagner H. (2005) Stimulating effect of adaptogens: an overview with particular reference to their efficacy following single dose administration. *Phytotherapy Research* 19(10): 819–838.

Panossian AG, Hamm R, Kadioglu O, Wikman G, Efferth T. (2013) Synergy and antagonism of active constituents of ADAPT-232 on transcriptional level of metabolic regulation of isolated neuroglial cells. *Frontiers in Neuroscience* 7: 16. doi: 10.3389/fnins.2013.00016.

Rasmussen CB, Glisson JK, Minor DS. (2012) Dietary supplements and hypertension: potential benefits and precautions. *Journal of Clinical Hypertension* 14(7): 467–471.

Schaffler K, Wolf OT, Burkart M. (2013) No benefit adding *Eleutherococcus senticosus* to stress management training in stress-related fatigue/weakness, impaired work or concentration, a randomized controlled study. *Pharmacopsychiatry* 46(5): 181–190.

Szołomicki J, Samochowiec L, Wójcicki J, Droździk M. (2000) The influence of active components of *Eleutherococcus senticosus* on cellular defence and physical fitness in man. *Phytotherapy Research* 14: 30–35.

Watanabe K, Kamata K, Sato J, Takahashi T. (2010) Fundamental studies on the inhibitory action of *Acanthopanax senticosus* Harms on glucose absorption. *Journal of Ethnopharmacology* 132: 193–199.

Weng S, Tang J, Wang G, Wang X, Wang H. (2007) Comparison of the Addition of Siberian Ginseng (*Acanthopanax senticosus*) versus Fluoxetine to Lithium for the Treatment of Bipolar Disorder in Adolescents: A Randomized, Double-Blind Trial. *Current Therapeutic Research* 68(4): 280–290.

WHO. (2004) WHO *Monographs on Selected Medicinal Plants*. Vol. 2. WHO, Geneva, 358 pp.

Wichtl M. (2008) *Teedrogen und Phytopharmaka: Ein Handbuch für die Praxis auf wissenschaftlicher Grundlage*. Wissenschaftliche Verlagsgesellschaft, Stuttgart.

Goldenrod   *Solidago virgaurea* L.

Family: Asteraceae (Compositae)

Other common names: European Goldenrod; virgaurea

Drug name: Solidaginis virgaureae herba

Botanical drug used: Flowering aerial parts (herb)

Indications/uses: Goldenrod has been used traditionally as a diuretic and for treating and preventing inflammatory conditions of the lower urinary tract and kidney stones. It is an ingredient of the popular formula Phytodolor, used for pain and inflammation.

Evidence: A small number of clinical studies support the use as a diuretic and for irritable bladder conditions. Goldenrod should not be used in conditions where a reduced fluid intake is recommended, e.g. severe cardiac or renal diseases (EMEA 2008).

Safety: Scientific data are lacking, but based on the long tradition of use, products complying with quality control guidelines seem to be safe.

Main chemical compounds: Oleanane-type triterpene saponins, the virgaurea saponins, based on polygalic acid; clerodane diterpenes known as solidagoic acids; flavonoids based on quercetin and kaempferol (the Ph. Eur. specifies 0.5–1.5%, expressed as hyperoside), and chlorogenic acid derivatives, have been isolated from the herb (Laurençon et al. 2013; Melzig 2004; Pharmaceutical Press Editorial Team 2013; Starks et al. 2010).

Clinical evidence:

There is limited clinical evidence available for some traditional uses.

Diuretic effects: In an open multi-centre study of chronic recurrent irritable bladder conditions (512 patients), a very large percentage (96%) of patients showed an improvement in the clinical global impression scale after taking 424.8 mg goldenrod extract three times daily. Urinary urgency and painful micturition were reduced significantly (Pfannkuch and Stammwitz 2002). In a small open study including 74 female patients with dysuria who received an extract of European goldenrod three times daily for 14 days, the frequency and pain of urination was reduced in 69% of patients (Pharmaceutical Press Editorial Team 2013).

Phytopharmacy: An evidence-based guide to herbal medicinal products, First Edition.
Sarah E. Edwards, Inês da Costa Rocha, Elizabeth M. Williamson and Michael Heinrich.
© 2015 John Wiley & Sons, Ltd. Published 2015 by John Wiley & Sons, Ltd.

Anti-inflammatory effects: Phytodolor, a product containing extracts of ash (*Fraxinus excelsior* L.) bark and aspen (*Populus tremula* L.) leaf and bark, in addition to goldenrod, is widely used for pain and inflammation and is reputed to be clinically effective and safe (Gundermann and Müller 2007).

Pre-clinical evidence and mechanisms of action:

Diuretic effects: Anti-muscarinic effects have been demonstrated in an *in vivo* rat bladder model which had been transfected with human M2 and M3 muscarinic receptors (Borchert et al. 2010). Diuretic effects have also been demonstrated in rats.

Anti-inflammatory effects: Antioxidant and anti-inflammatory effects have been shown *in vitro*, and analgesic and anti-inflammatory effects *in vivo* in a rat arthritis and other animal models (Melzig 2004; Pharmaceutical Press Editorial Team 2013).

Spasmolytic effects: These have been demonstrated in animal models (Melzig 2004).

Antimicrobial effects: Antibacterial effects, including against *Escherichia coli* (a main pathogen in bladder infections) and *Staphylococcus aureus* have been noted for the clerodane diterpenoids (Starks et al. 2010), and antifungal effects for the virgaurea saponins (Laurençon et al. 2013).

Interactions: No reports available.

Adverse effects: Hypersensitivity to *Solidago* species and the Asteraceae in general. In the absence of data, it is not recommended for use in infants and breastfeeding mothers.

Dosage: Based on EMEA (2008) a dose of 350–450 mg of dry extract, three times daily is recommended. Herb: 6–12 g/day in the form of a herbal tea in divided doses.

General plant information: *S. gigantea* Aiton (giant goldenrod) and *S. canadensis* L. (Canadian goldenrod) and their hybrids are also used for similar indications, but show differences in their chemical profile and reported pharmacological effects and have not been well investigated.

References

Borchert VE, Czyborra P, Fetscher C, Goepel W, Michel MC. (2010) Extracts from Rhois aromatic and Solidaginis virgaurea inhibit rat and human bladder contraction. *Naunyn-Schmiedeberg's Archives of Pharmacology* 369(3): 281–286.

EMEA. (2008) *Community herbal monograph on Solidago virgaurea L., herba.* European Medicines Agency http://www.ema.europa.eu/docs/en_GB/document_library/Herbal_-_Community_herbal_monograph/2009/12/WC500018159.pdf.

Gundermann KJ, Müller J. (2007) Phytodolor – effects and efficacy of a herbal medicine. *Wiener Medizinische Wochenschrift* 157(13–14): 343–347.

Laurençon L, Sarrazin E, Chevalier M, Prêcheur I, Herbette G, Fernandez X. (2013) Triterpenoid saponins from the aerial parts of *Solidago virgaurea alpestris* with inhibiting activity of *Candida albicans* yeast-hyphal conversion. *Phytochemistry* 86: 103–111.

Melzig M. (2004) [Goldenrod – a classical exponent in the urological phytotherapy] (in German). *Wiener Medizinische Wochenschrift* 154: 523–527.

Pfannkuch A, Stammwitz U. (2002) Effectiveness and tolerability of a monograph-conform goldenrod extract in patients with an irritable bladder [in German]. *Zeitschrift fuer Phytotherapie* 23: 20–25.

Pharmaceutical Press Editorial Team. (2013) *Herbal Medicines.* 4th Edition. Pharmaceutical Press, London, UK.

Starks CM, Williams RB, Goering MG, O'Neil-Johnson M, Norman VL, Hu JF, Garo E, Hough GW, Rice SM, Eldridge GR. (2010) Antibacterial clerodane diterpenes from Goldenrod (*Solidago virgaurea*). *Phytochemistry* 71(1): 104–109.

Goldenseal　　　　　　　　　*Hydrastis canadensis* L.

Synonyms: *H. trifolia* Raf.; *Warnera canadensis* Mill.

Family: Ranunculaceae (formerly in Berberidaceae)

Other common names: Orange root; yellow puccoon; yellow root

Drug name: Hydrastis rhizoma

Botanical drug used: Whole or cut dried rhizome and root

Indications/uses: Traditionally, goldenseal rhizome/root has been used to treat a wide range of ailments, mainly involving infection. Internally, it is used to treat colds and other upper respiratory tract infections, gastrointestinal disorders including infectious diarrhoea and dyspepsia, urinary tract infections, to improve appetite, and as an immune stimulant. It is also used externally to treat conjunctivitis, vaginitis, skin inflammations and cold sores. Historically, Native Americans also used goldenseal as a yellow dye. Goldenseal root is often combined with Echinacea in preparations used to treat colds.

Evidence: Clinical and pre-clinical evidence for goldenseal is lacking, although the active constituent berberine has been found effective against acute diarrhoea, and to improve symptoms of chronic cholecystitis. The antimicrobial, antiprotozoal, metabolic and other effects of berberine are well-documented.

Safety: Safety data for goldenseal are lacking. Long-term use and high dosages should be avoided. Evidence of liver carcinogenicity was found in rodents fed powdered goldenseal root in a standard 2-year bioassay. Safety of goldenseal during pregnancy and lactation has not been established and should be avoided, especially in view of the reported cytotoxicity and uterine stimulating effects of berberine.

Main chemical compounds: Isoquinoline alkaloids (2.5–6%): predominantly hydrastine (1.5–5%), berberine (0.5–6.0%), berberastine (2–3%) and canadine (0.5–1%), with lesser quantities of related alkaloids, including canadaline, corypalmine, hydrastidine and jatrorrhizine (Pharmaceutical Press Editorial Team 2013; WHO 2007; Williamson et al. 2013). The British and European Pharmacopoeias

Phytopharmacy: An evidence-based guide to herbal medicinal products, First Edition.
Sarah E. Edwards, Inês da Costa Rocha, Elizabeth M. Williamson and Michael Heinrich.
© 2015 John Wiley & Sons, Ltd. Published 2015 by John Wiley & Sons, Ltd.

define berberine content in goldenseal root as a minimum of 3% (dried drug), and hydrastine as a minimum of 2.5% (dried drug).

Clinical evidence:

Clinical data assessing the effects of goldenseal are limited. More information is available for berberine.

Metabolic effects: A recent systematic review found that berberine may have beneficial effects in the control of blood lipid levels, but that further studies are needed (Dong et al. 2013). Berberine also has antidiabetic properties and several studies have shown it is able to improve glycaemic control (Derosa et al. 2012).

Pre-clinical evidence and mechanisms of action:

The pharmacological activity of goldenseal is attributed to the alkaloids, which have been well studied, but may not necessarily apply to whole extracts of goldenseal rhizome. Well-documented bioactivities exhibited by berberine include antibacterial, antifungal, antidiabetic, anti-hyperlipidaemic and anticancer effects.

Antimicrobial effects: *H. canadensis* leaf extract was found to be a more potent antimicrobial against methicillin-resistant *Staphylococcus aureus* (MRSA) than berberine alone. The extract inhibited toxin production by MRSA and prevented damage by MRSA to keratinocyte cells *in vitro* (Cech et al. 2012). The antimicrobial action of berberine may be potentiated by non-antimicrobial compounds found in berberine-containing plants, suggesting a synergistic action (Pharmaceutical Press Editorial Team 2013).

Metabolic effects: The effects of berberine on glycaemic and lipid-profile control, and also some aspects of the anti-inflammatory activity, are mediated at least in part via PPAR gamma (Chen et al. 2008).

Anticancer effects: Ethanolic root extract of goldenseal was shown to induce apoptosis in HeLa cells *in vitro*, with a dose-dependent decrease in expression of NF-κβ via TNF-α-mediated pathway, indicating chemopreventive potential (Saha et al. 2013).

Carcinogenic activity: When goldenseal was evaluated by the US National Toxicology Program (NTP), there was clear evidence of carcinogenic activity of goldenseal root powder based on an increased incidence of liver tumours in rats and mice in standard 2-year feed studies (NTP 2010). In a study assessing the toxicity of five goldenseal alkaloids, berberine, followed by palmatine, appeared to be the most potent DNA damage inducers in human hepatoma HepG2 cells. Commercially available goldenseal extracts were also found to induce DNA damage, with a positive correlation to berberine content (Chen et al. 2013).

Effects on the central nervous system: Berberine derivatives have been advocated as potential treatments for nervous system disease, including dementias (Kulkarni and Dhir 2010).

Interactions: Evidence from human studies has shown that goldenseal may inhibit drug-metabolising isoenzymes CYP3A4 and CYP2D6, slowing the clearance of drugs metabolised by these enzymes and increasing plasma levels of these drugs. However, no significant clinical interactions have been reported (Gurley et al. 2012; Williamson et al. 2013).

Contraindications: As berberine has been shown to promote blood coagulation in mice and rats, use of goldenseal should be avoided by individuals with high blood pressure, bleeding or blood clotting disorders, or if using medications to control such a disorder (Pharmaceutical Press Editorial Team 2013; WHO 2007).

Goldenseal is contraindicated during pregnancy and lactation. The active constituents – berberine, canadine, hydrastine and hydrastinine – have all been reported to stimulate uterine contractions, and berberine was found to be present in the breast milk of lactating women who have taken berberine-containing plants. Reports that berberine can cause neonatal jaundice have been tested in a small clinical study and are now thought to be unfounded (Linn et al. 2012); however, goldenseal should not be given to children due to lack of safety data.

Adverse effects: Symptoms of goldenseal overdose include stomach upset, nervous symptoms and depression, and large quantities may be fatal (Pharmaceutical Press Editorial Team 2013). High doses of hydrastine are reported to cause exaggerated reflexes, convulsions, paralysis and respiratory failure (WHO 2007).

Dosage: Standard herbal reference texts give dosages for oral administration (adults) as dried rhizome: 0.5–1.0 g as a decoction three times daily; liquid extract of Hydrastis (BPC 1949): 0.3–1.0 ml; tincture of Hydrastis (BPC 1949): 2–4 ml (Pharmaceutical Press Editorial Team 2013).

General plant information: Goldenseal is endemic to North America, where it is primarily wild harvested. It is one of the most commercially important medicinal plants in North America, but habitat loss and over-exploitation have resulted in conservation concerns. Goldenseal has been listed in the Convention in Trade in Endangered Species of Wild Flora and Fauna (CITES) Appendix II since 1997 (Lonner 2007).

References

Cech NB, Junio HA, Ackermann LW, Kavanaugh JS, Horswill AR. (2012) Quorum quenching and antimicrobial activity of goldenseal (*Hydrastis canadensis*) against methicillin-resistant *Staphylococcus aureus* (MRSA). *Planta Medica* 78(14): 1556–1561.

Chen FL, Yang ZH, Liu Y, Li LX, Liang WC, Wang XC, Zhou WB, Yang YH, Hu RM. (2008) Berberine inhibits the expression of TNF-alpha, MCP-1, and IL-6 in AcLDL-stimulated macrophages through PPAR gamma pathway. *Endocrine* 33(3): 331–337.

Chen S, Wan L, Couch L, Lin H, Li Y, Dobrovolsky VN, Mei N, Guo L. (2013) Mechanism study of goldenseal-associated DNA damage. *Toxicology Letters* 221(1): 64–72.

Derosa G, Maffioli P, Cicero AF. (2012) Berberine metabolic and cardiovascular risk factors: an analysis from preclinical evidences to clinical trials. *Expert Opinion on Biological Therapy* 12(8): 1113–1124.

Dong H, Zhao Y, Zhao L, Lu F. (2013) The effects of berberine on blood lipids: a systemic review and meta-analysis of randomized controlled trials. *Planta Medica* 79(6): 437–446.

Linn YC, Lu J, Lim LC, Sun H, Sun J, Zhou Y, Ng HS. (2012) Berberine-induced haemolysis revisited: safety of Rhizoma coptidis and Cortex phellodendri in chronic haematological diseases. *Phytotherapy Research* 26(5): 682–686.

Gurley BJ, Fifer EK, Gardner Z. (2012) Pharmacokinetic herb-drug interactions (part 2): drug interactions involving popular botanical dietary supplements and their clinical relevance. *Planta medica* 78(13): 1490–1514.

Kulkarni SK, Dhir A. (2010) Berberine: a plant alkaloid with therapeutic potential for central nervous system disorders. *Phytotherapy Research* 24(3): 317–324.

Lonner J. (2007) Medicinal Plant Fact Sheet: *Hydrastis canadensis* / Goldenseal. A collaboration of the IUCN Medicinal Plant Specialist Group, PCA-Medicinal Plant Working

Group and North American Pollinator Protection Campaign. PCA-Medicinal Plant Working Group, Arlington, Virginia. http://pollinator.org/Resources/Hydrastis%20canadensis%20fact%20sheet.pdf (accessed 1 January 2014).

NTP. (2010) Toxicology and carcinogenesis studies of goldenseal root powder (*Hydrastis canadensis*) in F344/N rats and B6C3F1 mice (feed studies). *National Toxicology Program Technical Report Series* 562: 1–188.

Pharmaceutical Press Editorial Team. (2013) *Herbal Medicines*. 4th Edition. Pharmaceutical Press, London, UK.

Saha SK, Sikdar S, Mukherjee A, Bhadra K, Boujedaini N, Khuda-Bukhsh AR. (2013) Ethanolic extract of the Goldenseal, *Hydrastis canadensis*, has demonstrable chemopreventive effects on HeLa cells in vitro: Drug-DNA interaction with calf thymus DNA as target. *Environmental Toxicology and Pharmacology* 36(1): 202–214.

WHO. (2007) *WHO Monographs on Selected Medicinal Plants*. Vol. 3. WHO, Geneva, 390 pp.

Williamson EM, Driver S, Baxter K. (Eds.) (2013) *Stockley's Herbal Medicines Interactions*. 2nd Edition. Pharmaceutical Press, London, UK.

Grapeseed

Vitis vinifera L.

Family: Vitaceae

Other common names: Grapevine seeds

Drug name: Vitis viniferae semen

Botanical drug used: Seeds

Indications/uses: Grapeseed is promoted as a rich antioxidant supplement for the prevention of degenerative disorders such as cardiovascular disease.

Evidence: There is no convincing clinical evidence as yet to support a therapeutic effect on cardiovascular conditions. Further studies are needed to support its possible long-term benefits.

Safety: It is considered to be safe and well tolerated, but limited data are available.

Main chemical compounds: The main active components are flavonoids, which include esters of gallic acid, (+)-catechin, (−)-epicatechin and their galloylated derivatives (e.g. (−)-epicatechin 3-O-gallate) and proanthocyanidins. Polyphenolic stilbene derivative, resveratrol, tocopherols and tocotrienols are also present (Gabetta et al. 2000; Nassiri-Asl and Hosseinzadeh 2009; Williamson et al. 2013).

Clinical evidence:

Chronic venous insufficiency: Extracts of grapeseed proanthocyanidins (GSP) are widely available and have been clinically studied; however, results are inconclusive. A placebo-controlled, double-blind crossover study assessed the effect of daily intake of a wine-grape and grapeseed mix (6 capsules/day providing 800 mg of polyphenols expressed as gallic acid equivalents) on endothelial function and other cardiovascular health markers in healthy participants ($n = 35$; male) for 2 weeks. The grape polyphenols did not improve endothelial function (non-invasive flow-mediated dilation) and related cardiovascular health markers such as blood pressure, platelet function and serum lipids, but further studies are needed in populations with increased cardiovascular risk (van Mierlo et al. 2010).

Effect on hypertension and lipids: A systematic review and meta-analysis based on nine randomised clinical studies found that grapeseed extract had a

---

Phytopharmacy: An evidence-based guide to herbal medicinal products, First Edition.
Sarah E. Edwards, Inês da Costa Rocha, Elizabeth M. Williamson and Michael Heinrich.
© 2015 John Wiley & Sons, Ltd. Published 2015 by John Wiley & Sons, Ltd.

lowering effect on systolic blood pressure as well as heart rate, but this effect was lower than with antihypertensive drugs. The extract failed to show lipid or C-reactive protein (CRP) lowering effects (Feringa et al. 2011). Previous reports on grape juice had shown that its consumption (and also its derived polyphenols) reduced LDL-cholesterol susceptibility to oxidation and increased the plasma total antioxidant capacity in healthy subjects. A recent randomised, double-blind, placebo-controlled, pilot study was conducted on the antioxidant and lipid effects of a whole grape extract (which included seed, skin and pulp; 350 mg/day standardised to 60–70% proathocyanidins for 6 weeks) on pre-hypertensive, overweight, and/or pre-diabetic subjects (24 individuals). The study showed that the extract was able to significantly increase HDL-c while decreasing the total cholesterol/HDL-c ratio, as well as having a positive effect on oxidised LDL in serum (Evans et al. 2014).

Anti-inflammatory effect: A systematic review based on human intervention studies on the use of *V. vinifera* extracts (grapes, grape juice, seeds and grape skin) showed promising results for coping with inflammatory conditions. However, due to the heterogeneity of the extracts used, and the limited number of subjects, further studies are desirable to assess its efficacy (Dell'Agli et al. 2013).

Pre-clinical evidence and mechanisms of action: Grapeseed and GSP extracts have potent antioxidant and cardioprotective effects *in vitro* and *in vivo*. They have shown anti-carcinogenic activity in various tumour models (Nandakumar et al. 2008). Beneficial effects on the cardiovascular system have been reported in several pre-clinical studies via a number of different mechanisms, such as modulating anti-apoptotic genes and by modifying molecular targets (DNA damage/repair and lipid peroxidation) among others (Bagchi et al. 2003), and decreasing low-density lipoprotein–cholesterol oxidation and platelet aggregation (Leifert and Abeywardena 2008). Extracts minimised cardiotoxicity induced by doxorubicin and ciclosporin without attenuating their antineoplastic activity (Li et al. 2009; Ozkan et al. 2012).

Interactions: The concomitant use of grapeseed with vitamin C (ascorbic acid) appeared to increase both systolic and diastolic blood pressure in one clinical study (Ward et al. 2005), although no supporting mechanism was proposed. However, caution is advised in patients with poorly controlled blood pressure taking grapeseed supplements (Williamson et al. 2013).

Contraindications: Grapeseed is not recommended during pregnancy and lactation due to insufficient safety data.

Adverse effects: Grapeseed may cause allergic reactions in sensitive individuals. Adverse effects reported in one clinical trial were suggested to be unrelated to the test product (van Mierlo et al. 2010).

Dosage: Products should be used as recommended by the manufacturer. The dose used in clinical trials suggests up to 400 mg of polyphenol rich grapeseed extract per day (calculated as proanthocyanidin) (Brooker et al. 2006; Sano et al. 2007; Shenoy et al. 2007), or 800 mg gallic acid equivalents per day in products containing grape solids (van Mierlo et al. 2010).

General plant information: The grape vine is native to the Mediterranean region and western Asia, but is now cultivated in temperate regions worldwide. The seeds also contain fixed oil, which is rich in polyunsaturated fatty acids and vitamin E derivatives and has a culinary use.

References

Bagchi D, Sen CK, Ray SD, Das DK, Bagchi M, Preuss HG, Vinson JA. (2003) Molecular mechanisms of cardioprotection by a novel grape seed proanthocyanidin extract. *Mutation Research* 523–524: 87–97.

Brooker S, Martin S, Pearson A, Bagchi D, Earl J, Gothard L, Hall E, Porter L, Yarnold J. (2006) Double-blind, placebo-controlled, randomised phase II trial of IH636 grape seed proanthocyanidin extract (GSPE) in patients with radiation-induced breast induration. *Radiotherapy and Oncology* 79(1): 45–51.

Dell'Agli M, Di Lorenzo C, Badea M, Sangiovanni E, Dima L, Bosisio E, Restani P. (2013) Plant food supplements with anti-inflammatory properties: a systematic review (I). *Critical Reviews in Food Science and Nutrition* 53(4): 403–413.

Evans M, Wilson D, Guthrie N. (2014) A randomized, double-blind, placebo-controlled, pilot study to evaluate the effect of whole grape extract on antioxidant status and lipid profile. *Journal of Functional Foods* 7: 680–691.

Feringa HH, Laskey DA, Dickson JE, Coleman CI. (2011) The effect of grape seed extract on cardiovascular risk markers: a meta-analysis of randomized controlled trials. *Journal of the American Dietetic Association* 111(8): 1173–1181.

Gabetta B, Fuzzati N, Griffini A, Lolla E, Pace R, Ruffilli T, Peterlongo F. (2000) Characterization of proanthocyanidins from grape seeds. *Fitoterapia* 71(2): 162–175.

Leifert WR, Abeywardena MY. (2008) Cardioprotective actions of grape polyphenols. *Nutrition Research* 28(11): 729–737.

Li W, Xu B, Xu J, Wu XL. (2009) Procyanidins produce significant attenuation of doxorubicin-induced cardiotoxicity via suppression of oxidative stress. *Basic and Clinical Pharmacology and Toxicology* 104(3): 192–197.

van Mierlo LA, Zock PL, van der Knaap HC, Draijer R. (2010) Grape polyphenols do not affect vascular function in healthy men. *Journal of Nutrition* 140(10): 1769–1773.

Nandakumar V, Singh T, Katiyar SK. (2008) Multi-targeted prevention and therapy of cancer by proanthocyanidins. *Cancer Letters* 269(2): 378–387.

Nassiri-Asl M, Hosseinzadeh H. (2009) Review of the pharmacological effects of *Vitis vinifera* (Grape) and its bioactive compounds. *Phytotherapy Research* 23(9): 1197–1204.

Ozkan G, Ulusoy S, Alkanat M, Orem A, Akcan B, Ersoz S, Yulug E, Kaynar K, Al S. (2012) Antiapoptotic and antioxidant effects of GSPE in preventing cyclosporine A-induced cardiotoxicity. *Renal Failure* 34(4): 460–466.

Sano A, Uchida R, Saito M, Shioya N, Komori Y, Tho Y, Hashizume N. (2007) Beneficial effects of grape seed extract on malondialdehyde-modified LDL. *Journal of Nutritional Science and Vitaminology* 53(2): 174–182.

Shenoy SF, Keen CL, Kalgaonkar S, Polagruto JA. (2007) Effects of grape seed extract consumption on platelet function in postmenopausal women. *Thrombosis Research* 121(3): 431–432.

Ward NC, Hodgson JM, Croft KD, Burke V, Beilin LJ, Puddey IB. (2005) The combination of vitamin C and grape-seed polyphenols increases blood pressure: a randomized, double-blind, placebo-controlled trial. *Journal of Hypertension* 23(2): 427–434.

Williamson EM, Driver S, Baxter K. (Eds.) (2013) *Stockley's Herbal Medicines Interactions*. 2nd Edition. Pharmaceutical Press, London, UK.

Graviola   *Annona muricata* L.

Family: Annonaceae

Other common names: Brazilian paw paw; corossol; guanábana; prickly custard apple

Drug name: No authorised botanical drug products are available. Marketed on the Internet as 'Triamazon' or graviola

Botanical drug used: Leaves, seeds and bark. The fruit is eaten widely in the tropics and used to produce beverages.

Indications/uses: An extract is being advocated for treatment of cancer, mainly on the Internet. Traditionally, the leaves are taken as a herbal tea for a wide variety of illnesses, including cancer, hypertension, diabetes, fever, indigestion, nervousness, palpitations and skin diseases. It is also applied externally to kill ectoparasites.

Evidence: There is experimental evidence supporting many of the traditional uses, and selective cytotoxicity has been observed *in vitro* in various cancer lines. There is no clinical evidence available.

Safety: The fruit pulp is considered safe (see below) but there are concerns over the use of the leaf, seeds and bark.

Main chemical compounds: The leaves, seeds and bark contain annonaceous acetogenins, including annonacin, annonacinone, annomontacin, murisolin and a series of muricins (Liaw et al. 2010; McLaughlin 2008); and alkaloids including the benzylisoquinolines reticuline, annonaine, and *N*-methylcoculaurine and an aporphine, annonamine (Matsushige et al. 2012).

Clinical evidence: None, for any indication. Claims of potent anticancer effects are based on laboratory tests using cancer cell lines (McLaughlin 2008).

Pre-clinical evidence and mechanisms of action: The cytotoxic activity observed in cancer cell lines is mainly due to the presence of acetogenins (Liaw et al. 2010). Annonacin is an inhibitor of mitochondrial complex I and as such is

---

Phytopharmacy: An evidence-based guide to herbal medicinal products, First Edition.
Sarah E. Edwards, Inês da Costa Rocha, Elizabeth M. Williamson and Michael Heinrich.
© 2015 John Wiley & Sons, Ltd. Published 2015 by John Wiley & Sons, Ltd.

widely used as a biochemical tool to induce experimental Parkinsonism in animal models. The alkaloids are also known neurotoxins and have been shown to inhibit dopamine synthesis (Matsushige et al. 2012).

Interactions: None known.

Contraindications: Leaves, seeds and bark are not recommended for internal use. The fruit is considered safe if the seeds are removed, or remain whole when consumed (i.e. not ground or fragmented). The safety of topical application of the seeds, leaf or bark extract is unknown, but as these are not ingested, nor used over long periods of time, are likely to be relatively harmless.

Adverse effects: The regular high consumption of graviola leaves is associated with a high incidence of atypical Parkinson's disease (PD), or progressive supranuclear palsy, in countries where the leaves are taken as a herbal tea. This has been linked to the high concentration of the acetogenins and alkaloids present in the leaves, seeds and bark (Champy et al. 2004; Schapira 2010). There may be a genetic susceptibility in certain populations to this condition – but in view of the mounting epidemiological and experimental evidence demonstrating a link with PD, the herb cannot be considered safe.

Dosage: Not recommended, so a dose cannot be given.

General plant information: Soursop fruit is widely eaten in tropical areas without reported adverse effects, as the flesh contains low levels of the toxins and the seeds are indigestible.

References

Champy P, Höglinger GU, Féger J, Gleye, C, Hocquemiller R, Laurens A, Guérineau V, Laprévote O, Medja F, Lombès A, Michel PP, Lannuzel A, Hirsch EC, Ruberg M. (2004) Annonacin, a lipophilic inhibitor of mitochondrial complex I, induces nigral and striatal neurodegeneration in rats: possible relevance for atypical Parkinsonism in Guadeloupe. *Journal of Neurochemistry* 88(1): 63–69.
Liaw CC, Wu TY, Chang FR, Wu YC. (2010) Historic perspectives on Annonaceous acetogenins from the chemical bench to preclinical trials. *Planta Medica* 76(13): 1390–1404.
Matsushige A, Kotake Y, Matsunami K, Otsuka H, Ohta S, Takeda Y. (2012) Annonamine, a new aporphine alkaloid from the leaves of *Annona muricata*. *Chemical and Pharmaceutical Bulletin (Tokyo)*. 60(2): 257–259.
McLaughlin JL. (2008) Paw paw and cancer: annonaceous acetogenins from discovery to commercial products. *Journal of Natural Products* 71(7): 1311–1321.
Schapira AH. (2010) Complex I: inhibitors, inhibition and neurodegeneration. *Experimental Neurology* 224(2): 331–335.

Green Tea   *Camellia sinensis* (L.) Kuntze

Synonyms: *Thea sinensis* L.; *T. viridis* L.

Family: Theaceae

Other common names: Chinese tea

Drug name: Theae viridis folium

Botanical drug used: Unfermented dried leaf

Indications: Taken internally, green tea and its preparations have been widely acclaimed as beneficial for a wide range of indications, mainly weight loss, cognitive improvement and the prevention of degenerative diseases such as cancer and cardiovascular disease, and to maintain bone mineralisation. Green tea polyphenols have antibacterial and antiviral effects and an extract has been licensed for the topical treatment of condylomata acuminata.

Evidence: Limited evidence base data are available for most of the indications. Further clinical studies are warranted.

Safety: Considered to be safe: very many people consume green tea, but the safety of high doses of concentrated extracts has not been confirmed.

Main chemical compounds: The main constituents are polyphenols, gallic acid and catechins. The main catechins in green tea are epicatechin, epicatechin-3-gallate, epigallocatechin and epigallocatechin-3-gallate (EGCG). Other constituents present are caffeine, xanthines (theophylline and theobromine), flavonoids (theaflavin), tannins (proanthocyanidins), polysaccharides and vitamins (Williamson et al. 2013).

Clinical evidence:

Weight loss: The use of green tea preparations for weight loss and weight maintenance in overweight or obese adults has been assessed in a Cochrane review (Jurgens et al. 2012), but due to the heterogeneity of the studies no clinical relevant claims could be made for green tea preparations. These products do appear to induce a small, statistically non-significant, weight loss in overweight or obese adults.

Cancer prevention: A Cochrane review on green tea consumption for cancer prevention assessed all prospective, controlled interventional studies and observational

---

Phytopharmacy: An evidence-based guide to herbal medicinal products, First Edition.
Sarah E. Edwards, Inês da Costa Rocha, Elizabeth M. Williamson and Michael Heinrich.
© 2015 John Wiley & Sons, Ltd. Published 2015 by John Wiley & Sons, Ltd.

studies that either assessed the associations between green tea consumption and risk of cancer incidence, or reported on cancer mortality. The review concluded that there is insufficient data as yet to make recommendations as to whether green tea has cancer preventative properties (Boehm et al. 2009). Further studies are needed.

Cardiovascular health: Regular consumption of green tea has been associated with enhanced cardiovascular and metabolic health, but further clinical studies are needed. In a double-blind, placebo-controlled trial with obese, hypertensive subjects receiving a daily supplement of 1 capsule of green tea extract (379 mg) for 3 months, it reduced blood pressure compared to placebo, along with reductions in fasting serum glucose and insulin levels and insulin resistance (Bogdanski et al. 2012). Another study with habitual tea drinkers in China, green or oolong (120 ml/day or more for 1 year) also showed that tea drinking significantly reduced the risk of developing hypertension (Yang et al. 2004). The epigallocatechin-3-gallate (EGCG), the major catechin found in green tea, is thought to be beneficial in insulin resistance and other metabolic parameters (Brown et al. 2009).

Bone health: Epidemiological evidence has shown an association between tea consumption and the prevention of age-related bone loss in elderly women and men. Tea and its active components may decrease the risk of fracture by improving bone mineral density and supporting osteoblastic activities while suppressing osteoclastic activities (Shen et al. 2009). Green tea polyphenols, in combination with Tai Chi exercise, have been shown to mitigate oxidative damage in post-menopausal women with osteopenia (Qian et al. 2012), but clinical studies remain to be carried out to clarify whether supplementation with green tea is beneficial for maintaining bone mineralisation in osteoporosis.

External application for warts: Clinical studies have showed that the topical application of a green tea polyphenol extract was efficacious and safe to treat condylomata acuminata (external genital and perianal warts) (Stockfleth et al. 2008; Tatti et al. 2008).

Pre-clinical evidence and mechanisms of action: Green tea has been the subject of an immense amount of pharmacological research both *in vitro* and in animal models and many of the exact mechanisms of action remain to be elucidated. The following reviews can be consulted for further detail on activity and mechanisms: weight loss, Thavanesan (2011); anticancer effects, Singh et al. (2011); cardiovascular properties, Hodgson and Croft (2010); bone health, Shen et al. (2011); anti-infective, Steinmann et al. (2013).

Interactions: Tea contains significant amounts of caffeine, therefore the interactions of caffeine should be considered (e.g. with benzodiazepines and related drugs), unless the product is stated to be decaffeinated. An excess of caffeine consumption can cause adverse effects such as headache, jitteriness (anxiety), restlessness and insomnia. In those cases, reducing caffeine intake is recommended. Tea may have some antiplatelet effects that may be additive with those of conventional antiplatelet drugs (e.g. aspirin, clopidogrel). Patients should discuss any episode of prolonged bleeding with a health-care professional. Tea may interact with warfarin and related drugs as case reports suggested that tea may reduce the INR in response to warfarin, but modest consumption is unlikely to cause any problems (Williamson et al. 2013). Pre-clinical data indicates possible interactions with chemotherapeutic drugs such as tamoxifen (Shin and Choi 2009) and bortezomib (Golden et al. 2009); green tea supplements should probably be avoided in individuals taking these.

Contraindications: Any conditions in which caffeine is contraindicated; pregnant and lactating women should limit their intake as caffeine passes into breast milk and may cause sleep disorders or insomnia in nursing infants. Green tea consumption has been linked with iron deficiency effect and a recent review concluded that where people have adequate iron stores, there is no problem, but in populations with marginal iron status there appears to be a negative association between tea consumption and iron status (Temme and Van Hoydonck 2002).

Adverse effects: Although green tea is consumed globally and considered very safe, the effects of high doses and prolonged use of extracts is not known and concerns have been raised about the promotion of these; for review, see Schönthal (2011). Nausea, rash and gastrointestinal problems have been reported with high doses of supplements. Due to the caffeine content, insomnia, irritability and nervousness may also occur.

Dosage: Doses vary significantly ranging between 1 and 10 cups tea daily. Desirable green tea intake is from 3 to 5 cups per day (up to 1200 ml/day), providing a minimum of 250 mg/day catechins. Commercially available green tea capsules recommend a daily dose of 100–150 mg (Shord et al. 2009).

General plant information: Tea is one of the main beverages widely consumed in the world. It is produced from the leaves of *C. sinensis* prepared in different ways. Green tea (Theae viridis folium) is non-fermented, where the fresh leaves have been steamed and dried immediately after picking, to deactivate the enzymes; black tea (Theae nigrae folium) leaves are fermented prior to drying and steaming, Oolong tea is partly fermented and pu-erh tea is drastically fermented and aged (Cao et al. 2013). These are all used medicinally, mainly in Asia. Green tea is predominantly produced in China and Japan, while black tea is predominantly produced in India, Sri Lanka and Kenya (Williamson et al. 2013). 'White tea' and 'yellow tea' are specialist teas and have not been evaluated for their health benefits yet.

References

The literature on green tea is immense. Here we have included mainly clinical assessments and provided a selection of papers and reviews that may be consulted for further pre-clinical information.

Boehm K, Borrelli F, Ernst E, Habacher G, Hung SK, Milazzo S, Horneber M. (2009) Green tea (*Camellia sinensis*) for the prevention of cancer. *Cochrane Database of Systematic Reviews* 3: CD005004.

Bogdanski P, Suliburska J, Szulinska M, Stepien M, Pupek-Musialik D, Jablecka A. (2012) Green tea extract reduces blood pressure, inflammatory biomarkers, and oxidative stress and improves parameters associated with insulin resistance in obese, hypertensive patients. *Nutrition Research* 32(6): 421–427.

Brown A L, Lane J, Coverly J, Stocks J, Jackson S, Stephen A, Bluck L, Coward A, Hendrickx H. (2009) Effects of dietary supplementation with the green tea polyphenol epigallocatechin-3-gallate on insulin resistance and associated metabolic risk factors: randomized controlled trial. *British Journal of Nutrition* 101(6): 886–894.

Cao ZH, Yang H, He ZL, Luo C, Xu ZQ, Gu DA, Jia JJ, Ge CR, Lin QY. (2013) Effects of aqueous extracts of raw pu-erh tea and ripened pu-erh tea on proliferation and differentiation of 3T3-L1 preadipocytes. *Phytotherapy Research* 27(8): 1193–1199.

Golden EB, Lam PY, Kardosh A, Gaffney KJ, Cadenas E, Louie SG, Petasis NA, Chen TC, Schonthal AH. (2009) Green tea polyphenols block the anticancer effects of bortezomib and other boronic acid-based proteasome inhibitors. *Blood* 113(23): 5927–5937.

Hodgson JM, Croft KD. (2010) Tea flavonoids and cardiovascular health. *Molecular Aspects of Medicine* 31(6): 495–502.

Jurgens TM, Whelan AM, Killian L, Doucette S, Kirk S, Foy E. (2012) Green tea for weight loss and weight maintenance in overweight or obese adults. *Cochrane Database of Systematic Reviews* 12: CD008650.

Qian G, Xue K, Tang L, Wang F, Song X, Chyu MC, Pence BC, Shen CL, Wang JS. (2012) Mitigation of oxidative damage by green tea polyphenols and Tai Chi exercise in postmenopausal women with osteopenia. *PLoS One* 7(10): e48090.

Schönthal AH. (2011) Adverse effects of concentrated green tea extracts. *Molecular Nutrition & Food Research* 55(6): 874–885.

Shen CL, Yeh JK, Cao JJ, Wang JS. (2009) Green tea and bone metabolism. *Nutrition Research* 29(7): 437–456.

Shen CL, Yeh JK, Cao JJ, Chyu MC, Wang JS. (2011) Green tea and bone health: Evidence from laboratory studies. *Pharmacological Research* 64(2): 155–161.

Shin SC, Choi JS. (2009) Effects of epigallocatechin gallate on the oral bioavailability and pharmacokinetics of tamoxifen and its main metabolite, 4-hydroxytamoxifen, in rats. *Anticancer Drugs* 20(7): 584–588.

Shord SS, Shah K, Lukose A. (2009) Drug-botanical interactions: a review of the laboratory, animal, and human data for 8 common botanicals. *Integrative Cancer Therapies* 8(3): 208–227.

Singh BN, Shankar S, Srivastava RK. (2011) Green tea catechin, epigallocatechin-3-gallate (EGCG): mechanisms, perspectives and clinical applications. *Biochemical Pharmacology* 82(12): 1807–1821.

Steinmann J, Buer J, Pietschmann T, Steinmann E. (2013) Anti-infective properties of epigallocatechin-3-gallate (EGCG), a component of green tea. *British Journal of Pharmacology* 168(5): 1059–1073.

Stockfleth E, Beti H, Orasan R, Grigorian F, Mescheder A, Tawfik H, Thielert C. (2008) Topical Polyphenon E in the treatment of external genital and perianal warts: a randomized controlled trial. *British Journal of Dermatology* 158(6): 1329–1338.

Tatti S, Swinehart JM, Thielert C, Tawfik H, Mescheder A, Beutner KR. (2008) Sinecatechins, a defined green tea extract, in the treatment of external anogenital warts: a randomized controlled trial. *Obstetrics & Gynecology* 111(6): 1371–1379.

Temme EH, Van Hoydonck PG. (2002) Tea consumption and iron status. *European Journal of Clinical Nutrition* 56(5): 379–386.

Thavanesan N. (2011) The putative effects of green tea on body fat: an evaluation of the evidence and a review of the potential mechanisms. *British Journal of Nutrition* 106(9): 1297–1309.

Williamson EM, Driver S, Baxter K. (Eds.) (2013) *Stockley's Herbal Medicines Interactions.* 2nd Edition. Pharmaceutical Press, London UK.

Yang YC, Lu FH, Wu JS, Wu CH, Chang CJ. (2004) The protective effect of habitual tea consumption on hypertension. *Archives of Internal Medicine* 164(14): 1534–1540.

♦ ♦ ♦ ♦ ♦

Hawthorn   *Crataegus spp.* (mainly *C. laevigata* (Poir.) DC., *C. monogyna* Jacq., *C. rhipidophylla* Gand., and their hybrids)

Synonyms: *C. oxyacantha* L. (= *C. rhipidophylla* Gand.)

Family: Rosaceae

Other common names: Bread-and-cheese

Drug name: Crataegi folium cum flore; Crataegi fructus

Note: C. pinnatifida Bunge is used in Chinese medicine.

Botanical drug used: Flowering branch; fruit (berries)

Indications/uses: Improvement of blood circulation, mild heart failure, mild antihypertensive and antisclerotic, and as a general heart tonic.

Evidence: Overall, a large number of clinical studies have shown that hawthorn may represent an effective agent in the treatment of mild cardiovascular disease and ischaemic heart disease.

Safety: Overall, considered to be safe, although there is no data for use in pregnancy. Hawthorn preparations are not suitable for self-medication of heart failure and hypertension; consequently, currently there are no over-the-counter (OTC) preparations on the UK market, but such OTC drugs are available in some Continental European markets.

Main chemical compounds: The major constituents of the leaf and flower are flavonoid glycosides, and include kaempferol, quercetin, apigenin, luteolin, vitexin and orientin glycosides, schaftoside, isoschaftoside, vivenins and many others, with some procyanidins. Other compounds include triterpenes (ursolic acid, oleanolic acid and crataegolic acid), phenolic acids (chlorogenic acid and caffeic acid), amines (choline), xanthines and minerals.

The fruits (berries) contain lower levels of flavonoids and higher concentrations of procyanidins, incuding oligomeric compounds mainly composed of epicatechin units, and triterpene acids (ursolic, crataegolic and others) (Edwards et al. 2012; ESCOP 2003; Upton 1999a,b).

Phytopharmacy: An evidence-based guide to herbal medicinal products, First Edition.
Sarah E. Edwards, Inês da Costa Rocha, Elizabeth M. Williamson and Michael Heinrich.

Clinical evidence: A Cochrane review concluded that hawthorn extracts showed a significant benefit in symptom control (shortness of breath and fatigue) and physiological outcomes (exercise tolerance and pressure-heart rate), compared with placebo, as an adjunctive treatment for chronic heart failure (classes I–III) (Pittler et al. 2008). A more recent review points out that there has been a large consistent positive outcome from clinical trials using hawthorn extracts for cardiovascular disease but they have shown inconsistency in terms of criteria used (sample size, preparation, dosage, among others) (Koch and Malek 2011). Mild antihypertensive, but clinically significant, effects have been seen in small human trials (e.g. Walker et al. 2002).

Hawthorn preparations are not suitable for use as a self-medication for heart failure and hypertension. These indications should be managed by a qualified medical professional.

Pre-clinical evidence and mechanisms of action: The exact mechanisms of action are not clear but many relevant pharmacological effects *in vivo* and *in vitro* have been described. Inhibition of cAMP-phosphodiesterase, sodium and calcium-ATP-ase activity and thromboxane (TXA_2) synthesis, and stimulation of prostacyclin synthesis, antioxidant activity and inhibition of human neutrophil elastase have all been reported. Hypotensive activity via vasorelaxation is thought to result from an increase in nitrous oxide production and a tonic action on cardiac myocytes (ESCOP 2003; Koch and Malek 2011; Upton 1999a,b).

Interactions: None known, but in one clinical study hawthorn showed additive blood-pressure-lowering effects when taken with conventional anti-hypertensives. However, this effect was small and may not be clinically relevant (Williamson et al. 2013).

Contraindications: Not recommended during pregnancy and lactation, as safety has not been established.

Adverse effects: These are mild and transient, and include digestive complaints, nausea, dizziness, sweating and rash (Daniele et al. 2006; Pittler et al. 2008; Upton 1999a,b). A physician should be consulted if symptoms worsen, or remain unchanged for longer than 6 weeks, since this is indicative of more serious heart disease. If oedema occurs in the legs or if pain occurs in the region of the heart, spreading out to the arms, upper abdomen or neck area, or in cases of respiratory distress (e.g. dyspnoea), medical attention should be sought immediately (WHO 2004).

Dosage: Daily doses have been reported to be between 160 and 1800 mg of dried ethanol or methanol extracts for 3–24 weeks. Commercially available products, standardised to oligomeric procyanidins or flavonoids, are available in other parts of the European Union and manufacturers' instructions should be followed (WHO 2004; Williamson et al. 2013).

General plant information: Distributed throughout temperate areas of the Northern Hemisphere (WHO 2004). The leaves and flowers have been used as a remedy for heart disorders since the 19th century. Hawthorn berries are eaten in syrups and jams (Anon. 2010; Upton 1999a,b).

References

The literature on hawthorn is extensive. This is a selection of references and reviews in support of the information given above.

Anon. (2010) *Crataegus oxyacantha* (Hawthorn). Monograph. *Alternative Medicine Review* 15(2): 164–167.

Daniele C, Mazzanti G, Pittler MH, Ernst E. (2006) Adverse-event profile of *Crataegus spp.*: a systematic review. *Drug Safety* 29(6): 523–535.

Edwards JE, Brown PN, Talent N, Dickinson TA, Shipley PR. (2012) A review of the chemistry of the genus *Crataegus. Phytochemistry* 79: 5–26.

ESCOP. (2003) *ESCOP Monographs: The Scientific Foundation for Herbal Medicinal Products.* 2nd Edition. Thieme, Exeter and London, UK.

Koch E, Malek FA. (2011) Standardized extracts from hawthorn leaves and flowers in the treatment of cardiovascular disorders – preclinical and clinical studies. *Planta Medica* 77(11): 1123–1128.

Pittler MH, Guo R, Ernst E. (2008) Hawthorn extract for treating chronic heart failure. *Cochrane Database of Systematic Reviews* 1: CD005312.

Upton R. (Ed.) (1999a). Hawthorn Leaf with Flower. *American Herbal Pharmacopoeia and Therapeutic Compendium.* AHP, Santa Cruz, USA.

Upton R. (Ed.) (1999b). Hawthorn Berry. *American Herbal Pharmacopoeia and Therapeutic Compendium.* AHP, Santa Cruz, USA.

Walker AF, Marakis G, Morris AP, Robinson PA. (2002) Promising hypotensive effect of hawthorn extract: A randomized double-blind pilot study of mild, essential hypertension. *Phytotherapy Research* 16(1): 48–54.

WHO. (2004) WHO. *Monographs on Selected Medicinal Plants.* Vol. 2. WHO, Geneva, 358 pp.

Williamson EM, Driver S, Baxter K. (Eds.) (2013) *Stockley's Herbal Medicines Interactions.* 2nd Edition. Pharmaceutical Press, London, UK.

Holy Basil   *Ocimum tenuiflorum* L.

Synonyms: *O. sanctum* L; O. tomentosum Lam.

Family: Lamiaceae (Labiatae)

Other common names: Sacred basil; Tulsi

Drug name: Ocimi sancti folium

Botanical drug used: Most commonly the leaves, but the aerial parts or the seed oil are also used.

Indications/uses: The herb is used to combat stress and as a general tonic, for managing diabetes and its complications, as an anti-inflammatory agent, to treat respiratory conditions, especially cough, sore throat and asthma and for oral hygiene where the fresh leaves can be chewed with 'misiri' (crystallised sugar). The essential oil is used as a mosquito repellent.

Evidence: Overall, there is limited evidence for any clinical effectiveness for treating symptoms of stress conditions or diabetes. Holy basil products may be effective in oral hygiene due to their eugenol content, which as a well-known antiseptic and breath-freshener is already widely used in such products.

Safety: There is a lack of evidence in terms of its safety, but there are concerns about the constituents of the essential oil, especially methyl-eugenol. A study on commercial extracts of the plant, which may have low levels of essential oil, found it to be devoid of mutagenic, genotoxic or acute toxic effects in mice.

Main chemical compounds: The chemical composition varies depending on the cultivar, growing conditions, age of the plant and other factors. Holy basil contains essential oil composed of eugenol (ca 70%), methyl eugenol (approx. 20%), β-caryophyllene, carvacrol, methyl chavicol, linalool and others; flavonoids including apigenin, vicenin, cirsilineol and quercetin; polyphenolic acids including rosmarinic, protocatechuic, caffeic and chlorogenic acids; triterpenes including stigmasterol and ursolic acid; ocimarin (a coumarin), and ocimumosides A and B, which are glycoglycerolipids (Ahmad et al. 2012; WHO 2004; Williamson 2002).

Clinical evidence:

Overall, although there is a large body of pre-clinical work on this herb, it is not possible to draw clear conclusions on specific therapeutic effects from clinical studies (Engels and Brinckmann 2013).

Phytopharmacy: An evidence-based guide to herbal medicinal products, First Edition.
Sarah E. Edwards, Inês da Costa Rocha, Elizabeth M. Williamson and Michael Heinrich.
© 2015 John Wiley & Sons, Ltd. Published 2015 by John Wiley & Sons, Ltd.

Stress and related conditions: In a randomised, double-blind, placebo-controlled six-week study using an extract of the whole plant (1200 mg/day, one capsule after breakfast and two capsules after dinner, for a maximum of 6 weeks), the severity of self-reported stress-related symptoms at weeks 0, 2, 4 and 6 of the trial was reduced significantly. Forgetfulness, sexual problems, feelings of exhaustion and sleep problems of recent origin decreased significantly ($p \leq 0.05$) during this period (Saxena et al. 2012).

In a double-blinded randomised controlled cross-over trial in healthy volunteers, an ethanolic extract of Tulsi leaves (300 mg capsules) resulted in immunomodulatory effects (Mondal et al. 2011).

Oral hygiene: In a small cross over design clinical study in children, a holy basil preparation reduced salivary levels of *Streptococcus mutans*, comparable to that of chlorhexidine and Listerine treatment (Agarwal and Nagesh 2011).

Antidiabetic effects: A small randomised placebo-controlled, single blind trial of holy basil leaves in patients with non-insulin-dependent diabetes found a modest reduction in blood sugar levels (Agrawal et al. 1996).

Pre-clinical evidence and mechanisms of action:

Stress and related conditions: Ocimumosides A and B have anti-stress effects in a chronic unpredictable stress model in rats, produced via modulation of brain monoamines and antioxidant systems (Ahmad et al. 2012). The antioxidant properties have frequently been described (e.g. Williamson 2002).

Oral hygiene: The antibacterial and anti-inflammatory properties of essential oils, including that of holy basil, are well known. *Ex-vivo* and animal model studies have confirmed these properties for the essential oil extract, for its proposed use as a dental agent (Giridharan et al. 2011).

Antidiabetic effects: A tetracyclic triterpenoid [16-hydroxy-4,4,10,13-tetramethyl-17-(4-methyl-pentyl)-hexadecahydro-cyclopenta[a]phenanthren-3-one] was found to possess potent antidiabetic properties in alloxan-induced diabetic rats, ameliorating glucose and lipid parameters including total cholesterol, triglycerides, and low and high density lipoprotein cholesterol (Patil et al. 2011).

Anticancer effects: Several studies have shown potential anticancer properties, but there is no clinical evidence to support these effects (Baliga et al. 2013; Shimizu et al. 2013).

Interactions with other drugs: No robust data are available. Holy basil leaf can modestly reduce blood glucose levels so caution should be observed when taking hypoglycaemic drugs concurrently (Williamson et al. 2013)

Adverse effects: A study on a commercial extract of the plant (which may have low levels of essential oil), showed it to be devoid of mutagenic, genotoxic or acute toxic effects in mice (Chandrasekaran et al. 2013).

The presence of methyl-eugenol, often in relatively high concentrations, could potentially give rise to concerns. The compound is known to be genotoxic in *in vivo* models, for example, by inducing DNA damage in rat liver and causing oxidative DNA damage. This was at oral doses of 400 and 1000 mg/kg in male rats, unlikely to be encountered clinically (Ding et al. 2011). Higher doses of eugenol (400–600 mg/kg body weight) have been reported to produce liver damage in mice (WHO 2004). Insufficient data on carcinogenesis, mutagenesis and on the teratogenicity/impairment of fertility are available and thus the herb does not hold GRAS (Generally Recognised as Safe) status in the United States (Engels and Brinckmann 2013). The herb is contraindicated in lactating mothers, pregnant women and children due to insufficient information.

Dosage: A dose of 1–2.5 g daily of dried herb is used, in the form of a dried powder, or equivalent extract, or as a herbal tea.

General plant information: Within Hindu culture, tulsi ('the incomparable one') is a holy plant, the incarnation of the Hindu goddess Tulsi. It is commonly grown and adorned in British Indian households, where it is placed in the centre of the house or in the courtyard. It is native to India, parts of north and eastern Africa, and southern parts of China, but is widely cultivated.

References

Agrawal P, Rai V, Singh RB. (1996) Randomized placebo-controlled, single blind trial of holy basil leaves in patients with noninsulin-dependent diabetes mellitus. *International Journal of Clinical Pharmacology and Therapeutics* 34(9): 406–409.

Agarwal P, Nagesh L. (2011) Comparative evaluation of efficacy of 0.2% Chlorhexidine, Listerine and Tulsi extract mouth rinses on salivary *Streptococcus mutans* count of high school children-RCT. *Contemporary Clinical Trials* 32(6): 802–808.

Ahmad A, Rasheed N, Gupta P, Singh S, Siripurapu KB, Ashraf GM, Kumar R, Chand K, Maurya R, Banu N, Al-Sheeha M, Palit G. (2012) Novel Ocimumoside A and B as anti-stress agents: modulation of brain monoamines and antioxidant systems in chronic unpredictable stress model in rats. *Phytomedicine* 19(7): 639–647.

Baliga MS, Jimmy R, Thilakchand KR, Sunitha V, Bhat NR, Saldanha E, Rao S, Rao P, Arora R, Palatty PL. (2013) *Ocimum sanctum* L. (Holy Basil or Tulsi) and its phytochemicals in the prevention and treatment of cancer. *Nutrition and Cancer* 65(Suppl 1): 26–35.

Chandrasekaran CV, Srikanth HS, Anand MS, Allan JJ, Viji MM, Amit A. (2013) Evaluation of the mutagenic potential and acute oral toxicity of standardized extract of *Ocimum sanctum* (OciBest™). *Human & Experimental Toxicology* 32(9): 992–1004.

Ding W, Levy DD, Bishop ME, Lascelles ELC, Kulkarni R, Chang CW, Aidoo A, Manjanatha MG. (2011) Methyleugenol genotoxicity in the Fischer 344 rat using the comet assay and pathway-focused gene expression profiling. *Toxicological Sciences* 123(1): 103–112.

Engels, G, Brinckmann, J. (2013) Holy Basil. *HerbalGram* 98: 1–5.

Giridharan VV, Thandavarayan RA, Mani V, Ashok Dundapa T, Watanabe K, Konishi T. (2011) *Ocimum sanctum* Linn. leaf extracts inhibit acetylcholinesterase and improve cognition in rats with experimentally induced dementia. *Journal of Medicinal Food* 14(9): 912–919.

Mondal S, Varma S, Bamola VD, Naik SN, Mirdha BR, Padhi MM, Mehta N, Mahapatra SC. (2011) Double-blinded randomized controlled trial for immunomodulatory effects of Tulsi (*Ocimum sanctum* Linn.) leaf extract on healthy volunteers. *Journal of Ethnopharmacology* 136(3): 452–456.

Patil R, Patil R, Ahirwar B, Ahirwar D. (2011) Isolation and characterization of anti-diabetic component (bioactivity-guided fractionation) from *Ocimum sanctum* L. (Lamiaceae) aerial part. *Asian Pacific Journal of Tropical Medicine* 4(4): 278–282.

Saxena RC, Singh R, Kumar P, Singh Negi MP, Saxena VS, Geetharani P, Allan JJ, Venkateshwarlu KA. (2012) Efficacy of an Extract of *Ocimum tenuiflorum* (OciBest) in the Management of General Stress: A Double-Blind, Placebo-Controlled Study. *Evidence-Based Complementary and Alternative Medicine* 2012: 894509.

Shimizu T, Torres MP, Chakraborty S, Souchek JJ, Rachagani S, Kaur S, Macha M, Ganti AK, Hauke RJ, Batra SK. (2013) Holy Basil leaf extract decreases tumorigenicity and metastasis of aggressive human pancreatic cancer cells *in vitro* and *in vivo*: Potential role in therapy. *Cancer Letters* 336(2): 270–280.

WHO. (2004) *WHO monographs on medicinal plants*. Vol. 2. WHO, Geneva, 358 pp.
Williamson EM (Ed.) (2002). *Major Herbs of Ayurveda*. Dabur Research Foundation. Churchill Livingstone Elsevier, UK.
Williamson EM, Driver S, Baxter K. (Eds.) (2013) *Stockley's Herbal Medicines Interactions*. 2nd Edition. Pharmaceutical Press, UK.

♠ ♠ ♠ ♠ ♠

Hoodia *Hoodia gordonii* (Masson) Sweet ex Decne, and other *Hoodia spp.*

Synonyms: *Scytanthus gordonii* (Masson) Hook; *Stapelia gordonii* Masson

Family: Apocynaceae (previously in Asclepiadaceae)

Other common names: Bushman's hat; hoodia 'cactus'; Kalahari 'cactus'; xhoba

Botanical drug used: Stem

Indications/uses: Appetite suppressant for weight loss and, more recently, for treating diabetes.

Evidence: Limited clinical and pre-clinical data supports the traditional use of hoodia as an appetite suppressant. Based on the existing data and a risk-benefit assessment, its use cannot be recommended.

Safety: Limited safety or toxicity data are available. There are concerns about the safety profile of preparations derived from hoodia. Increases in blood pressure, pulse, heart rate and bilirubin were reported in one hoodia extract trial. Many Hoodia products have been found to be adulterated with other ingredients, which may pose potential safety risks. Clearly, hoodia should not be used during pregnancy and lactation. Long-term or excessive doses of hoodia are not recommended.

Main chemical compounds: *H. gordonii* extract has been extensively characterised; chemical profiles of other *Hoodia* species are less well known. Main constituents of *H. gordonii* are steroidal (pregnane) glycosides (>70% w/w), based on hoodigogenin A and calogenin, including the hoodigosides, the hoodistanalosides and the gordonosides. The isolated appetite-suppressant component, a minor constituent, was identified as a triglycoside of 12β-tigloyloxy-14β-hydroxypregn-5-en-20-one (known as 'P57'). Several other closely related glycosides have been characterised, all of them containing 6-deoxy- and 2,6-dideoxy sugars. *H. gordonii* extracts also contain fatty acids (e.g. myristic acid, palmitic acid, stearic acid, oleic acid, and linoleic acids); sterols (e.g. cholesterol, β-sitosterol, stigmasterol); α-tocopherol; and alcohols (van Heerden 2008; Russell and Swindells 2012; Shukla et al. 2009).

Phytopharmacy: An evidence-based guide to herbal medicinal products, First Edition.
Sarah E. Edwards, Inês da Costa Rocha, Elizabeth M. Williamson and Michael Heinrich.

Clinical evidence: The main focus of research into hoodia has been on the patented extract enriched with the pregnane glycoside P57, which has been the subject of controversial intellectual property rights and benefit sharing issues. Development of hoodia extracts by Unilever for use in functional food products was abandoned in 2008 due to safety and efficacy concerns (Vermaak et al. 2011). Few clinical studies on hoodia extracts have been published. A number of unpublished clinical studies (using crude extracts and concentrated active ingredients of *Hoodia*) undertaken by the pharmaceutical/functional foods industry are available from the South African Council for Scientific and Industrial Research. While evidence of appetite-suppressant effects was observed in some (but not all) of these trials, adverse effects, including hyperbilirubinaemia, and tolerability issues were reported in some individuals administered the concentrated active ingredient extracts (CSIR 2011).

In a published randomised controlled-trial in 49 healthy overweight women, administration of purified hoodia extract (1110 mg twice per day for 15 days) did not result in any significant reduction in energy intakes or body weights relative to placebo. However, the hoodia preparation resulted in a significant ($p < 0.05$) increase in adverse effects, including increase in blood pressure, pulse, heart rate, bilirubin and alkaline phosphatase, and was less tolerated than placebo (Blom et al. 2011).

Other published studies include an observational pilot study on obese patients administered a hoodia product (containing 400 mg *H. gordonii*) twice daily for 4 weeks. All subjects reported reduction in appetite (500–1000 calories per day) and weight loss (2–15 lb). In a double-blind, placebo-controlled study, treatment with *H. gordonii* resulted in a reduction of food intake up to 1000 calories per day and body weight loss of about 2 kg, with a decrease in blood glucose and triglycerides (Jain and Singh 2013).

Even though some clinical data are available, there are no products on the market that comply with European quality requirements and there are serious concerns about the level of adulteration in unlicensed 'hoodia' products.

Pre-clinical evidence and mechanisms of action: In a study in pregnant mice, administration of hoodia extract (15 mg/kg/day or 50 mg/kg/day) resulted in a dose-dependent decrease in feed consumption and body weight gain, which was also associated with reduced foetal and uterine weights. No foetal malformations were found associated with hoodia, although some effects on ossification parameters were seen. No observed adverse effects related to treatment were observed at 5 mg/kg/day (Dent et al. 2012a). In a similar study, pregnant rabbits administered 6 or 12 mg/kg/day hoodia extract resulted in a dose-dependent reduction in feed intake and bodyweight gain. No effects on embryonic or foetal development at 12 mg/kg/day were observed (Dent et al. 2012b).

In an *in vivo* study in rats, a dose-dependent reduction in food intake (12–26%) was observed at a dose of 100 and 150 mg/kg body weight ($p < 0.05$). A significant increase in hepatic glycogen levels, activity of mitochondrial CPT-1 and thyroid hormones, and changes in levels of appetite regulatory peptides were found in the hoodia-treated rats. While the precise mechanism of action of hoodia is uncertain, the authors postulated that the hormonal and metabolic changes induced by the hoodia extract may be responsible for its anorectic activity (Jain and Singh 2013).

H. gordonii extract was found to be non-genotoxic in three independent assays (a bacterial mutant test, a gene mutation assay using mouse lymphoma cells, and a bone marrow micronucleus assay in the mouse) (Scott et al. 2012). However, in a study in mice, doses of 200 mg/kg or more of hoodia extract were associated with acute hepatoxicity (Vermaak et al. 2011).

The absorption of P57 has been found to be higher when administered as part of the extract than when given alone (Vermaak et al. 2011b).

Interactions: None known.

Contraindications: Its use should not be recommended. Due to lack of specific safety data, hoodia should not be used during pregnancy or lactation, or in under 18s.

Adverse effects: Flatulence, nausea, emesis, headaches and disturbances of skin sensation have been reported. Hyperbilirubinaemia, and increases in blood pressure, pulse, heart rate, bilirubin and alkaline phosphatase have also been reported with some hoodia extracts (Blom et al. 2011; CSIR 2011). Deep vein thrombosis has been reported in one UK case associated with a hoodia product (MHRA 2012).

Dosage: There is insufficient data to determine the appropriate dose or safe duration of hoodia supplements.

General plant information: Traditionally used as an appetite suppressant by Khoi-San bushmen and other groups in the Kalahari Desert of southern Africa, the global popularity of hoodia for use in weight loss regimes (as a result of Internet marketing) has led to overharvesting of the species. *Hoodia spp.* are listed in CITES Appendix II, which regulates international trade in endangered species. Experimental cultivation of hoodia was started as part of the efforts to develop a food supplement, but was abandoned when these projects were dropped by Unilever. It is believed that considerable illegal trade in *Hoodia spp.* occurs. Due to the limited supply of hoodia, up to 75% of products purporting to contain *Hoodia* extracts are adulterated with other ingredients, including *Opuntia ficus-indica* (prickly pear), pharmaceuticals and other materials (Gathier et al. 2013; Vermaak et al. 2010).

References

Blom WA, Abrahamse SL, Bradford R, Duchateau GS, Theis W, Orsi A, Ward CL, Mela DJ. (2011) Effects of 15-d repeated consumption of *Hoodia gordonii* purified extract on safety, ad libitum energy intake, and body weight in healthy, overweight women: a randomized controlled trial. *American Journal of Clinical Nutrition* 94(5): 1171–1181.

CSIR. (2011) Summary reports for key Hoodia clinical studies. http://researchspace.csir.co .za/dspace/bitstream/10204/5375/1/Maharaj_2011_Hoodia%20summary%20reports.pdf (accessed 17 August 2013).

Dent MP, Wolterbeek ATM, Russell PJ, Bradford R. (2012a) Safety profile of *Hoodia gordonii* extract: Mouse prenatal developmental toxicity study. *Food and Chemical Toxicology* 50: S20–S25.

Dent MP, Wolterbeek ATM, Russell PJ, Bradford R. (2012b) Safety profile of *Hoodia gordonii* extract: Rabbit prenatal developmental toxicity study. *Food and Chemical Toxicology* 50: S26–S33.

Gathier G, van der Niet T, Peelen T, van Vugt RR, Eurlings MC, Gravendeel B. (2013) Forensic identification of CITES protected slimming cactus (Hoodia) using DNA barcoding. *Journal of Forensic Science* 58(6): 1467–1471.

van Heerden FR. (2008) *Hoodia gordonii*: A natural appetite suppressant. *Journal of Ethnopharmacology* 119(3): 434–437.

Jain S, Singh SN. (2013) Metabolic effect of short-term administration of *Hoodia gordonii*, an herbal appetite suppressant. *South African Journal of Botany* 86: 51–55.

MHRA. (2012) Drug Analysis Print: *Hoodia gordonii* Web All. http://www.mhra.gov.uk /home/groups/public/documents/sentineldocuments/dap_1374836419814.pdf (accessed 18 August 2013).

Russell PJ, Swindells C. (2012) Chemical characterisation of *Hoodia gordonii* extract. *Food and Chemical Toxicology* 50(Suppl. 1): S6–S13.

Scott AD, Orsi A, Ward C, Bradford R. (2012) Genotoxicity testing of a *Hoodia gordonii* extract. *Food and Chemical Toxicology* 50: S34–S40.

Shukla YJ, Pawar RS, Ding Y, Li XC, Ferreira D, Khan IA. (2009) Pregnane glycosides from *Hoodia gordonii*. *Phytochemistry* 70(5): 675–683.

Vermaak I, Hamman JH, Viljoen AM. (2010) High performance thin layer chromatography as a method to authenticate *Hoodia gordonii* raw material and products. *South African Journal of Botany* 76: 119–124.

Vermaak I, Hamman JH, Viljoen AM. (2011a) *Hoodia gordonii*: an up-to-date review of a commercially important anti-obesity plant. *Planta Medica* 77(11): 1149–1160.

Vermaak I, Viljoen AM, Chen W, Hamman JH. (2011b) *In vitro* transport of the steroidal glycoside P57 from *Hoodia gordonii* across excised porcine intestinal and buccal tissue. *Phytomedicine* 18(8–9): 783–787.

Hops   *Humulus lupulus* L.

Synonyms: *Lupulus humulus* Mill.; and others. Sometimes incorrectly referred to as *H. lupus* L.

Family: Cannabaceae

Other common names: Common hop(s); hop

Drug name: Lupuli flos; Lupuli strobilus. 'Lupulin' is the powder composed of the glandular trichomes separated from the strobiles (conical flowers).

Botanical drug used: Dried female inflorescence

Indications/uses: Traditionally used to relieve mild symptoms of stress and to aid sleep. Registered THR products containing hops usually contain other herbal sedatives, including valerian root or passionflower herb. Hops have oestrogenic properties and hop-containing preparations are reputed to reduce menopausal hot flushes. Hops are a major component of beer.

Evidence: Clinical data to support the traditional sedative uses of hops are lacking, although there is plenty of anecdotal evidence recorded and some animal studies which support this use. Oestrogenic, chemopreventive and anti-microbial activities of hop constituents have also been demonstrated *in vitro*.

Safety: Safety data are lacking, although the wide use of hops suggests their use is safe. As a potential sedative, hops may impair the ability to drive or use machinery (EMEA 2008).

Main chemical compounds: More than 1000 chemical constituents have been identified from hops. The most significant include the volatile constituents, which include humulene, myrcene, β-caryophyllene and 2-methly-3-buten-2-ol, the α- and β-bitter acids (di- and tri-prenylated phloroglucinol derivatives respectively), the prenylated flavonoids, and the polyphenols (e.g. gallocatechin). The main α-bitter acids are humulone, cohumulone and adhumulone, and the main β-acids are lupulone, colupulone and adlupulone. These isomerise naturally and on heating to produce isohumulones and isolupulones. The most important preny-lated flavonoids include 8-prenylnaringenin, xanthohumol, isoxanthohumol, 6-isopentenylnaringenin. Other flavonoids present include astragalin, kaempferol,

---

Phytopharmacy: An evidence-based guide to herbal medicinal products, First Edition.
Sarah E. Edwards, Inês da Costa Rocha, Elizabeth M. Williamson and Michael Heinrich.
© 2015 John Wiley & Sons, Ltd. Published 2015 by John Wiley & Sons, Ltd.

quercetin, quercitrin and rutin (Chadwick et al. 2006; Pharmaceutical Press Editorial Team 2013; Zanoli and Zavatti 2008).

Clinical evidence:

Sedative effects: Human studies supporting the traditional uses of hops are limited. Clinical trials have investigated the sedative effects of hops only in combination with other botanical drugs, mainly valerian. A randomised, double-blind controlled trial in subjects suffering from sleep disorders showed equivalent efficacy and tolerability between a hop-valerian preparation and a benzodiazepine (Zanoli and Zavatti 2008). Another study found that addition of hops to a valerian formula reduced sleep latency (time required to fall asleep), compared to valerian on its own (Koetter and Biendl 2010).

Antidiabetic effects: Isohumulones derived from hops have been shown to have beneficial effects in diabetes and obesity. In a study in subjects with mild type 2 diabetes, 100 mg of an isohumulone extract administered twice daily was shown to significantly reduce blood glucose, haemoglobin A1C and systolic blood pressure after 8 weeks. In a follow-up study in subjects with pre-diabetes, body mass index was significantly reduced after administration of 48 mg of isohumulones for 12 weeks (Koetter and Biendl 2010).

Pre-clinical evidence and mechanisms of action: 8-prenylnaringenin (8-PN) is one of the most potent _in vitro_ oestrogenic substances found in the plant kingdom. In humans, 8-PN has been shown to derive from isoxanthohumol either by the action of liver cytochrome P450 enzymes, or by intestinal flora. The identified sedative principle is 2-methly-3-buten-2-ol, which, although found in low amounts, may be formed _in vivo_ by metabolism of α-bitter acids. High concentrations of 2-methly-3-buten-2-ol may also be achieved in both tea and bath products containing hops. The sedative activity of hops is poorly understood and studies in rodents have also consistently failed to demonstrate anxiolytic action. Reduced locomotor activity, reduced body temperature and increased ketamine-induced sleeping time have been demonstrated with ethanolic and CO_2 hop extracts administered to mice by oral gavage. Another rodent study demonstrated a significant increase in pentobarbital-induced sleeping time and antidepressant effect following oral administration of both hop extracts and α-acids fractions and data suggests that several of the constituents of hops, including α-acids, β-bitter acids and hop oil, contribute to the overall sedative activity (Schiller et al. 2006; Zanoli and Zavatti 2008). A dried hop extract (which did not contain any bitter acids due the extraction method) was shown to bind to serotonergic 5-HT_6 receptors as well as melatonergic ML_1 receptors. It has been suggested that the sedative effects of hops probably involve GABA receptors (Zanoli and Zavatti 2008).

8-PN mimics the activity of 17β-oestradiol, but with far less potency, and has been shown to reduce raised skin temperature in a rat model of menopausal hot flushes. It may act as a selective oestrogen receptor modulator. It also stimulates alkaline phosphatase activity in a human endometrial adenocarcinoma epithelial cell line, and the growth of oestrogen dependent MCF7 breast cancer cells (Bowe et al. 2006; Chadwick et al. 2006; Zanoli and Zavatti 2008).

Numerous _in vitro_ studies have also demonstrated that the prenylated flavonoids found in hops have chemopreventive properties. Anti-proliferative activity against human breast, colon, ovarian and prostate cancers has been reported. Potential mechanisms of action include induction of quinone reductase, reduced expression of CYP enzymes that activate procarcinogens (e.g. CYP1A1), inhibition of nitric oxide synthase and inhibition of cyclooxygenase enzymes, COX1 and COX2 (Zanoli and Zavatti 2008). The prenylflavonoids have been shown to induce a caspase-independent form of cell death (Delmulle et al. 2008), and xanthohumol inhibits replication of human hepatocellular carcinoma cell lines (Ho et al. 2008).

Other studies have shown antimicrobial, antispasmodic, antimycobacterial and antioxidant activities for hop extracts (Chadwick et al. 2006; Pharmaceutical Press Editorial Team 2013; Stavri et al. 2004; Zanoli and Zavatti 2008).

Interactions: None confirmed. Studies in rodents indicate that hops may potentiate the effects of other sedatives. Theoretically, due to the phytoestrogenic activity of 8-prenylnaringenin, hops may interact with hormonal therapies, but there is no evidence to confirm this and it is not thought to be clinically significant (Williamson et al. 2013).

Contraindications: Hypersensitivity to hops. Prolonged use of high doses is not recommended during pregnancy and lactation or in under 18s due to lack of safety data. Antispasmodic activity on the uterus has been reported *in vitro* (Pharmaceutical Press Editorial Team 2013).

Adverse effects: Small doses of hops are non-toxic, although large doses administered by injection in animals have caused soporific effects and death. Allergies have been reported in sensitive individuals. Hops are known to cause contact dermatitis in hop-pickers, attributed to myrcene in fresh hop oil. In addition, patch test reactions have been documented for fresh hop oil and the constituents, humulone and lupulone. Contact dermatitis has also been attributed to the pollen. No clinical cases of allergy resulting from therapeutic use of hops have been reported (Anon. 2003; Pharmaceutical Press Editorial Team 2013; Zanoli and Zavatti 2008).

Dosage: For products, see manufacturers' instructions. Oral administration (adults) of the dried strobiles (flowers): 0.5 g as an infusion (tea) two to four times/day; liquid extract – 0.5–2.0 ml (1:1 in 45% alcohol) up to three times/day; tincture – 1–2 ml (1:5 in 60% alcohol) up to three times/day (Pharmaceutical Press Editorial Team 2013).

General plant information: *H. lupulus* is naturalised in Central Europe, and is widely cultivated in the temperate regions of the world (North and South America, South Africa and Australia). Hops were first cultivated in the mid-9th century, between 859 CE and 875 CE in Germany and other regions of Central Europe. By the 18th century, hops overtook the use of *Myrica gale* L. (bog myrtle) in brewing, possibly due to its better preserving properties (Zanoli and Zavatti 2008).

References

Anon. (2003) *Humulus lupus*. Monographs *Alternative Medicine Review* 8(2): 190–192.

Bowe J, Li XF, Kinsey-Jones J, Heyerick A, Brain S, Milligan S, O'Byrne K. (2006) The hop phytoestrogen, 8-prenylnaringenin, reverses the ovariectomy-induced rise in skin temperature in an animal model of menopausal hot flushes. *Journal of Endocrinology* 191(2): 399–405.

Chadwick LR, Pauli GF, Farnsworth NR. (2006) The pharmacognosy of *Humulus lupulus* L. (hops) with an emphasis on estrogenic activity. *Phytomedicine* 13(1–2): 119–131.

Delmulle L, Vanden Berghe T, De Keukeleire D, Vandenabeele P. (2008) Treatment of PC-3 and DU145 prostate cancer cells by prenylflavonoids from hop (*Humulus lupulus* L.) induces a caspase-independent form of cell death. *Phytotherapy Research* 22(2): 197–203.

EMEA. (2008) Community herbal monograph on *Humulus lupulus* L., flos. European Medicines Agency. http://www.ema.europa.eu/docs/en_GB/document_library/Herbal_-_Community_herbal_monograph/2010/01/WC500059222.pdf (accessed 22 March 2014).

Ho YC, Liu CH, Chen CN, Duan KJ, Lin MT. (2008) Inhibitory effects of xanthohumol from hops (*Humulus lupulus* L.) on human hepatocellular carcinoma cell lines. *Phytotherapy Research* 22(11): 1465–1468.

Koetter U, Biendl M. (2010) Hops (*Humulus lupulus*): A review of its historic and medicinal uses. *HerbalGram* 87: 44–57.

Pharmaceutical Press Editorial Team. (2013) *Herbal Medicines*. 4th Edition. Pharmaceutical Press, London, UK.

Schiller H, Forster A, Vonhoff C, Hegger M, Biller A, Winterhoff H. (2006) Sedating effects of *Humulus lupulus* L. extracts. *Phytomedicine* 13(8): 535–541.

Stavri M, Schneider R, O'Donnell G, Lechner D, Bucar F, Gibbons S. (2004)The antimycobacterial components of hops (*Humulus lupulus*) and their dereplication. *Phytotherapy Research* 18(9): 774–776.

Williamson EM, Driver S, Baxter K. (Eds.) (2013) *Stockley's Herbal Medicines Interactions*. 2nd Edition. Pharmaceutical Press, London, UK.

Zanoli P, Zavatti M. (2008) Pharmacognostic and pharmacological profile of *Humulus lupulus* L. *Journal of Ethnopharmacology* 116(3): 383–396.

Horny Goat Weed   *Epimedium spp.*

Mainly: *E. brevicornu* Maxim.; *E. koreanum* Nakai; *E. pubescens* Maxim.; *E. sagittatum* (Siebold & Zucc.) Maxim. Other species of *Epimedium* are used medicinally, including *E. acunatum* Franch and *E. wushanense* T.S. Ying (as unofficial substitutes of Epimedii herba)

Synonyms: *Aceranthus macrophyllus* Blume ex. K.Koch; *A. sagittatus* Siebold & Zucc. (=*E. sagittatum*)

Family: Berberidaceae

Other common names: Barrenwort; bishop's hat; epimedium; yin yang huo

Drug name: Epimedii herba

Botanical drug used: Aerial parts

Note: Based on the large number of species accepted in the Pharmacopoeia of China, in combination with the poor information that is generally available about the quality of products sold as horny goat weed, there can be no evidence-based use of a product with an acceptable quality. Therefore, it clearly cannot be recommended by a health care professional.

Indications/uses: To treat sexual dysfunction, particularly impotence in men and lack of libido in women. In traditional Chinese medicine, it is also used widely for osteoporosis. *Epimedium* species are used to treat a number of other conditions including rheumatic arthritis, menopausal symptoms and memory loss.

Evidence: There is limited evidence to support the use of epimedium to treat sexual dysfunction and osteoporosis, although it is known to contain potent phytoestrogenic compounds.

Safety: Tests in animal models have shown no serious toxicity issues, but human clinical safety has not been established. Epimedium should not be used concurrently with other phosphodiesterase-5 (PDE-5) inhibitors, or when PDE-5 inhibitors are contraindicated (e.g. cardiac arrhythmias), or during pregnancy and lactation.

Main chemical compounds: The main active constituents are the prenylated flavonoids, icariin and its metabolites, epimedins A, B, C, and the sagittatosides.

Phytopharmacy: An evidence-based guide to herbal medicinal products, First Edition.
Sarah E. Edwards, Inês da Costa Rocha, Elizabeth M. Williamson and Michael Heinrich.
© 2015 John Wiley & Sons, Ltd. Published 2015 by John Wiley & Sons, Ltd.

Other active compounds present include flavonol glycosides (of kaempferol, quercetin, myricetin); flavones (tricin, luteolin); biflavones (gingketin, isogingketin, bilobetin); lignans such as syringaresinol; alkaloids (e.g. epimediphine from *E. koreanum*); β-sitosterol (*E. sagittatum*, *E. brevicornu*); and a chalcone, isoliquiritigenin (from *E. koreanum*) (Ma et al. 2011; Pharmaceutical Press Editorial Team 2013; Zhang et al. 2013c).

Clinical evidence:

Sexual dysfunction: Limited clinical data are available on the effects of epimedium on erectile dysfunction. A small double-blind clinical trial was reported assessing an epimedium herbal complex supplement in 25 healthy men and 13 men who had used sildenafil (Viagra). The men were administered with the herbal complex for a minimum of 45 days. Daily use of the epimedium preparation was found to enhance sexual satisfaction to a greater extent than sildenafil (Ma et al. 2011).

Bone health and osteoporosis in menopause: A five-year follow-up study of a herbal preparation of epimedium for prevention of postmenopausal osteoporosis and fragility fractures found that it was able to reduce postmenopausal bone loss, and also showed some potential for reduction in fragility fracture incidence (Deng et al. 2012). A randomised, double-blind placebo-controlled clinical study in 85 postmenopausal women with osteopenia (lower than normal mineral bone density, considered a precursor to osteoporosis), were treated with icariin 60 mg, combined with the isoflavonoids daidzein 15 mg, and genistein 3 mg, daily for 24 months. All patients were also given calcium 300 mg daily. A statistically significant, though small, increase in bone density was found in the test group and bone resorption markers were significantly decreased, compared to the control group (Zhang et al. 2007). Isoflavones are oestrogenic and so also have osteogenic activities.

Other oestrogenic effects: A study evaluating the effects of Epimedii herba water extract on blood lipid and sex hormone levels in 90 postmenopausal women found that after 6 months of medication, the extract decreased total cholesterol and triglyceride levels ($p < 0.01$) and significantly increased serum levels of oestradiol, compared with the pre-treatment level (Yan et al. 2008).

Pre-clinical evidence and mechanisms of action:

Sexual dysfunction: Icariin has erectogenic properties in animals, via phosphodiesterase type 5 (PDE-5) inhibition, and also neurotrophic effects *in vitro* and *in vivo* (Shindel et al. 2010). A study in male rats found increased sexual behaviour after treatment with epimedium in combination with four other medicinal plants. However, individual assessment of the herbs showed no improvement, indicating a possible synergistic action between them (Zanoli et al. 2008). A study in isolated rabbit corpus cavernosum (CC) smooth muscle showed that epimedium extracts could elicit a relaxation effect through activation of multiple targets on NO/cGMP signalling pathways. The study also found that they potentiated the effects of PDE-5 inhibitors (commonly used for erectile dysfunction), such as sildenafil and vardenafil. The crude extracts were found to have a greater potency than the purified compounds (Chiu et al. 2006).

Bone health and osteoporosis: Icariin is involved in the regulation of multiple signalling pathways in osteogenesis, anti-osteoclastogenesis, chondrogenesis, angiogenesis and inflammation. It has been reported to have the potential to be used as a substitute for osteoinductive protein-bone morphogenetic proteins, or to enhance their therapeutic effects, for bone tissue engineering as well as treatment of osteoporosis (Zhang et al. 2013b).

Other relevant effects: Icariin has phytoestrogenic properties and has been shown to promote the biosynthesis of oestrogen by aromatase (Yang et al. 2013). It reduced the levels of serum total cholesterol and low-density lipoprotein cholesterol, and reduced platelet adhesiveness and aggregation in atherosclerotic rabbits, demonstrating lipid-lowering effects (Zhang et al. 2013a). The alkaloid epimediphine has anticholinesterase activity (Zhang et al. 2013c), which together with the oestrogenic effects of other constituents, may support the use for memory enhancement.

Interactions: There is conflicting evidence about a possible interaction with other PDE-5 inhibitors such as sildenafil (Viagra). *In vitro* evidence has suggested that epimedium potentiates PDE-5 inhibitors (Chiu et al. 2006), but a study of the effects of *E. sagittatum* extract on the pharmacokinetics of sildenafil demonstrated that the area under the concentration–time curve of sildenafil was significantly decreased in groups that received a high dose of epimedium extract, suggesting antagonistic effects (Hsueh et al. 2013). Co-administration with PDE-5 inhibitors should, therefore, be avoided until further information is available.

Another *in vitro* study found that epimedium is a potent inhibitor of cytochrome P450 isoforms (including CYP1A2, CYP2C19, CYP2E1, CYP2C9, CYP3A4, CYP2D6) and NADPH-CYP reductase, indicating a possible potential for interaction with other drugs, although the clinical significance is not known (Liu et al. 2006).

Contraindications: Epimedium should not be used during pregnancy and lactation as it has oestrogenic activity. Paediatric use is not appropriate. People with heart conditions should avoid using epimedium and long-term use is not recommended (Pharmaceutical Press Editorial Team 2013).

Adverse effects: In high doses (not specified), epimedium may have a stimulatory effect and cause sweating or a feeling of heat. Prolonged use of excessive amounts in animal studies was associated with decreased thyroid activity (Ma et al. 2011).

A case was reported of a 66-year-old man who developed heart arrhythmia and hypomania after taking a herbal sexual enhancement product containing epimedium. As it was a multi-ingredient preparation, and the man was predisposed to heart disease and possible mood disorder, it is unclear if this was caused by epimedium or another ingredient (Partin and Pushkin 2004).

Dosage: For products, see manufacturers' instructions. For the herb, the adult dose is 3–9 g dried aerial parts daily (Pharmaceutical Press Editorial Team 2013).

General plant information: According to a Chinese legend, a goat herder first noticed the qualities of the plant after he observed far more sexual activity in his goats after they ate it; hence the common name (Ma et al. 2011).

References

Chiu J-H, Chen K-K, Chien T-M, Chiou W-F, Chen C-C, Wang J-Y, Lui W-Y, Wu C-W. (2006) *Epimedium brevicornum* Maxim extract relaxes rabbit corpus cavernosum through multitargets on nitric oxide/cyclic guanosine monophosphate signaling pathway. *International Journal of Impotence Research* 18(4): 335–342.

Deng WM, Zhang P, Huang H, Shen YG, Yang QH, Cui WL, He YS, Wei S, Ye Z, Liu F, Qin L. (2012) Five-year follow-up study of a kidney-tonifying herbal Fufang for prevention of postmenopausal osteoporosis and fragility fractures. *Journal of Bone and Mineral Metabolism* 30(5): 517–524.

Hsueh TY, Wu YT, Lin LC, Chiu AW, Lin CH, Tsai TH. (2013) Herb-drug interaction of *Epimedium sagittatum* (Sieb. et Zucc.) maxim extract on the pharmacokinetics of sildenafil in rats. *Molecules* 18(6): 7323–7335.

Liu KH, Kim MJ, Jeon BH, Shon JH, Cha IJ, Cho KH, Lee SS, Shin JG. (2006) Inhibition of human cytochrome P450 isoforms and NADPH-CYP reductase in vitro by 15 herbal medicines, including Epimedii herba. *Journal of Clinical Pharmacology and Therapeutics* 31(1): 83–91.

Ma H, He X, Yang Y, Li M, Hao D, Jia Z. (2011) The genus *Epimedium*: An ethnopharmacological and phytochemical review. *Journal of Ethnopharmacology* 134(3): 519–541.

Pharmaceutical Press Editorial Team. (2013) *Herbal Medicines*. 4th Edition. Pharmaceutical Press, London, UK.

Partin JF, Pushkin YR. (2004) Tachyarrhythmia and hypomania with horny goat weed. *Psychosomatics* 45(6): 536–537.

Shindel AW, Xin ZC, Lin G, Fandel TM, Huang YC, Banie L, Breyer BN, Garcia MM, Lin CS, Lue TF. (2010) Erectogenic and neurotrophic effects of icariin, a purified extract of horny goat weed (*Epimedium spp.*) *in vitro* and *in vivo*. *Journal of Sexual Medicine* 4(Pt 1): 1518–1528.

Yan FF, Liu Y, Liu YF, Zhao YX. (2008) Herba Epimedii water extract elevates estrogen level and improves lipid metabolism in postmenopausal women. *Phytotherapy Research* 22(9): 1224–1228.

Yang L, Lu D, Guo J, Meng X, Zhang G, Wang F. (2013) Icariin from *Epimedium brevicornum* Maxim promotes the biosynthesis of estrogen by aromatase (CYP19). *Journal of Ethnopharmacology* 145(3): 715–721.

Zanoli P, Benellia A, Zavatti M, et al. (2008) Improved sexual behaviour in male rats treated with a Chinese herbal extract: hormonal and neuronal implications. *Asian Journal of Andrology* 10(6): 937–945.

Zhang G, Qin L, Shi Y. (2007) Epimedium-derived phytoestrogen flavonoids exert beneficial effect on preventing bone loss in late postmenopausal women: a 24-month randomized, double-blind and placebo-controlled trial *Journal of Bone and Mineral Metabolism* 22(7): 1072–1079.

Zhang WP, Bai XJ, Zheng XP, Xie XL, Yuan ZY. (2013a) Icariin attenuates the enhanced prothrombotic state in atherosclerotic rabbits independently of its lipid-lowering effects. *Planta Medica* 79(9): 731–736.

Zhang X, Liu T, Huang Y, Wismeijer D and Liu Y. (2013b) Icariin: does it have an osteo inductive potential for bone tissue engineering? *Phytotherapy Research* 28(4): 498–509.

Zhang X, Oh M, Kim S, Kim J, Kim H, Kim S, Houghton PJ, Whang W. (2013c) Epimediphine, a novel alkaloid from *Epimedium koreanum*, inhibits acetylcholinesterase. *Natural Product Research* 27(12): 1067–1074.

Horse Chestnut   *Aesculus hippocastanum* L.

Family: Sapindaceae (Hippocastanaceae)

Other common names: Aesculus; common horse chestnut; conker tree

Drug name: Hippocastani semen

Botanical drug used: Dried ripe seeds

Indications/uses: A range of products are used in the treatment of chronic venous insufficiency (CVI), mainly varicose veins, venous ulcers, swollen legs, feeling of heavy legs, pain, tiredness, itching, tension and cramps in the calves.

Evidence: Overall, a large number of randomised clinical studies provide evidence that horse chestnut preparations seem to be effective for CVI when compared with placebo and reference treatment.

Safety: Overall considered to be safe and the usual precautions are required (see below).

Main chemical compounds: The main active constituents of horse chestnut are the acylated triterpene glycosides (saponins). This mixture is known as 'aescin' (escin), with α-escin, β-escin and cryptoaescin as the major glycosides. Other constituents include flavonoids (di- and triglycosides of quercetin and kaempferol), sterols, essential oil and a high proportion of starch (ESCOP 2003), as well as coumarins (aesculetin, fraxin and scopoli) and tannins among others (Pharmaceutical Press Editorial Team 2013).

Clinical evidence: A Cochrane review assessing oral horse chestnut seed extract mono-preparations (standardised to 'aescin') versus placebo or reference treatment, showed that the extract was safe and efficacious for treating patients with CVI related signs and symptoms, i.e. leg pain, pruritus (itching), and oedema (swelling), as well as leg volume and leg circumference at ankle and calf). However, larger studies are needed to confirm its efficacy (Pittler and Ernst 2012).

Pre-clinical evidence and mechanisms of action: The exact mechanism of action is unknown, but based on pre-clinical and clinical studies it might involve an influence on venous tone and capillary filtration rate. It is unclear whether or not aescin is the therapeutically active substance (EMA 2009a).

Phytopharmacy: An evidence-based guide to herbal medicinal products, First Edition.
Sarah E. Edwards, Inês da Costa Rocha, Elizabeth M. Williamson and Michael Heinrich.
© 2015 John Wiley & Sons, Ltd. Published 2015 by John Wiley & Sons, Ltd.

Interactions: None reported. However, one *in vitro* study showed that a horse chestnut extract inhibited, to a minor extent, the transport of digoxin by P-glycoprotein and that some interaction may occur, although no recommendation was made (Williamson et al. 2013). Potential interaction with anticoagulants has been suggested, but no cases have been reported (EMA 2009b).

Contraindications: Not recommended during pregnancy and lactation, as safety has not been established. The use in children under 12 years of age is not recommended. Horse chestnut should not be used on broken skin, around the eyes or on mucous membranes. A doctor should be consulted if there is inflammation of the skin, thrombophlebitis or subcutaneous induration, severe pain, ulcers, sudden swelling of one or both legs, cardiac or renal insufficiency (EMA 2009a; Pittler and Ernst 2012).

Adverse effects: Hypersensitivity to the active substance. Mild and infrequent adverse events such as gastrointestinal complaints, dizziness, nausea, headache and pruritus were also reported in clinical trials (Pittler and Ernst 2012).

Dosage: Dried ethanolic extracts standardised to 50–100 mg of aescin twice a day (Pittler and Ernst 2012; Wichtl 2008).

General plant information: This tree is native to the Balkans, but is now widely cultivated in temperate countries. Despite the fact that there were two suspected cases of toxic nephropathy due to very high doses of aescin in the late 1970s (Pharmaceutical Press Editorial Team 2013), there are no safety concerns when horse chestnut extract is given orally, as the bioavailability of aescin is about 1.5% of the given dose (standardised to 50 mg of aescin) (EMA 2009b).

References

EMA. (2009a) Community herbal monograph on *Aesculus hippocastanum* L., semen. European Medicines Agency http://www.ema.europa.eu/docs/en_GB/document_library/Herbal_-_Community_herbal_monograph/2010/01/WC500059105.pdf.

EMA. (2009b) Assessment report on *Aesculus hippocastanum* L., semen. European Medicines Agency http://www.ema.europa.eu/docs/en_GB/document_library/Herbal_-_HMPC_assessment_report/2010/01/WC500059103.pdf.

ESCOP. (2003) *ESCOP Monographs: The Scientific Foundation for Herbal Medicinal Products.* 2nd Edition. Thieme, Exeter and London, UK.

Pharmaceutical Press Editorial Team. (2013) *Herbal Medicines.* 4th Edition. Pharmaceutical Press, London, UK.

Pittler MH, and Ernst E. (2012) Horse chestnut seed extract for chronic venous insufficiency. *Cochrane Database of Systematic Reviews* 11: CD003230.

Wichtl M. (2008) *Teedrogen und Phytopharmaka: Ein Handbuch für die Praxis auf wissenschaftlicher Grundlage.*. Wissenschaftliche Verlagsgesellschaft, Stuttgart.

Williamson EM, Driver SB, Baxter K. (Eds.) (2013) *Stockley's Herbal Medicines Interactions.* 2nd Edition. Pharmaceutical Press, London, UK.

Horsetail   *Equisetum arvense L.*

Family: Equisetaceae

Other common names: Field horsetail

Drug name: Equiseti herba

Botanical drug used: Aerial parts

Indications/uses: The aerial parts are used externally as an ingredient in shampoos and skincare products, to boost hair and nail strength due to the silicate content, to stem bleeding and to treat wounds. Internally it is used as a herbal tea and in diuretic products, often in combination with dandelion (*Taraxacum officinale*), to flush the urinary tract in minor urinary complaints, and to relieve premenstrual water retention. It has also been used to treat osteoporosis and other bone disorders.

Evidence: Limited information is available to support its use.

Safety: External use is safe, and overall, internal use is considered safe at normal doses. It should be avoided in pregnancy and lactation, in children, and in patients with cardiac, liver and kidney disorders.

Main chemical compounds: The herb contains high concentrations of silicic acid and silicates. Flavonoids such as apigenin, kaempferol, luteolin, and quercetin glycosides and their malonyl esters, and other polyphenolic compounds (caffeic acid derivatives) are also present. Trace amounts of the alkaloid nicotine have been reported but are not confirmed, as well as sterols including cholesterol, isofucosterol and campesterol. Thiaminase (an enzyme that breaks down thiamine) is present in the fresh plant but is usually inactivated during processing (EMEA 2008; Williamson et al. 2013).

Clinical evidence: No clinical studies have been performed using *E. arvense* preparations, although a study reported the use of another species of *Equisetum* (*E. bogotense* Kunth) as a diuretic (Lemus et al. 1996). Older *in vivo* studies support its use as a diuretic in different animal species (mice/rats, rabbits and dogs) (EMEA 2008). However, a clinical trial of two formulations containing *E. arvense* with other herbs showed a reduction in nocturia in benign prostatic hyperplasia (BPH)

Phytopharmacy: An evidence-based guide to herbal medicinal products, First Edition.
Sarah E. Edwards, Inês da Costa Rocha, Elizabeth M. Williamson and Michael Heinrich.
© 2015 John Wiley & Sons, Ltd. Published 2015 by John Wiley & Sons, Ltd.

patients (Tamaki et al. 2008). Even though the evidence based on clinical studies is very limited, its use as a diuretic is based on long-standing medical experience.

Pre-clinical evidence and mechanisms of action: Mechanism of action is unknown, but extracts of some species of *Equisetum* have shown diuretic activity in mice (Perez Gutierrez et al. 1985). *E. arvense* extracts have a negative effect on human osteoclastogenesis *in vitro*, supporting the traditional use for bone disorders (Costa-Rodrigues et al. 2012). The topical use may in part be supported by evidence that an ointment containing *E. arvense* exhibited significant diabetic wound healing activity in excision wounds in rats (Ozay et al. 2013).

Interactions: None known.

Contraindications: Not recommended during pregnancy and lactation, or in children under 12 years of age as safety has not been established. Not recommended in patients with liver disease or with conditions where a reduced fluid intake is recommended, such as in severe cardiac or renal disease. A qualified practitioner should be consulted if symptoms such as fever, dysuria, spasm, or blood in urine occur during its use (EMEA 2008).

Adverse effects: Mild gastrointestinal complaints and allergic reactions (e.g. rash) have been reported (EMEA 2008). One case of liver injury was reported in a patient after taking *E. arvense* 'juice' for a week (Kilincalp et al. 2012). Adverse effects are thought to be due mainly to the use of adulterated material, including *E. palustre* L. (marsh horsetail), which is reported to contain toxic spermidine-type alkaloids – but these are unconfirmed.

Dosage: According to manufacturers' specifications. As a single dose, 570 mg of comminuted herbal substance, or as a tea, 2 to 3 g of herbal substance into 250 ml boiling water, 3–4 times per day, is recommended over a period of 2 to 4 weeks (EMEA 2008).

General plant information: The herb is widely distributed throughout the temperate zones of the Northern Hemisphere, being mainly imported from China and eastern and south-eastern European countries. Two chemotypes, an Asian and North American type, and a European type which does not contain luteolin-5-glycoside, have been described (EMEA 2008). The ancient Greeks used horsetail in the treatment of wounds, while the Romans used it as food for animals and humans, as well as medicine (Stajner et al. 2009). It has a historic use in polishing metal, particularly pewter: as the herb dries, silica crystals form in the stems and branches and give it abrasive properties.

References

Costa-Rodrigues J, Carmo SC, Silva JC, Fernandes MH. (2012) Inhibition of human *in vitro* osteoclastogenesis by *Equisetum arvense*. *Cell Proliferation* 45(6): 566–576.
EMEA. (2008) Assessment report on *Equisetum arvense* L., herba. European Medicines Agency http://www.ema.europa.eu/docs/en_GB/document_library/Herbal_-_HMPC _assessment_report/2009/12/WC500018418.pdf.
Kilincalp S, Ekiz F, Basar O, Coban S, Yuksel O. (2012) *Equisetum arvense* (Field Horsetail)-induced liver injury. *European Journal of Gastroenterology and Hepatology* 24(2): 213–214.
Lemus I, Garcia R, Erazo S, Pena R, Parada M, Fuenzalida M. (1996) Diuretic activity of an *Equisetum bogotense* tea (Platero herb): evaluation in healthy volunteers. *Journal of Ethnopharmacololgy* 54(1): 55–58.

Ozay, Y, Kasim Cayci, M, Guzel-Ozay, S, Cimbiz, A, Gurlek-Olgun, E, Sabri Ozyurt, M. (2013) Effects of *Equisetum arvense* ointment on diabetic wound healing in rats. *Wounds* 25(9): 234–241.

Perez Gutierrez RM, Laguna GY, Walkowski A. (1985) Diuretic activity of Mexican *Equisetum*. *Journal of Ethnopharmacology* 14(2–3): 269–272.

Stajner D, Popovic BM, Canadanovic-Brunet J Anackov G. (2009) Exploring *Equisetum arvense* L., *Equisetum ramosissimum* L. and *Equisetum telmateia* L. as sources of natural antioxidants. *Phytotherapy Research* 23(4): 546–550.

Tamaki M, Nakashima M, Nishiyama R, Ikeda H, Hiura M, Kanaoka T, Nakano T, Hayashi T, Ogawa O. (2008) Assessment of clinical usefulness of Eviprostat for benign prostatic hyperplasia – comparison of Eviprostat tablet with a formulation containing two-times more active ingredients. *Hinyokika Kiyo* 54(6): 435–445.

Williamson EM, Driver S, Baxter K. (Eds.) (2013) *Stockley's Herbal Medicines Interactions.* 2nd Edition. Pharmaceutical Press, London, UK.

Ipecacuanha

Carapichea ipecacuanha (Brot.) L. Andersson

Synonyms: *Cephaelis acuminata* H.Karst; *C. ipecacuanha* (Brot.) Tussac; *Ipecacuanha officinalis* Arruda; *Psychotria ipecacuanha* (Brot.) Standl.; *Uragoga ipecacuanha* (Brot.) Baill.; and others

Family: Rubiaceae

Other common name: Ipecacuanha is often shortened to 'ipecac'

Drug name: Ipecacuanhae radix

Botanical drug used: Root

Indications/uses: A traditional herbal remedy for relief of coughs and colds, as an expectorant, generally used in combination products, e.g. with (a) liquorice root, *Glycyrrhiza glabra* L., and Indian squill bulb, *Drimia indica* (Roxb.) Jessop; *or* (b) senega root, *Polygala senega* L., and marshmallow root, *Althaea officinalis* L.

Evidence: There is a lack of clinical evidence on its use for the relief of coughs and colds.

Safety: Possible cardiotoxic effects; the alkaloids are also emetic.

Main chemical compounds: The main active compounds are the alkaloids emetine, cephaeline, emetamine, ipecacuanhic acid, psychotrine, and *O*-methyl-psychotrine (Itoh et al. 1999).

Clinical evidence: Ipecac-containing cough syrups (e.g. 'glycerine lemon honey and ipecac') have been used for centuries, but with very little evidence to support this use. No clinical data is available for ipecacuanha alone for coughs. One double-blind, three-arm, randomised placebo-controlled clinical trial assessed the effect of a multi-component plant-based formulation containing ipecacuanha (*C. ipecacuanha*, together with *Bryonia alba* L. and *Drosera peltata* Thunb.) in chronic obstructive pulmonary disease (COPD). Patients (105 participants) received either the herbal formulated product, or the drug combination salbutamol + theophylline + bromhexine, or placebo over a 6-month study period. The herbal formulation was effective in the management of COPD and equivalent to the conventional therapy, and without side effects (Murali et al. 2006).

Phytopharmacy: An evidence-based guide to herbal medicinal products, First Edition.
Sarah E. Edwards, Inês da Costa Rocha, Elizabeth M. Williamson and Michael Heinrich.
© 2015 John Wiley & Sons, Ltd. Published 2015 by John Wiley & Sons, Ltd.

Ipecac syrup has been used as an emetic in the management of acute poisonings and overdoses, but this is no longer recommended (Höjer et al. 2013).

Pre-clinical evidence and mechanisms of action: Mechanism of action is unknown, but the alkaloids have well-documented emetic properties (e.g. Höjer et al. 2013) in doses higher than for expectorant use. The herb has been used since the 18th century as an emetic, diaphoretic and expectorant (Júnior et al. 2012; Manzali de Sa and Elisabetsky 2012).

The alkaloids have amoebicidal properties and have been suggested as having potential use in the development of antimalarial or other antiprotozoal drugs (Wright and Phillipson 1990).

Interactions: None reported.

Contraindications: Not recommended during pregnancy and lactation, as safety has not been established. For use by adults and children over the age of 12 years (MHRA 2012). Ipecac-containing products have been consumed by patients suffering from eating disorders, to eject food already eaten. This is less common than the abuse of laxatives in such patients, but is still dangerous (Steffen et al. 2007).

Adverse effects: Toxic effects due to the alkaloids are possible. Short-term use of ipecac syrup may cause minor side effects such as drowsiness, while long-term ingestion of ipecac syrup has been reported to result in cardiotoxicity (Yamashita et al. 1986).

Dosage: Not recommended. However, if used in formulations for cough, follow manufacturers' instructions.

General plant information: The species is native to the Atlantic Forest of South America and is used for its anti-diarrhoeal and emetic properties by Native Americans (Brandao et al. 2012). It has a long and interesting history of use (see e.g. Júnior et al. 2012) and was described as a medicine by Portuguese Jesuit priests as early as 1625, and later included in the first edition of the '*Pharmacopêa Portugueza*' in 1876 (Manzali de Sa and Elisabetsky 2012).

References

Brandao MG, Pignal M, Romaniuc S, Grael CF, Fagg CW. (2012) Useful Brazilian plants listed in the field books of the French naturalist Auguste de Saint-Hilaire (1779–1853). *Journal of Ethnopharmacology* 143(2): 488–500.

Höjer J, Troutman WG, Hoppu K, Erdman A, Benson BE, Mégarbane B, Thanacoody R, Bedry R, Caravati EM. (2013) Position paper update: ipecac syrup for gastrointestinal decontamination. *Clinical Toxicology (Philadelphia, Pa.)* 51(3): 134–139.

Itoh A, Ikuta Y, Baba Y, Tanahashi T, Nagakura N. (1999) Ipecac alkaloids from *Cephaelis acuminata*. *Phytochemistry* 52(6): 1169–1176.

Júnior WSF, Cruz MP, Santos LL, Medeiros MFT. (2012) Use and importance of quina (*Cinchona spp.*) and ipeca (*Carapichea ipecacuanha* (Brot.) L. Andersson): Plants for medicinal use from the 16th century to the present. *Journal of Herbal Medicine* 2(4): 103–112.

Manzali de Sa I, Elisabetsky E. (2012) Medical knowledge exchanges between Brazil and Portugal: an ethnopharmacological perspective. *Journal of Ethnopharmacology* 142(3): 762–768.

MHRA. (2012) Lists of substances. Retrieved November 2012, from http://www.mhra.gov.uk /Howweregulate/Medicines/Licensingofmedicines/Legalstatusandreclassification/Listsof substances/index.htm.

Murali PM, Rajasekaran S, Paramesh P, Krishnarajasekar OR, Vasudevan S, Nalini K, Lakshmisubramanian S, Deivanayagam CN. (2006) Plant-based formulation in the management of chronic obstructive pulmonary disease: a randomized double-blind study. *Respiratory Medicine* 100(1): 39–45.

Steffen KJ, Mitchell JE, Roerig JL, Lancaster KL. (2007) The eating disorders medicine cabinet revisited: a clinician's guide to ipecac and laxatives. *International Journal of Eating Disorders* 40(4): 360–368.

Wright C, Phillipson JD. (1990) Natural products and the development of selective antiprotozoal drugs. *Phytotherapy Research* 4(4): 127–139.

Yamashita M, Nakamura K, Mizutani T, Koyama K, Naitou H. (1986) Effectiveness and side effects of ipecac syrup. *American Journal of Emergency Medicine* 4(5): 468.

Ispaghula Husk, Psyllium Husk

Plantago ovata Forssk.

Synonyms: *P. decumbens* Forssk.; *P. ispaghul* Roxb; and others.

Family: Plantaginaceae

Other common names: Blond psyllium; Indian psyllium; isabgol; pale psyllium; spogel

Drug name: Plantaginis ovatae seminis tegumentum, Plantaginis ovatae testa (both refer to the husk); Plantaginis ovatae semen (seed)

Botanical drug used: Husk; less often, whole seed

Note: Flea (or fleawort) seeds (Psyllii semen), from *P. indica* L. (syn.: *P. psyllium* L.; *P. arenaria* Waldst. & Kit.), *P. afra* L., and *P. orbignyana* Steinh. ex Decne., are used for the same conditions and have similar pharmacognostic and phytotherapeutic profiles.

Indications/uses: Ispaghula preparations are used for the treatment of habitual constipation, and where soft stools are desirable, for example, after rectal or anal surgery, anal fissures, during pregnancy, in patients with haemorrhoids and in those where an increased daily fibre intake may be advisable. It is also used for diarrhoea and dysentery, where it is taken with minimal liquid in order to absorb fluid from the colon, irritable bowel syndrome and in other conditions such as hyperlipidaemia, as an adjuvant.

Evidence: Ispaghula husk products are well-established as bulk-forming laxatives, and are widely prescribed. While there is limited evidence based on clinical studies, the empirical medical evidence is strong.

Safety: Overall, considered safe and high risk is not expected since constituents are largely unabsorbed. However, ispaghula husk should be taken with a sufficient amount of liquid, as abdominal distension and risk of intestinal obstruction and faecal impaction may occur. High doses may cause abdominal discomfort and flatulence.

Main chemical compounds: The main constituents are water-soluble fibres and mucilages, mainly highly branched acidic arabinoxylan polysaccharides like

---

Phytopharmacy: An evidence-based guide to herbal medicinal products, First Edition.
Sarah E. Edwards, Inês da Costa Rocha, Elizabeth M. Williamson and Michael Heinrich.
© 2015 John Wiley & Sons, Ltd. Published 2015 by John Wiley & Sons, Ltd.

D-xylose, L-arabinose, rhamnose and D-galacturonic acid (EMA 2013; ESCOP 2003). The husk also contains monoterpene alkaloids (indicaine and plantagonine), phenylethanoids forsythoside and acteoside. Sterols (campesterol, β-sitosterol, stigmasterol) and triterpenes (α- and β-amyrin) are also present (Williamson et al. 2013).

Clinical evidence:

Constipation: In cases of simple and chronic constipation and among the elderly, the use of ispaghula husk is reported to be effective (Ashraf et al. 1997; Cheskin et al. 1995; McRorie et al. 1998). Its use as a bulking agent is well established, but many of the reported studies were only single-blinded. A systematic review on the use comparing non-bulk-forming laxatives (lactulose, lactitol, docusate, magnesium salts and bisocodyl) with fibre (ispaghula and bran) on chronic constipation in adults found that there was little evidence to establish which class of laxative was superior to another. Both improved the symptoms of constipation compared to placebo, with no severe side effects being reported (Tramonte et al. 1997). Similar results were reported by another review on laxative therapies for treatment of chronic constipation in older adults (Fleming and Wade 2010). A study investigating the use of ispaghula after haemorrhoidectomy found that it reduced pain and tenesmus rate and shortened postoperative hospital stay (Kecmanovic et al. 2006).

A Cochrane review assessing different strategies for managing faecal incontinence and constipation in patients with neurological disease showed that ispaghula husk increased stool frequency in cases of Parkinson's disease, but did not alter colonic transit time (Coggrave et al. 2014). No robust conclusions could be drawn as the study included a limited number of patients. Therefore, the role of fibre supplementation in neurogenic bowel management in this type of patients is yet to be determined.

Hypercholesterolaemia: A randomised clinical study assessing the efficacy and safety of ispaghula husk (Isapgol) plus atorvastatin, versus atorvastatin alone, in subjects with hypercholesterolaemia found that although no significant differences were observed between the two groups in terms of cholesterol-lowering effects, there were less side effects in the combination group than in the atorvastatin alone after 8 weeks' treatment (Jayaram et al. 2007). However, in another study in patients taking simvastatin 10 and 20 mg, ispaghula (15 g daily for 8 weeks) produced a significant lowering effect in cholesterol in those taking 10 mg of simvastatin, equivalent to 20 mg of simvastatin alone (Moreyra et al. 2005).

Prevention of colorectal adenomas and carcinomas: A Cochrane review on fibre supplementation, including with ispaghula husk, showed that there is still no evidence from randomised clinical trials (RCTs) to support that an increase in dietary fibre intake may reduce the occurrence of adenomatous polyps (Asano and McLeod 2002). However, an RCT showed that ispaghula husk supplementation might have adverse effects on colorectal adenoma recurrence, mainly in patients with a high dietary intake of calcium (Bonithon-Kopp et al. 2000).

Pre-clinical evidence and mechanisms of action:

Laxative effects: Ispaghula husk acts as a laxative due to the content of water soluble fibre, which shortens gastrointestinal transit time and increases stool weight. It produces an increase in volume of intestinal content, leading to a stretch stimulus followed by defecation. Simultaneously, the mucilage forms a lubricant layer, which facilitates intestinal transit (EMA 2013). A recent *in vivo* study showed that psyllium husk had both a gut-stimulatory effect (mediated partially by muscarinic and 5-HT4 receptor activation) and inhibitory effects (by blockade of calcium channels and activation of NO-cyclic guanosine monophosphate pathways). This also helps to explain the dual efficacy in constipation and diarrhoea (Mehmood et al. 2011).

Hypocholesterolemic effects: Ispaghula husk showed hypocholesterolemic effects in two studies with adult male African green monkeys, the authors suggesting that it decreased plasma cholesterol by reducing low-density lipoprotein (LDL) apo B production and decreasing LDL synthesis, and not by increasing bile acid excretion. The mucilage may contribute to the effect by delaying absorption and reducing the circulating triglyceride-rich lipoprotein pool (McCall et al. 1992a; McCall et al. 1992b).

Interactions: Ispaghula should not be taken within at least half to one hour of any other medication, as it may delay their absorption (EMA 2013). In insulin-dependent patients, concomitant use with thyroid hormones requires medical supervision, as doses may need to be adjusted (ESCOP 2003). To decrease the risk of gastrointestinal obstruction (ileus), it may be advisable that use with drugs that inhibit peristaltic movement (e.g. opioids, loperamide) should be carried out only under medical supervision (EMA 2013). In theory, ispaghula should not be given with mesalazine (mesalamine), as it lowers colonic pH and may reduce the efficacy of mesalazine, although no interactions have been reported. The bioavailability of carbamazepine and lithium may be reduced when given with ispaghula (Williamson et al. 2013).

Contraindications: Not recommended for children under 6 years of age for constipation, or below 12 years for other indications, due to insufficient evidence of efficacy. Ispaghula husk should not be taken immediately before bedtime. Not recommended for patients with a sudden change in bowel habit that persists for more than 2 weeks, undiagnosed rectal bleeding and failure to defecate following the use of a laxative. Not recommended in patients suffering from abnormal constrictions in the gastrointestinal tract, oesophagus and cardiac disease, intestinal blockage, paralysis or megacolon, and diabetes mellitus when insulin adjustment is difficult. If symptoms of chest pain, vomiting or difficulty in swallowing or breathing occur (signs of potential or existing intestinal blockage), a health care professional should be consulted immediately. Not recommended in patients who have difficulty in swallowing. There have been rare reports of individual allergic and anaphylactic reactions. In debilitated and/or elderly patients, and for use as an adjuvant to diet in hypercholesterolemia, medical supervision is required (EMA 2013; ESCOP 2003).

Adverse effects: Flatulence may occur but usually disappears over the course of treatment. Overdosing and/or insufficient liquid intake may cause abdominal distension and the risk of oesophageal or intestine obstruction and faecal impaction (ESCOP 2003).

Dosage: According to manufacturers' instructions. The daily dosage ranges from 3–20 g daily in one to three doses. Usually the effect is observed within 12 to 24 hours after a single administration, with maximum effect after 2 to 3 days. Should be taken with adequate fluid intake (30 ml of water per 1 g of herbal substance) (EMA 2013).

General plant information: The plant is native to the Mediterranean region and West Asia but is now widely distributed throughout the world. Ispaghula/psyllium husk is the epidermis (episperm and collapsed adjacent layers) of the seed of *P. ovata*, and is capable of absorbing up to 40 times its own weight in water. It is obtained by milling the seed to remove the husk. The term 'ispaghula' is used to refer to *P. ovata* only, while 'psyllium' may be used for other species of the genus *Plantago* as well, e.g. *P. indica* L. The whole seeds are commercially available and in some studies are used instead of the husk. In some scientific publications, the broad term 'psyllium' is used, without defining which type of preparation is used, husk or seed (EMA 2013).

References

Asano T, McLeod RS. (2002) Dietary fibre for the prevention of colorectal adenomas and carcinomas. *Cochrane Database of Systematic Reviews* (2): CD003430.

Ashraf W, Pfeiffer RF, Park F, Lof J, Quigley E M. (1997) Constipation in Parkinson's disease: objective assessment and response to psyllium. *Movement Disorders* 12(6): 946–951.

Bonithon-Kopp C, Kronborg O, Giacosa A, Rath U, Faivre J. (2000) Calcium and fibre supplementation in prevention of colorectal adenoma recurrence: a randomised intervention trial. European Cancer Prevention Organisation Study Group. *Lancet* 356(9238): 1300–1306.

Cheskin LJ, Kamal N, Crowell MD, Schuster MM, Whitehead WE. (1995) Mechanisms of constipation in older persons and effects of fiber compared with placebo. *Journal of the American Geriatrics Society* 43(6): 666–669.

Coggrave M, Norton C, Cody JD. (2014) Management of faecal incontinence and constipation in adults with central neurological diseases. *Cochrane Database of Systematic Reviews* 1:CD002115.

EMA. (2013) Community Herbal monograph on *Plantago ovata* Forssk., Seminis tegumentum. European Medicines Agency http://www.ema.europa.eu/docs/en_GB/document_library/Herbal_-_Community_herbal_monograph/2013/07/WC500146508.pdf.

ESCOP. (2003) *ESCOP Monographs: The Scientific Foundation for Herbal Medicinal Products.* 2nd Edition. Thieme, Exeter and London, UK.

Fleming V, Wade WE. (2010) A review of laxative therapies for treatment of chronic constipation in older adults. *American Journal of Geriatric Pharmacotherapy* 8(6): 514–550.

Jayaram S, Prasad HB, Sovani VB, Langade DG, Mane PR. (2007) Randomised study to compare the efficacy and safety of isapgol plus atorvastatin versus atorvastatin alone in subjects with hypercholesterolaemia. *Journal of the Indian Medical Association* 105(3): 142–145, 150.

Kecmanovic DM, Pavlov MJ, Ceranic MS, Kerkez MD, Rankovic VI, and Masirevic VP. (2006) Bulk agent *Plantago ovata* after Milligan-Morgan hemorrhoidectomy with Ligasure™. *Phytotherapy Research* 20(8): 655–658.

McCall MR, Mehta T, Leathers CW, Foster DM. (1992a) Psyllium husk. I: Effect on plasma lipoproteins, cholesterol metabolism, and atherosclerosis in African green monkeys. *American Journal of Clinical Nutrition* 56(2): 376–384.

McCall MR, Mehta T, Leathers CW, Foster DM. (1992b) Psyllium husk. II: Effect on the metabolism of apolipoprotein B in African green monkeys. *American Journal of Clinical Nutrition* 56(2): 385–393.

McRorie JW, Daggy BP, Morel JG, Diersing PS, Miner PB, Robinson M. (1998) Psyllium is superior to docusate sodium for treatment of chronic constipation. *Alimentary Pharmacology and Therapeutics* 12(5): 491–497.

Mehmood MH, Aziz N, Ghayur MN, Gilani AH. (2011) Pharmacological basis for the medicinal use of psyllium husk (Ispaghula) in constipation and diarrhea. *Digestive Diseases and Sciences* 56(5): 1460–1471.

Moreyra AE, Wilson AC, Koraym A. (2005) Effect of combining psyllium fiber with simvastatin in lowering cholesterol. *Archives of Internal Medicine* 165(10): 1161–1166.

Ivy

Hedera helix L.

Family: Araliaceae

Other common names: Common ivy; English ivy; European ivy

Drug name: Hederae folium

Botanical drug used: Dried leaf

Indications/uses: Uses include for the treatment of upper respiratory tract infections (URTIs), as an expectorant in cough associated with cold, and for symptomatic treatment of acute and chronic inflammatory bronchial disorders. There are registered products available for use in the treatment of chesty coughs associated with the common cold.

Evidence: Clinical studies and a post-marketing surveillance study suggest that ivy preparations are effective for the treatment of URTIs, but further rigorously designed randomised controlled trials are necessary.

Safety: Overall, ivy preparations are considered to be safe in recommended doses, including for children, although allergic reactions have been reported.

Main chemical compounds: Saponins based on hederagenin, bayogenin and oleanolic acid, including α-hederin, β-hederin, hederacoside C; polyacetylenic compounds such as falcarinol and didehydrofalcarinol; flavonoids and other polyphenolic compounds (Heinrich et al. 2012; Paulsen et al. 2010).

Clinical evidence: Several studies have concluded that ivy leaf extracts are effective in reducing symptoms of acute URTI. A systematic review of clinical trials, however, found that many studies lacked a placebo control or contained other methodological flaws (Holzinger and Chenot 2011).

A double-blind randomised study in 590 patients with acute bronchitis compared the efficacy and tolerability of two ivy leaf extracts (a dry extract and a soft extract) and found both to be effective in improving symptoms; overall, there were few side effects which were not considered serious (Cwientzek et al. 2011).

Two formulations of an ivy leaf extract, syrup and cough drops, were tested for efficacy and safety in children with cough and bronchitis in two independent open, non-interventional studies of identical design. A total of 268 children, aged 0–12 years, were treated with either of the preparations for up to 14 days, and the effects on cough-related symptoms measured using a verbal rating scale. The major symptoms of rhinitis, cough and viscous mucus were found to be mild or absent (in 93%, 94.2% and 97.7% of cases, respectively) and the overall effect was rated as 'good' or 'very good' in 96.5% of cases. On completion of the study, tolerability and compliance were reported as 'good' or 'very good' in 99% (syrup) and 100% (drops) of patients. A subgroup analysis of four different age and dosing groups did not reveal any differences in response. Five adverse events, classified as mild (1.9%), were reported and the authors concluded that ivy leaf extract in the form of syrup or cough drops was an effective and safe treatment of cough in children (Schmidt et al. 2012).

A placebo-controlled, randomised clinical trial of a product containing both ivy and thyme extracts provided evidence of a faster recovery from symptoms in the intervention group (Kemmerich et al. 2006).

A post-marketing surveillance study that included adults and children (330 patients aged 11–85 years, of both genders) assessed the tolerability and safety of a product consisting of film-coated tablets containing ivy leaf dry extract. This study showed good to very good tolerability of the tablets when assessed by both practitioner and patient (Stauss-Grabo et al. 2011).

Pre-clinical evidence and mechanisms of action: The bronchiolytic and secretolytic effects of ivy extracts are thought to be due mainly to the effect of the saponin α-hederin, which increases β2-adrenergic responsiveness and elevates cyclic adenosine monophosphate (cAMP) levels, leading to a relaxation of airway smooth muscle cells (Sieben et al. 2009; Wolf et al. 2011). A pre-clinical study showed that a-hederin, but not hederacoside C, contributed to the contractile response of rat isolated stomach corpus and fundus strips (Mendel et al. 2011).

Interactions: None known (Williamson et al. 2013).

Contraindications: Not recommended during pregnancy and lactation, as safety has not been established. Hypersensitivity to plants of the Araliaceae family. There is risk of aggravation of respiratory symptoms in children less than 2 years of age. The saponins and the polyacetylenic compounds, falcarinol and didehydrofalcarinol, may cause irritant and allergic contact dermatitis (Paulsen et al. 2010).

Adverse effects: Gastrointestinal reactions (nausea, vomiting, diarrhoea) as well as allergic reactions (urticaria, skin rash, dyspnoea) have been reported, with gastrointestinal reactions being the most common. A qualified health care practitioner should be consulted if adverse reactions should occur. Overdose may cause nausea, vomiting, diarrhoea and agitation (EMA 2011b). Although contact with the fresh plant of *H. helix* is known to cause allergic contact dermatitis (Jones et al. 2009; Paulsen et al. 2010), and occasionally occupational asthma in gardeners exposed to the leaves (Hannu et al. 2008), no severe adverse events were reported in more than 17,000 subjects treated with ivy extracts, suggesting that there is considerable evidence for its safety (Holzinger and Chenot 2011).

Dosage: The maximum daily dose for finished ivy products is 67 mg extract, corresponding to 420 mg herbal substance (EMA 2011a).

General plant information: Ivy is a common evergreen climbing plant, native to Europe. It has a long tradition of use as a decoration in mid-winter and Christmas festivities.

References

Cwientzek U, Ottillinger B, Arenberger P. (2011) Acute bronchitis therapy with ivy leaves extracts in a two-arm study. A double-blind, randomised study vs. an other ivy leaves extract. *Phytomedicine* 18(13): 1105–1109.

EMA. (2011a) Community herbal monograph on *Hedera helix* L., folium. European Medicines Agency. http://www.ema.europa.eu/docs/en_GB/document_library/Herbal_-_Community_herbal_monograph/2011/04/WC500105313.pdf.

EMA. (2011b) Assessment report on *Hedera helix* L., folium. European Medicines Agency. http://www.ema.europa.eu/docs/en_GB/document_library/Herbal_-_HMPC_assessment_report/2012/01/WC500120648.pdf.

Hannu T, Kauppi P, Tuppurainen M, Piirila P. (2008) Occupational asthma to ivy (*Hedera helix*). *Allergy* 63(4): 482–483.

Heinrich M, Barnes J, Gibbons S, Williamson EM. (2012) *Fundamentals of Pharmacognosy and Phytotherapy*. 2ⁿᵈ Edition. Churchill Livingstone Elsevier, Edinburgh and London.

Holzinger F, Chenot JF. (2011) Systematic review of clinical trials assessing the effectiveness of ivy leaf (*Hedera helix*) for acute upper respiratory tract infections. *Evidence-Based Complementary and Alternative Medicine* 2011: 382789.

Jones JM, White IR, White JM, McFadden JP. (2009) Allergic contact dermatitis to English ivy (*Hedera helix*) – a case series. *Contact Dermatitis* 60(3): 179–180.

Kemmerich B, Eberhardt R, Stammer H. (2006) Efficacy and tolerability of a fluid extract combination of thyme herb and ivy leaves and matched placebo in adults suffering from acute bronchitis with productive cough. A prospective, double-blind, placebo-controlled clinical trial. *Arzneimittelforschung* 56(9): 652–660.

Mendel M, Chlopecka M, Dziekan N, Wiechetek M. (2011) The effect of the whole extract of common ivy (*Hedera helix*) leaves and selected active substances on the motoric activity of rat isolated stomach strips. *Journal of Ethnopharmacology* 134(3): 796–802.

Paulsen E, Christensen LP, Andersen KE. (2010) Dermatitis from common ivy (*Hedera helix* L. subsp. *helix*) in Europe: past, present, and future. *Contact Dermatitis* 62(4): 201–209.

Schmidt M, Thomsen M, Schmidt U. (2012) Suitability of ivy extract for the treatment of paediatric cough. *Phytotherapy Research* 26(12): 1942–1947.

Sieben A, Prenner L, Sorkalla T, Wolf A, Jakobs D, Runkel F, Haberlein H. (2009) Alpha-hederin, but not hederacoside C and hederagenin from *Hedera helix*, affects the binding behavior, dynamics, and regulation of beta 2-adrenergic receptors. *Biochemistry* 48(15): 3477–3482.

Stauss-Grabo M, Atiye S, Warnke A, Wedemeyer RS, Donath F, Blume HH. (2011) Observational study on the tolerability and safety of film-coated tablets containing ivy extract (Prospan® Cough Tablets) in the treatment of colds accompanied by coughing. *Phytomedicine* 18(6): 433–436.

Williamson EM, Driver S, Baxter K. (Eds.) (2013) *Stockley's Herbal Medicines Interactions*. 2ⁿᵈ Edition. Pharmaceutical Press, London, UK.

Wolf A, Gosens R, Meurs H, Haberlein H. (2011) Pre-treatment with alpha-hederin increases beta-adrenoceptor mediated relaxation of airway smooth muscle. *Phytomedicine* 18(2–3): 214–218.

Kalmegh   *Andrographis paniculata* (Burm.f.) Nees

Synonyms: *Justicia paniculata* Burm.f; and others.

Family: Acanthaceae

Other common names: Andrographis; green chiretta; Indian echinacea; kalamegha

Drug name: Andrographis herba

Botanical drug used: Aerial parts

Indications/uses: Mainly used to treat infection, inflammation and fever, and specifically the common cold, influenza type and other upper respiratory tract infections (URTIs). In both Chinese medicine and Ayurveda, it is used for a wide range of complaints including cardiovascular disorders, and as a general tonic.

Evidence: Randomised controlled trials suggest that *A. paniculata* extract alone or in combination with *Eleutherococcus senticosus* is superior to placebo in alleviating the symptoms of uncomplicated URTIs. There is also preliminary evidence of a preventative effect. There is no clinical evidence available for other indications, but there is pre-clinical evidence in support of some of these uses.

Safety: Overall, considered safe, but may cause minor gastrointestinal disturbances or allergic responses, and the usual precautions are required (see below).

Main chemical compounds: The main constituents are diterpenes (e.g. andrographic acid, diterpene glucoside (deoxyandrographolide-19-β-D-glucoside) and diterpene dimers (bis-andrographolides A, B, C and D), lactones (andrographolide, kalmeghin, chuanxinlian A (deoxyandrographolide), B (andrographolide), C (neoandrographolide) and D (14-deoxy-11,12-didehydroandrographolide) and flavonoids (e.g. 5,7,2′,3′-tetramethoxyflavanone and 5-hydroxy-7,2′,3′-trimethoxyflavone) (Akbar 2011).

Clinical evidence: Two systematic reviews and seven clinical trials (six in adults and one in children) were retrieved for the use of *A. paniculata* in URTI (Akbar 2011; Kligler et al. 2006). The majority of the clinical trials were conducted using a standardised commercial preparation of *A. paniculata* extract in a fixed combination with *E. senticosus* (Kan Jang). The authors concluded that the data available suggests that this herb is effective in reducing the severity and duration of the symptoms

---

Phytopharmacy: An evidence-based guide to herbal medicinal products, First Edition.
Sarah E. Edwards, Inês da Costa Rocha, Elizabeth M. Williamson and Michael Heinrich.
© 2015 John Wiley & Sons, Ltd. Published 2015 by John Wiley & Sons, Ltd.

in URTI, both in children and adults, but further research is warranted. Another randomised, double-blind placebo-controlled clinical trial was conducted, evaluating the efficacy of a specific extract of *A. paniculata* (200 mg/day for 5 days), in patients with uncomplicated URTI. The study showed that the extract was effective in reducing the symptoms of URTI (2.1 times or 52.7% higher than placebo with $p \leq 0.05$) when quantifying symptom scores of cough, expectoration, nasal discharge, headache, fever, sore throat, earache, malaise/fatigue and sleep disturbance (Saxena et al. 2010).

Pre-clinical evidence and mechanisms of action: The exact mechanism of action is unknown, but it may be based either on the anti-inflammatory properties or the immunomodulatory properties of this botanical drug. Extract of *A. paniculata* and its main component andrographolide enhanced the immune system by increasing the proliferation of lymphocytes and production of interleukin (IL)-2, and inhibited tumour cell proliferation (Kumar et al. 2004; Rajagopal et al. 2003). Andrographolide and neoandrographolide showed significant anti-inflammatory effects *in vivo* (Liu et al. 2007). Andrographolide and Kan Jang (a standardised fixed combination of *A. paniculata* extract SHA-10 and *E. senticosus* extract SHE-3) showed an effect *in vitro* on the activation and proliferation of immune-competent cells, as well as on the production of key cytokines and immune activation markers (Panossian et al. 2002).

Andrographolide and neoandrographolide have shown potent hypolipidemic effects in mice, reducing the serum total triglyceride, total cholesterol and low-density lipoprotein cholesterol, without significant liver damage, which may support the use of *Andrographis* in traditional Chinese medicine to protect the cardiovascular system (Yang et al. 2013). Extracts and andrographolide have antioedema and analgesic activities (Lin et al. 2009).

Interactions: No clinical cases reported (Williamson et al. 2013). Based on animal models, possible interaction with antiplatelet (inhibit platelet aggregation) (Amroyan et al. 1999) and antidiabetic drugs (possible hypoglycemic effect) (Husen et al. 2004). It might cause decrease of fertility at high doses; thus, patients undergoing infertility treatment should avoid this herb (Kligler et al. 2006).

Contraindications: None have been reported. Not recommended during pregnancy and lactation, as safety has not been established.

Adverse effects: One case of anaphylactic shock and two cases of anaphylactic reactions were reported to the WHO Collaborating Centre for International Drug Monitoring (Coon and Ernst 2004). From clinical trials, no serious side effects were reported. However, safety of longer term use is impossible to determine as most trials to date were of short duration (2 weeks or less in general) (Kligler et al. 2006). A case of unpleasant sensations in the chest and intensified headache (Melchior et al. 2000) and urticaria were associated with *A. paniculata* (Melchior et al. 1997; Poolsup et al. 2004) but in children, the preparation was well tolerated with no reported adverse effects (Spasov et al. 2004). In a phase I dose-escalating clinical trial of andrographolide from *A. paniculata*, one HIV-positive participant experienced an anaphylactic reaction at a dose of 10 mg/kg body weight. Others reported headache, fatigue, rash, bitter/metallic/decreased taste, loose stool/diarrhoea, or pruritus as being mild to moderate. Among the HIV-negative patients only one reported body rash and headache (Calabrese et al. 2000).

Dosage: Most of the clinical studies for common colds and URTI used a patented product (Kan Jang), in which andrographis is standardised to contain 4–6% of andrographolides. The dose varied from 60–72 mg/day to about 300 mg/day (Akbar 2011).

General plant information: The plant is native to India, China, and Southeast Asia. It is a commonly used plant in Chinese traditional medicine as well as the Unani and Ayurvedic systems (Akbar 2011). A couple of clinical trials reported positive effects for its use in mild-to-moderate active ulcerative colitis (Sandborn et al. 2012; Tang et al. 2011).

References

Akbar S. (2011) *Andrographis paniculata*: a review of pharmacological activities and clinical effects. *Alternative Medicine Review* 16(1): 66–77.

Amroyan E, Gabrielian E, Panossian A, Wikman G, Wagner H. (1999) Inhibitory effect of andrographolide from *Andrographis paniculata* on PAF-induced platelet aggregation. *Phytomedicine* 6(1): 27–31.

Calabrese C, Berman SH, Babish JG, Ma X, Shinto L, Dorr M, Wells K, Wenner CA, Standish LJ. (2000) A phase I trial of andrographolide in HIV positive patients and normal volunteers. *Phytotherapy Research* 14(5): 333–338.

Coon JT, Ernst E. (2004) *Andrographis paniculata* in the treatment of upper respiratory tract infections: a systematic review of safety and efficacy. *Planta Medica* 70(4): 293–298.

Husen R, Pihie AH, Nallappan M. (2004) Screening for antihyperglycaemic activity in several local herbs of Malaysia. *Journal of Ethnopharmacology* 95(2–3): 205–208.

Kligler B, Ulbricht C, Basch E, Kirkwood CD, Abrams TR, Miranda M, Singh Khalsa KP, Giles M, Boon H, Woods J. (2006) *Andrographis paniculata* for the treatment of upper respiratory infection: a systematic review by the natural standard research collaboration. *Explore (NY)* 2(1): 25–29.

Kumar RA, Sridevi K, Kumar NV, Nanduri S, Rajagopal S. (2004) Anticancer and immunostimulatory compounds from *Andrographis paniculata*. *Journal of Ethnopharmacology* 92(2–3): 291–295.

Lin FL, Wu SJ, Lee SC, Ng LT. (2009) Antioxidant, antioedema and analgesic activities of *Andrographis paniculata* extracts and their active constituent andrographolide. *Phytotherapy Research* 23(7): 958–964.

Liu J, Wang ZT, Ji LL. (2007) *In vivo* and *in vitro* anti-inflammatory activities of neoandrographolide. *American Journal of Chinese Medicine* 35(2): 317–328.

Melchior J, Palm S, Wikman G. (1997) Controlled clinical study of standardized *Andrographis paniculata* extract in common cold – a pilot trial. *Phytomedicine* 3(4): 315–318.

Melchior J, Spasov AA, Ostrovskij OV, Bulanov AE, Wikman G.. (2000) Double-blind, placebo-controlled pilot and phase III study of activity of standardized *Andrographis paniculata* Herba Nees extract fixed combination (Kan jang) in the treatment of uncomplicated upper-respiratory tract infection. *Phytomedicine* 7(5): 341–350.

Panossian A, Davtyan T, Gukassyan N, Gukasova G, Mamikonyan G, Gabrielian E, Wikman G. (2002) Effect of andrographolide and Kan Jang – fixed combination of extract SHA-10 and extract SHE-3 on proliferation of human lymphocytes, production of cytokines and immune activation markers in the whole blood cells culture. *Phytomedicine* 9(7): 598–605.

Poolsup N, Suthisisang C, Prathanturarug S, Asawamekin A, Chanchareon U. (2004) *Andrographis paniculata* in the symptomatic treatment of uncomplicated upper respiratory tract infection: systematic review of randomized controlled trials. *Journal of Clinical Pharmacy and Therapeutics* 29(1): 37–45.

Rajagopal S, Kumar RA, Deevi DS, Satyanarayana C, Rajagopalan R. (2003) Andrographolide, a potential cancer therapeutic agent isolated from *Andrographis paniculata*. *Journal of Experimental Therapeutics and Oncology* 3(3): 147–158.

Sandborn WJ, Targan SR, Byers VS, Rutty DA, Mu H, Zhang X, Tang T. (2012) *Andrographis paniculata* extract (HMPL-004) for active ulcerative colitis. *American Journal of Gastroenterology* 108: 90–98.

Saxena R C, Singh R, Kumar P, Yadav SC, Negi MP, Saxena VS, Joshua AJ, Vijayabalaji V, Goudar KS, Venkateshwarlu K, Amit A. (2010) A randomized double blind placebo controlled clinical evaluation of extract of *Andrographis paniculata* (KalmCold) in patients with uncomplicated upper respiratory tract infection. *Phytomedicine* 17(3–4): 178–185.

Spasov AA, Ostrovskij OV, Chernikov MV, Wikman G. (2004) Comparative controlled study of *Andrographis paniculata* fixed combination, Kan Jang® and an Echinacea preparation as adjuvant, in the treatment of uncomplicated respiratory disease in children. *Phytotherapy Research* 18(1): 47–53.

Tang T, Targan SR, Li ZS, Xu C, Byers VS, Sandborn WJ. (2011) Randomised clinical trial: herbal extract HMPL-004 in active ulcerative colitis – a double-blind comparison with sustained release mesalazine. *Alimentary Pharmacology & Therapeutics* 33(2): 194–202.

Williamson EM, Driver S, Baxter K (Eds.) (2013). *Stockley's Herbal Medicines Interactions*. 2nd Edition. Pharmaceutical Press, London, UK.

Yang T, Shi HX, Wang ZT, Wang CH. (2013) Hypolipidemic effects of andrographolide and neoandrographolide in mice and rats. *Phytotherapy Research* 27(4): 618–623.

♠ ♦ ♦ ♦ ♦

Lapacho   *Handroanthus impetiginosus* (Mart. ex DC.) Mattos

Synonyms: *Handroanthus avellanedae* (Lorentz ex Griseb.) Mattos; *Tabebuia avellanedae* Lorentz ex Griseb.; *T. impetiginosa* (Mart ex DC.) Standl.; and others

Family: Bignoniaceae

Other common names: Ipê roxo; pau d'arco; red or purple lapacho; taheebo

Drug name: Tabebuiae cortex

Botanical drug used: Bark (inner bark preferred)

Indications/uses: Lapacho is widely used, especially in South America, and often in the form of a herbal tea, for a range of conditions, including infections (bacterial, fungal, protozoal and viral), for digestive disorders, and as an anti-inflammatory agent. It is applied externally for skin ulcerations and boils, eczema and psoriasis, and even skin cancer. It is reputed to strengthen the immune system and is a popular herbal medicine for treating all forms of cancer. The FDA has registered the drug 'to alleviate conditions and symptoms of cancer' as a dietary supplement.

Evidence: There is limited evidence available to support the alleged health benefits of lapacho, and despite a considerable amount of pharmacological evidence of anticancer effects for the constituent β-lapachone, these were not borne out by clinical trials, which have now been discontinued.

Safety: While the tea is generally considered to be safe in moderate doses, there is also a lack of data regarding the quality of the products on the market, which needs to be taken into consideration.

Main chemical compounds: Naphthoquinones, the most important being lapachol and β-lapachone, iridoid, lignan, isocoumarin, phenolic and phenylethanoid glycosides (Gómez-Castellanosa et al. 2009). Several novel benzoyl apiosides have been isolated (Suo et al. 2012).

Clinical evidence:

The clinical evidence is limited to studies on the potential anticancer effects in phase I and II clinical trials (see Gómez-Castellanosa et al. 2009).

Phytopharmacy: An evidence-based guide to herbal medicinal products, First Edition.
Sarah E. Edwards, Inês da Costa Rocha, Elizabeth M. Williamson and Michael Heinrich.
© 2015 John Wiley & Sons, Ltd. Published 2015 by John Wiley & Sons, Ltd.

Anticancer effects: In the 1960s, an Investigational New Drug Application (IND) was filed in the United States, and phase I clinical trials were initiated, with doses up to 4000 mg/day administered orally. No therapeutic response was observed, and it was determined that satisfactory blood levels could not be obtained by oral administration; also, some toxicity was observed. The IND was closed in 1970 (Gómez-Castellanosa et al. 2009).

Pre-clinical evidence and mechanisms of action:

Anticancer effects: β-lapachone is reported to have significant activity against a range of tumour cell lines, including breast, leukaemia and prostate, as well as several multi-drug resistance cell lines, by interfering with topoisomerases. Other potential mechanisms have been reviewed by Gómez-Castellanosa et al. (2009). The anti-proliferative effects of lapacho correlated with down-regulated cell cycle regulatory and oestrogen-responsive genes, and up-regulated apoptosis specific and xenobiotic metabolism specific genes, in human breast carcinoma derived ER+MCF-7 cells (Mukherjee et al. 2009).

Anti-inflammatory effects: In the mouse hot plate and writhing tests, a 200 mg/kg dose of lapacho extract induced a significant antinociceptive effect and increased the pain threshold by approximately 30% compared with the control. In vascular permeability, phorbol ester or carrageenan-induced rat paw oedema and related tests, 200 mg/kg extract inhibited inflammation by 30–50% compared with the control (Lee et al. 2012). The water extract significantly suppressed the production of prostaglandin E_2 and nitric oxide in lipopolysaccharide-stimulated RAW264.7 cells. Oral administration of taheebo (100 mg/kg) for 1 week completely diminished mouse ear oedema induced by arachidonic acid, an activator of COX-II, but not croton oil, an activator of lipoxygenase (Byeon et al. 2008). The benzoyl apiosides have recently been found to have significant anti-inflammatory effects by directly suppressing inflammatory cytokine production (Suo et al. 2012). Multiple constituents and multiple mechanisms are therefore likely to account for its anti-inflammatory effects.

Effects on the GI tract: The ethanolic bark extract was shown to accelerate healing of acetic acid-induced gastric ulcer in rats, through increase of mucus content and cell proliferation (Pereira et al. 2013). The extract also inhibited the growth of *Helicobacter pylori*, which would contribute to an anti-ulcerative effect (Park et al. 2006).

Administration of lapacho tea led to a significant delay in the postprandial increase of plasma triglycerides in rats; however, lapachol had no lipase inhibitory effect *in vitro* and thus does not seem to mediate this effect (Kiage-Mokua et al. 2012).

Other effects: Weak antimicrobial properties (in addition to anti-*H. pylori* effects) have been reported (Gómez-Castellanosa et al. 2009).

Antidepressant-like action has been observed for the bark ethanolic extract in several mouse models: the forced swimming test, the tail suspension test, and in olfactory bulbectomised mice (Freitas et al. 2013).

Interactions: No clinical reports available, but in view of the wide range of biological activities exhibited by lapacho, concurrent administration with drugs for similar conditions should be avoided. Lapachol is a potent inhibitor of vitamin K metabolism, which explains the anticoagulant activity of the botanical drug, so it should be avoided in combination with warfarin (see Gómez-Castellanosa et al. 2009).

Contraindications: No specific contraindications recorded, but the range of biological activities exhibited by lapacho should be taken into account when deciding whether to take the botanical drug. In the United States, the tea has an FDA regulatory classification of 'generally recognised as safe' (GRAS) status.

Adverse effects: Very limited data are available. In rats, genotoxic effects were observed at a comparatively high dose range (Lemos et al. 2012). The herbal material available on the international market is of variable quality, giving rise to concerns of additional risks associated with adulterated or poor quality materials.

Dosage: 1–2 g inner bark, in the form of a tea, twice daily. Higher doses are often used but increase the likelihood of toxicity.

General plant information: Red lapacho is a canopy tree indigenous to the Amazonian rainforest and other parts of South America.

References

Byeon SE, Chung JY, Lee YG, Kim BH, Kim KH, Cho JY. (2008) *In vitro* and *in vivo* anti-inflammatory effects of taheebo, a water extract from the inner bark of *Tabebuia avellanedae*. *Journal of Ethnopharmacology* 119(1): 145–152.

Gómez-Castellanosa JR, Prieto JM, Heinrich M. (2009) Red Lapacho (*Tabebuia impetiginosa*) – a global ethnopharmacological commodity? *Journal of Ethnopharmacology* 121(1): 1–13.

Freitas AE, Machado DG, Budni J, Neis VB, Balen GO, Lopes MW, de Souza LF, Veronezi PO, Heller M, Micke GA, Pizzolatti MG, Dafre AL, Leal RB, Rodrigues AL. (2013) Antidepressant-like action of the bark ethanolic extract from *Tabebuia avellanedae* in the olfactory bulbectomized mice. *Journal of Ethnopharmacology* 145(3): 737–745.

Kiage-Mokua BN, Roos N, Schrezenmeir J. (2012) Lapacho tea (*Tabebuia impetiginosa*) extract inhibits pancreatic lipase and delays postprandial triglyceride increase in rats. *Phytotherapy Research* 26(12): 1878–1883.

Lee MH, Choi HM, Hahm DH, Her E, Yang HI, Yoo MC, Kim KS. (2012) Analgesic and anti-inflammatory effects in animal models of an ethanolic extract of Taheebo, the inner bark of *Tabebuia avellanedae*. *Molecular Medicine Reports* 6(4): 791–796.

Lemos OA, Sanches JCM, Silva IEF, Silva MLA, Vinhólis AHC, Felix,MAP, Santos RA, Cecchi AO. (2012) Genotoxic effects of *Tabebuia impetiginosa* (Mart. Ex DC.) Standl. (Lamiales, Bignoniaceae) extract in Wistar rats. *Journal of Ethnopharmacology* 35(2): 498–502.

Mukherjee B, Telang N, Wong GY. (2009) Growth inhibition of estrogen receptor positive human breast cancer cells by Taheebo from the inner bark of *Tabebuia avellanedae* tree. *International Journal of Molecular Medicine* 24(2): 253–260.

Park BS, Lee HK, Lee SE, Piao XL, Takeoka GR, Wong RY, Ahn YJ, Kim JH. (2006) Antibacterial activity of *Tabebuia impetiginosa* Martius ex DC (Taheebo) against *Helicobacter pylori*. *Journal of Ethnopharmacology* 105(1–2): 255–262.

Pereira IT, Burci LM, da Silva LM, Baggio CH, Heller M, Micke GA, Pizzolatti MG, Marques MC, Werner MF. (2013) Antiulcer effect of bark extract of *Tabebuia avellanedae*: activation of cell proliferation in gastric mucosa during the healing process. *Phytotherapy Research* 27(7): 1067–1073.

Suo M, Isao H, Kato H, Takano F, Ohta T. (2012) Anti-inflammatory constituents from *Tabebuia avellanedae*. *Fitoterapia* 83(8): 1484–1488.

♦ ♦ ♦ ♦ ♦

Lavender *Lavandula angustifolia* Mill.

Synonyms: *L. officinalis* Chaix; *L. spica* L.; *L. vulgaris* Lam.

Family: Lamiaceae (Labiatae)

Other common names: English lavender; common lavender; garden lavender; true lavender. (English lavender from France is often traded as 'French lavender', although the related *L. stoechas* L. is more usually referred to as French lavender.)

Drug name: Lavandulae flos; Lavandulae aetheroleum

Botanical drug used: Dried flower; essential oil (obtained by steam distillation from fresh flowering tops)

> **Indications/uses:** Traditional uses of lavender include treatment of restlessness, agitation, insomnia and nervous intestinal discomfort. Lavender flowers have also been used in infusions to treat asthma, whooping cough, laryngitis and influenza. Externally, lavender oil is used for scar and wound healing, in the treatment of insect bites and burns, and for its aroma, and antimicrobial and analgesic properties.
>
> **Evidence:** Clinical evidence supports the use of lavender oil (both administered orally in dosage form and nebulised in aromatherapy) for its anxiolytic effects and to improve associated symptoms such as disturbed sleep. There is some evidence that inhalation of lavender oil may reduce postoperative and menstrual pain. Antimicrobial and pesticidal activity of lavender oil has also been demonstrated.
>
> **Safety:** Lavender essential oil and dried lavender flowers are considered safe. Few adverse effects have been reported, although very rare cases of allergic dermatitis have been reported due to contact with lavender oil. If taken orally, the oil must be used very cautiously and the dose carefully controlled.

Main chemical compounds: Essential oil: monoterpene alcohols are the main constituents (60–65%) and include linalool (20–45%), and its acetate ester, linalyl acetate (25–46%). Other terpenoids include 1,8-cineole, terpinen-4-ol, lavendulyl acetate, α-terpineol; camphor; limonene; geraniol and β-caryophyllene and the non-terpenoid 3-octanone. Lavender flowers contain essential oil (1–3%); coumarin derivatives (e.g. umbelliferone); flavonoids; sterols (traces); triterpenes (traces); tannins (up to 13%); phenylcarboxylic acids including rosmarinic acid, ferulic acid, caffeic acid and others (EMA 2012; WHO 2007).

Phytopharmacy: An evidence-based guide to herbal medicinal products, First Edition.
Sarah E. Edwards, Inês da Costa Rocha, Elizabeth M. Williamson and Michael Heinrich.
© 2015 John Wiley & Sons, Ltd. Published 2015 by John Wiley & Sons, Ltd.

Clinical evidence:

Numerous studies have been undertaken investigating lavender, many of which are summarised in the European Medicine Agency's 2012 assessment report on *L. angustifolia* oil and flowers. These include examinations of the effects of lavender oil on the central nervous system (CNS), neuronal activity, antioxidant activity, and trials investigating anxiolytic, antidepressant, analgesic, sedative and anti-stress effects (EMA 2012). Some of the more recent clinical trials are summarised here:

Anxiety: A review of three randomised, double-blind clinical trials investigated the efficacy and tolerability of a proprietary oral lavender oil capsule preparation ('Silexan') in the treatment of anxiety disorders. Silexan (80 mg/day) ($n = 280$) was found to be superior to placebo ($n = 192$), while incidence of adverse events in patients treated with the preparation was comparable to placebo (Kasper et al. 2010). In another study, silexan was compared against lorazepam in the treatment of generalised anxiety disorder. The results showed that silexan ameliorated symptoms of generalised anxiety comparable to lorazepam. No sedative effects were associated with silexan. The authors concluded that silexan appears to be a well-tolerated alternative to benzodiazepines for amelioration of generalised anxiety (Woelk and Schläfke 2011). A phase II clinical trial investigated silexan (administered 1×80 mg/day for 6 weeks) in 50 patients with neurasthenia, and post-traumatic stress disorder. Patients showed statistically significant and clinically meaningful improvements of symptoms like restlessness, sleep-disturbances and sub-threshold anxiety (Uehleke et al. 2013).

In a systematic review of randomised clinical trials investigating anxiolytic activity of lavender oil, 15 trials met the inclusion criteria, of which 7 appeared to favour lavender over control for at least one clinical outcome. The authors concluded that methodological issues limit the extent to which conclusions can be drawn, but that the evidence suggests that lavender oil does have some therapeutic effects (Perry et al. 2012).

Sleep: A systematic review assessed available evidence on lavender aroma inhalation as a possible self-care intervention to improve sleep architecture (initiation, maintenance and quality). Initial results from eight eligible studies (including four randomised controlled trials) were promising, although they had small sample sizes and methodological limitations (Fismer and Pilkington 2012).

Coronary circulation: The effect of lavender aromatherapy on coronary circulation in 30 young healthy men in a single-blind (operator) study was assessed by measuring coronary flow velocity reserve (CFVR), using non-invasive transthoracic Doppler echocardiography. Lavender aromatherapy significantly decreased serum cortisol and improved CFVR, indicating its relaxation effects and beneficial effects on coronary circulation (Shiina et al. 2008).

Pain: A randomised controlled trial in 92 patients undergoing haemodialysis with arteriovenous fistulas found that inhalation of lavender significantly reduced pain compared to placebo following needle insertion into a fistula (Bagheri-Nesami et al. 2014). In another randomised controlled trial in 48 children aged 6–12 years, inhalation of essential lavender oil caused a statistically significant reduction in daily use of acetaminophen post-operatively following tonsillectomy (Soltani et al. 2013).

Dysmenorrhoea: In a randomised controlled trial, 96 students with dysmenorrhoea were treated either with inhalation of lavender in sesame base oil, or sesame oil (as placebo). Symptoms of dysmenorrhoea were significantly lower in the lavender group, and amount of menstrual bleeding was reduced (but was not statistically significant) (Dehkordi et al. 2014).

Postpartum care: A review of the role of lavender oil in relieving perineal trauma symptoms following childbirth identified 6 studies for inclusion. In a randomised blind controlled trial ($n = 635$), use of lavender oil as a bath additive found no significant difference between lavender group and controls, although the lavender group recorded lower mean discomfort score between days 3 and 5 (Jones 2011). In two Iranian studies (both randomised controlled trials) conducted on women following episiotomy ($n = 180$), lavender oil administered in baths was found to be effective at reducing perineal discomfort and redness (Sheikhan et al. 2012; Vakilian et al. 2011).

Head lice: In a randomised, assessor-blind, comparative parallel study with 123 subjects, a head lice treatment product containing lavender oil in combination with tea tree oil was found to be highly effective against head lice. Significantly more subjects (97.6%) were louse-free 1 day after end of treatment compared to subjects treated with a product containing pyrethrins and piperonyl butoxide (25.0%) (Barker and Altman 2010).

Pre-clinical evidence and mechanisms of action:

The mechanism of action has not been clearly established. Linalool is considered the primary active constituent, but many constituents of lavender oil have demonstrable pharmacological activity and are likely to contribute to the total therapeutic effect (Denner 2009; Woronuk et al. 2011).

CNS effects: Lavender exerts a number of actions on the CNS, including anti-convulsive and sedative effects, anti-conflict behaviour and anxiolytic activity (EMA 2012; Kumar 2013; Rahmati et al. 2013). Animal studies indicate that the anxiolytic effects of lavender oil are associated with the serotonergic system, but not $GABA_A$/benzodiazepine neurotransmission (Chioca et al. 2013; Silenieks et al. 2013). Neuroprotective effects of inhaled lavender oil have been shown in rats with scopolamine-induced dementia (Hancianu et al. 2013). While antinociceptive and anti-inflammatory effects have been demonstrated in animals, high doses of lavender oil were used, which cannot be extrapolated to human conditions (EMA 2012; Hajhashemi et al. 2003).

Antimicrobial effects: *In vitro* studies have shown that lavender oil has antimicrobial activity against *Escherichia coli*, *Bacillus subtilis*, *Candida albicans* and *Staphylococcus aureus* (but not *Pseudomonas aeruginosa*), and inhibits potential pathogenic intestinal bacteria at concentrations that had no effect on beneficial bacteria examined (EMA 2012). Insecticidal activity has been demonstrated with both *L. angustifolia* oil and linalool (Cavanagh and Wilkinson 2002).

Other effects: *In vitro* studies have demonstrated spasmolytic effect of lavender oil. Weak oestrogenic effects have been found in a study on MCF-7 human breast cancer cells in concentrations of 0.01 and 0.03% (V/V) treated for 18 h, and lavender oil was shown to be weakly anti-androgenic in concentrations between 0.0001 and 0.01% V/V in MDA-kb2 cells treated for 24 h (EMA 2012).

Interactions: None known (Gardner and McGuffin 2013). Caution is suggested with concurrent use of CNS depressants or anticonvulsants, due to the potential of a synergistic narcotic or sedative effect.

Contraindications: None known (Gardner and McGuffin 2013). Although there is no scientific evidence to support the historical belief that lavender oil is an emmenagogue, many sources recommend that the use of lavender during pregnancy and lactation should be restricted (Cavanagh and Wilkinson 2002; Denner 2009).

Adverse effects: Allergic contact dermatitis has been reported in individuals exposed to lavender essential oil. Case reports and an *in vitro* study suggested that lavender essential oil has oestrogenic activity, and that lavender as an ingredient in shampoo and soaps was the probable cause of gynaecomastia in several teenage boys. However, this has been disputed and is not supported by other research (Denner 2009; Gardner and McGuffin 2013). Pharmacokinetic data are limited; however, in one study, transdermal applications of lavender oil resulted in the accumulation of linalool and linalyl acetate in blood samples. Most lavender constituents are metabolised into carbon dioxide or excreted in conjugated form by the kidneys (Woronuk et al. 2011).

Dosage: Recommendations for use of lavender essential oil include both external and inhalational administration. Adults – as a bath additive: from 6 drops (120 mg) for a 20 L bath; inhalational (aromatherapy): 2–4 drops in 2–3 cups of boiling water, or by a diffuser. Massage therapy: 1–4 drops per 20 ml carrier/base oil (it may be mixed with other essential oils). There is insufficient data to recommend a suitable dose in children under 18, although some authors indicate that lavender is gentle and safe to use externally in children (Denner 2009). A medicinal product authorised in Germany for treatment of restlessness due to anxiety in adults: 80 mg of lavender oil in a soft capsule, one capsule to be taken orally once per day (EMA 2012).

General plant information: Lavender is endemic to the northern Mediterranean region, but is widely cultivated elsewhere (WHO 2007). The genus name '*Lavandula*' is derived from the Latin term '*lavare*', meaning 'to wash'. There is evidence for its use centuries ago as an antiseptic and a disinfectant by ancient Arabians, Greeks and Romans. Lavender essential oil was used as part of the 'mummification' process in ancient Egypt (Denner 2009), and was reported to be particularly effective in dermatological wound healing during World War I (Cavanagh and Wilkinson 2002).

References

Barker SC, Altman PM. (2010) A randomised, assessor blind, parallel group comparative efficacy trial of three products for the treatment of head lice in children – melaleuca oil and lavender oil, pyrethrins and piperonly butoxide, and a "suffocation" product. *BMC Dermatology* 10: 6 doi: 10.1186/1471-5945-10-6.

Bagheri-Nesami M, Espahbodi F, Nikkhah A, Shorofi SA, Charati JY. (2014) The effects of lavender aromatherapy on pain following needle insertion into a fistula in hemodialsysis patients. *Complementary Therapies in Clinical Practice* 20(1): 1–4.

Cavanagh HMA, Wilkinson JM. (2002) Bioactivities of lavender essential oil. *Phytotherapy Research* 16: 301–308.

Chioca LR, Ferro MM, Baretta IP, Oliveira SM, Silva CR, Ferreira J, Loss EM, Andreatini R. (2013) Anxiolytic-like effect of lavender essential oil inhalation in mice: participation of serotonergic but not GABA$_A$/benzodiazepene neurotransmission. *Journal of Ethnopharmacology* 147: 412–418.

Dehkordi ZR, Baharanchi FSH, Bekhradi R. (2014) Effect of lavender inhalation on the symptoms of primary dysmenorrhea and the amount of menstrual bleeding: a randomized clinical trial. *Complementary Therapies in Medicine*, 22(2): 212–219.

Denner SS. (2009) *Lavandula angustifolia* Miller. *Holistic Nursing Practice* 23(1): 57–64.

EMA (2012) Assessment report on *Lavandula angustifolia* Mill., aetheroleum and *Lavandula angustifolia* Mill., flos. European Medicines Agency http://www.ema.europa.eu/docs/en_GB/document_library/Herbal_-_HMPC_assessment_report/2012/06/WC500128642.pdf.

Fismer KL, Pilkington K. (2012) Lavender and sleep: A systematic review of the evidence. *European Journal of Integrative Medicine* 4(4): e436–e447.

Gardner Z, McGuffin M. (Eds.) (2013) *American Herbal Product Association's Botanical Safety Handbook.* 2nd Edition. CRC Press, USA, 1072 pp.

Hajhashemi V, Ghannadi A, Sharif B. (2003) Anti-inflammatory and analgesic properties of the leaf extract and essential oil of *Lavandula angustifolia* Mill. *Journal of Ethnopharmacology* 89(1): 67–71.

Hancianu M, Cioanca O, Mihasan M, Hritcu L. (2013) Neuroprotective effects of inhaled lavender oil on scopolamine-induced dementia *via* anti-oxidative activities in rats. *Phytomedicine* 20(5): 446–452.

Jones C. (2011) The efficacy of lavender oil on perineal trauma: a review of the evidence. *Complementary Therapies in Clinical Practice* 17(4): 215–220.

Kasper S, Gastpar M, Müller WE, Volz HP, Möller HJ, Dienel A, Schläfke S. (2010) Efficacy and safety of silexan, a new, orally administered lavender oil preparation, in subthreshold anxiety disorder – evidence from clinical trials. *Wiener Medizinsche Wochenschrift* 160(21–22): 547–556.

Kumar V. (2013) Characterization of anxiolytic and neuropharmacological activities of Silexan. *Wiener Medizinsche Wochenschrift* 163(3–4): 89–94.

Perry R, Terry R, Watson LK, Ernst E. (2012) Is lavender an anxiolytic drug? A systematic review of randomised clinical trials. *Phytomedicine* 19(8–9): 825–835.

Rahmati B, Khalili M, Roghani M, Ahgari P. (2013) Anti-epileptogenic and antioxidant effect of *Lavandula officinalis* aerial part extract against pentylenetetrazol-induced kindling in male mice. *Journal of Ethnopharmacology* 148(1): 152–157.

Sheikhan F, Jahdi F, Khoei EM, Shamsalizadeh N, Sheikhan M, Haghani H. (2012) Episiotomy pain relief: use of lavender oil essence in primiparous Iranina women. *Complementary Therapies in Clinical Practice* 18(1): 66–70.

Shiina Y, Funabashi N, Lee K, Toyoda T, Sekine T, Honjo S, Hasegawa R, Kawata T, Wakatsuki Y, Hayashi S, Murakami S, Koike K, Daimon M, Komuro I. (2008) Relaxation effects of lavender aromatherapy improve coronary flow velocity reserve in healthy men evaluated by transthoracic Doppler echocardiography. *International Journal of Cardiology* 129(2): 193–197.

Silenieks LB, Koch E, Higgins GA. (2013) Silexan, an essential oil from flowers of *Lavandula angustifolia*, is not recognized as benzodiazepine-like in rats trained to discriminate a diazepam cue. *Phytomedicine* 20(2): 172–177.

Soltani R , Soheilipour S, Hajhashemi V, Asghari G, Bagheri M, Molavi M. (2013) Evaluation of the effect of aromatherapy with lavender essential oil on post-tonsillectomy pain in pediatric patients: a randomized controlled trial. *International Journal of Pediatric Otorhinolaryngology* 77(9): 1579–1581.

Uehleke B, Schaper S, Dienel A, Schlaefke S, Stange R. (2013) Phase II trial on the effects of Silexan in patients with neurasthenia, post-traumatic stress disorder or somatization disorder. *Phytomedicine* 19(8–9): 665–671.

Vakilian K, Atarha M, Bekhradi R, Chaman R. (2011) Healing advantages of lavender essential oil during episiotomy recovery: a clinical trial. *Complementary Therapies in Clinical Practice* 17(1): 50–53.

WHO. (2007) *WHO Monographs on Selected Medicinal Plants.* Vol. 3. WHO, Geneva, 390 pp.

Woelk H, Schläfke S. (2011) A multi-center, double-blind, randomised study of the Lavender oil preparation Silexan in comparison to Lorazepam for generalized anxiety disorder. *Phytomedicine* 17(2): 94–99.

Woronuk G, Demissie Z, Rheault M, Mahmoud S. (2011) Biosynthesis and therapeutic properties of *Lavandula* essential oil constituents. *Planta Medica* 77(1): 7–15.

Lemon Balm   *Melissa officinalis* L.

Family: Lamiaceae (Labiatae)

Other common names: Balm; balm mint; bee balm; honey plant; sweet balm

Drug name: Melissae folium

Botanical drug used: Dried leaf

Indications/uses: Traditionally used in oral preparations to relieve temporary symptoms of stress including mild anxiety and insomnia. Lemon balm is also used for the symptomatic relief of mild gastrointestinal complaints such as bloating and flatulence. In recent years, it has attracted interest as a topical treatment for herpes simplex labialis (cold sores) as a result of infection with herpes simplex virus type 1. *M. officinalis* is most commonly sold in combination with other herbs, notably *Valeriana officinalis* (valerian), as a mild sedative.

Evidence: Some clinical evidence supports the internal use of lemon balm as an anxiolytic, an antioxidant, and as a topical treatment against cold sores. *In vitro* studies have demonstrated antiviral and antispasmodic effects for *M. officinalis* extracts.

Safety: Safety and toxicity data are lacking. Genotoxic tests reported to date have all been negative. However, in lieu of lack of data, use during pregnancy and lactation and in children and adolescents under 18 years of age is not recommended. As a potential sedative, use of *M. officinalis* may impair the ability to drive or use machinery.

Main chemical compounds: Essential oil 0.06–0.8% containing monoterpene aldehydes, mainly geranial and its isomer neral, with citronellal, neryl acetate, β-caryophyllene, germacrene-D, geraniol and β-ocimene (Tisserand and Young 2014); monoterpene glycosides; flavonoids (0.5%) including glycosides of luteolin, quercetin, apigenin and kaempferol; polyphenols, including hydroxycinnamic derivatives, caffeic acide, chlorogenic acid, and in particular rosmarinic acid (up to 6%); tannins; and triterpenes including ursolic and oleanolic acids (EMA 2013; Pharmaceutical Press Editorial Team 2013).

Clinical evidence:

Sedative, cognitive and anti-stress effects: Preparations containing *M. officinalis* extracts have been evaluated in several trials, although some of these investigated

Phytopharmacy: An evidence-based guide to herbal medicinal products, First Edition.
Sarah E. Edwards, Inês da Costa Rocha, Elizabeth M. Williamson and Michael Heinrich.
© 2015 John Wiley & Sons, Ltd. Published 2015 by John Wiley & Sons, Ltd.

combination preparations, mainly with valerian (e.g. Kennedy et al. 2006; Müller and Klement 2006). Studies that have investigated the action of mono-preparations of *M. officinalis* suggest that it may mitigate symptoms of stress. For example, in one double-blind, placebo-controlled, crossover trial, 18 healthy volunteers received two separate single doses of a standardised methanolic lemon balm leaf extract (300 mg, 600 mg) and a placebo, on separate days separated by a 7-day 'washout' period. The results demonstrated that 600 mg had a significant effect on ameliorating negative moods artificially induced using a 'Defined Intensity Stress Stimulation' (DISS) computerised battery. The 600 mg dose also significantly increased self-ratings of calmness and reduced self-ratings of alertness. After ingestion of the 300 mg dose, a significant increase in speed of mathematical processing, with no reduction in accuracy, was observed (Kennedy et al. 2004). A previous study by the same research group also demonstrated that extracts of *M. officinalis* modulate cognitive performance and mood in healthy participants, with lower doses (300 mg) raising self-ratings of calmness and high doses (600 mg), significantly reducing self-ratings of alertness (Kennedy et al. 2002).

In a 4-month randomised, placebo-controlled study in patients with mild-to-moderate Alzheimer's disease (aged 65–80), a fixed dose of 60 drops/day of a 50% ethanolic extract (1:4) of *M. officinalis* or placebo was administered. At the end of 16 weeks, the lemon balm extract resulted in a significantly better outcome on cognitive function than placebo, with agitation more common in the placebo group (Akhondzadeh et al. 2003). In another double-blind, placebo-controlled randomised study in patients with Alzheimer's disease, lemon balm extract was compared with donepezil, an anti-cholinesterase drug, to assess its efficacy as an aromatherapy external application (to hands and upper arms) in the treatment of agitation. No evidence was found that lemon balm aromatherapy was superior to either donepezil or placebo (Burns et al. 2011).

Topical antiviral effects: Studies have demonstrated that topical applications of a 1% lyophilised aqueous extract of *M. officinalis* in a cream base reduced the healing time of cold sores (from 10–14 days to 6–8 days), and significantly reduced the size of herpetic lesions in comparison to placebo within 5 days of treatment (WHO 2004).

Other clinical effects: A study in 55 radiology staff demonstrated that a herbal tea of *M. officinalis* taken twice daily (1.5 g/100 ml), resulted in significantly improved oxidative stress status and DNA damage (Zeraatpishe et al. 2011).

Pre-clinical evidence and mechanisms of action:

Sedative, cognitive and anti-stress effects: A study in mice demonstrated that a proprietary extract of *M. officinalis* (Cyracos®) exhibited dose-dependent anxiolytic-like effects under moderate stress conditions, without altering activity levels (Ibarra et al. 2010). Inhalation of the essential oil has also been shown to have a weak tranquilising effect in mice (WHO 2004).

Antiviral effects: *In vitro* studies have demonstrated that aqueous extracts of lemon balm can inhibit replication of herpes simplex virus type 2 (HSV-2), and have activity against Semliki Forest, influenza and vaccinia viruses. Tannin constituents are able to inhibit haemagglutination induced by Newcastle disease virus or mumps virus, and protect cell cultures from infection by Newcastle disease virus. A tannin-free polyphenol fraction of an aqueous extract was also found to be active against HSV and vaccinia virus in cell culture (WHO 2004). One proposed mechanism of action is the inhibition of attachment of the virus (Astani et al. 2012).

Antispasmodic effects: An ethanolic extract and essential oil of *M. officinalis* have been shown to exhibit antispasmodic effects in guinea pig ileum and other animal models *in vitro* (WHO 2004).

Interactions: No interactions have been reported. However, given its putative sedative properties, lemon balm should not be used concomitantly with synthetic sedatives such as lorazepam or phenobarbital, as it may increase the risk of drowsiness (Williamson et al. 2013).

Contraindications: Hypersensitivity to active ingredients. Not recommended during pregnancy and lactation, or in children and adolescents under 18 years of age due to lack of safety data (Tisserand and Young 2014).

Adverse effects: None reported.

Dosage: (a) Oral administration (adults): dried herb – 1.5–4.5 g as an infusion in 150 ml water several times daily; 45% alcohol extract (1:1) – 2–4 ml three times daily; tincture – (1:5 in 45% ethanol), 2–6 ml three times daily; (b) topical application – cream containing 1% of a lyophilised aqueous extract of dried leaves of *M. officinalis* (70:1) 2–4 times daily (Pharmaceutical Press Editorial Team 2013; WHO 2004).

General plant information: Native to regions of southern and eastern Europe, northern Africa and western Asia, lemon balm is naturalised and widely cultivated in most temperate regions. The genus name *Melissa* is derived from Greek, the word meaning 'bee', since the plants are very attractive to bees (GRIN 2013). Lemon balm is an important ingredient in the herb liqueurs Chartreuse and Bénédictine.

References

Akhondzadeh S, Noroozian N, Mohammadi M, Ohadinia S, Jamshidi AH, Khani M. (2003) *Melissa officinalis* extract in the treatment of patients with mild to moderate Alzheimer's disease: a double blind, randomised, placebo controlled trial. *Journal of Neurology, Neurosurgery, and Psychiatry* 74(7): 863–866.

Astani A, Reichling J, Schnitzler P. (2012) *Melissa officinalis* extract inhibits attachment of herpes simplex virus *in vitro*. *Chemotherapy* 58(1): 70–77.

Burns A, Perry E, Holmes C, Francis P, Morris J, Howes MJ, Chazot P, Lees G, Ballard C. (2011) A double-blind placebo-controlled randomized trial of *Melissa officinalis* oil and donepezil for the treatment of agitation in Alzheimer's disease. *Dementia and Geriatric Cognitive Disorders* 31(2): 158–164.

EMA (2013) Assessment report on *Melissa officinalis* L., folium. European Medicines Agency http://www.ema.europa.eu/docs/en_GB/document_library/Herbal_-_HMPC_assessment _report/2013/08/WC500147187.pdf.

GRIN. (2013) Germplasm Resources Information Network (GRIN) Taxonomy for Plants: *Melissa officinalis* L. http://www.ars-grin.gov/cgi-bin/npgs/html/taxon.pl?24036 (accessed 4 September 2013).

Ibarra A, Feuillere N, Roller M, Lesburgere E, Beracochea D. (2010) Effects of chronic administration of *Melissa officinalis* L. extract on anxiety-like reactivity and on circadian and exploratory activities in mice. *Phytomedicine;* 17(6): 397–403.

Kennedy DO, Little W, Haskell, CF, Scholey AB. (2006) Anxiolytic effects of a combination of *Melissa officinalis* and *Valeriana officinalis* during laboratory induced stress. *Phytotherapy Research* 20(2): 96–102.

Kennedy DO, Little W, Scholey AB. (2004) Attenuation of laboratory-induced stress in humans after acute administration of *Melissa officinalis* (Lemon Balm). *Psychosomatic Medicine* 66(4): 607–613.

Kennedy DO, Scholey AB, Tildesley NTJ, Perry EK, Wesnes KA. (2002) Modulation of mood and cognitive performance following acute administration of *Melissa officinalis* (lemon balm). *Pharmacology, Biochemistry and Behavior* 72(4): 953–964.

Müller SF, Klement S. (2006) A combination of valerian and lemon balm is effective in the treatment of restlessness and dyssomnia in children. *Phytomedicine* 13(6): 383–387.

Pharmaceutical Press Editorial Team. (2013) *Herbal Medicines*. 4th Edition. Pharmaceutical Press, London, UK.

Tisserand R, Young R. (2014) *Essential Oil Safety*. 2nd Edition. Churchill Livingstone Elsevier, UK, pp. 350–351.

WHO. (2004) *WHO Monographs on Selected Medicinal Plants*. Vol. 2. WHO, Geneva, 358 pp.

Williamson EM, Driver S, Baxter K. (Eds.) (2013) *Stockley's Herbal Medicines Interactions*. 2nd Edition. Pharmaceutical Press, London, UK.

Zeraatpishe A, Oryan S, Banheri MH, Pilevarian AA, Malekirad AA, Baeeri M, Abdollahi M. (2011) Effects of *Melissa officinalis* L. on oxidative status and DNA damage in subjects exposed to long-term low-dose ionizing radiation. *Toxicology and Industrial Health* 27(3): 205–212.

Linseed (Flaxseed)   *Linum usitatissimum* L.

Family: Linaceae

Other common names: Flax; flaxseed oil

Drug name: Lini semen; Lini oleum virginale

Botanical drug used: Seeds, expressed fixed oil

Indications/uses: Traditional linseed extracts are often mucilaginous preparations, used as demulcents; they are taken internally for the symptomatic relief of constipation, gastrointestinal discomfort, bronchitis and coughs, and applied externally as a soothing emollient agent for burns. Flaxseed oil is available in capsules as a food supplement, as a source of omega-3 essential fatty acids and phytoestrogens. Linseeds and linseed flour are used as an ingredient of functional foods (such as bread) where the phytoestrogenic properties of the lignans are intended to help to alleviate menopausal symptoms.

NB: 'Linseed oil' is a term normally given to the oil when used for industrial purposes (wood treatment etc.); it may have been boiled or otherwise further treated and is not usually suitable for human consumption.

Evidence: Limited clinical studies are available to support its use as an emollient, although there is a long history of such use, as well as for the treatment of habitual constipation due to the fibre content. Linseed flour may be beneficial in hypertension. A recent systematic review suggests that flaxseed products do not have harmful effects in patients with, or at risk of, breast cancer and may be beneficial.

Safety: Overall, linseed is considered to be safe. Overdose may cause abdominal discomfort, flatulence and possibly intestinal obstruction.

Main chemical compounds: The seed contains a fixed oil, the main constituents of which are glycerides of linoleic and linolenic acid and mucilage composed of a rhamnogalacturonan backbone. It also contains the lignans secoisolariciresinol and its diglucoside (SDG), the cyanogenetic glycosides linamarin, lotaustralin, linustatin and neolinustatin, proteins, minerals and vitamins (EMEA 2006b; Kaewmanee et al. 2014; Williamson et al. 2013).

Phytopharmacy: An evidence-based guide to herbal medicinal products, First Edition.
Sarah E. Edwards, Inês da Costa Rocha, Elizabeth M. Williamson and Michael Heinrich.
© 2015 John Wiley & Sons, Ltd. Published 2015 by John Wiley & Sons, Ltd.

Clinical evidence:

Even though the evidence based on controlled clinical studies is somewhat limited, for some of the key indications, especially for the management of gastrointestinal discomfort, use of linseed is well established based on medical experience. In other cases, the evidence is much more limited.

Gastrointestinal discomfort including constipation and irritable bowel syndrome: In a randomised investigator-blinded trial, patients (55 subjects) received either linseed (roughly ground, partly defatted) or psyllium seed 6–24 g a day for 3 months for the treatment of constipation with irritable bowel syndrome. Linseed treatment showed a significant decrease in constipation and abdominal symptoms (bloating and pain) compared to psyllium. After the blinding period, 40 out of 55 patients continued in an open period of further 3 months with only flaxseed treatment and constipation and abdominal symptoms were further reduced. (Tarpila et al. 2003).

Effects on blood lipids: A meta-analysis on the effects of flaxseed and flaxseed-derived products (flaxseed oil or lignans) on blood lipids, which included 28 randomised clinical trials (1539 subjects) concluded that the consumption of flaxseed and its derivatives reduced blood total and LDL-cholesterol concentrations, with no effect on HDL-cholesterol levels and triglycerides. The effect was more evident when whole flaxseed was used in (particularly postmenopausal) women and in subjects with initial higher cholesterol levels (Pan et al. 2009). In children, however, the use of flaxseed for hypercholesterolemia might not be a viable option. A study of paediatric patients (given flaxseed in muffins and bread) found that it was associated with adverse changes in the lipid profile of these children, although a potential benefit of low-density lipoprotein cholesterol lowering could not be excluded (Wong et al. 2013). Although no other safety concerns have been noted, further evidence is needed to support its use for the prevention of hypercholesterolaemia.

Effects on hypertension: In a recent prospective, double-blinded, placebo-controlled, randomised trial, patients (110 in total) ingested a variety of foods containing 30 g of milled flaxseed or placebo each day over 6 months. Plasma levels of the α-linolenic acid and enterolignans increased 2- to 50-fold in the flaxseed-fed group, but not in the placebo group. Systolic blood pressure (SBP) was ≈10 mm Hg lower, and diastolic blood pressure (DBP) was ≈7 mm Hg lower in the flaxseed group compared with placebo after 6 months. Patients who entered the trial with a SBP ≥140 mm Hg at baseline obtained a significant reduction of 15 mm Hg in SBP and 7 mm Hg in DBP from flaxseed ingestion. The antihypertensive effect was achieved selectively in hypertensive patients. Circulating α-linolenic acid levels correlated with SBP and DBP and lignan levels correlated with changes in DBP. The authors have suggested that flaxseed may induce one of the most potent antihypertensive effects possible by a dietary intervention (Rodriguez-Leyva et al. 2013).

Effects on blood glucose: The effects of a flaxseed-derived lignan supplement, 3 lignan capsules a day (0.6 g/capsule, equivalent to 360 mg/day of secoisolariciresinol diglucoside (SDG)) was studied in a randomised, double-blind, cross-over trial in type 2 diabetic patients (73 participants with mild hypercholesterolaemia) for 12 weeks. The study showed that there was a modest improvement in the glycaemic control without affecting fasting glucose, lipid profiles and insulin sensitivity in these patients (Pan et al. 2007).

Phytoestrogenic effects: A phase III, randomised, placebo-controlled, double-blind trial was carried out to determine the efficacy of a standardised flaxseed product (providing 410 mg of lignans) for the treatment of hot flashes in post-menopausal women (188 participants) with or without breast cancer for 6 weeks, versus a placebo. The study showed that flaxseed did not reduce hot flashes more than a placebo, despite preliminary data suggesting its effects on hot flashes and other menopausal symptoms (Pruthi et al. 2012).

Use of flaxseed products as an adjuvant treatment in breast cancer: A systematic review of flax for efficacy in improving menopausal symptoms in women living with breast cancer, and for potential impact on risk of breast cancer incidence or recurrence, suggests that flax may be associated with decreased risk of breast cancer. Flax demonstrates antiproliferative effects in breast tissue of women at risk of breast cancer and may protect against primary breast cancer (Flower et al. 2014).

Pre-clinical evidence and mechanisms of action:

Laxative effects: Laxative effects are due to the bulk forming capacity of the seeds, mainly due to the mucilage, which after binding with water swell to form a demulcent gel in the intestine. As water is being held back in the intestine, it makes the faeces softer and causes a stretch stimulus. This swollen mass of mucilage forms a lubrication layer that facilitates the transit in the intestinal tract. Part of the bulk material is defecated while the other part is fermented by the bacteria in the colon (EMEA 2006a).

Effects on blood glucose and lipids: Studies have shown that effects on blood glucose and lipid levels are partly due to a delay in food absorption, as well as oestrogenic effects (EMEA 2006a; ESCOP 2003). Flaxseed, the lignan extract and SDG, but not flaxseed oil, has been shown to suppress atherosclerosis in animal models, slowing progression of atherosclerosis but with no effect on regression. Flaxseed oil suppressed oxygen radical production by white blood cells, prolonged bleeding time and in higher doses suppressed serum levels of inflammatory mediators, but does not lower serum lipids (Prasad 2009).

Phytoestrogenic effects: The lignans are metabolised to the mammalian lignans enterodiol and enterolactone by human intestinal microflora (EMEA 2006a; Rodriguez-Leyva et al. 2013).

Anti-inflammatory effects: The anti-arthritic and immunomodulatory activity of flaxseed oil has been demonstrated in several experimental models. The oil produced a reduction in joint swelling and circulating TNF-α levels in preventive and curative protocols of arthritis induced by complete Freund's adjuvant. TNF-R1 and Interleukin (IL)-6 expression was also significantly reduced in the treated animals; and in the cotton pellet-induced granuloma model, the oil significantly reduced the dry granuloma weight (Singh et al. 2012).

Interactions: Flaxseed oil may increase bleeding times so caution is recommended in patients taking aspirin and anticoagulants (Williamson et al. 2013). Concomitant use of linseed with other drugs may delay their enteral absorption thus linseed should be taken at least half an hour to 1 hour before other drugs. Linseed may interact with drugs that inhibit peristaltic movement (e.g. opioids and loperamide), thus its concomitant use should be avoided to decrease the risk of gastrointestinal obstruction (ileus) (EMEA 2006b). Diabetic patients may experience delay in glucose absorption (ESCOP 2003).

Contraindications: Flaxseed is not recommended during pregnancy and lactation due to lack of data and is also not recommended for children under 12 years of age. A health-care professional should be consulted if the symptoms persist for more than 1 week. It should be avoided in patients with potential or existing intestinal blockage or paralysis of the intestine, and by patients with a sudden change in bowel habit that continues for more than 2 weeks. Its use should be discontinued in cases of abdominal pain or if any irregularity of faeces occurs and medical support is advisable (EMEA 2006b).

Adverse effects: Flaxseed should be taken along with plenty of fluids to avoid it blocking the throat or oesophagus, which may cause choking. Intestinal obstruction may also occur if adequate amount of fluid is not maintained. Flatulence (meteorism) is a very common side effect. Anaphylaxis-like reactions may occur but are very rare. Overdosing may cause abdominal discomfort, flatulence and possibly intestinal obstruction (EMEA 2006b). The presence of cyanogenic glycosides in the seeds is a concern as they release cyanide upon hydrolysis; however, cyanide levels produced as a result of autolysis have been shown to be below the limits that are known to be harmful to humans (Wanasundara and Shahidi 1998).

Dosage: For products, the manufacturer's indications should be followed. Usually 10–15 g of seeds with 150 ml of water or any other beverage can be taken 2–3 times a day. To obtain a mucilaginous preparation soak 5–10 g of whole or broken seeds in water (250 mL), preferably by the evening before, and take it half an hour before eating up to three times a day as a single dose. It should not to be taken immediately prior to bedtime. It may be consumed with or without the seeds. Its effect is usually observed within 12–24 hours with maximum effect after 2–3 days (EMEA 2006a). The usual dose of flaxseed oil is at least 1 g three times daily.

General plant information: Flax is widely cultivated in Europe, North- and South-America and also used for the production of linen from the fibres. Linseed oil is a drying oil, meaning it polymerizes easily, and as such has many uses, including as a wood treatment, and especially for cricket bats. It is the source of the traditional floor covering, linoleum and used as a paint carrier and in putty, as a sealant.

References

EMEA. (2006a) Assessment report on *Linum usitatissimum* L., Semen. European Medicines Agency http://www.ema.europa.eu/docs/en_GB/document_library/Herbal_-_HMPC _assessment_report/2010/01/WC500059156.pdf.

EMEA. (2006b) Community herbal monograph on *Linum usitatissimum* L., Semen. European Medicines Agency http://www.ema.europa.eu/docs/en_GB/document_library /Herbal_-_Community_herbal_monograph/2010/01/WC500059157.pdf.

ESCOP. (2003) *ESCOP Monographs: The Scientific Foundation for Herbal Medicinal Products.* 2nd Edition. Thieme, Exeter and London, UK.

Flower G, Fritz H, Balneaves LG, Verma S, Skidmore B, Fernandes R, Kennedy D, Cooley K, Wong R, Sagar S, Fergusson D, Seely D. (2014) Flax and breast cancer: a systematic review. *Integrative Cancer Therapy* 13(3): 181–192.

Kaewmanee T, Bagnasco L, Benjakul S, Lanteri S, Morelli CF, Speranza G, Cosulich ME. (2014) Characterisation of mucilages extracted from seven Italian cultivars of flax. *Food Chemistry* 148: 60–69.

Pan A, Sun J, Chen Y, Ye X, Li H, Yu Z, Wang Y, Gu W, Zhang X, Chen X, Demark-Wahnefried W, Liu Y, Lin X. (2007) Effects of a flaxseed-derived lignan supplement in type 2 diabetic patients: a randomized, double-blind, cross-over trial. *PLoS One* 2(11): e1148.

Pan A, Yu D, Demark-Wahnefried W, Franco O H and Lin X. (2009) Meta-analysis of the effects of flaxseed interventions on blood lipids. *American Journal of Clinical Nutrition* 90(2): 288–297.

Prasad K. (2009) Flaxseed and cardiovascular health. *Journal of Cardiovascular Pharmacology* 54(5): 369–377.

Pruthi S, Qin R, Terstreip SA, Liu H, Loprinzi CL, Shah TR, Tucker KF, Dakhil SR, Bury MJ, Carolla RL, Steen PD, Vuky J, Barton DL. (2012) A phase III, randomized, placebo-controlled, double-blind trial of flaxseed for the treatment of hot flashes: North Central Cancer Treatment Group N08C7. *Menopause* 19(1): 48–53.

Rodriguez-Leyva D, Weighell W, Edel AL, Lavallee R, Dibrov E, Pinneker R, Maddaford TG, Ramjiawan B, Aliani M, Guzman R, Pierce GN. (2013) Potent antihypertensive action of dietary flaxseed in hypertensive patients. *Hypertension* 62(6): 1081–1089.

Singh S, Nair V, Gupta YK. (2012) Linseed oil: an investigation of its antiarthritic activity in experimental models. *Phytotherapy Research* 26(2): 246–252.

Tarpila S, Tarpila A, Gröhn P, Silvennoinen T, Lindberg L. (2003) Efficacy of ground flaxseed on constipation in patients with irritable bowel syndrome. *Nutritional Genomics and Functional Foods* 1(1): 1–7.

Wanasundara PK, Shahidi F. (1998) Process-induced compositional changes of flaxseed. *Advances in Experimental Medicine and Biology* 434: 307–325.

Williamson EM, Driver S, Baxter K. (Eds.) (2013) *Stockley's Herbal Medicines Interactions*. 2nd Edition. Pharmaceutical Press, London UK.

Wong H, Chahal N, Manlhiot C, Niedra E, McCrindle BW. (2013) Flaxseed in pediatric hyperlipidemia: a placebo-controlled, blinded, randomized clinical trial of dietary flaxseed supplementation for children and adolescents with hypercholesterolemia. *JAMA Pediatrics* 167(8): 708–713.

Liquorice *Glycyrrhiza glabra* L., *G. inflata* Batalin, *G. uralensis* Fisch. ex DC.

Synonyms: *Liquiritia officinalis* Moench (= *G. glabra* L.); and others

Family: Fabaceae (Leguminosae)

Other common name: Gancao; glycyrrhiza; licorice; sweet root

Drug name: Glycyrrhizae radix; Liquiritiae radix

Botanical drug used: Root and stolon

Indications/uses: Liquorice has long been used for the relief of digestive symptoms such as dyspepsia and to aid the healing of ulcers, both gastric and apthous types. It is commonly used for respiratory tract infections as an ingredient in cough mixtures and lozenges, as an expectorant for the relief of chesty coughs, catarrh and sore throats, and as an anti-inflammatory agent. It is used for many other indications, including skin conditions, and in most types of traditional medicine, especially in combination formulae (cf., e.g. Wang et al. 2013).

Evidence: Some clinical trial evidence is available to support the use as an anti-ulcer treatment, although studies are small, and newer treatments are more effective with fewer side effects. Limited clinical studies are available to support its traditional use as an expectorant, although it has a very long history of use.

Safety: Liquorice is widely eaten in confectionary and is overall considered to be safe. However, in high doses (equivalent to more than 100 mg glycyrrhizin content) and/or prolonged use (longer than 4–6 weeks), it has corticosteroid effects which include hypokalaemia, hypertension and oedema, and more rarely cardiac rhythm disorders.

Main chemical compounds: The main active constituents are the triterpene glycosides such as glycyrrhizin (also known as glycyrrhizic or glycyrrhizinic acid) and its aglycone glycyrrhetinic acid, liquiritic acid, glycyrretol, glabrolide, isoglaborlide and others; flavonoids including liquiritin apioside, liquiritin, isoliquiritin and glabrol; isoflavonoids such as glabrene and glabridin, coumarins (e.g. liqcoumarin, glabrocoumarones A and B, herniarin and glycyrin), and many other compounds including polysaccharides (glycyrrhizan GA), and an essential oil containing

Phytopharmacy: An evidence-based guide to herbal medicinal products, First Edition.
Sarah E. Edwards, Inês da Costa Rocha, Elizabeth M. Williamson and Michael Heinrich.
© 2015 John Wiley & Sons, Ltd. Published 2015 by John Wiley & Sons, Ltd.

anethole, linallol, fenchone (Asl and Hosseinzadeh 2008; EMA 2013; ESCOP 2003; Williamson et al. 2013).

Clinical Evidence:

Gastrointestinal effects: Several clinical trials using liquorice extract in gastrointestinal and ulcerative conditions have been performed, but no meaningful conclusions can be drawn due to the small size of the studies (EMA 2013).

Clinical studies assessing the effect of deglycyrrhizinated liquorice extract (i.e. from which the glycyrrhizinic acid has been removed) on gastric and duodenal ulcers have found positive effects (Engqvist et al. 1973; Feldman and Gilat 1971; Hollanders et al. 1978), and a commercial product of liquorice (a flavonoid-rich root extract of *G. glabra*) in patients with functional dyspepsia showed beneficial effects on relief of the symptoms (Raveendra et al. 2012).

A study investigating the efficacy of bioadhesive patches containing liquorice extract to control the pain and reduce the healing time of recurrent apthous ulcers (Moghadamnia et al. 2009) showed that the patches were effective in the reduction of pain and inflammation of the ulcers, but the study was small.

Respiratory tract effects: No clinical studies are available to support the use of liquorice as an expectorant, despite its very wide usage for this purpose. Liquorice is present in combination products with ivy (*Hedera helix* L.) and thyme (*Thymus vulgaris* L.) aerial parts for cough relief (EMA 2013).

A prospective, randomised, single-blind clinical study assessed the efficacy of liquorice gargle for attenuating post-operative sore throat (POST) in 49 participants. A decoction of liquorice powder (5 g in 300 ml of water), used as a gargle just before induction of anesthesia, was found to be an effective method for attenuating both the incidence and severity of POST, although further studies are warranted (Agarwal et al. 2009).

Skin conditions: A Cochrane review assessing interventions for the treatment of *Pityriasis rosea* reported a small clinical trial (23 participants) where an active ingredient of liquorice root (glycyrrhizin) was compared to an anaesthetic (procaine) injected intravenously. No significant difference was found, but the authors concluded that further studies are needed since this study was of poor methodological quality (Chuh et al. 2007). Other clinical evidence of efficacy in skin disorders is lacking, although liquorice has been used to treat hyper-pigmentation, eczema and psoriasis, applied topically and taken orally.

Pre-clinical evidence and mechanisms of action: Liquorice extracts and derivatives have shown a very wide range of biological activities, which include anti-inflammatory, anti-allergic, antioxidant, antibacterial, antiviral, anticancer, antithrombotic, antidiabetic, antispasmodic, hepatoprotective and neuroprotective activities. Studies mainly focus on the pharmacological effects of the saponins, especially glycyrrhizin and its aglycone, 18β-glycyrrhetinic acid, and the many flavonoids, both of which have gastroprotective effects in animals. The anti-ulcer activity is thought to be due to the reduction of gastric secretions via the inhibition of gastrin release and the promotion of gastric mucosa healing. The anti-inflammatory and anti-allergic activities have been linked to the corticosteroid-like activity of glycyrrhizin and 18β-glycyrrhetinic acid, although the flavonoids also possess anti-inflammatory and anti-allergic properties. The antitussive and expectorant properties have been attributed to the increase in tracheal mucus secretion caused by glycyrrhizin (Asl and Hosseinzadeh 2008; EMA 2013). Antitussive properties have also been attributed to the flavonoids liquiritin apioside, liquiritin and liquiritigenin, which were able to reduce capsaicin-induced cough in guinea pigs (Kamei et al. 2003).

Interactions: May have an additive hypokalaemia effect (potassium depletion) with loop and thiazide diuretics, corticosteroids and laxatives if given in large quantities. Caution should be exercised with patients who are taking digitalis glycosides and who regularly use/abuse laxatives including liquorice and/or anthraquinone-containing substances such as rhubarb, as there was one report of digoxin toxicity (Williamson et al. 2013). Concomitant use with other products containing liquorice is not recommended due to possible additive effects.

Contraindications: Not recommended in patients suffering from hypertension, hypokalaemia, kidney disease, liver or cardiovascular disorders. Liquorice extracts in high doses should be avoided in pregnancy and lactation and are not recommended in children and adolescents under 18 due to lack of data, although *moderate* ingestion of liquorice-containing confectionery is not thought to be harmful. A qualified health care professional should be consulted if symptoms persist for longer than 1 to 2 weeks, or if dyspnoea, fever or purulent sputum is present (EMA 2012; EMA 2013).

Adverse effects: Adverse events such as water retention, hypokalaemia, hypertension, cardiac rhythm disorders and hypertensive encephalopathy may occur in cases of liquorice overdosing (high dose and/or prolonged use) (EMA 2012).

Dosage: As a herbal tea, 1.5–2 g in 150 ml of boiling water two to four times a day and as an extract not more than 160 mg a day for the relief of digestive symptoms. As an expectorant, the dose should be lower, 1.5 g of herbal substance in boiling water two times a day and 1.2–1.5 g three to four times a day as an extract (EMA 2012). Commercial products should be taken in accordance with the manufacturers' instructions.

General plant information: The use of liquorice dates back to ancient Egypt, with traces being found in Pharaohs' tombs. It has many uses in addition to its medical applications and in confectionery, such as flavouring agent, sweetener, and in cosmetics (see Fiore et al. 2005 and Asl and Hosseinzadeh 2008 for more comprehensive information).

References

The primary literature on liquorice is immense. The references cited here are reports and reviews which support the most important clinical uses of liquorice.

Agarwal A, Gupta D, Yadav G, Goyal P, Singh PK, Singh U. (2009) An evaluation of the efficacy of licorice gargle for attenuating postoperative sore throat: a prospective, randomized, single-blind study. *Anesthesia and Analgesia* 109(1): 77–81.

Asl MN, Hosseinzadeh H. (2008) Review of pharmacological effects of *Glycyrrhiza sp.* and its bioactive *compounds. Phytotherapy Ressearch* 22(6): 709–724.

Chuh AA, Dofitas BL, Comisel GG, Reveiz L, Sharma V, Garner SE, Chu F. (2007) Interventions for *pityriasis rosea. Cochrane Database of Systematic Reviews* 2: CD005068.

EMA. (2012) Community herbal monograph on *Glycyrrhiza glabra* L. and/or *Glycyrrhiza inflata* Bat. and/or *Glycyrrhiza uralensis* Fisch., radix. European Medicines Agency http://www.ema.europa.eu/docs/en_GB/document_library/Herbal_-_Community_herbal_monograph/2012/08/WC500131287.pdf.

EMA. (2013) Assessment report on *Glycyrrhiza glabra* L.and/or *Glycyrrhiza inflata* Bat. and/or *Glycyrrhiza uralensis* Fisch., radix. European Medicines Agency http://www.ema .europa.eu/docs/en_GB/document_library/Herbal_-_HMPC_assessment_report/2012/08 /WC500131285.pdf.

Engqvist A, von Feilitzen F, Pyk E, Reichard H. (1973) Double-blind trial of deglycyrrhizinated liquorice in gastric ulcer. *Gut* 14(9): 711–715.

ESCOP. (2003) *ESCOP Monographs: The Scientific Foundation for Herbal Medicinal Products*. 2nd Edition. Thieme, Exeter and London, UK.

Feldman H, Gilat T. (1971) A trial of deglycyrrhizinated liquorice in the treatment of duodenal ulcer. *Gut* 12(6): 449–451.

Fiore C, Eisenhut M, Ragazzi E, Zanchin G, Armanini D. (2005) A history of the therapeutic use of liquorice in Europe. *Journal of Ethnopharmacology* 99(3): 317–24.

Hollanders D, Green G, Woolf IL, Boyes BE, Wilson RY, Cowley DJ, Dymock IW. (1978) Prophylaxis with deglycyrrhizinised liquorice in patients with healed gastric ulcer. *British Medical Journal* 1(6106): 148.

Kamei J, Nakamura R, Ichiki H, Kubo M. (2003) Antitussive principles of Glycyrrhizae radix, a main component of the Kampo preparations Bakumondo-to (Mai-men-dong-tang). *European Journal of Pharmacology* 469(1–3): 159–163.

Moghadamnia AA, Motallebnejad M, Khanian M. (2009) The efficacy of the bioadhesive patches containing licorice extract in the management of recurrent aphthous stomatitis. *Phytotherapy Research* 23(2): 246–250.

Raveendra KR, Jayachandra, SV, Sushma KR, Allan JJ, Goudar KS, Shivaprasad HN, Venkateshwarlu K, Geetharani P, Sushma G, Agarwal A. (2012) An extract of *Glycyrrhiza glabra* (GutGard) alleviates symptoms of functional dyspepsia: a randomized, double-blind, placebo-controlled study. *Evidence-Based Complementary and Alternative Medicine* 2012: 216970.

Wang X, Zhang H, Chen L, Shan L, Fan G, Gao X. (2013) Liquorice, a unique "guide drug" of traditional Chinese medicine: A review of its role in drug interactions. *Journal of Ethnopharmacology* 150(3): 781–790.

Williamson EM, Driver S, Baxter K. (Eds.) (2013) *Stockley's Herbal Medicines Interactions*. 2nd Edition. Pharmaceutical Press, London, UK.

Lobelia   *Lobelia inflata* L.

Synonyms: *L. michauxii* Nutt; and others

Family: Campanulaceae

Other common name: Indian tobacco; puke weed

Drug name: Lobeliae herba

Botanical drug used: Herb

Indications/uses: Traditionally used as a respiratory stimulant in asthma, for the relief of coughs and as an expectorant. It is sometimes used as an aid to smoking cessation, although the clinical evidence does not support this.

Evidence: No clinical data available to validate its traditional use.

Safety: Safe in recommended doses. Overdose can produce severe side effects, including nausea, tremor, tachycardia and other nicotine-like effects. Deaths have been recorded.

Main chemical compounds: The active constituents are the piperidine alkaloids, mainly lobeline, but also lobelanine, lobelanidine, norlobelanine, lelobanidine, norlelobanidine, norlobelanidine and lobinine (Felpin and Lebreton 2004; Williamson et al. 2013).

Clinical evidence: No data available to support the traditional use for respiratory tract disorders. A Cochrane review suggests that there is no evidence from long-term trials to show that lobeline can aid smoking cessation, and the short-term evidence also suggests there is no benefit (Stead and Hughes 2012).

Pre-clinical evidence and mechanisms of action: Lobeline and its analogues have nicotine-like effects and have been suggested to have some potential as a pharmacotherapy for psychostimulant abuse (Crooks et al. 2011; Dwoskin and Crooks 2002; Felpin and Lebreton 2004). Lobeline is a neuronal nicotinic acetylcholine receptor (nAChR) antagonist, with antidepressant properties in animals (Roni and Rahman 2013). It does, however, have dual effects: in a cell model of expressed rat $\alpha 4\beta 2$ nicotinic receptors, lobeline caused inhibition when applied alone, whereas it caused potentiation by binding to a second agonist site when the first one was occupied by acetylcholine, leading to channel opening (Kaniaková et al. 2011).

Phytopharmacy: An evidence-based guide to herbal medicinal products, First Edition.
Sarah E. Edwards, Inês da Costa Rocha, Elizabeth M. Williamson and Michael Heinrich.
© 2015 John Wiley & Sons, Ltd. Published 2015 by John Wiley & Sons, Ltd.

Interactions: None reported.

Contraindications: Pregnancy and lactation, as safety has not been established.

Adverse effects: May cause nausea, vomiting, diarrhoea, coughing, tremors and dizziness. Lobeline has nicotine-like effects, which include tachycardia and hypertension (Dwoskin and Crooks 2002). Symptoms of over-dosage have been reported such as profuse diaphoresis, tachycardia, convulsions, hypothermia, hypotension and coma, resulting in fatalities (Bradley 1992; Williamson et al. 2013).

Dosage: For products, follow manufacturers' instructions. Lobelia herb has been given as an infusion (tea) or decoction, 50–200 mg dried herb three times daily (Bradley 1992).

General plant information: Lobelia herb is subject to sales restrictions under SI 2130 Parts II & III: it is a 'P' medicine; but for internal use, a maximum dose of 200 mg with a maximum daily dose of 600 mg can be supplied following a one-to-one consultation with a practitioner. Lobelia has been used in cough mixtures in combination with squill (*Drimia maritima*, see p. 366) extract. Lobeline is a regulated substance in high-performance sports. Competitors should check with the appropriate authorities before using products containing lobeline (Williamson et al. 2013), although they should first consider whether there is enough evidence to justify its use.

References

Bradley PR. (Ed.) (1992) *Lobelia. British Herbal Compendium* Vol 1. British Herbal Medicine Association, UK, pp. 150–151.

Crooks PA, Zheng G, Vartak AP, Culver JP, Zheng F, Horton DB, Dwoskin LP. (2011) Design, synthesis and interaction at the vesicular monoamine transporter-2 of lobeline analogs: potential pharmacotherapies for the treatment of psychostimulant abuse. *Current Topics in Medicinal Chemistry* 11(9): 1103–1127.

Dwoskin LP, Crooks PA. (2002) A novel mechanism of action and potential use for lobeline as a treatment for psychostimulant abuse. *Biochemical Pharmacology* 63(2): 89–98.

Felpin F-X, Lebreton J. (2004) History, chemistry and biology of alkaloids from *Lobelia inflata. Tetrahedron* 60(45): 10127–10153.

Kaniaková M, Lindovský J, Krušek J, Adámek S, Vyskočil F. (2011) Dual effect of lobeline on α4β2 rat neuronal nicotinic receptors. *European Journal of Pharmacology* 658(2–3): 108–113.

Roni MA, Rahman S. (2013) Antidepressant-like effects of lobeline in mice: behavioral, neurochemical, and neuroendocrine evidence. *Progress in Neuro-Psychopharmacology & Biological Psychiatry* 41: 44–51.

Stead LF, Hughes JR. (2012) Lobeline for smoking cessation. *Cochrane Database of Systematic Reviews* 2: CD000124.

Williamson EM, Driver S, Baxter K. (Eds.) (2013) *Stockley's Herbal Medicines Interactions*. 2nd Edition. Pharmaceutical Press, London, UK.

Maca   *Lepidium meyenii* Walp.

Synonyms: *L. gelidum* Wedd.; *L. peruvianum* G.Chacón; and others

Family: Brassicaceae (Cruciferae)

Other common names: Chilque; macaia; Peruvian ginseng

Botanical drug used: Dried powdered 'root' (hypocotyl)

Indications/uses: Popularly, it is used as a tonic and as an aphrodisiac to enhance sexual drive and female fertility.

Evidence: The evidence from clinical studies to support its use as an aphrodisiac is very limited, although there is some evidence for its use in managing menopausal symptoms. Further studies are needed before any recommendation can be made.

Safety: Overall, considered to be safe, as the root is also used as a vegetable, but further studies are needed to confirm its safety. Despite the limited clinical studies on its efficacy and safety, it is widely marketed over the Internet for its purported beneficial effects on sexual function and menopausal symptoms.

Main chemical compounds: Unsaturated fatty acids, mainly macaene and macamide, as well as linoleic and oleic acids, the glucosinolates glucotropaeolin and m-methoxyglucotropaeolin, alkaloids including lepidiline A, lepidiline B and macaridine, and sterols such as β-sitosterol, campesterol and stigmasterol. The root is rich in minerals and trace elements (Wang et al. 2007).

Clinical evidence: A systematic review on the use of maca for treatment of menopausal symptoms, which included four double-blind randomised placebo-controlled clinical trials in healthy peri-menopausal, early post-menopausal and late post-menopausal women, showed that there is limited evidence to support its use (Lee et al. 2011), and that further studies are needed. However, based on these studies, maca was considered to be more effective than a placebo.

Its effect on improving sexual function was also assessed in a systematic review. Again, the evidence available to support its effects on the treatment of sexual function in healthy volunteers (men and women) and men with sexual dysfunction was limited (Shin et al. 2010). Similar findings were reported in an overview of clinical studies by Ernst et al. (2011) for its use in older patients with sexual dysfunction and/or erectile dysfunction.

Phytopharmacy: An evidence-based guide to herbal medicinal products, First Edition.
Sarah E. Edwards, Inês da Costa Rocha, Elizabeth M. Williamson and Michael Heinrich.
© 2015 John Wiley & Sons, Ltd. Published 2015 by John Wiley & Sons, Ltd.

Pre-clinical evidence and mechanisms of action: The mechanisms of action are unknown. It has been proposed that the polyunsaturated fatty acids (macaene and macamide), the glucosinolates, and the alkaloid macaridine are the compounds responsible for its biological activity. Pre-clinical studies suggest that maca may improve sperm count and motility and enhance fertility, and it improved sexual performance in rats, although the mechanism is not clear (Lee et al. 2011; Shin et al. 2010; Wang et al. 2007). Maca has antioxidant and free radical scavenging effects and a high nutritional content, which might help to promote vitality. The phytosterols may have an effect on menopause symptoms, as shown in ovariectomised rats studies (Lee et al. 2011; Wang et al. 2007).

Interactions: None known.

Contraindications: Not recommended during pregnancy and lactation due to lack of safety data.

Adverse effects: None known.

Dosage: For products, as recommended by the manufacturer. The optimum dose of maca is unknown (Lee et al. 2011) but doses up to 3500 mg/day have been used (Shin et al. 2010).

General plant information: This plant is cultivated in the Andes, mainly in the highlands of Peru and has been used for over 2000 years in the region as food and medicine (Wang et al. 2007). It has been used traditionally in the Andes as a general tonic, for anaemia, infertility and female hormone balance. There are different types of maca (eight or more different ecotypes) with different colours ranging from white to black, with the yellow one being the most common cultivar in the Andes. Recently, it has been introduced to Western countries, but there is considerable concern about the quality and composition of maca products (Valerio and Gonzales 2005) as different maca ecotypes (with differently coloured hypocotyls) appear to differ in their biological effects (Clement et al. 2012). However, the planting site appears to be the most important factor affecting chemical variation (Gonzales et al. 2009; Wang et al. 2007; Zhao et al. 2012).

References

Clement C, Witschi U, Kreuzer M. (2012) The potential influence of plant-based feed supplements on sperm quantity and quality in livestock: a review. *Animal Reproduction Science* 132(1–2): 1–10.

Ernst E, Posadzki P, Lee MS. (2011) Complementary and alternative medicine (CAM) for sexual dysfunction and erectile dysfunction in older men and women: an overview of systematic reviews. *Maturitas* 70(1): 37–41.

Gonzales GF, Gonzales C, Gonzales-Castaneda C. (2009) *Lepidium meyenii* (Maca): a plant from the highlands of Peru – from tradition to science. *Forschende Komplementaermedizin* 16(6): 373–380.

Lee MS, Shin BC, Yang EJ, Lim HJ, Ernst E. (2011) Maca (*Lepidium meyenii*) for treatment of menopausal symptoms: a systematic review. *Maturitas* 70(3): 227–233.

Shin BC, Lee MS, Yang EJ, Lim HS, Ernst E. (2010) Maca (*L. meyenii*) for improving sexual function: a systematic review. *BMC Complementary and Alternative Medicine* 10: 44.

Valerio LG Jr., Gonzales GF. (2005) Toxicological aspects of the South American herbs cat's claw (Uncaria tomentosa) and Maca (*Lepidium meyenii*) : a critical synopsis. *Toxicological Reviews* 24(1): 11–35.

Wang Y, Wang Y, McNeil B, Harvey LM. (2007) Maca: An Andean crop with multi-pharmacological functions. *Food Research International* 40(7): 783–792.

Zhao J, Avula B, Chan M, Clement C, Kreuzer M, Khan IA. (2012) Metabolomic differentiation of maca (*Lepidium meyenii*) accessions cultivated under different conditions using NMR and chemometric analysis. *Planta Medica* 78(1): 90–101.

Mallow   *Malva sylvestris* L.

Synonyms: *M. ambigua* Guss.; *M. erecta* C.Presl; *M. mauritiana* L.; and others

Family: Malvaceae

Other common names: Common mallow; malva

Drug name: Malvae folium; Malvae flos

Botanical drug used: Whole or fragmented dried leaf (commonly); dried flower (occasionally)

Indications/uses: Mallow herbal teas and hydroalcoholic extracts are most commonly used for upper respiratory tract inflammation, due to their antibacterial and demulcent properties. They are formulated into cough mixtures and mouth-washes, and applied topically as emollients for a range of skin conditions.

Evidence: No controlled clinical studies are available. Pre-clinical data support the use as a demulcent.

Safety: There are no known safety concerns, and the plant is also used widely as a food ingredient, but there is a lack of data about herbal extracts of the plant.

Main chemical compounds: Mucilage polysaccharides (6% to >10%) composed of neutral and acidic monosaccharide residues including rhamnose, galactose, arabinose, galacturonic acid and glucuronic acid, anthocyanins (6–7%), mainly malvidin 3,5-diglucoside, malvidin 3-glucoside, malvidin 3-(6''-malonylglucoside)-5-glucoside and delphinidin 3-glucoside with traces of petunidin and cyanidin glycosides, as well as scopoletin, ursolic acid and phytosterols (ESCOP 2009). The essential oil of the flowers contains the aroma compounds β-damascenone, phenylacetaldehyde, (E)-β-ocimene, (E)-β-ionone, and decanal (Usami et al. 2013).

Clinical evidence: No clinical data are available.

Pre-clinical evidence and mechanisms of action: Studies have demonstrated antioxidant free-radical scavenging, antibacterial, and anti-inflammatory effects (DellaGreca et al. 2009; Gasparetto et al. 2012), which are relevant to the respiratory tract uses and the topical uses in treating chronic and acute skin conditions.

Phytopharmacy: An evidence-based guide to herbal medicinal products, First Edition.
Sarah E. Edwards, Inês da Costa Rocha, Elizabeth M. Williamson and Michael Heinrich.
© 2015 John Wiley & Sons, Ltd. Published 2015 by John Wiley & Sons, Ltd.

The anti-inflammatory action of the hydroalcoholic extract (HE) in the mouse ear inflammation model induced by 12-O-tetradecanoylphorbol-acetate has been ascribed mainly to the malvidin 3-glucoside content: topical application reduced oedema, polymorphonuclear cells influx and interleukin (IL)-1β levels in the tissue (Prudente et al. 2013). A range of other properties have been shown, including anti-cholinesterase effects, which are being explored for their potential application in anti-ageing and anti-hair loss products (Gasparetto et al. 2012). None of these have, however, been substantiated sufficiently to recommend their use.

Interactions: No data are available and significant interactions are not expected.

Adverse effects: There is a lack of data, but due to the wide use of mallow species there are no major concerns about safety.

Dosage: The plant is generally used as an infusion, and often mixed with other demulcents. A common dose range is 2–5 g dried herb per dose.

General plant information: The rootstock of *Althaea officinalis* L. (marshmallow) is used in a very similar way and has similar constituents and effects. It is part of some THR preparations.

Common mallow is commonly consumed in Southern Europe in stews, soups and vegetable dishes.

References

DellaGreca M, Cutillo F, D'Abrosca B, Fiorentino A, Pacifico S, Zarrelli A. (2009) Antioxidant and radical scavenging properties of *Malva sylvestris*. *Natural Product Communications* 4(7): 893–896.

ESCOP. (2009) *ESCOP Monographs: The Scientific Foundation for Herbal Medical Products.* 2nd Edition Supplement 2009. Thieme, Stuttgart, New York.

Gasparetto JC, Martins CA, Hayashi SS, Otuky MF, Pontarolo R. (2012) Ethnobotanical and scientific aspects of *Malva sylvestris* L.: a millennial herbal medicine. *Journal of Pharmacy and Pharmacology* 64(2): 172–189.

Prudente AS, Loddi AM, Duarte MR, Santos AR, Pochapski MT, Pizzolatti MG, Hayashi SS, Campos FR, Pontarolo R, Santos FA, Cabrini DA, Otuki MF. (2013) Pre-clinical anti-inflammatory aspects of a cuisine and medicinal millennial herb: *Malva sylvestris* L. *Food and Chemical Toxicology* 58: 324–331.

Usami A, Kashima Y, Marumoto S, Miyazawa M. (2013) Characterization of aroma-active compounds in dry flower of *Malva sylvestris* L. by GC-MS-O analysis and OAV calculations. *Journal of Oleo Science* 62(8): 563–570.

Maritime Pine (Bark) *Pinus pinaster* subsp. *pinaster*

Synonyms: *P. maritima* Lam.; *P. pinaster* subsp. *atlantica* Villar; and others

Family: Pinaceae

Other common names: Cluster pine; pinaster pine

Botanical drug used: Bark (generally, a specific bark extract derived from a population of French maritime pine grown on the Atlantic coast)

Indications/uses: French maritime pine bark has been recommended for a wide range of conditions, especially those relating to chronic venous insufficiency and neurodegeneration, most of which are attributed to its antioxidant effects. These include improvements to cardiovascular health, including prophylaxis of air flight oedema and thrombosis, treatment of venous ulcers, improvement of risk factors in metabolic syndrome and even erectile dysfunction. It is also used for chronic inflammatory disorders such as osteo-arthritis and asthma.

Evidence: Many clinical trials have been carried out showing benefits of French maritime pine bark in conditions of chronic venous insufficiency and other conditions. Although positive and mainly well conducted, there exist only one or two fairly small trials per condition.

Safety: Generally safe, even in children, but it has been suggested that there may be an additive effect with antiplatelet drugs.

Main chemical compounds: French maritime pine bark contains a range of procyanidins, flavonoids and polyphenols. The procyanidins account for 65–75% of the extract and are composed of catechin and epicatechin subunits of varying chain lengths. It contains other phenolic acids, including cinnamic acids and their glycosides, and flavonoids including taxifolin (D'Andrea 2010; Schoonees et al. 2012; Williamson et al. 2013).

Clinical evidence: Various pine bark extracts are available (see Maimoona et al. 2011) but almost all of the clinical research has been conducted on French maritime pine bark, most notably Pycnogenol®.

Phytopharmacy: An evidence-based guide to herbal medicinal products, First Edition.
Sarah E. Edwards, Inês da Costa Rocha, Elizabeth M. Williamson and Michael Heinrich.
© 2015 John Wiley & Sons, Ltd. Published 2015 by John Wiley & Sons, Ltd.

Clinical studies have shown benefits in several conditions. These have recently been systematically reviewed and 15 trials (with a total of 791 participants) were evaluated for the treatment of seven different chronic disorders (Schoonees et al. 2012): asthma (2 studies, 1 in children; $n = 86$), attention deficit hyperactivity disorder (1 study in children; $n = 61$), chronic venous insufficiency (2 studies; $n = 60$), diabetes mellitus (4 studies; $n = 201$), erectile dysfunction (1 study; $n = 21$), hypertension (2 studies; $n = 69$) and osteoarthritis (3 studies; $n = 293$). The trials provided positive results, but due to the limited numbers of trials per condition and variation in outcome measures used, the authors concluded that it is not possible to draw definitive conclusions as yet, and that larger, well-designed, adequately powered trials are still needed. Other clinical studies have shown the benefits of French pine bark for varicose ulcers, for long haul air flight travel and tinnitus. The uses of French pine bark for these and other conditions involving chronic venous insufficiency have been reviewed and mechanisms identified (Gulati 2014).

Recent clinical trials have found that French maritime pine bark is able to improve health risk factors in subjects with metabolic syndrome (Belcaro et al. 2013); reduce vasomotor and insomnia/sleep problems in perimenopausal women (Kohama and Negami 2013); improve cognitive function, attention and mental performance in students (Luzzi et al. 2011); and benefit skin condition in women, by increasing hydration and skin elasticity (Marini et al. 2012).

Pre-clinical evidence and mechanisms of action: French maritime pine bark protects against oxidative stress by increasing the intracellular synthesis of antioxidative enzymes and by acting as a potent scavenger of free radicals via the regeneration and protection of vitamins C and E. Anti-inflammatory activity has been demonstrated *in vitro* and *in vivo* in animals, including protection against UV-radiation-induced erythema. Immunomodulation has been observed in animal models and also in patients with lupus. French maritime pine bark antagonises the vasoconstriction caused by adrenaline and noradrenaline by increasing the activity of endothelial nitric oxide synthase (eNOS) and prevents smoking-induced platelet aggregation. It has been shown to relieve some premenstrual symptoms, including abdominal pain. This action may be associated with the spasmolytic action of some phenolic acids (D'Andrea 2010; Schoonees et al. 2012). It has many other pharmacological activities, including protection against ovariectomy-induced bone loss in rats (Mei et al. 2012).

Interactions: French maritime pine bark inhibits ADP-induced platelet aggregation in *ex-vivo* studies, so there is potential for it to interact with antiplatelet drugs due to its effects on COX-1 and COX-2 (Williamson et al. 2013). No clinically significant interactions have been recorded. French maritime pine bark appears to modulate the immune system so it may be wise to avoid with immunosuppressant drugs.

Contraindications: None known

Adverse effects: Mild gastrointestinal discomfort, headache, nausea and dizziness have been reported rarely.

Dosage: 100–360 mg daily, in divided doses; for specific products and for asthma in children, the manufacturers' instructions should be followed.

General plant information: *P. pinaster* Aiton is native to the western and south-western Mediterranean region. It is fast growing and widely planted for its timber.

In many regions, including South Africa it has become an invasive species, reducing the local biodiversity.

References

Belcaro G, Cornelli U, Luzzi R, Cesarone MR, Dugall M, Feragalli B, Errichi S, Ippolito E, Grossi MG, Hosoi M, Cornelli M, Gizzi G. (2013) Pycnogenol® supplementation improves health risk factors in subjects with metabolic syndrome. *Phytotherapy Research* 27(10): 1572–1578.

D'Andrea G. (2010) Pycnogenol: a blend of procyanidins with multifaceted therapeutic applications? *Fitoterapia* 81(2010) 724–736.

Gulati OP. (2014) Pycnogenol® in chronic venous insufficiency and related venous disorders. *Phytotherapy Research* 28(3): 348–362.

Kohama T, Negami M. (2013) Effect of low-dose French maritime pine bark extract on climacteric syndrome in 170 perimenopausal women: a randomized, double-blind, placebo-controlled trial. *Journal of Reproductive Medicine* 58(1–2): 39–46.

Lee OH, Seo MJ, Choi HS, Lee BY. (2012) Pycnogenol® inhibits lipid accumulation in 3T3-L1 adipocytes with the modulation of reactive oxygen species (ROS) production associated with antioxidant enzyme responses. *Phytotherapy Research* 26(3); 403–411.

Luzzi R, Belcaro G, Zulli C, Cesarone MR, Cornelli U, Dugall M, Hosoi M, Feragalli B. (2011) Pycnogenol® supplementation improves cognitive function, attention and mental performance in students. *Panminerva Medica* 53(3 Suppl 1): 75–82.

Maimoona A, Naeem I, Saddique Z, Jameel K. (2011) A review on biological, nutraceutical and clinical aspects of French Maritime pine bark extract. *Journal of Ethnopharmacology* 133(2): 261–277.

Marini A, Grether-Beck S, Jaenicke T, Weber M, Burki C, Formann P, Brenden H, Schönlau F, Krutmann J. (2012) Pycnogenol® effects on skin elasticity and hydration coincide with increased gene expressions of collagen type I and hyaluronic acid synthase in women. *Skin Pharmacology and Physiology* 25(2): 86–92.

Mei L, Mochizuki M, Hasegawa N. (2012) Protective effect of Pycnogenol® on ovariectomy-induced bone loss in rats. *Phytotherapy Research* 26(1): 153–155.

Schoonees A, Visser J, Musekiwa A, Volmink J. (2012) Pycnogenol® (extract of French maritime pine bark) for the treatment of chronic disorders. *Cochrane Database of Systematic Reviews* 4:CD008294.

Williamson EM, Driver S, Baxter K. (Eds.) (2013) *Stockley's Herbal Medicines Interactions.* 2nd Edition. Pharmaceutical Press, London, UK.

Milk Thistle   *Silybum marianum* (L.) Gaertn.

Synonyms: *Carduus marianus* L.; *S. maculatum* (Scop.) Moench; and others

Family: Asteraceae (Compositae)

Other common names: Blessed thistle; holy thistle; lady's milk; lady's thistle; St. Mary's thistle; and others

Drug name: Silybi mariani fructus

Botanical drug used: Dried ripe fruits, freed from the pappus

---

Indications/uses: Traditionally used to relieve the symptoms associated with overindulgence of food and drink, including indigestion and upset stomach. Known as a liver protectant, milk thistle is also used to treat gallstones and as supportive treatment for alcoholic cirrhosis and hepatitis. An intravenous preparation containing silibinin (Legalon® SIL) is licensed in a number of European countries, where it is used in emergency rooms to counteract *Amanita* mushroom poisoning.

Evidence: Clinical data to support the traditional uses of milk thistle are lacking. Despite a number of trials, clinical evidence remains inconclusive for the effectiveness of milk thistle in treating liver disease except in *Amanita* poisoning. However, constituents of milk thistle are being investigated for potential treatment in hepatitis and for their anticancer, antidiabetic and cardioprotective properties.

Safety: Appears safe at recommended doses, including for long-term use. Animal studies indicate that milk thistle is low in toxicity. Although historically used as a galactagogue, there is a lack of safety data for the use of milk thistle during pregnancy and lactation or in paediatric use.

Main chemical compounds: Flavonolignans (1.5–3.0%) which, together with a flavonoid – taxifolin (a 2,3-dihydroflavonol) - are collectively known as silymarin, comprising 65–80% of the crude extract. The major components of silymarin include two pairs of a 1:1 mixture of diastereoisomers, silybin A and silybin B (previously thought to be a single compound, silibinin), with isosilybin A, isosilybin B, silychristin, silydianin, 2,3-dehydrosilybin and 2,3-dihydrosilychristin (Lee and Liu 2003; WHO 2004).

Phytopharmacy: An evidence-based guide to herbal medicinal products, First Edition.
Sarah E. Edwards, Inês da Costa Rocha, Elizabeth M. Williamson and Michael Heinrich.
© 2015 John Wiley & Sons, Ltd. Published 2015 by John Wiley & Sons, Ltd.

Clinical evidence:

Clinical data to support the main traditional uses of milk thistle are sparse and conflicting.

Functional dyspepsia: A double-blind placebo-controlled study in which 60 patients diagnosed with functional dyspepsia were treated with a proprietary herbal preparation containing milk thistle, in combination with eight other herbs, showed that the herbal preparation significantly improved symptoms compared with placebo (Madisch et al. 2001). However, no data on the effectiveness of mono-preparations of milk thistle for indigestion were found.

Liver disease: Data to support the use of milk thistle preparations to treat liver diseases are mixed (Abenavoli et al. 2010). A Cochrane review assessed 13 randomised controlled trials in patients ($n = 915$) with alcoholic and/or hepatitis B or C liver diseases, and found that milk thistle versus placebo or no intervention had no significant effect on mortality (Rimbaldi et al. 2007). In a double-blind, randomised controlled trial, 154 patients with chronic hepatitis C were administered higher than customary doses of standardised silymarin preparation for 24 weeks. Silymarin did not significantly reduce serum alanine aminotransferase (ALT) compared with placebo (Fried et al. 2012). However, other clinical trials in hepatitis patients found that administration of silymarin significantly improved mortality, decreased serum levels of liver enzymes, improved liver function and improved histopathological findings compared to placebo groups (Abenavoli et al. 2010; WHO 2004). Inconsistent results of various studies may be attributed to heterogeneous populations of patients with liver disease and use of non-standardised silymarin preparations.

Amanita poisoning: For ethical reasons, there are no controlled clinical studies available, but uncontrolled trials and case reports describe successful treatment with intravenous silibinin (Legalon® SIL). In nearly 1500 documented cases, the overall mortality in patients treated with Legalon® SIL was less than 10%, in comparison with over 20% when using penicillin. Silibinin has been shown to interact with specific hepatic transport proteins blocking cellular amatoxin re-uptake and thus interrupting enterohepatic circulation of the toxin (Mengs et al. 2012).

Antidiabetic effects: Silymarin treatment was found to have a significant beneficial effect in type 2 diabetic patients, reducing levels of glycosylated haemoglobin, fasting blood glucose, total cholesterol, LDL, triglyceride, SGOT and SGPT compared with placebo and values at the beginning of a 4-month randomised double-blind controlled trial (Huseini et al. 2006).

Pre-clinical evidence and mechanisms of action:

Liver protection and detoxification effects: Silymarin also promotes hepatocyte regeneration, reduces inflammatory reaction and inhibits fibrogenesis in the liver. The major mechanism in hepatoprotective activity appears to be inhibition of intrahepatic NF-κB activation, which prevents synthesis of TNF, INF-g, IL-2 and iNOS (Abenavoli et al. 2010). Silymarin may also inhibit binding of toxins to the hepatocyte cell membrane receptors.

Other effects: Silymarin acts as an antioxidant by reducing free radical production and lipid peroxidation, and these properties may be responsible for many of its effects. A number of *in vitro* and *in vivo* studies have demonstrated anti-inflammatory, antidiabetic, anticancer, antifibrotic and cardioprotective properties of milk thistle extracts (Abenavoli et al. 2010; Milić et al. 2013; Salamone et al. 2011; WHO 2004), However, the therapeutic benefits of silymarin are severely limited by poor bioavailability. A number of pharmaceutical preparations have been developed in the past few years to overcome this, and show promise in several areas (Cufi et al. 2013).

Interactions: No confirmed reports of interactions are available (Williamson et al. 2013). Milk thistle constituents have been shown to inhibit CYP isozymes, including CYP3A *in vitro* although the clinical significance of this is unknown (Brantley et al. 2013).

Contraindications: Hypersensitivity to plants from the Asteraceae family (WHO 2004).

Adverse effects: In clinical trials, adverse effects were similar to those seen with placebo. The most common adverse effects reported after oral ingestion are minor gastrointestinal disturbances, headache and allergic reactions (urticaria, skin rash, pruritis), although frequency is unknown. Anaphylactic shock was reported in a patient ingesting a tea made from milk thistle (WHO 2004). Exacerbation of haemochromatosis (iron overload caused by excessive absorption) has been associated with ingestion of milk thistle, although this is unproven and the patient was also taking paracetamol (acetaminophen) regularly (Whittington 2007).

Dosage: Daily oral dosage: 12–15 g crude drug; 200–400 mg silymarin, calculated as silybin, in standardised preparations (WHO 2004).

General plant information: Native to the Mediterranean region, *S. marianum* has naturalised throughout much of Europe, including southern Britain. It is considered a noxious weed in some areas where it has been introduced (e.g. North America and Australia).

References

Abenavoli L, Capasso R, Milic N, Capasso F. (2010) Milk thistle in liver diseases: past, present and future. *Phytotherapy Research* 24(10): 1423–1432.

Brantley SJ, Graf TN, Oberlies NH, Paine MF. (2013) A systematic approach to evaluate herb-drug interaction mechanisms: investigation of milk thistle extracts and eight isolated constituents as CYP3A inhibitors. *Drug Metabolism and Disposition* 41(9): 1662–1670.

Cufi S, Bonavia R, Vazquez-Martina A, Corominas-Fajaa B, Oliveras-Ferraros C, Cuyàs E, Martin-Castillo B, Barrajón-Catalán E, Visa J, Segura-Carretero A, Bosch-Barrera J, Joveni J, Micol V, Menendez JA. (2013) Silibinin meglumine, a water-soluble form of milk thistle silymarin, is an orally active anti-cancer agent that impedes the epithelial-to-mesenchymal transition (EMT) in EGFR-mutant non-small-cell lung carcinoma cells. *Food and Chemical Toxicology* 60: 360–368.

Fried MW, Navarro VJ, Afdhal N, Belle SH, Wahed AS, Hawke RL, Doo E, Meyers CM, Reddy KR. (2012) Effect of silymarin (milk thistle) on liver disease in patients with chronic hepatitis C unsuccessfully treated with interferon therapy: a randomized controlled trial. *Journal of the American Medical Association (JAMA)* 308(3): 274–282.

Huseini HF, Larijani B, Heshmat R, Fakrhzadeh H, Radjabipour B, Toliat T, Raza M. (2006) The efficacy of *Silybum marianum* (L.) Gaertn. (silymarin) in the treatment of type II diabetes: a randomized, double-blind, placebo-controlled, clinical trial. *Phytotherapy Research* 20(12): 1036–1039.

Lee DY, Liu Y. (2003) Molecular structure and stereochemistry of silybin A, silybin B, isosilybin A, and isosilybin B, Isolated from *Silybum marianum* (milk thistle). *Journal of Natural Products* 66(9): 1171–1174.

Madisch A, Melderis H, Mayr G, Sassin I, Hotz J. (2001) A plant extract and its modified preparation in functional dyspepsia. Results of a double-blind placebo controlled comparative study. *Zeitschrift für Gastroenterologie* 39(7): 511–517 [Article in German].

Mengs U, Pohl RT, Mitchell T. (2012) Legalon® SIL: The antidote of choice in patients with acute hepatoxicity from amatoxin poisoning. *Current Pharmaceutical Biotechnology* 13(10): 1964–1970.

Milić N, Milosević N, Suvajdzić L, Zarkov M, Abenavoli L. (2013) New therapeutic potentials of milk thistle (*Silybum marianum*). *Natural Product Communication* 8(12): 1801–1810.

Rimbaldi A, Jacobs BP, Gluud C. (2007) Milk thistle for alcoholic and/or hepatitis B or C virus liver diseases. *Cochrane Database of Systematic Reviews* (4): CD003620.

Salamone F, Cappello F, La Delia F, Mangiameli A, Galvano F, Bugianesi E, Li Volti G. (2011) P.1.10: Cardioprotective effect of silibinin in mice with non-alcoholic fatty liver disease. *Digestive and Liver Disease* 43(Suppl. 3): S151.

Whittington C.. (2007) Exacerbation of hemochromatosis by ingestion of milk thistle. *Canadian Family Physician* 53(10): 1671–1673.

WHO. (2004) *WHO Monographs on Selected Medicinal Plants*. Vol. 2. WHO, Geneva, 358 pp.

Williamson EM, Driver S, Baxter K. (Eds.) (2013) *Stockley's Herbal Medicines Interactions*. 2nd Edition. Pharmaceutical Press, London, UK.

Neem

Azadirachta indica A.Juss.

Synonyms: *Melia azadirachta* L.; *M. indica* (A.Juss.) Brandis; and others

Family: Meliaceae

Other common names: Arista; bakam; margosa; nim

Drug names: Azadirachti folium/cortex/semen/oleum

Botanical drug used: Most parts of the neem tree are used medicinally, commonly the leaves, bark, seeds and seed oil (WHO 2007).

Indications/uses: Neem is a natural insecticide, and leaf and seed extracts are applied topically to treat infestations of lice, a common use in Europe. Traditional Ayurvedic uses are numerous and include oral as well as topical administration. It is widely used as a treatment for skin diseases, inflammation, gingivitis (gum disease), cancer, fevers, malaria, diabetes, controlling gastric hyperacidity and as a spermicide.

Evidence: Clinical evidence is lacking for most internal uses but as a topical insecticide, *in vivo* and *in vitro* studies have demonstrated the effectiveness of neem seed extracts against head and body lice. Evidence for the internal uses is weak.

Safety: While neem appears to be safe when used topically, internal use cannot yet be widely recommended. Large doses or long-term internal use should be avoided due to the risk of kidney and liver damage. Children should not take neem internally. Neem has genotoxic, cytotoxic, spermicidal and abortifacient properties, so it should be avoided in women who are (or wish to be) pregnant and their partners, and also during breastfeeding.

Main chemical compounds: The active components are considered to be several different types of limonoids and include the azadirachtin, azadiradione, gedunin and nimbin types. There are many other types of compounds present, including flavonoids, coumarins, glycoproteins and tannins (Biswas et al. 2002l; Kikuchi et al. 2011; WHO 2007).

Phytopharmacy: An evidence-based guide to herbal medicinal products, First Edition.
Sarah E. Edwards, Inês da Costa Rocha, Elizabeth M. Williamson and Michael Heinrich.
© 2015 John Wiley & Sons, Ltd. Published 2015 by John Wiley & Sons, Ltd.

Clinical evidence:

External application as an insecticide: Some studies on the clinical efficacy of external application against head lice are available to support this use (Abdel-Ghaffar and Semmler 2007; Abdel-Ghaffar et al. 2012).

Internal application and other uses: There is a lack of clinical trial evidence to support many of the traditional medicinal uses of neem, but the plant is used so widely in Asia that there is a great deal of anecdotal evidence.

Pre-clinical evidence and mechanisms of action:

Insecticidal use: *In vitro* studies have demonstrated the effectiveness of neem seed extract preparations against head lice and their eggs (Abdel-Ghaffar et al. 2012; Heukelbach et al. 2006; Mehlhorn et al. 2011). Azadirachtin is a feeding deterrent and disrupts the insect moulting process, preventing maturation of larvae, as well as having direct insecticidal activity (Biswas et al. 2002).

Other uses: Isolated constituents and extracts exhibit a very wide spectrum of pharmacological activities, such as anti-inflammatory, antibacterial (Biswas et al. 2002), anti-ulcer (Maity et al. 2009) and antidiabetic (Perez-Guttierez and Damian-Guzman 2012) properties. Mechanisms of action are likely to be complex, with the possibility of synergistic or antagonistic effects occurring within extracts.

There is considerable *in vitro* evidence that neem extracts contain compounds that have potent cytotoxic and other anticancer properties (Elumalia et al. 2012; Gunadharini et al. 2011; Kikuchi et al. 2011; Mahapatra et al. 2011). For this reason, the internal use of neem as a herbal medicine cannot be recommended and such uses are outside the scope of herbal self-treatment.

Interactions: An aqueous extract interacted with chloroquine, causing slower absorption and elimination, as well as increasing its half-life (Nwafor et al. 2003). As neem has hypoglycaemic effects (i.e. reduces blood sugar levels), there is the potential that it may interact with antidiabetic drugs if taken internally, causing blood sugar levels to become too low. Neem should be avoided with immuno-suppressant and anticancer therapy, due to its reported immunomodulatory and cytotoxic activities.

Contraindications: Internal use is not recommended for self-treatment. Neem should only be used internally, if at all, by adults and under the guidance of a knowledgeable qualified health professional, and not at all by children, since Reye's syndrome and infant deaths have been associated with its use (Sinniah and Baskaran 1981). It should not be used during pregnancy and lactation, due to its cytotoxicity. Neem should be avoided by patients suffering from autoimmune diseases, such as multiple sclerosis, lupus and rheumatoid arthritis, due to its reported immunomodulating properties.

Adverse effects: Internal use may cause gastrointestinal irritation, infertility, oliguria (decreased urine excretion), acute renal failure, jaundice and anaemia (Boeke et al. 2004).

Dosage: For external use, follow manufacturers' instructions. There is not enough scientific data to provide a safe dose range of neem for internal use.

General plant information: Neem has been utilised for centuries in India, the earliest recorded uses dating over 4500 years ago as an ingredient in health and beauty preparations. Today, in rural India, the neem tree is known as the 'Village Pharmacy', due to its many medicinal uses (Dasgupta et al. 2004). It is also used to prevent insect infestation in stored food crops. An interesting review of this plant from pre-history to the present day is available (Kumar and Navaratnam 2013).

References

Abdel-Ghaffar F, Al-Quaraishy S, Al-Rasheid KA, Mehlhorn H. (2012) Efficacy of a single treatment of head lice with a neem seed extract: an *in vivo* and *in vitro* study on nits and motile stages. *Parasitology Research* 110(1): 277–280.

Abdel-Ghaffar F, Semmler M. (2007) Efficacy of neem seed extract shampoo on head lice of naturally infected humans in Egypt. *Parasitology Research* (2): 329–332.

Biswas K, Chattopadhyay I, Banerjee RK, Bandyopadhyay U. (2002) Biological activities and medicinal properties of neem (*Azadirachta indica*). *Current Science* 82(11): 1336–1345.

Boeke SJ, Boersma MG, Alink GM, van Loon JJ, van Huis A, Dickie M, Rietjens IM. (2004) Safety evaluation of neem (*Azadirachta indica*) derived pesticides. *Journal of Ethnopharmacology* 94(1): 25–41.

Dasgupta T, Banerjee S, Yadava PK, Rao AR. (2004) Chemopreventive potential of *Azadirachta indica* (Neem) leaf extract in murine carcinogenesis model systems. *Journal of Ethnopharmacology* 92(1): 23–26.

Elumalia P, Gunadharini DK, Senthilkumar K, Banudevi S, Arunkumar R, Benson CS, Sharmila G, Arunakaran J. (2012) Ethanolic neem (*Azadirachta indica* A.Juss) leaf extract induces apoptosis and inhibits the IGF signaling pathway in breast cancer cell lines. *Biomedicine and Preventive Nutrition* 2(1): 59–68.

Gunadharini DN, Elumalai P, Arunkumar R, Senthilkumar K, Arunakaran J. (2011) Induction of apoptosis and inhibition of PI3K/Akt pathway in PC-3 and LNCaP prostate cancer cells by ethanolic neem leaf extract. *Journal of Ethnopharmacology* 134(3): 644–650.

Heukelbach J, Oliveira FA, Speare R. (2006) A new shampoo based on neem (*Azadirachta indica*) is highly effective against head lice *in vitro*. *Parasitology Research* 99(4): 353–356.

Kikuchi T, Ishii K, Noto T, Takahashi A, Tabata K, Suzuki T, Akihisa T. (2011) Cytotoxic and apoptosis-inducing activities of limonoids from the seeds of *Azadirachta indica* (neem). *Journal of Natural Products* 74(4): 866–870.

Kumar VS, Navaratnam V. (2013) Neem (*Azadirachta indica*): Prehistory to contemporary medicinal uses to humankind. *Asian Pacific Journal of Tropical Biomedicine* 3(7): 505–514.

Mahapatra S, Karnes RJ, Holmes MW, Young CY, Cheville JC, Kohli M, Klee EW, Tindall DJ, Donkena KV. (2011) Novel molecular targets of *Azadirachta indica* associated with inhibition of tumor growth in prostate cancer. *AAPS Journal* 13(3): 365–377.

Maity P, Biswas K, Chattopadhyay I, Banerjee RK, Bandyopadhyay U. (2009) The use of neem for controlling gastric hyperacidity and ulcer. *Phytotherapy Research* 23(6): 747–755.

Mehlhorn H, Abel-Ghaffar F, Al-Rasheid KA, Schmidt J, Semmler M. (2011) Ovicidal effects of a neem seed extract preparation on eggs of body and head lice. *Parasitology Research* 109(5): 1299–1302.

Nwafor SV, Akah PA, Okoli CO, Onyirioha AC, Nworu CS. (2003) Interaction between chloroquine sulphate and aqueous extract of *Azadirachta indica* A. Juss. (Meliaceae) in rabbits. *Acta Pharmaceutica* 53(4): 305–311.

Perez-Guttierez RM, Damian-Guzman M. (2012) Meliacinolin: a potent α-amylase inhibitor isolated from *Azadirachta indica* leaves and in vivo antidiabetic property in streptozocin-nicotonimide-induced type 2 diabetes in mice. *Biological and Pharmaceutical Bulletin* 35(9): 1516–1524.

Sinniah D, Baskaran G. (1981) Margosa oil poisoning as a cause of Reye's syndrome. *The Lancet* 1(8218): 487–489.

WHO. (2007) *WHO Monographs on Medicinal Plants*. Vol. 3. WHO, Geneva, 390 pp.

Nettle   *Urtica dioica* L.; *U. urens* L.

Family: Urticaceae

Other common names: Stinging nettle (*U. dioica*); dwarf nettle (*U. urens*); Urtica

Drug name: Urticae folium/herba; Urticae radix et rhizoma

Botanical drug used: Leaf/herb; root and rhizome

Indication/uses: The modern use of nettle root (*U. dioica* and *U. urens*) is mainly to relieve symptoms of urinary tract discomfort in men with enlargement of the prostate (benign prostatic hyperplasia, BPH), where cancer has been ruled out. Nettle leaf and herb (*U. dioica*) preparations have been used traditionally to relieve minor aches and pains in the joints, and for allergic reactions, especially rhinitis. The herb has a reputation in folk medicine for improving glycaemic control in patients with diabetes mellitus.

Evidence: Some clinical studies are available to support the therapeutic claims for BPH and inflammatory conditions, but larger clinical studies are needed. Combination products with saw palmetto may prove to be more effective in the treatment of BPH.

Safety: Considered to be safe. Young nettle leaves are often eaten, especially in soups.

Main chemical compounds: The main constituents of the leaf are caffeic acid esters, mainly caffeoylmalic acid (not present in *U. urens*), chlorogenic, neochlorogenic, with free caffeic acids. Flavonoids such as kaempferol, isorhamnetin and quercetin, scopoletin, β-sitosterol and glycoprotein are also present. Histamine, serotonin, leukotriene and acetylcholine, present in the fresh plant and producing the 'sting', are denatured by drying (Anon. 2007; ESCOP 2003). The root contains a lectin (*Urtica dioica* agglutinin) composed of six isolectins, lignans including pinoresinol, (+)-neoolivil, (-)-secoisolariciresinol, dehydrodiconiferyl alcohol, 3,4-divanillyltetrahydrofuran and isolariciresinol, with polysaccharides, scopoletin, β-sitosterol and other sterols and their glucosides; caffeic acid esters, ceramides, hydroxyl fatty acids, and monoterpene diols and their glucosides (Anon. 2007; ESCOP 2003; Schöttner et al. 1997).

Phytopharmacy: An evidence-based guide to herbal medicinal products, First Edition.
Sarah E. Edwards, Inês da Costa Rocha, Elizabeth M. Williamson and Michael Heinrich.
© 2015 John Wiley & Sons, Ltd. Published 2015 by John Wiley & Sons, Ltd.

Clinical evidence:

Benign prostate hyperplasia: A comprehensive review on the clinical effectiveness of nettle root extract in the treatment of BPH concluded that there is some evidence to support its use in the short term (Chrubasik et al. 2007). Most of the studies cited were open and uncontrolled, and only a small number were randomised controlled clinical trials. Since then, a small clinical study (100 patients) found that nettle extract (plant part not stated) had a better effect in relieving clinical symptoms in BPH patients compared to placebo (Ghorbanibirgani et al. 2013). More randomised trials have been performed with nettle root extracts in combination with other herbs (see e.g. saw palmetto), and these may prove more effective than nettle root extracts alone in treatment of lower urinary tract symptoms (LUTS) (Wilt et al. 2000). An open-label extension of a randomised, double-blind clinical trial assessing the effect of a fixed combination product containing saw palmetto fruit (*Serenoa repens*) and nettle root extract (160:120 mg per capsule, twice a day for 24 weeks) in elderly men with moderate or severe LUTS caused by BPH, showed that there was a clinically relevant benefit over a follow-up period of 96 weeks. There was a reduction in the International Prostate Symptom Score (I-PSS) self-rating questionnaire, a decrease of residual urine volume, increase in urinary flow and very few adverse events after treatment (Lopatkin et al. 2007). A combination of nettle root (*U. dioica* 120 mg), saw palmetto (320 mg) and pine bark (*Pinus pinaster* Aiton, 5 mg) extract was effective in patients suffering from LUTS when treated for periods between 1 month and a year, with or without concomitant therapy (antibiotics or alpha-blockers). In this prospective study, it was observed that patients did not present changes in flow rate and prostate volume, but a marked reduction in pain and discomfort (Pavone et al. 2010).

Anti-inflammatory activity: A systematic review of herbal medicines which included *U. dioica* for effects on inflammatory conditions showed that there was some evidence of effectiveness in pain and the physical impairment caused by arthritis, and it reduced concurrent intake of analgesic drugs (Di Lorenzo et al. 2013). Oral preparations of nettle caused a reduction of symptoms in rheumatic conditions and were able to potentiate the efficacy of diclofenac (Anon. 2007). However, studies were limited and larger trials are needed.

Anti-allergic effects in rhinitis: A randomised, double-blind study of freeze-dried *U. dioica* in the treatment of allergic rhinitis involved 69 patients. Assessment was based on daily symptom diaries, and global responses recorded at the follow-up visit after 1 week of therapy. *U. dioica* was rated higher than placebo in the global assessments, but comparing the diary data, *U. dioica* was rated only slightly higher (Mittman, 1990).

Antidiabetic effects: In a randomised, double-blind, placebo-controlled clinical trial of patients with advanced type 2 diabetes who also needed insulin therapy, improved glycaemic control was observed after taking *U. dioica* leaf extract (Kianbakht et al. 2013).

Pre-clinical evidence and mechanisms of action:

Effects on the prostate: The mechanism of action by which nettle exerts its effect on BPH is likely to be related to interactions with sex hormone-binding globulin (SHBG), aromatase, epidermal growth factor and prostate steroid membrane receptors (Schöttner et al. 1997), but 5α reductase or androgen receptors do not appear to be involved (Chrubasik et al. 2007). For example, an aqueous nettle extract was found to inhibit the binding to the SHBG receptor *in vitro* (Hryb et al. 1995) and the lignans have been shown to interfere with SHBG binding of steroids (Gansser et al. 1995). It has been suggested that nettle exerts its effect by displacing steroid hormones from SHBG binding sites and preventing the interaction of prostate receptors with SHGB (Chrubasik et al. 2007).

Anti-inflammatory and antinociceptive effects: Extracts of *U. dioica* leaf, and its constituent caffeic malic acid, showed inhibitory effects on the arachidonic pathway *in vitro*, as well as antihistamine effects (Obertreis et al. 1996; Roschek et al. 2009). An *U. urens* extract was found to possess significant antinociceptive activity in chemically induced mouse pain models, the writhing test and the late phase of the formalin test, and anti-inflammatory activity in carrageenan-induced rat hind paw oedema, but not in a thermal model of pain, nor a topical inflammation model. The major component of this extract was chlorogenic acid (Marrassini et a 2010).

Allergic rhinitis: There is some evidence that *U. urens* extracts interact with receptors and enzymes associated with allergic rhinitis (Roschek et al. 2009).

Antidiabetic effects: Dried *U. dioica* leaf alcoholic and aqueous extracts promoted the repair of pancreatic tissue in a streptozocin-induced diabetic experimental model (Qujeq et al. 2013).

Interactions: No interactions recorded (Williamson et al. 2013). May have an inhibitory effect on drugs metabolised by cytochrome P450 enzymes (Ozen et al. 2003), or an additive effect with diuretics and hypotensives (Tahri et al. 2000). Excessive use of nettle may interact with concomitant therapy for diabetes, high or low blood pressure and may potentiate drugs with central nervous system (CNS)-depressant action (EMA 2011; EMA 2012).

Contraindications: Despite the food use of nettle leaves, medicinal preparations of nettle are not recommended during pregnancy and lactation, in patients with cardiac or renal oedema, nor for children and adolescents under 12 years of age, as safety has not been established (EMA 2011; EMA 2012; ESCOP 2003).

Adverse effects: Gastrointestinal complains (nausea, heartburn, feeling of fullness, flatulence, diarrhea) and allergic skin reactions (pruritus, rash) have been reported (ESCOP 2003). A case of gynaecomastia in a man and a case of galactorrhoea in a woman have been reported after consumption of nettle tea (Sahin et al. 2007).

Dosage: As recommend by the manufacturer. For the leaf, 4–6 g for preparation of an herbal tea, three to six times daily (equivalent to 8–12 g of herbal substance) or 150–460 mg dry extract one to three times daily (EMA 2011). For the root, 1.5 g as a decoction three to four times daily or 450–750 mg of dry extract, two to three times daily (EMA 2012).

General plant information: Nettles are a common weed native to Eurasia. The leaves have stinging hairs, which cause a painful skin eruption, blisters, itchiness and numbness that might last from minutes to days. Stings inflicted by fresh nettle leaves have been used to treat chronic knee pain and it has been reported that, in theory, this is acceptable to patients and GPs (Randall et al. 2008).

References

Anon. (2007) *Urtica dioica; Urtica urens* (nettle). Monograph. *Alternative Medicine Review* 12(3): 280–284.

Chrubasik JE, Roufogalis BD, Wagner H, Chrubasik S. (2007) A comprehensive review on the stinging nettle effect and efficacy profiles. Part II: Urticae radix. *Phytomedicine* 14(7–8): 568–579.

Di Lorenzo C, Dell'Agli M, Badea M, Dima L, Colombo E, Sangiovanni E, Restani P, Bosisio E. (2013) Plant food supplements with anti-inflammatory properties: a systematic review (II). *Critical Reviews in Food Science and Nutrition* 53(5): 507–516.

EMA. (2011) Community herbal monograph on *Urtica dioica* L.; *Urtica urens* L., folium. European Medicines Agency http://www.ema.europa.eu/docs/en_GB/document _library/Herbal_-_Community_herbal_monograph/2011/01/WC500100762.pdf.

EMA. (2012) Community herbal monograph on *Urtica dioica* L., *Urtica urens* L., their hybrids or their mixtures, radix. European Medicines Agency http://www.ema.europa.eu /docs/en_GB/document_library/Herbal_-_Community_herbal_monograph/2012/11 /WC500134486.pdf.

ESCOP. (2003) *ESCOP Monographs: The Scientific Foundation for Herbal Medicinal Products.* 2nd Edition. Thieme, Exeter and London, UK.

Gansser D, Spiteller G. (1995) Aromatase inhibitors from *Urtica dioica* roots. *Planta Medica* 61(2): 138–140.

Ghorbanibirgani A, Khalili A, Zamani L. (2013) The efficacy of stinging nettle (*Urtica dioica*) in patients with benign prostatic hyperplasia: a randomized double blind study in 100 patients. *Iranian Red Crescent Medical Journal* 15(1): 9–10.

Hryb DJ, Khan MS, Romas NA, Rosner W. (1995) The effect of extracts of the roots of the stinging nettle (*Urtica dioica*) on the interaction of SHBG with its receptor on human prostatic membranes. *Planta Medica* 61(1): 31–32.

Kianbakht S, Khalighi-Sigaroodi F, Dabaghian FH. (2013) Improved glycemic control in patients with advanced type 2 diabetes mellitus taking *Urtica dioica* leaf extract: a randomized double-blind placebo-controlled clinical trial. *Clinical Laboratory* 59(9–10): 1071–1076.

Lopatkin N, Sivkov A, Schlafke S, Funk P, Medvedev A, Engelmann U. (2007) Efficacy and safety of a combination of Sabal and Urtica extract in lower urinary tract symptoms – long-term follow-up of a placebo-controlled, double-blind, multicenter trial. *International Urology and Nephrology* 39(4): 1137–1146.

Marrassini C, Acevedo C, Miño J, Ferraro G, Gorzalczany S. (2010) Evaluation of antinociceptive, antinflammatory activities and phytochemical analysis of aerial parts of *Urtica urens* L. *Phytotherapy Research* 24(12): 1807–1812.

Mittman P. (1990) Randomized, double-blind study of freeze-dried *Urtica dioica* in the treatment of allergic rhinitis. *Planta Medica* 56(1): 44–47.

Obertreis B, Giller K, Teucher T, Behnke B, Schmitz H. (1996) Anti-inflammatory effect of *Urtica dioica* folia extract in comparison to caffeic malic acid. *Arzneimittelforschung* 46(1): 52–56.

Ozen T., Korkmaz H. (2003) Modulatory effect of *Urtica dioica* L. (Urticaceae) leaf extract on biotransformation enzyme systems, antioxidant enzymes, lactate dehydrogenase and lipid peroxidation in mice. *Phytomedicine* 10(5): 405–415.

Pavone C, Abbadessa D, Tarantino ML, Oxenius I, Lagana A, Lupo A, Rinella M. (2010) Associating *Serenoa repens, Urtica dioica* and *Pinus pinaster*. Safety and efficacy in the treatment of lower urinary tract symptoms. Prospective study on 320 patients. *Urologia* 77(1): 43–51.

Qujeq D, Tatar M, Feizi F, Parsian H, Sohan Faraji A, Halalkhor S. (2013) Effect of *Urtica dioica* leaf alcoholic and aqueous extracts on the number and the diameter of the islets in diabetic rats. *International Journal of Molecular Medicine* 2(1): 21–26.

Randall C, Dickens A, White A, Sanders H, Fox M, Campbell J. (2008) Nettle sting for chronic knee pain: a randomised controlled pilot study. *Complementary Therapies in Medicine* 16(2): 66–72.

Roschek B Jr., Fink RC, McMichael M, Alberte RS. (2009) Nettle extract (*Urtica dioica*) affects key receptors and enzymes associated with allergic rhinitis. *Phytotherapy Research* 23(7): 920–926.

Sahin M, Yilmaz H, Gursoy A, Demirel AN, Tutuncu NB, Guvener ND. (2007) Gynaecomastia in a man and hyperoestrogenism in a woman due to ingestion of nettle (*Urtica dioica*). *New Zealand Medical Journal* 120(1265): U2803.

Schöttner M, Gansser D, Spiteller G. (1997) Lignans from the roots of *Urtica dioica* and their metabolites bind to human sex hormone binding globulin (SHBG). *Planta Medica* 63(6): 529–532.

Tahri A, Yamani S, Legssyer A, Aziz M, Mekhfi HH, Bnouham M, Ziyyat A. (2000) Acute diuretic, natriuretic and hypotensive effects of a continuous perfusion of aqueous extract of *Urtica dioica* in the rat. *Journal of Ethnopharmacology* 73(1–2): 95–100.

Williamson EM, Driver S, Baxter K. (Eds.) (2013) *Stockley's Herbal Medicines Interactions*. 2nd Edition. Pharmaceutical Press, London, UK.

Wilt, TJ., Ishani, A, Rutks, I and MacDonald, R. (2000) Phytotherapy for benign prostatic hyperplasia. *Public Health Nutrition* 3(4A): 459–472.

Noni

Morinda citrifolia L.

Synonyms: *M. littoralis* Blanco; *M. macrophylla* Desf.; *M. tomentosa* B.Heyne ex Roth; and others

Family: Rubiaceae

Other common names: Canary wood; cheese fruit; hog apple; Indian mulberry

Botanical drug used: Fruit juice

Indications/uses: Traditionally, noni fruit has been used internally as a general tonic and to treat a wide range of chronic conditions, including hypertension, diabetes, cancer, asthma, gastrointestinal disorders and musculoskeletal disorders. Noni fruit and leaves have also been used topically to treat skin ailments and arthritic swellings.

Evidence: A range of *in vitro* and animal studies have indicated the potential beneficial effects of noni fruit juice, but clinical evidence is lacking.

Safety: Despite some reports of hepatotoxicity, probably due to the presence of anthraquinones (AQs) originating from the seeds, the European Food Safety Authority (EFSA) approved noni juice as a novel food ingredient in 2003 for use in pasteurised fruit drinks. Like other fruit juices, noni juice may increase potassium blood levels. Safety of consumption of noni during pregnancy and lactation has not been established and should be avoided.

Main chemical compounds: Noni fruit contains several classes of compounds: polysaccharides; glycosides of fatty acids and alcohols (including noniosides E-H and 3-methyl-3-butenyl glucoside); iridoids (e.g. asperuloside and its derivatives); flavonoids (e.g. narcissoside, kaempferol, quercetin and rutin); coumarins (e.g. scopoletin); lignans; phytosterols; carotinoids and volatile constituents including monoterpenes and short chain fatty acids and fatty acid esters (Dalsgaard et al. 2006; Pawlus and Kinghorn 2007; Potterat and Hamburger 2007). The seeds contain toxic AQs, including lucidin, alizarin and rubiadin; however, products made from ripe fruits without seeds do not contain these compounds. High levels of potassium (56 mEq/l) have been found in some noni juice products (Pharmaceutical Press Editorial Team 2013).

Phytopharmacy: An evidence-based guide to herbal medicinal products, First Edition.
Sarah E. Edwards, Inês da Costa Rocha, Elizabeth M. Williamson and Michael Heinrich.
© 2015 John Wiley & Sons, Ltd. Published 2015 by John Wiley & Sons, Ltd.

Clinical evidence:

Effects in cancer patients: Clinical data assessing the effects of noni are limited. In a phase I clinical trial investigating the optimal and maximum tolerated oral dosage of noni juice in 51 patients with advanced cancer, the maximum tolerated dose was found to be 12 g daily. Dose-related differences in self-reported physical functioning, pain and fatigue control were self-reported, but no tumour response attributable to noni was observed (Issell et al. 2009).

Antioxidant and anti-hyperlipidaemic effects in smokers: A randomised, double-blind, placebo-controlled trial investigated the effect of daily consumption of noni juice on blood lipid levels in 132 adult heavy smokers for 1 month. Results showed that noni juice was able to mitigate cigarette smoke-induced dyslipidemia, attributed to presence of iridoids (Wang et al. 2012). An earlier study by the same group found that 29.5 ml or 118 ml of noni juice taken daily by heavy smokers for 30 days reduced plasma superoxide anion radicals by 27% and 31%, respectively, and reduced plasma lipid hydroperoxide levels by 24.5% and 27%, while placebo juice did not significantly affect levels (Wang et al. 2009).

Lack of effects in dysmenorrhoea: A randomised, double-blind, placebo-controlled clinical trial in 100 women over 18 years, given 400 mg noni capsules or placebo over three menstrual cycles, found no reduction in menstrual pain or bleeding in the noni group compared to placebo (Fletcher et al. 2013).

Pre-clinical evidence and mechanisms of action:

A number of pharmacological studies on noni juice and its constituents have been undertaken, indicating anticancer, anti-inflammatory, hepatoprotective and hypoglycaemic activities (Brown 2012; Pawlus and Kinghorn 2007; Potterat and Hamburger 2007; Wang et al. 2008). However, little mechanistic work has been done and these properties may be due at least in part to non-specific antioxidant effects.

Anti-inflammatory and immune system effects via cannabinoid receptors: An *in vitro* study found that noni juice concentrate and a proprietary noni juice product (Tahitian Noni® Juice) potently activated cannabinoid 2 (CB_2) receptors, but inhibited cannabinoid 1 (CB_1) receptors in a concentration-dependent manner. *In vivo* (in mice) oral administration of noni juice *ad libitum* for 16 days was found to decrease the production of interleukin 4 (IL-4), but increase the production of interferon gamma (IFN-γ), suggesting that noni modulates the immune system via activation of the CB_2 receptors, suppression of IL-4 and increasing the production of IFN-γ cytokines (Palu et al. 2008).

Interactions: Limited information is available on interactions of noni with other medicines or supplements. While noni juice has been shown to inhibit CYP3A4 *in vitro*, it is not thought to be clinically relevant. A case report of a patient taking warfarin indicated that daily consumption of a product containing several plant ingredients, including noni, resulted in faster blood clotting. This was attributed to the presence of vitamin K, which the product may have been fortified with (Gardner and McGuffin 2013; Pharmaceutical Press Editorial Team 2013). In a randomised, open crossover study with 20 healthy adult volunteers, aqueous noni extract was administered orally 30 min. prior to a single oral dose (300 mg) of ranitidine (a histamine H_2-receptor antagonist that inhibits stomach acid production). Pre-treatment with the noni extract was found to increase the bioavailability of ranitidine, increasing the rate and extent of its absorption. The authors suggested that this was due to noni or its constituent scopoletin modifying upper gastrointestinal motility (Nima et al. 2012). Acute liver injury was described in another case report of a patient on long-term (9 months) anticonvulsive therapy with phenobarbitone after consuming a daily dose (60 ml) of noni juice for 1 week. The authors speculated that this was due to a synergistic idiosyncratic herb-drug-induced liver injury (Mrzljak et al. 2013). Noni products may contain high levels of potassium, so its use with medicines that increase blood potassium levels should be avoided (e.g. ACE inhibitors, angiotensin receptor blockers and potassium-sparing diuretics.

Contraindications: Noni should be avoided in people with liver disease as there have been some reports of hepatoxicity. Due to variable amounts of potassium in noni products, people with kidney disease or on potassium-restricted diets should also avoid consuming noni juice.

Safety of noni juice during pregnancy and lactation has not been established. Ethnobotanical studies have reported that use of large amounts of noni fruit can induce abortion, although this has not been confirmed experimentally (Pawlus and Kinghorn 2007). Conflicting data has been reported in animal studies, with one study showing no adverse effects on foetal development at very large (6 g/kg) doses, while another indicated that noni juice caused a delay in foetal bone formation. A slight mutagenic effect, attributed to the presence of flavonoids, was observed in strain TA1537 in the *Salmonella* microsome assay, but not in other strains (Gardner and McGuffin 2013; Pharmaceutical Press Editorial Team 2013).

Adverse effects: Hepatoxicity has been reported in several cases of noni juice ingestion, although in some of these, patients were also taking other medications or supplements. One case involved a multiple sclerosis patient also being treated with IFN-B1a, which is known to commonly cause hepatic dysfunction. A 2006 EFSA review of toxicological data and case reports found no convincing evidence for a causal relationship between noni consumption and the reported cases of hepatoxicity. However, the possibility of rare idiosyncratic hepatoxicity due to noni has not been ruled out (Gardner and McGuffin 2013; Mrzljak et al. 2013).

One case of hyperkalaemia was reported in a man with chronic renal insufficiency, who had consumed noni juice daily for an unspecified period of time. Noni contains potassium levels similar to those of other common fruit juices, which are usually restricted in diets of patients with kidney insufficiency (Gardner and McGuffin 2013; Pharmaceutical Press Editorial Team 2013).

Dosage: Standard herbal reference texts do not provide a recommended dose for oral administration of noni juice. However, a dose escalation study indicated that noni fruit is well tolerated up to 10 g daily for 28 days (Gardner and McGuffin 2013; Pharmaceutical Press Editorial Team 2013).

General plant information: *M. citrifolia* is a small evergreen tree, widely distributed in coastal regions of the Indo-Pacific region, including India, Southeast Asia, northern Australia and Polynesia. It is perhaps the most important medicinal plant in Polynesian culture, and is believed to be one of the so-called 'canoe plants' intentionally distributed by ancient voyagers as they colonised the Pacific Islands. The fruit has a pungent odour (resembling rotten cheese), and its consumption was probably restricted to times of famine. The root and stem bark were also traditionally used to make yellow and red dyes, due to their anthraquinone content (McClutchey 2002; Potterat and Hamburger 2007). Commercial products derived from noni fruit have been heavily promoted in the United States via the so-called 'pyramid schemes', often with unsubstantiated health claims (including the presence of a probably fictitious alkaloid 'xeronine'). Noni products are now widely available elsewhere (Pawlus and Kinghorn 2007).

References

Brown AC. (2012) Anticancer activity of *Morinda citrifolia* (noni) fruit: a review. *Phytotherapy Research* 26(10): 1427–1440.
Bussmann RW, Hennig L, Giannis A, Ortwein J, Kutchan TM, Feng X. (2013) Anthraquinone Content in Noni (*Morinda citrifolia* L.). *Evidence Based Complementary and Alternative Medicine* 2013: 208378.
Dalsgaard PW, Potterat O, Dieterle F, Paululat T, Kühn T, Hamburger M. (2006) Noniosides E–H, new trisaccharide fatty acid esters from the fruit of *Morinda citrifolia* (noni). *Planta Medica* 72(14): 1322–1327.
Fletcher HM, Dawkins J, Rattray C, Wharfe G, Reid M, Gordon-Strachan G. (2013) *Morinda citrifolia* (noni) as an anti-inflammatory treatment in women with primary dysmenorrhoea: a randomised double-blind placebo-controlled trial. *Obstetrics and Gynecology International* 2013: 195454.
Gardner Z, McGuffin M. (Eds.) (2013) *American Herbal Product Association's Botanical Safety Handbook*. 2nd Edition. CRC Press, USA, 1072 pp.
Issell BG, Gotay CC, Pagano I, Franke AA. (2009) Using quality of life measures in a Phase I clinical trial of noni in patients with advanced cancer to select a Phase II dose. *Journal of Dietary Supplements* 6(4): 347–359.
Mahattanadul S, Ribtitid W, Nima S, Phdoongsombut N, Ratanasuwon P, Kasiwong S. (2012) Effects of *Morinda citrifolia* aqueous fruit extract and its biomarker scopoletin on reflux esophagitis and gastric ulcer in rats. *Journal of Ethnopharmacology* 134(2): 243–250.
McClutchey W. (2002) From Polynesian healers to health food stores: changing perspectives of *Morinda citrifolia*. *Integrative Cancer Therapies* 1(2): 110–120.
Mrzljak A, Kosuta I, Skrtic A, Kanizaj TF, Vrhovac R. (2013) Drug-induced liver injury associated with noni (*Morinda citrifolia*) juice and phenobarbital. *Case Reports in Gastroenterology* 7(1): 19–24.
Nima S, Kasiwong S, Ridtitid W, Thaenmanee N, Mahattanadul S. (2012) Gastrokinetic activity of *Morinda citrifolia* aqueous fruit extract and its possible mechanism of action in human and rat models. *Journal of Ethnopharmacology* 142(2): 354–361.
Palu AK, Kim AH, West BJ, Deng S, Jensen J, White L. (2008) The effects of *Morinda citrifolia* L. (noni) on the immune system: its molecular mechanisms of action. *Journal of Ethnopharmacology* 115(3): 502–506.
Pawlus AD, Kinghorn AD. (2007) Review of the ethnobotany, chemistry, biological activity and safety of the botanical dietary supplement *Morinda citrifolia* (noni). *Journal of Pharmacy and Pharmacology* 59(12): 1587–1609.
Pharmaceutical Press Editorial Team. (2013) *Herbal Medicines*. 4th Edition. Pharmaceutical Press, London, UK.
Potterat O, Hamburger M. (2007) *Morinda citrifolia* (noni) fruit – phytochemistry, pharmacology, safety. *Planta Medica* 73(3): 191–199.
Wang MY, Nowicki D, Anderson G, Jenson J, West B. (2008) Liver protective effects of *Morinda citrifolia* (noni). *Plant Foods for Human Nutrition* 63(2): 59–63.
Wang MY, Lutfiyya MN, Weidenbacher-Hoper V, Anderson G, Su CX, West BJ. (2009) Antioxidant activity of noni juice in heavy smokers. *Chemistry Central Journal* 3: 13.
Wang MY, Peng L, Weidenbacher-Hoper V, Deng S, Anderson G, West BJ. (2012) Noni juice improves serum lipid profiles and other risk markers in cigarette smokers. *The Scientific World Journal* 2012: 594657.

Norway Spruce   *Picea abies* (L.) H.Karst.

Synonyms: *Abies excelsa* (Lam.) Poir.; *A. picea* Mill.; *P. excelsa* (Lam.) Link; *Pinus abies* L., *P. excelsa* Lam.; and others

Family: Pinaceae

Other common name: European spruce

Drug name: Piceae turiones recentes (fresh spruce tips); Piceae aetheroleum (pine needle oil)

Botanical drug used: Fresh shoots; essential oil; resin

Indications/uses: Relief of cough, especially chesty coughs and catarrh. The fresh shoot tips are used traditionally in the form of a tea, and the essential oil is used in cough syrups and throat pastilles. Externally, the oil and resin have also been used to treat wounds and skin infections and as an ingredient in salves and plasters.

Evidence: The use of Norway spruce preparations in the relief of coughs is based on traditional use only. There is some clinical evidence for their use to treat wounds and skin infections.

Safety: Overall considered to be safe in recommended doses.

Main chemical compounds: The main constituents of the essential oil and resin are the monoterpenes pinene, limonene, linalool, 1,8-cineole, terpinyl acetate, camphor, bornyl acetate, and the diterpenes levopimaradiene, abietadiene, neoabietadiene and isopimara-7,15-diene, and to a lesser extent the sesquiterpenes farnesene, bisabolene, longifolene and nerolidol (Martin et al. 2003; Martin et al. 2004).

Clinical evidence:

External use in wound healing: A pilot clinical trial on the use of resin salve from *Picea abies* was carried out in 23 participants. The resin salve, applied to the wound, was an effective and well-tolerated treatment for promoting the healing of chronic, complicated surgical wounds (Sipponen et al. 2012). Another prospective, randomised, controlled multi-centre clinical trial (37 patients) assessed the effect of resin salve compared to a cellulose polymer treatment on pressure ulcers in a

---

Phytopharmacy: An evidence-based guide to herbal medicinal products, First Edition.
Sarah E. Edwards, Inês da Costa Rocha, Elizabeth M. Williamson and Michael Heinrich.
© 2015 John Wiley & Sons, Ltd. Published 2015 by John Wiley & Sons, Ltd.

6-month study. The resin salve was found to be the more effective in the treatment of infected and non-infected severe pressure ulcers (Sipponen, et al. 2008); nevertheless, more randomised clinical trials are needed.

Pre-clinical evidence and mechanisms of action: Essential oils have well-known antiseptic and decongestant properties. The purified resin from Norway spruce has strong antifungal effects against dermatophytic fungi (*Trichophyton*) (Rautio et al. 2012) and antibacterial effects against both gram-positive (*Staphylococcus aureus,* including MRSA, *Bacilllus subtilis*) and gram-negative (*Escherichia coli, Pseudomonas aeruginosa*) bacteria as well as against *Candida albicans* (Sipponen and Laitinen 2011). Another *in vitro* study demonstrated antibacterial activity from a 'home-made' resin against gram-positive (genus *Streptococcus, Staphylococcus* and *Enterococcus*) and gram-negative bacteria (*Proteus vulgaris*) (Rautio et al. 2007). 7-Hydroxymatairesinol, a lignan extracted from *P. abies*, effectively inhibited some functions of both human monocytes (THP-1 cells) and human polymorphonuclear leukocytes by reducing the production of TNF-α and IL-8 , as well as of reactive oxygen species, showing anti-inflammatory and antioxidant activity (Cosentino et al. 2010).

Interactions: None known.

Contraindications: Not recommended during pregnancy and lactation, as safety has not been established. The use in children under 12 years of age is not recommended. Pine oils and resins may cause allergic reactions, thus avoid in sensitive individuals.

Adverse effects: Rarely, skin and subcutaneous tissue disorders such as rashes and pruritis.

Dosage: According to manufacturer indications. As a syrup, 2–5 ml (5 ml are equivalent to 1445 mg of extract from fresh spruce) two to four times a day.

General plant information: Norway spruce is native to Europe and widely cultivated as a Christmas tree for decoration. It has many other uses, for example, the resin is a source of Burgundy pitch (used as a varnish and in medicinal plasters; it is also a strong adhesive) and turpentine, which is a water-proofer and wood preservative. An essential oil from the leaves is used in perfumery and the bark has been used as a source of tannin (it contains up to 13%). The timber is used for general carpentry, to make musical instruments, and to make paper (PFAF 2014).

References

Cosentino M, Marino F, Maio RC, Delle Canne MG, Luzzani M, Paracchini S and Lecchini S. (2010) Immunomodulatory activity of the lignan 7-hydroxymatairesinol potassium acetate (HMR/lignan) extracted from the heartwood of Norway spruce (*Picea abies*). *International Immunopharmacology* 10(3): 339–343.

Martin DM, Faldt J, Bohlmann J. (2004) Functional characterization of nine Norway Spruce TPS genes and evolution of gymnosperm terpene synthases of the TPS-d subfamily. *Plant Physioliolgy* 135(4): 1908–1927.

Martin DM, Gershenzon J, Bohlmann J. (2003) Induction of volatile terpene biosynthesis and diurnal emission by methyl jasmonate in foliage of Norway spruce. *Plant Physiology* 132(3): 1586–1599.

PFAF. (2014) Plants for a Future Database. http://www.pfaf.org/user/Plant.aspx?LatinName =Picea+abies (accessed February 2014).

Rautio M, Sipponen A, Lohi J, Lounatmaa K, Koukila-Kahkola P, Laitinen K. (2012) *In vitro* fungistatic effects of natural coniferous resin from Norway spruce (*Picea abies*). *European Journal of Clinical Microbiology & Infectious Diseases* 31(8): 1783–1789.

Rautio M, Sipponen A, Peltola R, Lohi J, Jokinen JJ, Papp A, Carlson P, Sipponen P. (2007) Antibacterial effects of home-made resin salve from Norway spruce (*Picea abies*). *APMIS* 115(4): 335–340.

Sipponen A, Jokinen JJ, Sipponen P, Papp A, Sarna S, Lohi J. (2008) Beneficial effect of resin salve in treatment of severe pressure ulcers: a prospective, randomized and controlled multicentre trial. *British Journal of Dermatology* 158(5): 1055–1062.

Sipponen A, Kuokkanen O, Tiihonen R, Kauppinen H, Jokinen JJ. (2012) Natural coniferous resin salve used to treat complicated surgical wounds: pilot clinical trial on healing and costs. *International Journal of Dermatology* 51(6): 726–732.

Sipponen A, Laitinen K. (2011) Antimicrobial properties of natural coniferous rosin in the European Pharmacopoeia challenge test. *APMIS* 119(10): 720–724.

Sipponen A, Rautio M, Jokinen J J, Laakso T, Saranpaa P, Lohi J. (2007) Resin-salve from Norway spruce – a potential method to treat infected chronic skin ulcers? *Drug Metabolism Letters* 1(2): 143–145.

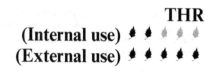
Oats

Avena sativa L.

Family: Poaceae (Gramineae)

Other common name: Common oat

Drug names: Avenae herba; Avenae fructus

Botanical drug used: Dried aerial parts, harvested before flowering; fruits (=seeds); colloidal oatmeal

---

Indications/uses: Aerial parts of the oat plant, both the herb and the extract of the fruits ('oats'), are reputed to have stimulant, neurotonic, antispasmodic and diuretic effects, and are traditionally used in oral preparations (especially tinctures) for the relief of mild symptoms of mental stress and nervous exhaustion. Oat seed extract is used in registered polyherbal preparations to relieve indigestion. Nutritional uses of oats include the reduction of cholesterol levels and associated cardiovascular complications.

Powdered oats, known as 'colloidal oatmeal', is widely used in topical preparations such as creams, lotions and bath additives, for the treatment of skin irritation and inflammation (e.g. eczema, sunburn).

Evidence: Clinical data to support traditional uses of oats as a sedative and 'nervine' are limited. Pre-clinical studies have highlighted anti-inflammatory, antioxidant, anti-atherogenic and chemopreventive properties of oat constituents.

The external use of colloidal oat preparations is well established and a number of licensed products are available to buy, and on prescription.

Safety: Safety data for internal use of oat herb extracts are lacking, although the widespread use of oats as food is safe. Internal use of the herb extract is not recommended during pregnancy and lactation or in children under 18 years of age due to lack of data. External use of colloidal oatmeal is safe in all age groups, including babies.

Main chemical compounds:

Aerial parts: beta-glucans, steroidal saponins, for example, avenacoside A and B, nitrogen-containing carboxylic acids (avenic acid A and B); flavonoids, for example, vitexin derivatives; minerals.

Phytopharmacy: An evidence-based guide to herbal medicinal products, First Edition.
Sarah E. Edwards, Inês da Costa Rocha, Elizabeth M. Williamson and Michael Heinrich.
© 2015 John Wiley & Sons, Ltd. Published 2015 by John Wiley & Sons, Ltd.

Fruits: Oat beta-glucan, proteins including glutelins and avenin, avenothionin (a lipoprotein); the alkaloid gramine; polyphenols including the avenanthramides, caffeic and ferulic acids and flavonoids; triterpene saponins and phytosterols (e.g. avenasterin and stigmasterin (Kutz and Wallo 2007; Singh et al. 2013).

Clinical evidence:

Stimulant and cognitive effects: A few studies have assessed the clinical effectiveness of a proprietary oat *herb* preparation, Neuravena® (a wild green oat extract) on cognition. A double-blind, randomised cross-over study demonstrated a beneficial effect in elderly volunteers with below-average cognitive performance administered single weekly doses of 1600 mg of extract, compared with placebo or 2400 mg (Berry et al. 2011). A study monitoring changes in healthy subjects administered 1250 mg or 2500 mg of extract, during resting and performance of a concentration test, found changes mainly in the regions of the brain known to be involved in cognitive tasks (Dimpfel et al. 2011). However, another trial did not show any effects on cognitive measurements in 37 healthy adults aged 67 ± 0.8 years when given 1500 mg/day for 12 weeks (Wong et al. 2012). The same group found that systemic and cerebral vasodilator responsiveness significantly increased (by 41% and 42%, respectively) over 24 weeks with 1500 mg/day of Neuravena® compared to placebo, supporting the use of oat herb extract in the maintenance of cardiovascular health (Wong et al. 2013).

Nutritional effects on cardiovascular health: The majority of clinical trials involving *A. sativa* have investigated the health benefits of consumption of oats, oat bran, or the constituent polysaccharide beta-glucan as part of the diet. There is some evidence (including from a Cochrane review) that oats and/or oat beta-glucan consumed in sufficient quantities (i.e. 3 g of beta-glucan per day) can have a beneficial effect on cardiovascular health, significantly lowering total cholesterol and low-density lipoprotein (LDL) cholesterol levels, improving liver functions and helping to reduce abdominal fat (Chang et al. 2013; Kelly et al. 2007).

Dermatological effects: Colloidal oatmeal is used to soothe and ameliorate atopic dermatitis and other pruritic, dry and/or scaling skin conditions. The effects of colloidal oatmeal as adjunct treatment in the management of atopic dermatitis (AD) have been reviewed in patients from 3 months to 60 years with mild-to-moderate atopic dermatitis, who were allowed to continue their prescribed topical medication. Daily use of moisturisers and/or cleansers containing colloidal oatmeal significantly improved many clinical outcomes of atopic dermatitis: investigator's assessment, eczema area and severity index, itch, dryness, and quality of life indices. The formulations were well tolerated in babies, children and adults with AD (Fowler et al. 2012).

Pre-clinical evidence and mechanisms of action:

Pre-clinical studies have demonstrated anti-inflammatory, antioxidant, anti-atherosclerosis and chemopreventive properties for oats, oat extracts and constituents, in particular the polyphenolic avenanthramides. Mechanisms of action include inhibition of pro-inflammatory cytokines (including IL-6, IL-8, NF-kappaB), inhibition of proliferation of vascular smooth muscle cells and increase in nitric oxide production (Andersson and Hellstrand 2012).

Stimulant, anti-stress and cognitive effects: *In vivo* studies in rats orally administered Neuravena® suggested stimulation of dopaminergic neurotransmission

(EMA 2008a). An extract of wild green oat, tested *in vivo* in rats using the elevated plus maze, forced swimming, conditioned avoidance response and tetradic encounter tests, found that apart from a slightly decreased food and fluid intake in the higher dose group (100 g/kg extract-admixed food), there were no side effects observed during the treatment. The low dose (10 g/kg), but not the high dose, led to an improvement of active stress response, an enhancement of shock avoidance learning and an increased synchrony in social behaviour (Schellekens et al. 2009).

Dermatological effects: The many clinical properties of colloidal oatmeal in skin conditions derive from its chemical polymorphism. The high concentration in starches and beta-glucan is responsible for the protective and water-holding functions, the phenolic constituents confer antioxidant and anti-inflammatory activity and some are also strong ultraviolet absorbers. The cleansing activity of oat is mostly due to saponins (Kurtz and Wallo 2007).

Interactions: The US Food and Drug Administration (FDA) recorded (as of 7 August 2013) five cases of interaction of oat *herb extract* with allopurinol (eHealthMe 2013). Three cases of stent thrombosis have been reported in Turkey in patients taking a herbal combination containing *A. sativa* (with *Tribulus terrestris* L. and *Panax ginseng* C.A.Mey.) along with anti-aggregant drugs (Vatankulu et al. 2013).

Contraindications: Sensitivity to oats. Patients with coeliac disease should consult a medical professional before using oat herb extracts internally, and they are not recommended during pregnancy and lactation, or in patients under 18 years of age due to lack of safety data. Eating oats as a cereal and the external use of oat products is safe.

Adverse effects: Acute toxicity tests of oat herb extract found no cytotoxic effects in human aortic endothelial cells at concentrations up to 40 µg/ml. Cosmetic products containing colloidal oats are well tolerated. Skin reactions may occur rarely in atopic individuals or in patients with contact dermatitis (EMA 2008a).

Dosage: Adults, including the elderly: powdered herb 3 g (for the production of herbal teas) or preparations containing the equivalent. For commercial products, follow manufacturers' instructions. Expressed juice from the fresh herb: 10 ml three to four times/day (EMA 2008b). Oats as food: no dose limit found.

General plant information: The German 12th century mystic and visionary, Hildegard von Bingen, described the use of oats as follows " ... *may everyone who is exhausted with an empty mind take steam bath by pouring water wherein oats have been cooked over hot stone. If this treatment is repeated, the patient will get to himself again and regain the capacity of thinking.*" (EMA 2008a). Oatmeal has been used externally for centuries to relieve itch and irritation associated with various dermatoses. In 1945, colloidal oatmeal was produced by finely grinding the oat and boiling it to extract the colloidal material, and became available in formulations including powders for the bath, shampoos and moisturising creams (Kurtz and Wallo 2007).

References

Andersson KE, Hellstrand P. (2012) Dietary oats and modulation of atherogenic pathways. *Molecular Nutrition and Food Research* 56(7): 1003–13.

Berry NM, Robinson MJ, Bryan J, Buckley JD, Murphy KJ, Howe PR. (2011) Acute effects of an *Avena sativa* herb extract on responses to the Stroop Color-Word test. *Journal of Alternative and Complementary Medicine* 17(7): 635–637.

Chang HC, Huang CN, Yeh DM, Wang SJ, Peng CH, Wang CJ. (2013) Oat prevents obesity and abdominal fat distribution, and improves liver function in humans. *Plant Foods for Human Nutrition* 68(1): 18.

Dimpfel W, Storni C, Verbruggen M. (2011) Ingested oat herb extract (*Avena sativa*) changes EEG spectral frequencies in healthy subjects. *Journal of Alternative and Complementary Medicine* 17(5): 427–434.

eHealthMe. (2013) From FDA reports: drug interactions between Allopurinol and *Avena sativa*. http://www.ehealthme.com/drug-interactions/allopurinol-and-avena-sativa (accessed August 2013).

EMA. (2008a) Assessment report on *Avena sativa* L., herba and *Avena sativa* L., fructus. European Medicines Agency. http://www.ema.europa.eu/docs/en_GB/document_library/Herbal_-_HMPC_assessment_report/2009/12/WC500017998.pdf.

EMA. (2008b) Community herbal monograph on *Avena sativa* L., herba. European Medicines Agency. http://www.ema.europa.eu/docs/en_GB/document_library/Herbal_-_Community_herbal_monograph/2009/12/WC500018067.pdf.

Fowler JF, Nebus J, Wallo W, Eichenfield LF. (2012) Colloidal oatmeal formulations as adjunct treatments in atopic dermatitis. *Journal of Drugs in Dermatology.* 11(7): 804–807.

Kelly SAM, Summerbell CD, Brynes A, Whittaker V, Frost G. (2007) Wholegrain cereals for coronary heart disease. *Cochrane Database of Systematic Reviews* (2): CD005051.

Kurtz ES, Wallo W. (2007) Colloidal oatmeal: history, chemistry and clinical propertiesJournal of *Drugs in Dermatology* 6(2): 167–170.

Schellekens C, Perrinjaquet-Moccetti T, Wullschleger C, Heyne A. (2009) An extract from wild green oat improves rat behaviour. *Phytotherapy Research* 23(10): 1371–1377.

Singh R, De S, Belkheir A. (2013) *Avena sativa* (Oat), a potential neutraceutical and therapeutic agent: an overview. *Critical Reviews in Food Science and Nutrition* 53(2): 126–144.

Vatankulu MA, Tasal A, Erdoğan E, Göktekin Ö. (2012) Three case reports of the use of herbal combinations resulted in stent thrombosis: herbal combinations; friend or foe? *Türk Kardiolyoji Derneği arşivi* 40(3): 265–268 [article in Turkish].

Wong RH, Howe PR, Bryan J, Coates AM, Buckley JD, Berry NM. (2012) Chronic effects of a wild green oat extract supplementation on cognitive performance in older adults: a randomised, double-blind, placebo-controlled, crossover trial. *Nutrients* 4: 331–342.

Wong RH, Howe PR, Coates AM, Buckley JD, Berry NM. (2013) Chronic consumption of a wild green oat extract (Neuravena®) improves brachial flow-mediated dilatation and cerebrovascular responsiveness in older adults. *Journal of Hypertension* 31(1): 192–200.

Passionflower

Passiflora incarnata L.; *P. edulis* Sims

Synonyms: *P. edulis* var. *keri* (Spreng.) Mast. = *P. incarnata*; *P. incarnata* Ker Gawl. = *P. edulis*

Studies generally consider *P. incarnata* and *P. edulis* as two separate species, but this is sometimes disputed. They have different distributions and are distinguishable phytochemically, with the former usually considered the 'medicinal' species and the latter the source of the cultivated passionfruit (although both are used medicinally). Published studies usually refer to *P. incarnata* but may not chemically characterise the plant enough for authentication.

Family: Passifloraceae

Other common names: Apricot vine; grenadille; maypop; passiflora; passion vine

Drug name: Passiflorae herba

Botanical drug used: Dried aerial parts

Indications/uses: Traditionally used to relieve mild anxiety and to aid sleep.

Evidence: Clinical data to support traditional uses are limited, although preliminary evidence of anxiolytic activity exists. Pre-clinical studies also provide evidence for anxiolytic, sedative, anti-convulsant and anti-inflammatory properties. Passiflora is mostly used in combination products with hops and valerian (see page 211, 383). The fruits of passion fruit are edible.

Safety: Safety and toxicity data are lacking. Although passiflora is not known to contain toxic compounds, and no side-effects have been reported, long term administration (>4 weeks), use during pregnancy and lactation or in children/adolescents under 18 years of age, cannot be recommended. As a sedative, passionflower may impair the ability to drive or operate machinery.

Main chemical compounds: The active constituents include the flavonoids apigenin and its C-glycosides (vitexin, isovitexin, schaftoside); chrysin, luteolin (and its C-glycosides orientin and iso-orientin), quercetin and kaempferol. β-carboline alkaloids (which are MAO inhibitors), including harman, harmol, harmine, harmalol and harmaline may be present as minor constituents, but are not always detectable.

Phytopharmacy: An evidence-based guide to herbal medicinal products, First Edition.
Sarah E. Edwards, Inês da Costa Rocha, Elizabeth M. Williamson and Michael Heinrich.
© 2015 John Wiley & Sons, Ltd. Published 2015 by John Wiley & Sons, Ltd.

Other constituents include a cyanogenic glycoside gynocardin; maltol and ethylmaltol; a polyacetylene, passicol; and essential oil containing carvone, hexanol, benzylalcohol, linalool, eugenol and α-bergamolol (Pharmaceutical Press Editorial Team 2013; Williamson et al. 2013).

Analysis of *P. incarnata* samples cultivated in Australia demonstrated two distinct chemotypes – one characterised by isovitexin and schaftoside/isoschaftoside, the other by a high level of swertisin and low levels of schaftoside/isoschaftoside (Wohlmuth et al. 2010).

Clinical evidence:

Most clinical studies investigating passionflower have investigated proprietary products, some of which are polyherbal preparations.

Use in anxiety: A Cochrane review identified two trials ($n = 198$) investigating the effectiveness of 'passiflora' (species not identified) for treating anxiety. The findings from one study indicated that passiflora was as effective as benzodiazepines. One study suggested an improvement in job performance (post-hoc outcome), and one found a lower rate of drowsiness compared with mexazolam, but neither of these findings were statistically significant. However, no conclusion could be drawn as to the effectiveness of passiflora due to the limited number of randomised controlled trials available (Myasaka et al. 2007). In two subsequent, randomised controlled trials, patients ($n = 60$ in both) were administered *P. incarnata* or placebo as premedication prior to operation (inguinal herniorrhaphy in one study and spinal anaesthesia in the other). Both studies found that *P. incarnata* significantly suppressed increase in anxiety compared to placebo without altering psychomotor function test results, and one found no alteration in sedation level or haemodynamics (Aslanargun et al. 2012; Movafegh et al. 2008).

Effects on opiate withdrawal: Passiflora extract in combination with clonidine was found to significantly reduce opiate withdrawal symptoms in a 14-day double-blind clinical trial compared to clonidine alone (the standard treatment). While both protocols (i.e. passiflora plus clonidine versus clonidine plus placebo) were equally effective in treating physical symptoms of withdrawal, passiflora plus clonidine was superior in the management of mental symptoms (Akhondzadeh et al. 2001).

Effects on sleep quality: A preliminary double-blind controlled study assessing the effects of low dose of *P. incarnata* (taken for one week as tea) on subjective sleep quality in 41 adults (18–35 years) found a significantly better rating for passionflower compared with placebo ($p<0.01$) (Ngan and Conduit 2011).

Pre-clinical evidence and mechanisms of action:

CNS depressant effects: Studies in mice have demonstrated the anti-convulsant activity of passionflower extracts and flavonoid fractions, protecting against pentylenetetrazol (PZT)-induced seizures in a dose-dependent manner (Elsas et al. 2010; Nassiri-Asl et al. 2007). A dose of 0.4 mg/kg of a proprietary passionflower preparation ('Pasipay') was found to be similar to diazepam 1 mg/kg. Data suggested gamma-amino butyric acid (GABA)-benzodiazepine and opioid receptor involvement in the response induced by passionflower (Nassiri-Asl et al. 2007). In an *in vitro* study, *Passiflora* extracts were found to contain a high amount of GABA, and were able to induce direct $GABA_A$ currents in hippocampal neurons. Interestingly, unlike anxiolytic effects that have been observed elsewhere, the authors found that mice administered the same extracts in drinking water for one week demonstrated anxiogenic effects, with no correlation to GABA content (Elsas et al. 2010). The specific constituents in passionflower responsible for its activity are still not clearly identified, although the flavonoid and possibly alkaloid content are considered important. The bioactivity may be the result of synergistic action of several compounds, e.g. a combination of GABA with other compounds that facilitate membrane permeation, in addition to modulation of $GABA_A$ receptors by flavonoids (Elsas et al. 2010; Pharmaceutical Press Editorial Team 2013).

Anti-inflammatory activity: Anti-inflammatory activity of *P. edulis* has been demonstrated in a mouse model of pleurisy induced by carrageenan. Results indicated that *P. edulis* inhibits pro-inflammatory cytokines (TNFα, IL-1β), enzyme (myeloperoxidase) and releases and/or induces mediators (bradykinin, histamine, substance P and nitric oxide), which may account for the anti-inflammatory action (Montanher et al. 2007).

Interactions: A case of a patient self-medicating with *Valeriana officinalis* L. and *P. incarnata* L. while he was on lorazepam treatment was reported. The patient developed hand tremor, dizziness, throbbing and muscular fatigue (Carrasco et al. 2009). It is thought that an additive or synergistic effect might increase the inhibitory activity of benzodiazepines binding to the GABA receptors, resulting in severe secondary effects. Use of passionflower concomitantly with other sedatives is not recommended (Williamson et al. 2013).

Contraindications: Hypersensitivity to passionflower.

Adverse effects: One case of hypersensitivity (vasculitis) and one case of nausea and tachycardia have been reported (EMEA 2007).

Dosage: Oral administration (adults): for tenseness, restlessness and irritability with difficulty falling asleep – 0.5–2 g of drug 3–4 times/day; 2.5 g of drug as an infusion 3–4 times/day; 1–4 ml of tincture (1:8) 3–4 times/day (Pharmaceutical Press Editorial Team 2013).

General plant information: *Passiflora* species are native to tropical and subtropical areas of North and South America, although have naturalised elsewhere. Spanish explorers 'discovered' the passionflower in Peru in the 16th century, who perceived the flowers as symbolic of the passion of Christ. The corona is said to resemble the crown of thorns.

References

Akhondzadeh S, Kashani L, Mobaseri M, Hosseini SH, Nikzad S, Khani M. (2001) Passionflower in the treatment of opiates withdrawal: a double-blind randomized controlled trial. *Journal of Clinical Pharmacy and Therapeutics* 26(5): 369–373.

Aslanargun P, Cuvas O, Dikmen B, Aslan E, Yuksel MU. (2012) *Passiflora incarnata* Linneaus as an anxiolytic before spinal anesthesis. *Journal of Anesthesia* 26(1): 39–44.

Carrasco MC, Vallejo JR, Pardo-de-Santayana M, Peral D, Martin MA, Altimiras J. (2009) Interactions of *Valeriana officinalis* L. and *Passiflora incarnata* L. in a patient treated with lorazepam. *Phytotherapy Research* 23(12): 1795–1796.

Elsas S-M, Rossi DJ, Raber J, White G, Seeley C-A, Gregory WL, Mohr C, Pfankuch T, Soumyanath A. (2010) *Passiflora incarnata* L. (Passionflower) extracts elicit GABA currents in hippocampal neurons *in vitro*, and show anxiogenic and anticonvulsant effects *in vivo*, varying with extraction method. *Phytomedicine* 17(12): 940–949.

EMEA. (2007) Community herbal monograph on *Passiflora incarnata* L., herba. European Medicines Agency. http://www.ema.europa.eu/docs/en_GB/document_library/Herbal_-_Community_herbal_monograph/2010/01/WC500059213.pdf.

Montanher AB, Zucolotto SM, Schenkel EP, Fröde TS. (2007) Evidence of anti-inflammatory effects of *Passiflora edulis* in an inflammation model. *Journal of Ethnopharmacology* 109(2): 282–288.

Movafegh A, Alizadeh R, Hajimohamadi F, Esfehani F, Nejatfar M. (2008) Preoperative oral *Passiflora incarnata* reduces anxiety in ambulatory surgery patients: a double-blind, placebo-controlled study. *Anesthesia and Analgesia* 106(6): 1728–1732.

Myasaka LS, Atallah AN, Soares B. (2007) Passiflora for anxiety disorder. *Cochrane Database of Systematic Reviews* (1): CD004518.

Nassiri-Asl M, Shariati-Rad S, Zamansoltani F. (2007) Anticonvulsant effects of aerial parts of *Passiflora incarnata* extract in mice: involvement of benzodiazepine and opiod receptors. *BMC Complementary and Alternative Medicine* 7: 26.

Ngan A, Conduit R. (2011) A double-blind, placebo-controlled investigation of the effects of *Passiflora incarnata* (passionflower) herbal tea on subjective sleep quality. *Phytotherapy Research* 25(8): 1153–1159.

Pharmaceutical Press Editorial Team. (2013) *Herbal Medicines*. 4th Edition. Pharmaceutical Press, London, UK.

Williamson EM, Driver S, Baxter K. (Eds.) (2013) *Stockley's Herbal Medicines Interactions*. 2nd Edition. Pharmaceutical Press, London, UK.

Wohlmuth H, Penman KG, Pearson T, Lehmann RP. (2010) Pharmacognosy and chemotypes of passionflower (*Passiflora incarnata* L.). *Biological and Pharmaceutical Bulletin* 33(6): 1015–1018.

Pelargonium	*Pelargonium sidoides* DC.; *P. reniforme* (Andrews) Curtis

Synonyms: *Geranium sidifolium* Thunb.; and others

Family: Geraniaceae

Other common names: Geranium; South African geranium; umckaloabo

Drug name: Pelargonii radix

Botanical drug used: Roots

Indications/uses: Acute bronchitis, tonsillitis and upper respiratory tract infections.

Evidence: Overall, the evidence for well-studied Pelargonium preparations is sound and its uses in alleviating symptoms of acute rhinosinusitis and the common cold in adults are relatively well established.

Safety: Overall, it is considered to be safe.

Main chemical compounds: Proanthocyanidin oligomers based on epigallo- and gallo-catechins. A series of ellagitannins known as pelargoniins and a novel series of O-galloyl-C- glucosylflavones have been isolated from *P. reniforme*. Sulphated and oxygenated coumarins have been found in the roots of *P. sidiodes* (Kolodziej 2000; Kolodziej et al. 2003; Williamson et al. 2013).

Clinical evidence: Several randomised, double-blind, placebo-controlled clinical trials support its use in adults and children with acute bronchitis (Agbabiaka et al. 2008) as well as a multi-centre, prospective, open observational study including adults, children and infants outside the strict indication for antibiotic treatment (Matthys et al. 2007). A systematic review based on randomised controlled trials examining the efficacy of *P. sidoides* preparations in acute respiratory infections (ARIs) showed that *P. sidoides* may be effective in alleviating symptoms of acute rhinosinusitis and the common cold in adults, but doubt exists. It may be effective in relieving symptoms in acute bronchitis in adults and children, and sinusitis in adults. Reliable data on treatment for other ARIs were not identified (Timmer et al. 2013).

A recent study showed that EPs 7630® (*P. sidoides* extract) interfered with the replication of different respiratory viruses including seasonal influenza A virus

---

Phytopharmacy: An evidence-based guide to herbal medicinal products, First Edition.
Sarah E. Edwards, Inês da Costa Rocha, Elizabeth M. Williamson and Michael Heinrich.
© 2015 John Wiley & Sons, Ltd. Published 2015 by John Wiley & Sons, Ltd.

strains, respiratory syncytial virus (RSV), human coronavirus, parainfluenza virus and coxsackie virus (Michaelis et al. 2011).

A study with athletes submitted to intense physical activity taking *P. sidoides* extract showed increased levels of secretory IgA, while levels of IL-15 and IL-6 were decreased, suggesting that it can exert a strong modulating influence on the immune response associated with the upper airway mucosa (Luna et al. 2011).

Pre-clinical evidence and mechanisms of action: The exact mechanism of action is unclear. *P. sidoides* extract has antimicrobial, antiviral and immunomodulatory effects. It was shown to increase natural killer cell formation, tumour necrosis factor alpha (TNFα), inhaled nitric oxide (iNO) and interferon-beta release and also demonstrated anti-adhesive effects (Wittschier et al. 2007), including HIV-1 attachment (Helfer et al. 2014). The extract improved peripheral blood phagocytes by enhancing the oxidative burst and intracellular killing *in vitro* (Conrad et al. 2007).

Interactions: No cases of concern have been reported but it might interact with warfarin and related drugs (based on *in vivo* studies with rats) (Koch and Biber 2007; Williamson et al. 2013).

Contraindications: Pelargonium is not recommended during pregnancy and lactation and in children below 6 years old. Pelargonium should be avoided in patients with severe hepatic and renal diseases, as well as those on anticoagulant and antiplatelet drugs, due to lack of adequate data.

Adverse effects: Gastrointestinal complaints (stomach pain, heartburn, nausea or diarrhoea) are possible side effects. Rare cases of bleeding from gum and nose and hypersensitivity reactions were reported (Kraft and Hobbs 2004).

Dosage: Dosage varies according to formulation; see manufacturer's instructions.

General plant information: Umckaloabo (*P. sidoides/P. reniforme*) is native to South Africa, where it has a long tradition of use. It was introduced into European herbal practice early on. Famously, "Stevens' consumption cure" was advertised as a treatment for a wide range of respiratory diseases including tuberculosis, and products containing *P. sidoides* have been sold in Europe for many decades.

References

Agbabiaka TB, Guo R, Ernst E. (2008) *Pelargonium sidoides* for acute bronchitis: a systematic review and meta-analysis. *Phytomedicine* 15(5): 378–385.

Conrad A, Hansmann C, Engels I, Daschner FD, Frank U. (2007) Extract of *Pelargonium sidoides* (EPs 7630) improves phagocytosis, oxidative burst, and intracellular killing of human peripheral blood phagocytes *in vitro*. *Phytomedicine* 14(Suppl 6): 46–51.

Helfer M, Koppensteiner H, Schneider M, Rebensburg S, Forcisi S, Müller C, Schmitt-Kopplin P, Schindler M, Brack-Werner R. (2014) The root extract of the medicinal plant *Pelargonium sidoides* is a potent HIV-1 attachment inhibitor. *PLoS One* 9(1): e87487.

Koch E, Biber A. (2007) Treatment of rats with the *Pelargonium sidoides* extract EPs 7630 has no effect on blood coagulation parameters or on the pharmacokinetics of warfarin. *Phytomedicine* 14(Suppl 6): 40–45.

Kolodziej H. (2000) Traditionally used *Pelargonium* species: chemistry and biological activity of umckaloabo extracts and their constituents. *Current Topics in Phytochemistry* 3: 77–93.

Kolodziej H, Kayser O, Radtke OA, Kiderlen AF, Koch E. (2003) Pharmacological profile of extracts of *Pelargonium sidoides* and their constituents. *Phytomedicine* 10(Suppl 4): 18–24.

Kraft K, Hobbs C. (2004) *Pocket Guide to Herbal Medicine*. Thieme, Stuttgart.

Luna LA Jr., Bachi AL, Novaes e Brito RR, Eid RG, Suguri VM, Oliveira PW, Gregorio LC, Vaisberg M. (2011) Immune responses induced by *Pelargonium sidoides* extract in serum and nasal mucosa of athletes after exhaustive exercise: modulation of secretory IgA, IL-6 and IL-15. *Phytomedicine* 18(4): 303–308.

Matthys H, Kamin W, Funk P, Heger M. (2007) *Pelargonium sidoides* preparation (EPs 7630) in the treatment of acute bronchitis in adults and children. *Phytomedicine* 14(Suppl 6): 69–73.

Michaelis M, Doerr HW, Cinatl J Jr. (2011) Investigation of the influence of EPs® 7630, a herbal drug preparation from *Pelargonium sidoides*, on replication of a broad panel of respiratory viruses. *Phytomedicine* 18(5): 384–386.

Timmer A, Günther J, Motschall E, Rücker G, Antes G, Kern WV. (2013) *Pelargonium sidoides* extract for treating acute respiratory tract infections. *Cochrane Database of Systematic Reviews* 10: CD006323.

Williamson EM, Driver S, Baxter K. (Eds.) (2013) *Stockley's Herbal Medicines Interactions.* 2nd Edition. Pharmaceutical Press, London, UK.

Wittschier N, Faller G, Hensel A. (2007) An extract of *Pelargonium sidoides* (EPs 7630) inhibits *in situ* adhesion of *Helicobacter pylori* to human stomach. *Phytomedicine* 14(4): 285–288.

Peony   *Paeonia lactiflora* Pall.

Synonyms: *P. edulis* Salisb.; *P. officinalis* Thunb.; and others

Family: Paeoniaceae

Other common names: Bai shao (root without bark); chi shao (*P. lactiflora* Pall. and *P. anomala* subsp. *veitchii* (Lynch) D.Y.Hong & K.Y.Pan, root with bark); Chinese peony; red peony; white peony

Drug name: Paeoniae radix

Botanical drug used: Roots

Indications/uses: Peony is used in traditional Chinese medicine (TCM), mainly in combination with other herbs, for the relief of gynaecological symptoms including hot flushes associated with the menopause, regulation of the menstrual cycle and alleviating anaemia caused by menorrhagia. It is also used in the treatment of high blood pressure, inflammation, diabetes and other metabolic disorders, and depression.

Evidence: No clinical data reported for single-herb products, but small clinical studies on the formulation Toki-shakuyaku-san, which is used for gynaecological symptoms in TCM, have shown positive results. A four-herb formula containing peony, PHY906, is in development as an adjuvant for cancer chemotherapy and has shown useful results in several clinical studies.

Safety: Overall considered to be safe.

Main chemical compounds: The main constituent of peony root is the monoterpene glycoside, paeoniflorin, together with 2'-*O*-benzoylpaeoni-florin, albiflorin R2 and albiflorin R3 (Fu et al. 2013; He et al. 2004; WHO 1999). The root also contains flavonoids, proanthocyanidins, tannins, complex polysaccharides, terpenoids and triterpenoids (Anon. 2001).

Clinical Evidence:

General: No clinical data on *P. lactiflora* root as a single herb extract are available, although studies have been carried out in China on its use in cardiovascular, gynaecological and inflammatory disorders. In combination with other herbs, peony has some other important applications for which clinical data are available.

Phytopharmacy: An evidence-based guide to herbal medicinal products, First Edition.
Sarah E. Edwards, Inês da Costa Rocha, Elizabeth M. Williamson and Michael Heinrich.
© 2015 John Wiley & Sons, Ltd. Published 2015 by John Wiley & Sons, Ltd.

Menstrual irregularities and dysmenorrhoea:

Toki-shakuyaku-san has been used to treat gynaecological symptoms in TCM. It contains *P. lactiflora* radix, *Atractylodes lancea* (Thunb.) DC. (red atractylodes) rhizome, *Alisma plantago-aquatica* L. (alisma) rhizome, *Wolfiporia extensa* (Peck) Ginns (syn. *Poria cocos* F.A. Wolf, hoelen) sclerotium, *Cnidium monnieri* (L.) Cusson (cnidium) rhizome and *Angelica dahurica* (Hoffm.) Benth. & Hook.f. ex Franch. & Sav. (Chinese angelica) radix. The formulation normalised the irregular menstrual cycle, healed cervical pseudo-erosion and reduced leukorrhagia in young women with insufficient luteal function (Sakamoto et al. 1996).

In some patients with mild or moderate anaemia, it was found to be a potentially useful alternative to iron supplementation in cases of severe side effects caused by oral iron supplements (Akase et al. 2003). In a study comparing its effects in women with luteal insufficiency with women with normal menstrual cycles, Toki-shakuyaku-san improved luteal insufficiency with no adverse effects on the hormonal levels of prolactin, gonadotropins, steroids, angiotensin II, ANP and renin in blood or urine (Usuki et al. 2002).

The analgesic effect of Toki-shakuyaku-san was also assessed in women suffering from dysmenorrhoea, over two menstrual cycles. This double-blind study showed that it was effective in the treatment of dysmenorrhoea compared to control (Kotani et al. 1997).

Kuei-chih-fu-ling-wan, another Chinese formulation, containing *P. lactiflora* (root), *Cinnamomum cassia* (L.) J.Presl (bark), *Prunus persica* (L.) Batsch (seeds), *W. extensa* (carpophores) and *P.* × *suffruticosa* Andrews (root bark), was used in older premenopausal patients with uterine myomas. Symptoms of hypermenorrhoea and dysmenorrhoea were greatly improved, with the uterine myomas shrinking in some patients (60%) (Sakamoto et al. 1992).

Cancer chemotherapy adjuvant activity:

PHY906 – is a decoction of *P. lactiflora* with *Scutellaria baicalensis* Georgi, *Glycyrrhiza uralensis* Fisch. ex DC. and *Ziziphus jujuba* Mill., which has been in use in TCM for over 1800 years for treating a variety of gastrointestinal disorders such as diarrhoea, cramps, nausea and vomiting and is now being developed as a cancer chemotherapy adjuvant. Ongoing clinical studies in a number of different types of cancer, and with different chemotherapy regimens, show beneficial effects in alleviating symptoms and improving quality of life (Liu and Cheng 2012).

Even though some clinical data are available for complex mixtures, these cannot be considered to be of direct relevance for providing some level of clinical evidence for *P. lactiflora* preparations.

Pre-clinical evidence and mechanisms of action:

Anti-inflammatory and cardiovascular effects: The monoterpene glycosides (especially paeoniflorin) have shown a wide range of pharmacological activities including anticoagulant, antithrombotic, anti-inflammatory, analgesic, anti-allergic and immunomodulatory effects in a wide variety of animal models and cell systems, and provide a basis for the clinical use in rheumatoid diseases (He and Dai 2011; Zhang and Dai 2012). Anti-complement effects have also recently been described for similar glucosides extracted from *P. suffruticosa* (Song et al. 2014).

Depression: Peony extracts have shown antidepressant activity in a number of animal models. It has been suggested that they are most likely mediated by the inhibition of monoamine oxidase activity, neuroprotection, modulation of the hypothalamic-pituitary-adrenal axis, inhibition of oxidative stress and the upregulation of neurotrophins (Mao et al. 2012).

Interactions: Peony should not be combined with *Fritillaria verticillata* Willd., *Cuscuta japonica* Choisy and *Rheum officinale* Baill. (WHO 1999). Although there are no reports, caution is warranted as peony may potentiate the effects of anticoagulant (e.g. warfarin) and antiplatelet (e.g. aspirin) drugs (Anon. 2001).

Contraindications: Peony is not recommended during pregnancy and lactation, as safety has not been established. It has been suggested to have abortifacient effects (WHO 1999) but this has not been confirmed by clinical reports.

Adverse effects: None known.

Dosage: For specific products, manufacturer's instructions should be followed. The usual dose of white peony is 1.5–4 g; while for red peony it is 1–3g, both three times daily (Anon. 2001).

General plant information: Peony species have been used as ornamental flowers in China and Europe for centuries. Four different peony species have been used in TCM under the name of peony: *P. suffruticosa* Andrews (tree peony), *P. lactiflora* (Chinese peony), *P. obovata* Maxim. (Chinese peony), *P. anomala* subsp. *veitchii* (Lynch) D.Y.Hong & K.Y.Pan (Anon. 2001). The drug Paeoniae radix from *P. lactiflora*, a traditional Chinese herbal drug, has been classified into two types, Paeoniae Radix Alba (Bai Shao or white peony), which comes from the dried root without bark, and Paeoniae Radix Rubra (Chi Shao or red peony), which comes from the dried root with bark. In China, Paeoniae Radix Alba has been reported to be more widely used (Zhang and Dai 2012). The roots of peony are used in TCM for several conditions such as amenorrhoea and dysmenorrhoea, inflammation, spasms and high blood pressure. It is also found in formulations for the treatment of symptoms related to metabolic syndrome and treatment and prevention of type-II diabetes mellitus (Baumgartner et al. 2010; Fu et al. 2013).

References

Akase T, Akase T, Onodera S, Jobo T, Matsushita R, Kaneko M, Tashiro S. (2003) A comparative study of the usefulness of toki-shakuyaku-san and an oral iron preparation in the treatment of hypochromic anemia in cases of uterine myoma. *Yakugaku Zasshi* 123(9): 817–824.

Anon. (2001) Peony (*Paeonia spp.*). Monograph. *Alternative Medicine Review* 6(5): 495–499.

Baumgartner RR, Steinmann D, Heiss EH, Atanasov AG, Ganzera M, Stuppner H, Dirsch VM. (2010) Bioactivity-guided isolation of 1,2,3,4,6-Penta-O-galloyl-D-glucopyranose from *Paeonia lactiflora* roots as a PTP1B inhibitor. *Journal of Natural Products* 73(9): 1578–1581.

Fu Q, Wang SB, Zhao SH, Chen XJ, Tu PF. (2013) Three new monoterpene glycosides from the roots of *Paeonia lactiflora*. *Journal of Asian Natural Product Research* 15(7): 697–702.

He DY, Dai SM. (2011) Anti-inflammatory and immunomodulatory effects of *Paeonia lactiflora* Pall., a traditional Chinese herbal medicine. *Frontiers in Pharmacology* 2: 10. doi: 10.3389/fphar.2011.00010.

He X, Xing D, Ding Y, Li Y, Xu L, Du L. (2004) Effects of cerebral ischemia-reperfusion on pharmacokinetic fate of paeoniflorin after intravenous administration of Paeoniae Radix extract in rats. *Journal of Ethnopharmacology* 94(2–3): 339–344.

Kotani N, Oyama T, Sakai I, Hashimoto H, Muraoka M, Ogawa Y, Matsuki A. (1997) Analgesic effect of a herbal medicine for treatment of primary dysmenorrhea – a double-blind study. *American Journal of Chinese Medicine* 25(2): 205–212.

Liu SH, Cheng YC. (2012) Old formula, new Rx: the journey of PHY906 as cancer adjuvant therapy. *Journal of Ethnopharmacology* 140(3): 614–623.

Mao QQ, Ip SP, Xian YF, Hu Z, Che CT. (2012) Anti-depressant-like effect of peony: a mini-review. *Pharmaceutical Biology* 50(1): 72–77.

Sakamoto S, Kudo H, Suzuki S, Sassa S, Yoshimura S, Nakayama T, Maemura M, Mitamura T, Qi Z, Liu XD, Yagishita Y, Asai A. (1996) Pharmacotherapeutic effects of toki-shakuyaku-san on leukorrhagia in young women. *American Journal of Chinese Medicine* 24(2): 165–168.

Sakamoto S, Yoshino H, Shirahata Y, Shimodairo K, Okamoto R. (1992) Pharmacotherapeutic effects of kuei-chih-fu-ling-wan (keishi-bukuryo-gan) on human uterine myomas. *American Journal of Chinese Medicine* 20(3-4): 313–317.

Song WH, Cheng ZH, Chen DF. (2014) Anticomplement Monoterpenoid Glucosides from the Root Bark of *Paeonia suffruticosa*. *Journal of Natural Products* 77(1): 42–48.

Usuki S, Higa TN, Soreya K. (2002) The improvement of luteal insufficiency in fecund women by tokishakuyakusan treatment. *American Journal of Chinese Medicine* 30(2–3): 327–338.

WHO. (1999) *WHO Monographs on Selected Medicinal Plants*. Vol. 1. WHO, Geneva, 295 pp.

Zhang W, Dai SM. (2012) Mechanisms involved in the therapeutic effects of *Paeonia lactiflora* Pallas in rheumatoid arthritis. *International Immunopharmacology* 14(1): 27–31.

Peppermint	*Mentha* × *piperita* L.

Synonyms: *M. officinalis* Hull; and others

Family: Lamiaceae (Labiatae)

Other common names: Balm mint; lamb mint

Drug name: Menthae piperitae folium; Menthae piperitae aetheroleum (peppermint oil)

Botanical drug used: Whole dried leaves; essential oil, obtained through steam distillation from flowering aerial parts

---

Indications/uses: Peppermint leaf is traditionally used internally, especially in the form of a herbal tea, for the symptomatic relief of digestive disorders such as dyspepsia, flatulence and nausea. Peppermint oil is taken internally, well-diluted with water, for the symptomatic relief of gastrointestinal spasm, flatulence and abdominal pain. Traditionally, peppermint oil is inhaled for the relief of symptoms associated with coughs and colds; and applied topically for the symptomatic relief of localised muscle pain, mild tension headache, and for the symptomatic relief of pruritic conditions in intact skin. Peppermint oil enteric-coated capsules are licensed in the United Kingdom for irritable bowel syndrome (IBS) or spastic colon syndrome.

Evidence: Considerable pharmacological and clinical evidence supports the use of peppermint oil as an antispasmodic, notably in the treatment of IBS. However, human studies on peppermint leaf are limited and clinical trials are lacking.

Safety: Peppermint is generally considered safe at recommended doses and is widely used as a deodorising agent and flavouring in toothpastes, mouthwashes, food and confectionery. Peppermint oil should be used with caution and only in small amounts. Safety has not been established during pregnancy and lactation and should be avoided. Topical use of peppermint oil is not recommended in young children, due to isolated reports of serious respiratory distress.

Main chemical compounds: The major constituent of peppermint leaves is the essential oil (0.5–4 %), which contains the monoterpenes menthol (30–55%) and menthone (14–32 %). Menthol is found mainly in the free alcohol form, with

Phytopharmacy: An evidence-based guide to herbal medicinal products, First Edition.
Sarah E. Edwards, Inês da Costa Rocha, Elizabeth M. Williamson and Michael Heinrich.
© 2015 John Wiley & Sons, Ltd. Published 2015 by John Wiley & Sons, Ltd.

small quantities as the acetate (3–5%) and valerate esters. Other monoterpene constituents include isomenthone (2–10%), 1,8-cineole (6–14%), α-pinene (1.0–1.5%), β-pinene (1–2%), limonene (1–5%), pulegone (<4%), neomenthol (2.5–3.5%) and menthofuran (1–9%). There is a maximum level of pulegone allowed due to its toxicity (EMEA 2008a,b; WHO 2004; Williamson et al. 2013).

Clinical evidence:

General: Clinical evidence for the use of peppermint leaf is lacking (McKay and Blumberg 2006). Human studies on the gastrointestinal, respiratory tract and analgesic effects of peppermint oil have been reported, although clinical trials are limited.

Antispasmodic effects of peppermint oil: A number of clinical trials on peppermint oil in the treatment for IBS have been evaluated in systematic reviews, which have concluded that peppermint is an effective antispasmodic (Ford et al. 2008; Ruepert et al. 2011).

Analgesic effects of peppermint oil: Studies investigating the analgesic effect of topically applied peppermint oil (10% in ethanol) have found that it reduced pain significantly more than placebo, and as much as paracetamol (Pharmaceutical Press Editorial Team 2013).

Pre-clinical evidence and mechanisms of action:

Antispasmodic effects of peppermint oil: The antispasmodic effects of peppermint oil have been largely attributed to its menthol content, which has calcium channel modifying activity (Grigoleit and Grigoleit 2005). The mechanism of action of peppermint oil has not been entirely elucidated, although both peppermint oil and menthol have been shown to act as 5-HT_3 receptor channel antagonists. This suggests that peppermint may positively influence the disturbed gastrointestinal motility and visceral sensitivity during symptoms of IBS and emesis, and thus contribute to the overall effect of peppermint oil (Heimes et al. 2011).

Analgesic effects: Menthol is a ligand of the 'cold receptor', the TRPM8 ion channel. Activation of TRPM8 is linked to cooling, analgesic and many other effects (Journigan and Zaveri 2013).

Antimicrobial effects: Numerous studies have assessed the antibacterial activity of peppermint oil, which inhibits both human and plant pathogens, including both gram-positive and gram-negative bacteria (Jeyakumar et al. 2011; McKay and Blumberg 2006). In one study, peppermint oil was also found to be effective against 11 out of 12 fungi tested, including *Candida albicans*, *Trichophyton mentagrophytes*, *Aspergillus fumigatus* and *Cryptococcus neoformans* at an MIC range of 0.25-10 μl/ml (McKay and Blumberg 2006).

Other effects: *In vitro* studies have shown that peppermint has choleretic action, exhibits strong antioxidant and anti-tumour effects and has some antiallergenic potential (McKay and Blumberg 2006). Genotoxicity studies have shown that peppermint oil induces mutations in a dose-independent manner and both peppermint oil and menthol are considered 'high toxicity clastogens'. Peppermint oil induces chromosome aberrations by secondary mechanism associated with cytotoxicity (Lazutka et al. 2000).

Interactions: Although no case reports of interactions have been reported, animal and *in vitro* pharmacological studies suggest that peppermint or peppermint oil may interact with antacids, ciclosporin and felodipine. Peppermint oil has been shown to inhibit CYP3A4, CYP2C8, CYP2C9, CYPC2C19 and CYP2D6 (Dresser et al. 2002; Wacher et al. 2002; Williamson et al. 2013). Peppermint oil has been demonstrated to slow intestinal transit, which may affect the absorption rate or total absorption of co-administered medications (Gardner and McGuffin 2013; Williamson et al. 2013). An *in vivo* study showed that pentobarbitone-induced sleeping time was significantly increased by acute pretreatment with peppermint oil (0.2 mg/kg), but significantly shortened by chronic pretreatment at the same dose. The effect of midozalam was enhanced and prolonged significantly at chronic pretreatment at the same dose. Gut motility was also increased by acute pretreatment of peppermint oil at this dose (Samojlik et al. 2012).

Peppermint leaf tea was found to reduce the activity of the CYP1A2 and CYP2E and to reduce the absorption of iron by 84%, similar to the effect associated with black tea (Gardner and McGuffin 2013).

Contraindications: Hypersensitivity to peppermint or menthol. Patients with gallstones or other biliary disorders should be cautious using preparations with peppermint leaf. Menthol can induce reflex apnoea and laryngospasm, so should not be used by children (Gardner and McGuffin 2013).

Adverse effects: No case reports of adverse events associated with peppermint leaf have been reported. In adults, adverse events are usually associated with a high intake of menthol from other products, including pharmaceuticals and confectionary. In clinical trials, side effects of peppermint oil that have been reported include heartburn and anal or perianal burning or discomfort. Allergic reactions to peppermint oil, including contact dermatitis have also been reported. Cases of mucosal ulcers have been associated with oral ingestion of products such as mouthwashes and confectionary that contain peppermint (Gardner and McGuffin 2013).

Dosage: Oral administration (adults): a) dried leaf – 2–4 g by infusion 3 times per day; b) concentrated peppermint water BP (1973) – 0.25–1 ml 3 times per day; c) peppermint oil – 0.05–0.2 ml 3 times per day. In the United Kingdom, licensed products of peppermint oil are not recommended for use in children (Pharmaceutical Press Editorial Team 2013).

General plant information: *M. × piperita* L. is a hybrid of water mint, *M. aquatica* L. and spearmint, *M. spicata* L. It is cultivated for commercial use in eastern and northern Europe and the United States of America (WHO 2004) and is widely used as a culinary herb.

References

Dresser GK, Wacher V, Wong S, Wong HT, Bailey DG. (2002) Evaluation of peppermint oil and ascorbyl palmitate as inhibitors of cytochrome P4503A4 activity *in vitro* and *in vivo*. *Clinical Pharmacology and Therapeutics* 72(3): 247–255.

EMEA. (2008a) Assessment report on *Mentha × piperita* L., folium. European Medicines Agency http://www.ema.europa.eu/docs/en_GB/document_library/Herbal_-_HMPC_assessment_report/2010/01/WC500059394.pdf (accessed August 2013).

EMEA. (2008b) Assessment report on *Mentha × piperita* L., aetheroleum. European Medicines Agency http://www.ema.europa.eu/docs/en_GB/document_library/Herbal_-_HMPC_assessment_report/2010/01/WC500059311.pdf (accessed August 2013).

Ford AC, Talley NJ, Spiegel BM, Foxx-Orenstein AE, Schiller L, Quigley EM, Moayyedi P. (2008) Effect of fibre, antispasmodics, and peppermint oil in the treatment of irritable bowel syndrome: systematic review and meta-analysis. *BMJ* 337: a2313.

Gardner Z, McGuffin M. (Eds.) (2013) *American Herbal Product Association's Botanical Safety Handbook*. 2nd Edition. CRC Press, USA, 1072 pp.

Grigoleit H-G, Grigoleit P. (2005) Pharmacology and preclinical pharmacokinetics of peppermint oil. *Phytomedicine* 12(8): 612–616.

Heimes K, Hauk F, Verspohl EJ. (2011) Mode of action of peppermint oil and (-)-menthol with respect to 5-HT_3 receptor subtypes: binding studies, cation uptake by receptor channels and contraction of isolated rat ileum. *Phytotherapy Research* 25(5): 702–708.

Jeyakumar E, Lawrence R, Pal T. (2011) Comparative evaluation in the efficacy of peppermint (*Mentha piperita*) oil with standards antibiotics against selected bacterial pathogens. *Asian Pacific Journal of Tropical Biomedicine* 1(2): S253–S257.

Journigan VB, Zaveri NT. (2013) TRPM8 ion channel ligands for new therapeutic applications and as probes to study menthol pharmacology. *Life Sciences* 92(8–9): 425–37.

Lazutka JR, Mierauskienė J, Slapšytė G, Dedonytė V. (2000) Genotoxicity of dill (*Anethum graveolens* L.), peppermint (*Mentha × piperita* L.) and pine (*Pinus sylvestris* L.) essential oils in human lymphocytes and *Drosophila melanogaster*. *Food and Chemical Toxicology* 39(5): 485–492.

McKay DL, Blumberg JB. (2006) A review of the bioactivity and potential health benefits of peppermint tea (*Mentha piperita* L.). *Phytotherapy Research* 20(8): 619–633.

Pharmaceutical Press Editorial Team. (2013) *Herbal Medicines*. 4th Edition. Pharmaceutical Press, London, UK.

Ruepert L, Quartero AO, de Wit NJ, van der Heijden GJ, Rubin G, Muris JWM. (2011) Bulking agents, antispasmodics and antidepressants for the treatment of irritable bowel syndrome. *Cochrane Database of Systematic Reviews* (8): CD003460.

Samojlik I, Petković S, Mimica-Dukić N, Božin B. (2012) Acute and chronic pretreatment with essential oil of peppermint (*Mentha × piperita* L., Lamiaceae) influences drug effects. *Phytotherapy Research* 26(6): 820–815.

Wacher VJ, Wong S, Wong HT. (2002) Peppermint oil enhances cyclosporine oral bioavailability in rats: comparison with D-alpha-tocopheryl poly(ethylene glycol 1000) succinate (TPGS) and ketoconazole. *Journal of Pharmaceutical Sciences* 91(1): 77–90.

WHO. (2004) *WHO Monographs on Selected Medicinal Plants*. Vol. 2. WHO, Geneva, 358 pp.

Williamson EM, Driver S, Baxter K. (Eds.) (2013) *Stockley's Herbal Medicines Interactions*. 2nd Edition. Pharmaceutical Press, London, UK.

Prickly Pear

Opuntia ficus-indica (L.) Mill., *O. streptacantha* Lem., and related species

Family: Cactaceae

Other common names: Indian fig; nopal; paddle cactus; tuna

Botanical drug used: Cladode

Indications/uses: The 'nopalitos', the thick, chlorophyll-rich stems (cladodes), are used widely in Mexico as a food and in the management of diabetes. In recent years, its use has become popular in Europe, including the United Kingdom, and in North America. In 2012, a 'fibre complex' containing *O. ficus-indica* was registered as a 'medical device' for use in weight loss, and the extract is also promoted in sports medicine to aid muscle recovery.

Evidence: The evidence is conflicting. Early studies in Mexico suggested good clinical effects but more recent studies have given more variable results.

Safety: Despite the long use as a food, limited evidence is available. Concerns have recently been raised due to a range of adverse reaction reports with the product registered as a medical device in the United Kingdom.

Main chemical compounds: The genus is rich in fibre and also in polysaccharides, especially the younger cladodes. A mucilage constituting up to 14% of the cladode dry weight is produced, especially by younger cladodes and also by the fruits, and consists of chains of (14)-linked β-D-galacturonic acid and R(1-2)-linked L-rhamnose residues (Stinzing and Carle 2005).

Clinical evidence:

Overall, some - so far still limited - evidence points to the potential of a chemically well characterised extract in the management of pre-diabetic conditions and in glucose management.

Antidiabetic effects: Initial studies in Mexico showed encouraging hypoglycaemic effects in human diabetic and normal subjects (Frati-Munari et al. 1991; Frati-Munari et al. 1988). More recently, a small (verum – *n* = 15, placebo – *n* = 14) double-blind, placebo-controlled study used 200 mg of a chemically characterised

---

Phytopharmacy: An evidence-based guide to herbal medicinal products, First Edition.
Sarah E. Edwards, Inês da Costa Rocha, Elizabeth M. Williamson and Michael Heinrich.
© 2015 John Wiley & Sons, Ltd. Published 2015 by John Wiley & Sons, Ltd.

extract over a 16-week period. An oral glucose tolerance test (OGTT) was also included, with a 400 mg dose of an *O. ficus-indica* preparation given 30 min before orally ingesting a 75 g glucose drink. The authors demonstrated acute blood-glucose-lowering effects and the long-term safety of the proprietary product used (Godard et al. 2010).

Stimulation of glycogen synthesis: An *O. ficus-indica* extract, together with carbohydrates, stimulated post-exercise muscle glycogen resynthesis (Pischel et al. 2012).

Pre-clinical evidence and mechanisms of action:

Much of the earlier research was done with related species (often *O. streptacantha* Lem.) or poorly defined material.

Antidiabetic effects: There is considerable evidence for its hypoglycaemic effects (e.g. Andrade-Cetto and Heinrich 2005; Butterweck et al. 2011).

Anti-inflammatory and cytoprotective effects: Anti-inflammatory and chondroprotective effect (Panico et al. 2007), prevention of chemically induced liver damage (Ncibi et al. 2008), cytoprotective effects on gastric mucosa (rat, Galati et al. 2007) and antioxidant activity (Lee et al. 2002) have also been described.

Interactions: No relevant data are available.

Adverse effects: A number of adverse drug reactions have been reported for products sold in the United Kingdom (MHRA 2013) and since currently there is no clear picture in terms of the pharmaceutical quality of the material used and of its composition, the use of *O. ficus-indica* products cannot be recommended. Food uses of the fruit and cladodes seem generally to be considered safe (Lopez 1995).

Dosage: Based on the study by Godard et al. (2010) a daily dose 200 mg/day is recommended, but further dose determining studies are urgently needed.

General plant information: This cactus is the source of the cactus pear fruit, which is widely cultivated. It is rich in betalaines (Stintzing and Carle 2004) and has been used for centuries in Mexico both medicinally and as a vegetable (Andrade-Cetto and Heinrich 2005). This and other species have now become noxious weeds in many drier, hot regions outside of its native habitat (e.g. Australia). The genus is also the host plant for the cochineal beetle, used to produce the intense red colour dye carmine (the aluminium salt of carminic acid).

References

Andrade-Cetto A, Heinrich M. (2005) Mexican plants with hypoglycaemic effect used in the treatment of diabetes. *Journal of Ethnopharmacology* 99(3): 325–348.
Butterweck V, Semlin L, Feistel B, Pischel I, Bauer K, Verspohl EJ. (2011) Comparative evaluation of two different *Opuntia ficus-indica* extracts for blood sugar lowering effects in rats. *Phytotherapy Research* 25(3): 370–5.
Frati-Munari AC, Gordillo BE, Altamirano P, Ariza CR. (1988) Hypoglycemic effect of *Opuntia streptacantha* Lemaire in NIDDM. *Diabetes Care* 11(1): 63–66.
Frati-Munari AC, Gordillo BE, Altamirano P, Ariza CR, Cortés-Franco R, Chávez-Negrete A, Islas-Andrade S. (1991) Influence of nopal intake upon fasting glycemia in type II diabetics and healthy subjects. *Archivo Investigaciones Medicas (Mex)*. 22(1): 51–56.
Galati EM, Monforte MT, Miceli N, Mondello MR, Taviano MF, Galluzzo M, Tripodo MM. (2007) *Opuntia ficus indica* (L.) Mill. mucilages show cytoprotective effect on gastric mucosa in rat. *Phytotherapy Research* 21(4): 344–346.

Ginestra G, Parker ML, Bennett RN, Robertson J, Mandalari G, Narbad A, Lo Curto RB, Bisignano G, Faulds CB, Waldron KW. (2009) Anatomical, chemical, and biochemical characterization of cladodes from prickly pear [*Opuntia ficus-indica* (L.) Mill.]. *Journal of Agricultural and Food Chemistry* 57(21): 10323–10330.

Godard MP, Ewing BA, Pischel I, Ziegler A, Benedek B, Feistel B. (2010) Acute blood glucose lowering effects and long-term safety of OpunDia supplementation in pre-diabetic males and females. *Journal of Ethnopharmacology* 130(3): 631–634.

Lee JC, Kim HR, Kim J, Jang YS. (2002) Antioxidant property of an ethanol extract of the stem of *Opuntia ficus-indica* var. *saboten*. *Journal of Agricultural and Food Chemistry* 50(22): 6490–6496.

Lopez, AD. (1995) Review: Use of the fruits and stems of the prickly pear cactus (*Opuntia* spp.) into human food. *Food Science and Technology International* 1(2–3): 65–74.

MHRA. (2013) Drug Analysis Print: *Opuntia ficus-indica*. http://www.mhra.gov.uk/home/groups/public/documents/sentineldocuments/dap_4869624197041.pdf.

Ncibi S, Ben Othman M, Akacha A, Krifi MN, Zourgui L. (2008) *Opuntia ficus indica* extract protects against chlorpyrifos-induced damage on mice liver. *Food and Chemical Toxicology* 46(2): 797–802.

Panico AM, Cardile V, Garufi F, Puglia C, Bonina F, Ronsisvalle S. (2007) Effect of hyaluronic acid and polysaccharides from *Opuntia ficus indica* (L.) cladodes on the metabolism of human chondrocyte cultures. *Journal of Ethnopharmacology* 111(2): 315–321.

Pischel I, van Proeyen, K, Hespel P. (2012) Dose finding of OpunDia™ (*O. ficus-indica* extract) for its effect on oral glucose tolerance and plasma insulin. *Journal of the International Society of Sports Nutrition* 29(Suppl 1): P25.

Stintzing, FC; Carle, R. (2004) Cactus stems (*Opuntia spp.*): A review on their chemistry, technology, and uses. *Molecular Nutrition and Food Research* 49(2): 175–194.

Pumpkin (Seed)　　　*Cucurbita pepo* L., and related cultivars

Synonyms: *Cucumis pepo* (L.) Dumort.; *Cucurbita elongata* Bean ex Schrad; and others

Family: Cucurbitaceae

Other common names: Bitter bottle gourd; pepita; pumpkin; summer pumpkin; squash

Drug name: Cucurbitae peponis semen

Botanical drug used: Dried seed

Indications/uses: Pumpkin seed extract is used for the symptomatic treatment of difficulties with urination in men due to benign prostatic hyperplasia (BPH). Women are often advised to eat pumpkin seeds for their phytoestrogen content. Historically, the seeds were used as a vermifuge and male anti-fertility agent, but these uses are no longer relevant.

Evidence: Overall only a few older but reasonably large (for herbal medicine) clinical trials provide some evidence for its use.

Safety: There is only limited data available on repeat dose or reproductive toxicity, but the data available do not point to any potential risks of registered products. It is important that the possibility of prostate cancer is eliminated before recommending pumpkin seed extract for BPH. The phytochemicals in the seeds are also common ingredients in other food plants.

Main chemical compounds: The seeds are rich in oil (approximately 35%) and protein (38%), and contain alpha-tocopherols (approximately 3 mg/100 g). The oil is composed of free fatty acids, mainly the unsaturated linoleic (35–68%), oleic acids (15–48%) and gadoleic acids, with the saturated palmitic and stearic acids; phytosterols (Δ^7-sterols including spinasterol, avenasterol, ergostenol and stigmastenol), together with smaller amounts of Δ^5-sterols such as campesterol, stigmasterol, clerosterol and isofucosterol, and terpenoids (especially squalene) (Stevenson et al. 2007; Yadav et al. 2010; Younis et al. 2000).The seeds also contain significant amounts of the phytoestrogenic lignan secoisolariciresinol (in one study, 21 mg/100 g of dry weight), with traces of lariciresinol (Sicilia et al. 2003).

Phytopharmacy: An evidence-based guide to herbal medicinal products, First Edition.
Sarah E. Edwards, Inês da Costa Rocha, Elizabeth M. Williamson and Michael Heinrich.
© 2015 John Wiley & Sons, Ltd. Published 2015 by John Wiley & Sons, Ltd.

Clinical evidence: Two comparatively large studies in men with BPH, (a 12-month randomised, placebo-controlled, multi-centre study with 476 patients and a 3-month open multi-centre study involving 2245 patients) showed an improvement in the International Prostate Symptom Score (IPSS). Most importantly, the frequency of urination decreased during the day from 6.7 to 4.8, and at night from 2.7 to 1.1 times (Bach 2000; Sabo et al. 1999; WHO 2009; Yadav et al. 2010).

A small trial was performed over 12 months on 47 BPH patients with an IPSS of over 8 to compare the effects of pumpkin seed oil (320 mg/day) with saw palmetto oil (320 mg/day), pumpkin seed oil plus saw palmetto oil (320 mg each/day), and placebo. In all treatment groups, including the pumpkin seed oil group, the IPSS were reduced after 3 months' treatment, and the quality of life score improved after 3 months with the pumpkin oil (and also the saw palmetto oil). No difference was observed in prostate volume in any of the treatment groups. Maximal urinary flow rates were gradually improved in the pumpkin oil group, with statistical significance after 6 months (12 months for saw palmetto). The combination showed no advantage over the pumpkin seed oil alone (Hong et al. 2009).

Thirty-five women who had undergone natural or surgical menopause were given 2 g daily of wheat germ oil (placebo; $n = 14$) or pumpkin seed oil ($n = 21$) in a pilot clinical study over 12 weeks. Serum lipids, fasting plasma glucose and blood pressure were measured and an 18-point questionnaire regarding menopausal symptoms was administered. Women receiving pumpkin seed oil showed a modest but significant increase in high-density lipoprotein cholesterol concentrations, and a decrease in diastolic blood pressure. There was significant improvement in the menopausal symptom scores and the authors suggested that further trials in postmenopausal women would be worthwhile (Gossell-Williams et al. 2011).

Pre-clinical evidence and mechanisms of action: Evidence points to inhibition of growth of the prostate gland, caused by a reduction in binding of dihydrotestosterone (measured *in vitro* using labelled cultured human prostate fibroblasts) as well as the inhibition of 5α-reductase, which prevents the conversion of testosterone to the more potent androgen dihydrotestosterone. The lignans are also phytoestrogenic. Pumpkin seed extract has anti-inflammatory effects in both acute and chronic models (WHO 2009; Yadav et al. 2010). Pumpkin seed oil has antihypertensive and cardioprotective effects in rats (El-Mosallamy et al. 2012), and in females, supplementation resulted in lower systolic and diastolic blood pressures in both non-ovariectomised and ovariectomised rats, thus preventing changes in plasma lipids and blood pressure associated with inadequate oestrogen availability (Gossell-Williams et al. 2008).

Interactions: No relevant information available.

Adverse effects: Limited data available but, based on its composition, registered products are unlikely to pose any major risk. Due to the lack of safety data, extracts of the crude drug should not be used during breastfeeding, by pregnant women or by children, but there is no need to avoid eating pumpkin seeds as food.

Dosage: 15–30 g (1–2 tablespoons) of crushed seeds daily. Commercially available extracts may vary in their dose; e.g. 500 mg per capsule with a drug-extract ratio of 15–25:1 and a daily dose of two capsules.

General plant information: The related *C. maxima* Duchesne has also been shown to be therapeutically beneficial in urinary problems, most notably in urinary dysfunction in human overactive bladder (Nishimura et al. 2014). Pumpkin is indigenous to North America but cultivated globally in temperate zones. The fruit, seeds and (to a lesser extent) the flowers, are used widely as food.

References

Bach D. (2000) Placebokontrollierte Langzeittherapiestudie mit Kurbissamen-extrakt bei BPH-bedingten Miktionsbeschwerden [Placebo-controlled longterm study with pumpkin seeds in BPH-induced problems with micturition]. *Urologe [B]* 40(5): 437–443.

El-Mosallamy AE, Sleem AA, Abdel-Salam OM, Shaffie N, Kenawy SA. (2012) Antihypertensive and cardioprotective effects of pumpkin seed oil. *Journal of Medicinal Foods* 15(2): 180–189.

Gossell-Williams M, Lyttle K, Clarke T, Gardner M, Simon O. (2008) Supplementation with pumpkin seed oil improves plasma lipid profile and cardiovascular outcomes of female non-ovariectomized and ovariectomized Sprague-Dawley rats. *Phytotherapy Research* 22(7): 873–877.

Gossell-Williams M, Hyde C, Hunter T, Simms-Stewart D, Fletcher H, McGrowder D, Walters CA. (2011) Improvement in HDL cholesterol in postmenopausal women supplemented with pumpkin seed oil: pilot study. *Climacteric* 14(5): 558–564.

Hong H, Kim CS, Maeng SH. (2009) Effects of pumpkin seed oil and saw palmetto oil in Korean men with symptomatic benign prostatic hyperplasia. *Nutrition Research and Practice* 3(4): 323–327.

Nishimura M, Ohkawara T, Sato, H, Takeda H, Nishihira J. (2014) Pumpkin seed oil extracted from *Cucurbita maxima* improves urinary disorder in human overactive bladder. *Journal of Traditional and Complementary Medicine*, 4(1): 72–74.

Sabo E, Berenji, J, Stoikov J, Bogdanovic J. (1999) Pharmacodynamic effect of pumpkin seed oil (Oleum cucurbitae pepo) in patients with adenoma prostate. *Fundamental and Clinical Pharmacology* 13(Suppl 1): 360.

Sicilia T, Heike B, Honig DM, Metzler M. (2003) Identification and stereochemical characterization of lignans in flaxseed and pumpkin seeds. *Journal of Agricultural and Food Chemistry* 51(5): 1181–1188.

Stevenson DG, Eller FJ, Wang L, Jane JL, Wang T, Inglett GE. (2007) Oil and tocopherol content and composition of pumpkin seed oil in 12 cultivars. *Journal of Agricultural and Food Chemistry* 55(10): 4005–4013.

WHO. (2009) *WHO Monographs on Medicinal Plants*. Vol. 4. WHO, Geneva, 456 pp.

Yadav M, Jain S, Tomar R, Prasad GB, Yadav H. (2010) Medicinal and biological potential of pumpkin: an updated review. *Nutrition Research Reviews* 23(2): 184–190.

Younis YM, Ghirmay S, Al-Shihry SS. (2000) African *Cucurbita pepo* L.: properties of seed and variability in fatty acid composition of seed oil. *Phytochemistry* 54(1): 71–75.

Raspberry Leaf   *Rubus idaeus* L.

Family: Rosaceae

Other common names: European raspberry; red raspberry

Drug name: Rubi idaei folium

Botanical drug used: Leaves (usually as a herbal tea)

Indications/uses: Raspberry leaf is very commonly taken as a tea during the later stages of pregnancy, as it is reputed to facilitate labour. No THR products for such a purpose are available since it is not deemed an appropriate indication for over-the-counter (OTC) use. Therefore, this cannot be recommended, although it is a widespread practice and without apparent harm. Raspberry leaf is also used for the relief of menstrual cramps, for its astringent properties in diarrhoea and disorders of the gastrointestinal tract, as a mouthwash for inflammation of the mouth and throat, and as an eye lotion for conjunctivitis.

Evidence: There is very limited clinical or other evidence to support the use of raspberry leaf to treat any indications, although pharmacological data do not conflict with traditional use.

Safety: Safety and toxicological data are limited for raspberry leaf preparations, including safety for when taken during pregnancy, but no serious adverse reports have been recorded.

Main chemical compounds: Tannins (including ellagitannins, gallotannins); flavonoids (including kaempferol, quercetin, hyperoside); phenolic acids (including caffeic acid, chlorogenic acid, *p*-coumaric acid); magnesium, zinc, vitamin C, tocopherol; alcohol; aldehydes; terpenoids (EMA 2013; Williamson et al. 2013).

Clinical evidence:

Use in pregnancy and birth: Evidence from clinical studies is sparse, despite the wide usage. For example, a study of 600 Norwegian women found that raspberry leaves were consumed by 5.7% of pregnant women; the most commonly reported reason for use was to 'prepare the uterus for labour' (Nordeng et al. 2011).

One study compared 96 pregnant women given 2 × 1.2 g a day of raspberry leaf, from 32 weeks of pregnancy until delivery, with a control group of 96 pregnant

---

Phytopharmacy: An evidence-based guide to herbal medicinal products, First Edition.
Sarah E. Edwards, Inês da Costa Rocha, Elizabeth M. Williamson and Michael Heinrich.
© 2015 John Wiley & Sons, Ltd. Published 2015 by John Wiley & Sons, Ltd.

women given a placebo. No significant difference was found between the groups. While the second stage of labour was 9.6 min shorter in the raspberry leaf group, and a few more women required forceps or vacuum-assisted birth in the placebo group, this was not statistically significant (Simpson et al. 2001). In a retrospective study, 57 women who used raspberry leaf during pregnancy (various doses) were compared with 51 randomly selected women who had not taken raspberry leaf, but had given birth during the same time period. No clinically significant difference was found between the groups. Although the mean time in first-stage labour was lower for the raspberry leaf group compared to control, there was no statistically significant difference in time for the three stages of labour (Parsons et al. 1999). A recent review of the clinical and pre-clinical evidence has concluded that raspberry leaf is not harmful in pregnancy (Smeriglio et al. 2014).

Other uses: No clinical studies could be found that investigated the effects of raspberry leaf for dysmenorrhea, diarrhoea or other indications (EMA 2013).

Pre-clinical evidence and mechanisms of action:

Effects on gestation and pregnancy: One study in three generations of pregnant rats found that administration of raspberry leaf resulted in a statistically significant increase in gestational length, and tended towards *lower* pregnancy success rates compared with controls and groups administered quercetin or kaempferol. Female offspring (F1 generation) showed precocious puberty, and their offspring (F2 generation) showed a significant increase in the proportion that were growth restricted (Holst et al. 2009).

In vitro and *in vivo* studies have shown variable effects on smooth muscle; the leaf extract showed relaxant activity in guinea-pig ileum *in vitro* (Rojas-Vera et al. 2002) but, overall, results from studies on raspberry leaf to support the traditional indications are very limited (EMA 2013).

Detoxification effects: It has been suggested from *in vitro* experiments that raspberry leaf may reduce paracetamol (acetaminophen) induced hepatotoxicity by inhibiting the multiple metabolic pathways of NAPQI formation, although animal and clinical studies have not yet been carried out (Langhammer and Nilsen 2014).

Interactions: None reported (Williamson et al. 2013), although raspberry leaf extract inhibited CYP1A2, 2D6, and 3A4 enzymes *in vitro* (Langhammer and Nilsen 2013), which may be clinically relevant (EMA 2013). The potential for interaction with other medicines should therefore be considered.

Contraindications: Safety has not been established for the use of raspberry leaf during pregnancy or lactation. Although there has been a long-established traditional use of raspberry leaf during pregnancy, it is not recommended for pregnant women, except under medical supervision. Raspberry leaf is not recommended for use during lactation or in children or adolescents under 18 years of age due to a lack of safety and efficacy data. Long-term use of raspberry leaf tea is not recommended (EMA 2013).

Adverse effects: Few adverse events have been recorded for raspberry leaf. Of two cases reported to the WHO's Uppsala Monitoring Centre, one was of a multi-herb preparation which could not be evaluated, and the other was a single-herb preparation that had been taken orally for 2 months to 'precipitate labour' (one dose weekly for 1 month). In the latter case, convulsions occurred in a 2-day-old boy. In a separate case, skin lesions were reported in a newborn infant after the mother drank raspberry leaf tea and took 6.5 g of evening primrose oil (as 500 mg capsules orally and vaginally) for 1 week prior to delivery. In general, minor side effects reported are likely to be those commonly found in pregnancy (nausea, sickness, diarrhoea, constipation, blood pressure changes, Braxton Hicks contractions) (Holst et al. 2009; EMA 2013), so causality is difficult to establish.

Dosage: Recommended oral dosages of raspberry leaf in adults are dried leaf – 4–8 g as an infusion daily; liquid extract – 4–8 ml (1:1 in 25% alcohol) 3 times daily (Pharmaceutical Press Editorial Team 2013).

General plant information: *R. idaeus* is native to Europe and northern Asia, and commonly cultivated elsewhere for the fruit.

References

EMA. (2013) Assessment report on *Rubus idaeus* L., folium. European Medicines Agency http://www.ema.europa.eu/docs/en_GB/document_library/Herbal_-_HMPC_assessment _report/2013/05/WC500142992.pdf (accessed July 2013).

Holst L, Haavik S, Nordeng H. (2009) Raspberry leaf – should it be recommended to pregnant women? *Complementary Therapies in Clinical Practice* 15(4): 204–208.

Langhammer AJ, Nilsen OG. (2013) *In vitro* inhibition of human CYP1A2, CYP2D6, and CYP3A4 by six herbs commonly used in pregnancy. *Phytotherapy Research* 28(4): 603–610.

Langhammer AJ, Nilsen OG. (2014) Fennel and Raspberry leaf as possible inhibitors of acetaminophen oxidation. *Phytotherapy Research* 28(10): 1573–1576.

Nordeng H, Bayne K, Havnen GC, Paulsen BS. (2011) Use of herbal drugs during pregnancy among 600 Norwegian women in relation to concurrent use of conventional drugs and pregnancy outcome. *Complementary Therapies in Clinical Practice* 17(3): 147–151.

Parsons M, Simpson M, Ponton T. (1999) Raspberry leaf and its effect on labour: safety and efficacy. *Australian College of Midwives Incorporated Journal* 12(3): 20–25.

Pharmaceutical Press Editorial Team. (2013) *Herbal Medicines.* 4th Edition. Pharmaceutical Press, London, UK.

Rojas-Vera J, Patel AV, Dacke CG. (2002) Relaxant activity of raspberry (*Rubus idaeus*) leaf extract in guinea-pig ileum *in vitro*. *Phytotherapy Research* 16(7): 665–668.

Simpson M, Parsons M, Greenwood J, Wade K. (2001) Raspberry leaf in pregnancy: its safety and efficacy in labor. *Journal of Midwifery and Women's Health* 46(2): 51–59.

Smeriglio A, Tomaino A, Trombetta D. (2014) Herbal products in pregnancy: experimental studies and clinical reports. *Phytotherapy Research* 28(8): 1107–1116.

Williamson EM, Driver S, Baxter K. (Eds.) (2013) *Stockley's Herbal Medicines Interactions.* 2nd Edition. Pharmaceutical Press, London, UK.

Red Clover   *Trifolium pratense* L.

Synonyms: *T. sativum* (Schreb.) Crome ex Boenn; and others

Family: Fabaceae (Leguminosae)

Other common names: Cow clover; creeping clover; meadow clover; peavine clover; purple clover; trefoil

Drug name (two drugs are used): Trifolii rubri flos; Trifolii pratensis herba

Botanical drug used: Dried inflorescence; aerial parts

Indications/uses: The isoflavone fraction of red clover extracts is used to relieve menopausal symptoms; traditionally, red clover has been used for skin conditions.

Evidence: The overall evidence for benefit in menopausal symptoms is positive but not unequivocal. There is no clinical evidence for any benefits on skin health.

Safety: Generally safe, but avoid where oestrogenic effects (e.g. in women with a history of hormone-dependent cancers, pregnancy and breast-feeding) may be contraindicated.

Main chemical compounds: Isoflavones, the major compound being biochanin A (together with afrormosin, daidzein, formononetin, genistein, orobol, pratensein, trifoside and others), and their glycoside conjugates. Other constituents include coumestrol, medicagol and coumarin, clovamides, soyasaponins, flavonoids and others (Sabudak and Guler 2009; Kolodziejczyk-Czepas 2012).

Clinical Evidence: A systematic review was carried out in 2007 to assess the efficacy of supplements containing red clover isoflavones in the reduction of hot flush frequency in menopausal women. The review concluded that there is a marginally significant effect for treating hot flushes in menopausal women. Seventeen potentially relevant articles were retrieved for further evaluation, but only five were suitable for inclusion in the meta-analysis, which indicated a reduction in hot flush frequency in the active treatment group (40–82 mg daily) compared with the placebo group (Coon et al. 2007).

Since then, further clinical studies have also shown positive but modest effects. A randomised, placebo-controlled, double-blind crossover trial of 199 post-menopausal women aged 40 or over who received 80 mg red clover isoflavones or

---

Phytopharmacy: An evidence-based guide to herbal medicinal products, First Edition.
Sarah E. Edwards, Inês da Costa Rocha, Elizabeth M. Williamson and Michael Heinrich.
© 2015 John Wiley & Sons, Ltd. Published 2015 by John Wiley & Sons, Ltd.

placebo for 90 days were monitored daily for 187 days. Daily hot flush/night sweat frequency and overall menopausal symptom intensity (Kupperman Index) were similar in both groups at baseline. Red clover isoflavone supplementation was more effective than placebo in reducing hot flushes and overall menopausal intensity. No side effects were observed with either active treatment or placebo (Lipovac et al. 2012).

A prospective, randomised, double-blind, placebo-controlled study evaluating the effects of red clover extract on menopausal symptoms and sexual satisfaction was carried out in 120 women with menopausal symptoms, who were given 40 mg isoflavone extract per day for 12 months. They were assessed before treatment and at 4, 8 and 12 months of treatment using the Kupperman Index and the Golombok Rust Inventory of Sexual Satisfaction. An improvement in menopausal symptoms was noted after 4 months of treatment, especially in relation to hot flashes, but no improvement in sexual satisfaction was reported (del Giorno et al. 2010).

Red clover isoflavones produced a positive effect on the lipid profile of women with increased body mass index, shown by a significant decrease in levels of total cholesterol, triglycerides, low-density lipoprotein cholesterol and lipoprotein A (Chedraui et al. 2008).

Pre-clinical evidence and mechanisms of action: Isoflavones are usually described as phytoestrogens but in certain circumstances, such as a high oestrogen environment (e.g. in premenopausal women), they can also act as anti-oestrogens. Isoflavones are SERMs (selective oestrogen receptor modulators), binding to certain oestrogen receptors and mimicking the effects of oestrogens, but they can also prevent the binding of natural, more potent, oestrogens such as oestriol. A diet with a high intake in isoflavone-containing foods (e.g. soya) is thought to protect against some hormone-dependent cancers, including those of the prostate and breast.

Interactions: Many studies have looked at the potential interaction of isoflavones with tamoxifen, an oestrogen receptor antagonist used widely in the treatment of breast cancer, and results have been conflicting – in some cases, the effect of tamoxifen was enhanced, and in others, inhibited. Current advice is to avoid the combination (Williamson et al. 2013). No other clinically significant interactions have been recorded.

Contraindications: Oestrogen-dependant cancers. The effect of isoflavones in breast tissue is complex, and currently not enough is known to recommend their use in patients with a history of breast or other hormone-dependent cancers.

Adverse effects: None reported, but see given contraindications. While there is no evidence of adverse events during short-term use, there are no data available on the safety of long-term administration (Coon et al. 2007).

Dosage: For products, use according to manufacturers' instructions. Herbal use is normally the extractive equivalent of 12 g of dried flower heads per day; isoflavone supplements 40–80 mg/day.

General plant information: Red clover is native to Europe, Western Asia and North Africa, but is naturalised in many other parts of the world. It is also grown as a fodder crop, and as 'green manure' due to its nitrogen fixation properties, which increase soil fertility.

References

Coon JT, Pittler MH, Ernst E. (2007) *Trifolium pratense* isoflavones in the treatment of menopausal hot flushes: a systematic review and meta-analysis. *Phytomedicine* 14(2–3): 153–159.

Chedraui P, San Miguel G, Hidalgo L, Morocho N, Ross S. (2008) Effect of *Trifolium pratense*-derived isoflavones on the lipid profile of postmenopausal women with increased body mass index. *Gynecological Endocrinology* 24(11): 620–624.

del Giorno C, Fonseca AM, Bagnoli VR, Assis JS, Soares JM Jr, Baracat EC. (2010) Effects of *Trifolium pratense* on the climacteric and sexual symptoms in postmenopause women. *Revista da Associação Médica Brasileira* 56(5): 558–562.

Kolodziejczyk-Czepas J. (2012) Trifolium species-derived substances and extracts – biological activity and prospects for medicinal applications. *Journal Of Ethnopharmacology* 143(1): 14–23.

Lipovac M, Chedraui P, Gruenhut C, Gocan A, Kurz C, Neuber B, Imhof M. (2012) The effect of red clover isoflavone supplementation over vasomotor and menopausal symptoms in postmenopausal women. *Gynecological Endocrinology* 28(3): 203–207.

Sabudak T, Guler N. (2009) *Trifolium* L.-a review of its phytochemical and pharmacological profile. *Phytotherapy Research* 23(3): 439–446.

Williamson EM, Driver S, Baxter K. (Eds.) (2013) *Stockley's Herbal Medicines Interactions.* 2nd Edition. Pharmaceutical Press, London, UK.

Red Vine Leaf — *Vitis vinifera* L.

Synonyms: *V. vinifera* subsp. *sativa* Hegi; *V. vinifera* var. *tinctorialis* Risso; and others (Note - *V. vinifera* var. *tinctoria* often appears in the literature, but is not considered a valid scientific name by botanists)

Family: Vitaceae

Drug name: Vitis viniferae folium

Botanical drug used: Leaves

Indications/uses: Red vine leaf is used for relieving symptoms associated with non-complicated chronic venous insufficiency (CVI), including mild oedema, and sensations of heaviness and tingling in the legs. Traditionally, red vine leaf extracts have been used both internally and externally to improve blood circulation for the treatment of symptoms associated with haemorrhoids, skin capillary fragility and for relief of dry eye.

Evidence: Limited clinical studies support its use for cardiovascular disease. The clinical evidence for its use in CVI is only available with a specific, chemically well characterised, proprietary red vine leaf extract. However, the available evidence on this specific red vine leaf extract suggests that it may help to maintain healthy leg-vein function.

Safety: Red vine leaf is considered to be safe and well tolerated.

Main chemical compounds: The main active components are flavonoids (kaempferol-3-*O*-glucoside, quercetin-3-*O*-glucoside, quercetin and myricetin) and tannins such as (+)-catechins and (−)-epicatechins, procyanidins and proanthocyanidins. Anthocyanins and stilbene derivatives, resveratrol and viniferins are also present. The red colour of the leaves is mainly due to the presence of anthocyanins (mainly glucosides of malvidin, delphinidin, cyanidin and pertunidin). Fruit acids such as tartaric acid, malic acid, succinic acid, citric acid and oxalic acid have also been reported, as well as phenylacrylic acid derivatives (*p*-coumaroyl acid, caffeoyl acid and feruloylsuccinic acid) (EMA 2010; ESCOP 2003; Nassiri-Asl and Hosseinzadeh 2009; Schneider et al. 2008; Williamson et al. 2013).

Clinical evidence:

Chronic venous insufficiency: Randomised clinical studies and open observational studies have been conducted with a specific, proprietary red vine leaf extract

Phytopharmacy: An evidence-based guide to herbal medicinal products, First Edition.
Sarah E. Edwards, Inês da Costa Rocha, Elizabeth M. Williamson and Michael Heinrich.
© 2015 John Wiley & Sons, Ltd. Published 2015 by John Wiley & Sons, Ltd.

(AS 195, 360 mg, 540 mg and 720 mg/day) for the treatment of CVI. The extract showed positive effects on improving the symptoms of CVI (tired, heavy and swollen legs or pain and tension in the legs) and reducing leg oedema compared to placebo (EMA 2010). More recently, a multicentre, randomised double-blind placebo-controlled trial showed that this specific extract of vine leaf (AS 195, 720 mg/day over 12 weeks) was well tolerated and significantly reduced the leg volume in patients with moderate to severe CVI (248 participants) than placebo (Rabe et al. 2011), but more robust studies are still needed to support its use in the treatment of CVI.

Pre-clinical evidence and mechanisms of action:

Anti-inflammatory effects: Red vine leaf extract exerts anti-inflammatory and anti-oedematous effects *in vitro* and *in vivo* (EMA 2010), due to the presence of flavonols and proanthocyanidins (Rabe et al. 2011). Resveratrol, also present in *V. vinifera* leaf, may also play an important role, inhibiting the release of chemokines and other biochemical mediators involved in the inflammatory process (Leifert and Abeywardena 2008).

Interactions: No interactions with red vine leaf have been reported (Williamson et al. 2013).

Contraindications: Although vine leaves are eaten as food, extracts cannot yet be recommended during pregnancy and lactation, or in children and adolescents under 18 years of age, due to insufficient safety data. If there is inflammation of the skin, thrombophlebitis, varicose or subcutaneous induration ulcers, sudden swelling of one or both legs, cardiac or renal insufficiency a health-care professional should be consulted immediately (EMA 2010).

Adverse effects: Very rare. Nausea, gastrointestinal complaints, headache and vertigo have been reported (EMA 2010; ESCOP 2003).

Dosage: Commercial products of red vine leaf extracts should be used as recommended by the manufacturer. Usually they have been standardised to 90% polyphenols and 5% astilbine with a daily dose of 360 mg (Heinrich et al. 2012). The dose used of red vine leaf extract varies from 360 to 720 mg/day of a dried water extract. As a herbal tea, infuse 5–10 g of dried leaves in 250 ml of water twice a day (EMA 2010; ESCOP 2003).

General plant information: The grape vine is native to the Mediterranean region and western Asia, but is now cultivated in temperate regions worldwide.

References

EMA. (2010) Assessment report on *Vitis vinifera* L., folium. European Medicines Agency http://www.ema.europa.eu/docs/en_GB/document_library/Herbal_-_HMPC_assessment _report/2011/01/WC500100390.pdf.

ESCOP. (2003) *ESCOP Monographs: The Scientific Foundation for Herbal Medicinal Products.* 2nd Edition. Thieme, Exeter and London, UK.

Heinrich M, Barnes J, Gibbons S, Williamson EM. (2012) *Fundamentals of Pharmacognosy and Phytotherapy.* 2nd Edition. Churchill Livingstone Elsevier, Edinburgh and London.

Leifert W R, Abeywardena M Y. (2008) Cardioprotective actions of grape polyphenols. *Nutrition Research* 28(11): 729–737.

Nassiri-Asl M, Hosseinzadeh H. (2009) Review of the pharmacological effects of *Vitis vinifera* (Grape) and its bioactive compounds. *Phytotherapy Research* 23(9): 1197–1204.

Rhodiola	*Sedum roseum* (L.) Scop. (better known under the synonym *Rhodiola rosea* L.)

Synonyms: *R. arctica* Boriss.; *R. odorata* Lam.; *R. rosea* L.; *S. arcticum* (Boriss.) Rønning; *S. rhodiola* Vill.; and others

Family: Crassulaceae

Other common names: Arctic root; golden root; roseroot; rosenrot

Drug name: Rhodiola rhizoma et radix

Botanical drug used: Rhizome and root

Indications: For the temporary relief of symptoms associated with stress such as fatigue, exhaustion and mild anxiety.

Evidence: Limited robust clinical trials exist supporting its use, but several small trials have shown positive effects on symptoms of stress and athletic performance.

Safety: Overall, considered to be safe.

Main chemical compounds: The main active constituents of the root are the monoterpene and phenylpropanoid glycosides rosin, rosarin, rosavin, rosiridin; and *p*-tyrosol; and the rhodiolosides (salidrosides), which are hydroxylated and methoxylated octenyl and octadienyl derivatives. It also contains polyphenolic acids such as gallic acid, caffeic acid and chlorogenic acid, as well as flavonoids (catechins, proanthocyanidins, kaempferol and its glycoside derivatives). Sterols (β-sitosterol), tannins, and essential oil are also present (EMA 2011; Panossian et al. 2010; Peschel et al. 2013; Williamson et al. 2013).

Clinical evidence:

General: A systematic review, including 11 randomised controlled trials (RCTs) which tested the efficacy or effectiveness of mono-preparations of Rhodiola, was conducted. Rhodiola as sole treatment was administered orally against a control group (placebo treatment, no treatment or active controls), in human individuals suffering from any condition and in healthy human volunteers. The study concluded that Rhodiola mono-preparation might have a positive effect on

---

physical performance, mental performance and certain mental health conditions such as stress. However, further research is warranted (Hung et al. 2011).

Effects on stress and fatigue: A multi-centre, non-randomised, open-label, single-arm trial investigated the therapeutic effects and safety over 4 weeks' treatment of a specific extract of Rhodiola (200 mg twice daily) in subjects with life-stress symptoms. The study found it to be effective in relieving symptoms associated with stress, such as fatigue, exhaustion and anxiety, and it was safe and well tolerated (Edwards et al. 2012).

A pilot study was conducted to assess the effect of a single dose of a mixed-herb combination product containing dried Rhodiola extract (each tablet standardised to rhodioloside (0.32%), rosavin, (0.5%), tyrosol (0.05%) schizandrin (0.37%), g-schizandrin (0.24%) and eleutherosides B and E (0.15%)) on mental performance in tired women. The mental performance (attention, speed and accuracy) was studied whilst patients performed stressful cognitive tasks under time pressure. Two hours after treatment, attention, speed and accuracy in situations of decreased performance caused by stress and tiredness was improved, compared to placebo (Aslanyan et al. 2010).

Effects on athletic performance: In a study conducted among a group of competitive athletes, the effects of Rhodiola supplementation on the physical performance and redox status was studied during endurance exercise for 4 weeks. The study concluded that this supplementation was able to increase physical performance as it reduced lactate levels and parameters of skeletal muscle damage after intense physical exercise (Parisi et al. 2010).

Effects on altitude sickness: The consumption of Rhodiola capsules for 15 days by soldiers helped reduce the symptoms (fatigue, drowsiness, chest tightness, palpitations, vertigo, lack of attention and memory loss) of high altitude sickness (Shi et al. 2011).

Pre-clinical evidence and mechanisms of action:

Adaptogenic activity: The effects of Rhodiola have been correlated to the content of the rosavins and rhodiolosides. They have been shown to have an effect on the levels and activity of serotonin, dopamine, and noradrenaline (norepinephrine) as well as β-endorphins. Rhodiola extracts were found to inhibit monoamine oxidases A and B, and catechol-o-methyl transferase, and may also facilitate the transport of neurotransmitters.

Other therapeutic effects have also been described for Rhodiola extract such as cardio- and hepato-protective and anticancer effects via several mechanisms (Kelly 2001; Panossian et al. 2010).

Interactions: Concomitant use of Rhodiola with pepper may diminish its antidepressant effect (Williamson et al. 2013). It should be avoided with antidepressants and central nervous system (CNS) stimulant drugs due to its effects on monoamine oxidases (van Diermen et al. 2009). Rhodiola products may interact with drugs that are metabolised via cytochrome P450 3A4 and P-glycoprotein, as the extract was found to inhibit their activity (Hellum et al. 2010).

Contraindications: Not recommended during pregnancy and lactation, as safety has not been established. The use in children and adolescents under 18 years of age is not recommended. If symptoms persist for longer than 2 weeks or worsen, a health care professional should be consulted (EMA 2011).

Adverse effects: May cause irritability and insomnia when taken in high doses (1.5–2.0 g or higher). Dizziness, dry mouth and gastrointestinal and nervous system disorders have also been reported (Bystritsky et al. 2008; Edwards et al. 2012; EMA 2011).

Dosage: According to manufacturers' indications. A daily dose of 100–600 mg Rhodiola extract, standardised to 1–3.6% rosavin (EMA 2011).

General plant information: Rhodiola is considered an adaptogen (a substance that enhances the state of non-specific resistance to stress). It has been used for centuries as a tonic in Eastern Europe and Asia to increase physical endurance, work productivity and longevity, enhance energy levels, increase resistance to high-altitude sickness, as well as for the treatment of fatigue, depression, impotence and infections. It was approved as an adaptogen for medical use in 1969 by the Ministry of Health of the Union of Soviet Socialist Republics (USSR) for fatigue (Dwyer et al. 2011; Kelly 2001; Panossian et al. 2010).

Due to the continued collection from the wild and high demand, a large number of adulterated non-regulated products can be found on some markets (Booker et al., unpublished).

References

Aslanyan G, Amroyan E, Gabrielyan E, Nylander M, Wikman G, Panossian A. (2010) Double-blind, placebo-controlled, randomised study of single dose effects of ADAPT-232 on cognitive functions. *Phytomedicine* 17(7): 494–499.
Bystritsky A, Kerwin L, Feusner JD. (2008) A pilot study of *Rhodiola rosea* (Rhodax) for generalized anxiety disorder (GAD). *Journal of Alternative and Complementary Medicine* 14(2): 175–180.
van Diermen D, Marston A, Bravo J, Reist M, Carrupt PA, Hostettmann K. (2009) Monoamine oxidase inhibition by *Rhodiola rosea* L. roots. *Journal of Ethnopharmacology* 122(2): 397–401.
Dwyer AV, Whitten DL, Hawrelak JA. (2011) Herbal medicines, other than St. John's Wort, in the treatment of depression: a systematic review. *Alternative Medicine Review* 16(1): 40–49.
Edwards D, Heufelder A, Zimmermann A. (2012) Therapeutic effects and safety of *Rhodiola rosea* extract WS(R) 1375 in subjects with life-stress symptoms–results of an open-label study. *Phytotherapy Research* 26(8): 1220–1225.
EMA. (2011) Community herbal monograph on *Rhodiola rosea* L., rhizoma et radix. European Medicines Agency http://www.ema.europa.eu/docs/en_GB/document_library /Herbal_-_Community_herbal_monograph/2011/09/WC500112677.pdf.
Hellum BH, Tosse A, Hoybakk K, Thomsen M, Rohloff J, Georg Nilsen O. (2010) Potent in vitro inhibition of CYP3A4 and P-glycoprotein by *Rhodiola rosea*. *Planta Medica* 76(4): 331–338.
Hung SK, Perry R, Ernst E. (2011) The effectiveness and efficacy of *Rhodiola rosea* L.: a systematic review of randomized clinical trials. *Phytomedicine* 18(4): 235–244.
Kelly GS. (2001) *Rhodiola rosea*: a possible plant adaptogen. *Alternative Medicine Review* 6(3): 293–302.
Panossian A, Wikman G, Sarris J. (2010) Rosenroot (*Rhodiola rosea*): traditional use, chemical composition, pharmacology and clinical efficacy. *Phytomedicine* 17(7): 481–493.
Parisi A, Tranchita E, Duranti G, Ciminelli E, Quaranta F, Ceci R, Cerulli C, Borrione P, Sabatini S. (2010) Effects of chronic *Rhodiola Rosea* supplementation on sport performance and antioxidant capacity in trained male: preliminary results. *Journal of Sports Medicine and Physical Fitness* 50(1): 57–63.
Peschel W, Prieto MJ, Karkour C, Williamson EM. (2013) Effect of provenance, plant part and processing on extract profiles from cultivated European *Rhodiola* for medicinal use. *Phytochemistry* 86: 92–102.
Shi ZF, Zhou QQ, Xiang L, Ma SD, Yan CJ, Luo H. (2011) Three preparations of compound Chinese herbal medicines for de-adaptation to high altitude: a randomized, placebo-controlled trial. *Zhong Xi Yi Jie He Xue Bao* 9(4): 395–401 [in Chinese].
Williamson EM, Driver S, Baxter K. (Eds.) (2013) *Stockley's Herbal Medicines Interactions*. 2nd Edition. Pharmaceutical Press, London, UK.

Ribwort Plantain   *Plantago lanceolata* L.

Family: Plantaginaceae

Other common names: English plantain; lance-leaf plantain; narrow-leaf plantain; plantain

Drug name: Plantaginis lanceolatae folium

Botanical drug used: Leaf, herb

Indications/uses: Plantain is used mainly to treat inflammatory upper respiratory tract infections, as a demulcent for the symptomatic relief of irritation and associated dry cough. It has also been used topically for inflammatory skin conditions and allergic rashes.

Evidence: There is pharmacological evidence in support of the traditional use, but clinical studies are lacking.

Safety: Overall, it is considered safe but there is a lack of toxicity data. The oral use is not recommended in children under 3 years of age and during pregnancy and lactation.

Main chemical compounds: The main constituents are iridoid glycosides including aucubin and catalpol, mucilage polysaccharides, flavonoids such as apigenin and luteolin, the phenylethanoids (acteoside, isoacteoside and plantamajoside), the phenol carboxylic acids, chlorogenic acid and vanillic acid, saponins, volatile oils and inorganic constituents (Beara et al. 2012; EMA 2011a; ESCOP 2003).

Clinical evidence: No clinical trials have been carried out. One post-marketing study assessed the tolerability of a *P. lanceolata* syrup (100 ml syrup containing 20 g ethanol extract of herb 1:1) in 593 patients with unspecific acute respiratory diseases for approximately 10 days. Patients were given 30 ml syrup daily, which corresponds to about 6 g of the herbal substance. A reduction in symptoms (e.g. cough frequency and intensity) with administration of syrup was found. The preparation was considered safe in elderly, adults, adolescents and children between 3 and 12 years of age (EMA 2011a).

Pre-clinical evidence and mechanisms of action: The pharmacologically active constituents are considered to be the mucilagous polysaccharides, the iridoid glycosides and the phenylethanoids. Extracts of *P. lanceolata* have shown

---

Phytopharmacy: An evidence-based guide to herbal medicinal products, First Edition.
Sarah E. Edwards, Inês da Costa Rocha, Elizabeth M. Williamson and Michael Heinrich.
© 2015 John Wiley & Sons, Ltd. Published 2015 by John Wiley & Sons, Ltd.

anti-inflammatory, antioxidant, antibacterial, immunostimulant, epithelising and procoagulant effects *in vitro* (EMA 2011a). The topical application of *P. lanceolata* extract on wound healing in rats was shown to be beneficial as it reduced the microbial count in the wound site and accelerate closure (Farahpour et al. 2012).

Interactions: None known.

Contraindications: It is not recommended during pregnancy and lactation, as safety has not been established (ESCOP 2003). Oral use in children under 3 years of age and the oromucosal use in children and adolescents (3 to 18 years of age) is not recommended due to lack of adequate data (EMA 2011b).

Adverse effects: None known.

Dosage: For specific products, manufacturers' instructions should be followed. As a herbal infusion, 2 g of the powdered herb in 150 ml of boiling water 2–3 times daily or 233–804 mg dry extract, 3–4 times daily (EMA 2011b).

General plant information: Plantain is a perennial weed widely distributed throughout Europe, North and Central Asia. Adulteration may occur with other species of plantain such as with the leaves of *P. major*, *P. media* or with *Digitalis lanata* (EMA 2011a). The leaf can be eaten in salads. Despite reports on allergies caused by the pollen from *P. lanceolata* (Granel et al. 1993; Mehta and Wheeler 1991), allergenic reactions due to the leaf extracts of *P. lanceolata* are unlikely (EMA 2011a). The fresh leaves have been used in Scotland for healing fresh wounds, either as a poultice or as an ointment (Lightfoot 1977).

References

Beara IN, Lesjak MM, Orcic DZ, Simin ND, Cetojevic-Simin DD, Bozin BN, Mimica-Dukic NM. (2012) Comparative analysis of phenolic profile, antioxidant, anti-inflammatory and cytotoxic activity of two closely-related Plantain species: *Plantago altissima* L. and *Plantago lanceolata* L. *Food Science and Technology* 47(1): 64–70.

EMA. (2011a) Assessment report on *Plantago lanceolata* L., folium. European Medicines Agency   http://www.ema.europa.eu/docs/en_GB/document_library/Herbal_-_HMPC_assessment_report/2012/02/WC500123351.pdf (accessed August 2014).

EMA. (2011b) Community herbal monograph on *Plantago lanceolata* L., folium. European Medicines Agency http://www.ema.europa.eu/docs/en_GB/document_library/Herbal_-_Community_herbal_monograph/2012/02/WC500123352.pdf (accessed August 2014).

ESCOP. (2003) *ESCOP Monographs: The Scientific Foundation for Herbal Medicinal Products*. 2nd Edition. Thieme, Exeter and London, UK.

Farahpour MR, Amniattalab A, Najad HI. (2012) Histological evaluation of *Plantago lanceolata* L. extract in accelerating wound healing. *Journal of Medicinal Plants Research* 6(33): 4844–4847.

Granel C, Tapias G, Valencia M, Randazzo L, Olive A. (1993) Plantain allergy (Plantago lanceolata): assessment of diagnostic tests. *Allergologia et immunopathologia (Madrid)* 21(4): 158–160.

Lightfoot J. (1977) *Flora Scotica Or a Systematic Arrangement, in the Linnaean Method, of the Native Plants of Scotland and the Hebrides*. B. White, London.

Mehta V, Wheeler AW. (1991) IgE-mediated sensitization to English plantain pollen in seasonal respiratory allergy: identification and partial characterisation of its allergenic components. *International Archives of Allergy and Immunology* 96(3): 211–217.

Rosehip *Rosa canina* L. & other allied *Rosa spp.*

Family: Rosaceae

Other common names: Dog rose; rose haw; rose hep

Botanical drug used: Ripe, fresh or dried whole 'fruits' (botanically they are, in fact, pseudofruits), ripe, fresh or dried receptacle, freed from 'seed' (which are botanically fruit) and trichomes, rosehip 'seed' (fruit)

Drug name: Rosae caninae pseudofructus cum fructibus; Rosae caninae pseudofructus (sine fructibus); Rosae caninae fructus

---

Indications/uses: The main use of rosehip is as an anti-inflammatory agent in rheumatic and arthritic diseases, including sciatica. It has also been used traditionally as a source of vitamin C (especially for babies), for colds and influenza, gastrointestinal diseases including gastric ulcers and many other conditions. At the time of writing, there are no UK registered (THR) products containing rosehip available.

Evidence: Systematic reviews and a meta-analysis have found clinical evidence for effectiveness of a proprietary rosehip powder in the treatment of osteoarthritis, where it decreased pain. However, clinical evidence for other indications is lacking.

Safety: Rosehip appears safe at recommended doses, with mild adverse effects reported infrequently.

Main chemical compounds: Biologically active compounds isolated from rosehips include lipids (including linoleic, oleic, linolenic and arachidonic acids), a galactolipid (2S)-1,2-di-O-[(9Z,12Z,15Z)-octadeca-9,12,15-trienoyl]-3-O-beta-d-galactopyranosyl glycerol (GLGPG), triterpene acids such as oleanolic acid, betulinic acid and ursolic acid, carotenoids, tocopherols and vitamins (particularly vitamin C, which is amongst the highest known in any fruit or vegetables, and also vitamins B1, B2, K, B3/niacin and E). Rosehip is also well known for its high phenolic content (flavonoids, ellagic acid, etc.) (Barros et al. 2011; Larson et al. 2003; Roman et al. 2013; Saaby et al. 2011).

Phytopharmacy: An evidence-based guide to herbal medicinal products, First Edition.
Sarah E. Edwards, Inês da Costa Rocha, Elizabeth M. Williamson and Michael Heinrich.
© 2015 John Wiley & Sons, Ltd. Published 2015 by John Wiley & Sons, Ltd.

Clinical evidence: A meta-analysis of randomised controlled trials identified three studies involving 287 patients with osteoarthritis, and found that a standardised rosehip powder (made from the seeds and husks of the fruits of a sub-type of *R. canina*) significantly ($p = 0.0019$) reduced pain scores more than placebo, with a combined effect size (ES) of 0.37 (95% CI: 0.13–0.60). Reduction in the use of 'rescue medication' was also found to be significant ($p = 0.0018$) in the rosehip group compared with placebo, with a combined ES of 0.28 (95% CI: 0.05–0.51). The trials had a median duration of 3 months. The authors concluded that dry rosehip powder seems to have a consistent, small-to-moderate efficacy in reducing pain in osteoarthritis patients, although efficacy was only observed in short-term trials (3–4 months) (Christenson et al. 2008).

In a longer, pilot surveillance study, 152 patients with acute exacerbations of chronic pain were treated with a propriety rosehip and seed powder (Litozin®) at a dose providing up to 3 mg of galactolipid/day for up to 54 weeks. Only 77 patients completed the year-long surveillance, but an appreciable overall improvement was observed, similar to that seen in surveillances of patients receiving an aqueous extract of *Harpagophytum sp.* (Devil's claw). Changes from baseline tended to be greater in patients with greater degrees of pain and disability (Chubrasik 2008a). Another study found that 10.5 g/day of the same rosehip powder preparation given to female rheumatoid arthritis patients and healthy volunteers for 28 days had no effect on clinical symptoms or laboratory measurements (Kirkeskov et al. 2011).

Systematic reviews (e.g., Chubrasik et al. 2006; Chubrasik 2008b) have found that while there is clinical evidence for effectiveness of rosehip in the field of osteoarthritis, it is still lacking for other indications. One small study found, however, that a proprietary rosehip drink given to 22 healthy volunteers for 3 weeks significantly increased faecal levels of bifidobacteria and lactobacilli, and abdominal pain was reduced after 4 weeks of treatment with rosehip in a double-blind randomised study in 60 patients with irritable bowel syndrome (Chubrasik 2008b).

Pre-clinical evidence and mechanisms of action: A number of *in vitro* and *in vivo* studies have demonstrated the antioxidant, anti-inflammatory, antinociceptive, anti-obesity and immunomodulatory properties of rosehip (e.g., see Chubrasik 2008b; Lattanzio et al. 2011; Saaby et al. 2011). Extracts derived from powdered rosehip without fruits were found to be more effective in assays carried out than powdered rosehip with fruits (Wenzig et al. 2008).

Rosehip powder and its constituent galactolipid GLGPG have been shown *in vitro* to attenuate inflammatory responses in macrophages, peripheral blood lymphocytes and chondrocytes; they have been shown to inhibit inflammatory mediators (nitric oxide and prostaglandin), and reduce secretion of inflammatory cytokines including TNF-α, IFN-g, IL-1b, IL-6, IL-12, and chemokines (CCL5/RANTES, CXCL10/IP-10). The effects of the galactolipid were weaker than those of the rosehip powder (Schwager et al. 2011), but rosehip contains other anti-inflammatory constituents such as the triterpene acids (Saaby et al. 2011).

Interactions: None known.

Contraindications: None known. During pregnancy and lactation, dosage should not exceed amounts usually found in foods or nutritional supplements due to lack of safety data (ESCOP 2003).

Adverse effects: In clinical trials, adverse effects were similar to those observed with placebo (Christensen et al. 2008). In rare cases, allergy may occur, for example, generalised exanthema (rash) and gastrointestinal complaints have been reported after drinking rosehip tea (Chrubasik et al. 2008a; Chrubasik et al. 2008b). At high doses (e.g., 45 g per day), mild gastrointestinal upset has been reported (ESCOP 2003).

Dosage: As for supplements in general, follow manufacturers' instructions.

General plant information: Children have used rosehip fruits as 'itching powder', which irritates through its fibres to provoke a prickle sensation. During World War II, collection of rosehips from hedgerows was encouraged in Britain to make rosehip syrup as a source of vitamin C, in order to replace citrus fruits which were in short supply at the time.

References

Barros L, Carvalho AM, Ferreira ICFR. (2011) Exotic fruits as a source of important phyto-chemicals: Improving the use of traditional *Rosa canina* fruits in Portugal. *Food Research International* 44(7): 2233–2236.

Christensen R, Bartels EM, Altman RD, Astrup A, Bliddal H. (2008) Does the hip powder of *Rosa canina* (rosehip) reduce pain in osteoarthritis patients? – a meta-analysis of random-ized controlled trials. *Osteoarthritis Cartilage* 16(9): 965–972.

Chrubasik C, Duke RK, Chrubasik S. (2006) The evidence for clinical efficacy of rose hip and seed: a systematic review. *Phytotherapy Research* 20(1): 1–3.

Chrubasik C, Wiesner L, Black A, Müller-Ladner U, Chrubasik S. (2008a) A one-year sur-vey on the use of a powder from *Rosa canina lito* in acute exacerbations of chronic pain. *Phytotherapy Research* 22(9): 1141–1148.

Chrubasik E, Rougofalis BD, Müller-Ladner U, Chrubasik S. (2008b) A systematic review on the *Rosa canina* effect and efficacy profiles. *Phytotherapy Research* 22(6): 725–733.

ESCOP. (2003) *ESCOP Monographs: The Scientific Foundation for Herbal Medicinal Prod-ucts.* 2nd Edition. Thieme, Exeter and London, UK.

Kirkeskov B, Christensen R, Bügel S, Bliddal H, Danneskiold-Samsøe B, Christensen LP, Andersen JR. (2011) The effects of rose hip (*Rosa canina*) on plasma antioxidative activity and C-reactive protein in patients with rheumatoid arthritis and normal controls: a prospec-tive cohort study. *Phytomedicine* 18(11): 953–958.

Larson E, Kharazmi A, Christensen LP, Christensen SB. (2003) An anti-inflammatory galac-tolipid from rose hip (*Rosa canina*) that inhibits chemostaxis of human peripheral blood neutrophils *in vitro. Journal of Natural Products* 66(7): 994–995.

Lattanzio F, Greco E, Carretta D, Cervellati R, Govoni P, Speroni E. (2011) *In vivo* anti-inflammatory effect of *Rosa canina* L. extract. *Journal of Ethnopharmacology* 137(1): 880–885.

Roman I, Stănilă A, Stănilă S. (2013) Bioactive compounds and antioxidant activity of *Rosa canina* L. biotypes from spontaneous flora of Transylvania. *Chemistry Central Journal* 7(1): 73.

Saaby L, Jäger AK, Moesby L, Hansen EW, Christensen SB. (2011) Isolation of immunomod-ulatory triterpene acids from a standardized rose hip powder (*Rosa canina* L.). *Phytotherapy Research* 25(2): 195–201.

Schwager J, Hoeller U, Worlfram S, Richard N. (2011) Rose hip and its constituent galac-tolipids confer cartilage protection by modulating cytokine, and chemokine expression. *BMC Complementary and Alternative Medicine* 11: 105.

Wenzig EM, Widowitz U, Kunert O, Chrubasik S, Bucar F, Knauder E, Bauer R. (2008) Phytochemical composition and in vitro pharmacological activity of two rose hip (*Rosa canina* L.) preparations. *Phytomedicine* 15(10): 826–835.

Roselle   *Hibiscus sabdariffa* L.

Synonyms: *Abelmoschus cruentus* (Bertol.) Walp.; *Sabdariffa rubra* Kostel.

Family: Malvaceae

Other common names: Hibiscus; Jamaica sorrel; karkade; red sorrel; soborodo; sour tea; Sudan tea

Drug name: Hibisci sabdariffae flos

Botanical drug used: Flowers, including calyx

Indications/uses: Traditionally an infusion of roselle flowers is used for colds and other upper respiratory tract infection, diabetes, as a gentle laxative and diuretic, to treat high blood pressure and to lower blood cholesterol levels. The leaves may be applied to boils and ulcers and a lotion made from them applied to wounds.

Evidence: Limited clinical evidence is available to support its antihypertensive and hypolipidaemic effects.

Safety: Overall roselle is considered safe and is commonly taken as a tea and in cold drinks; it is made into jams and jellies and may be added to foods and beverages to add colour and flavour.

Main chemical compounds: The main compounds present are anthocyanins based on delphinidin and cyanidin, such as delphinidin-3-sambubioside (delphinidin-3-glucoxyloside or hibiscin), cyanidin-3-sambubioside (gossypicyanin), cyanidin-3, 5-diglucoside, and others; organic acids including ascorbic acid, hydroxycitric acid and hibiscus acid; the flavonoids hibiscitrin (hibiscetin 3-glucoside), sabdaritrin, gossypitrin, gossytrin and many others; phenolic acids (protocatechuic acid, chlorogenic acid and others); polysaccharides composed of arabinose, galactose, glucose and rhamnose; glycine, betaine and trigonelline; and in the seeds, the phytosterols cholesterol, campasterol, stigmasterol, β-sitosterol, α-spinasterol and ergosterol (Da-Costa-Rocha et al. 2014; Hopkins et al. 2013; Williamson E et al. 2013).

Phytopharmacy: An evidence-based guide to herbal medicinal products, First Edition.
Sarah E. Edwards, Inês da Costa Rocha, Elizabeth M. Williamson and Michael Heinrich.
© 2015 John Wiley & Sons, Ltd. Published 2015 by John Wiley & Sons, Ltd.

Clinical evidence:

Hypertension, hypercholesterolaemia and diabetes: A systematic review on the use of Hibiscus for the treatment of hypertension based on four randomised clinical trials (sour tea vs black tea; aqueous extract (9.6 mg anthocyanins content) vs captopril (25 mg twice a day); or standardised extract (250 mg of total anthocyanins per dose) vs lisinopril (10 mg)) concluded that there was no reliable evidence to support its use in hypertension in adults despite the fact that a reduction of blood pressure was observed. The studies showed that Hibiscus caused a greater reduction of blood pressure when compared with black tea, but less than that of the ACE-inhibitors (Wahabi, et al. 2010). A study was carried out to determine the effect of a water *H. sabdariffa* extract on the lipid levels of metabolic syndrome patients alone or combined with diet where patients took a daily dose of 100 mg of extract (in capsule) for 1 month. A significant reduction in glucose and total cholesterol levels, an increase of HDL-c levels and an improved TAG/HDL-c ratio (an insulin resistance marker) were observed after taking the extract alone, and an additional triglyceride-lowering effect when taken in combination with a low-fat diet (Gurrola-Diaz et al. 2010). An extract of roselle significantly reduced blood pressure in both humans and rats. Diuresis and inhibition of the angiotensin I-converting enzyme were found to be less important mechanisms than those related to the antioxidant, anti-inflammatory, and endothelium-dependent effects of the polyphenols (Joven et al. 2014). Recent reviews have concluded that although there is promising data for its use in the treatment of hypertension and hyperlipidemia, further clinical studies need to be conducted before any clinical recommendations can be made, and the other traditional indications need further validation (Da-Costa-Rocha et al. 2014; Hopkins et al. 2013).

Pre-clinical evidence and mechanisms of action: The mechanism of action is unclear but *H. sabdariffa* extracts have been shown to have antioxidant, diuretic, hypotensive, hypocholesterolaemic, antidiabetic and anticancer effects *in vivo* and *in vitro*. The extract has a strong scavenging effect on reactive oxygen and free radicals (Olalye and Rocha 2007; Tseng et al. 1997). Several *in vivo* studies have reported lipid lowering activity (low-density lipoprotein cholesterol, triglycerides, total cholesterol) and reduced lipid peroxidation by extracts of *H. sabdariffa*, and especially the anthocyanins and protocatechuic acid. The reduced systolic and diastolic heart pressure, lowered heart rate and produced vasodilation, and the antihypertensive activity have been suggested to be linked to inhibition of angiotensin-converting enzymes, acetylcholine-like and histamine-like mechanisms, and the diuretic effect. The antidiabetic activity may be due to inhibition of α-glucosidase and α-amylase enzymes. The extracts also showed anticancer activity due to the protocatechuic acid and anthocyanins, through reduction of reactive oxygen species, DNA fragmentation, cell cycle arrest and apoptosis (Da-Costa-Rocha et al. 2014; Hopkins et al. 2013; Lin et al. 2011).

Interactions: None known.

Contraindications: Not recommended during pregnancy and lactation, as safety has not been established.

Adverse effects: Few reported. The extract has been reported to be generally safe. A single report has suggested that excessive doses for relatively long periods could have a deleterious effect on the testes of rats, although this has been disputed (Ali et al. 2005).

Dosage: Should be taken according to manufacturers' instructions. For the herbal drug, a clinical study reports that one tisane bag (serving) containing 1.25 g *H. sabdariffa* per 240 ml boiled water, three servings a day, is effective (McKay et al. 2010).

General plant information: *H. sabdariffa* is widely used in traditional medicine and as hot and cold beverage. It is widely cultivated in both tropical and subtropical regions (Ali et al. 2005).

References

Ali BH, Wabel NA, Blunden G. (2005) Phytochemical, pharmacological and toxicological aspects of *Hibiscus sabdariffa* L.: a review. *Phytotherapy Research* 19(5): 369–375.

Da Costa Rocha I, Bonnlaender B, Sievers H, Pischel I and Heinrich M. (2014) *Hibiscus sabdariffa* L. – a phytochemical and pharmacological Review. *Food Chemistry* 165: 424–443.

Gurrola-Diaz CM, Garcia-Lopez PM, Sanchez-Enriquez S, Troyo-Sanroman R, Andrade-Gonzalez I, Gomez-Leyva JF. (2010) Effects of *Hibiscus sabdariffa* extract powder and preventive treatment (diet) on the lipid profiles of patients with metabolic syndrome. *Phytomedicine* 17(7): 500–505.

Hopkins AL, Lamm MG, Funk JL, Ritenbaugh C. (2013) *Hibiscus sabdariffa* L. in the treatment of hypertension and hyperlipidemia: a comprehensive review of animal and human studies. *Fitoterapia* 85: 84–94.

Joven J, March I, Espinel E, Fernández-Arroyo S, Rodríguez-Gallego E, Aragonès G, Beltrán-Debón R, Alonso-Villaverde C, Rios L, Martin-Paredero V, Menendez JA, Micol V, Segura-Carretero A, Camps J (2014) *Hibiscus sabdariffa* extract lowers blood pressure and improves endothelial function. *Molecular Nutrition & Food Research*. 58(6): 1374–1378.

Lin HH, Chen JH, Wang CJ. (2011) Chemopreventive properties and molecular mechanisms of the bioactive compounds in *Hibiscus sabdariffa* Linne. *Current Medicinal Chemistry* 18(8): 1245–1254.

McKay DL, Chen CY, Saltzman E, Blumberg JB. (2010) *Hibiscus sabdariffa* L. tea (tisane) lowers blood pressure in prehypertensive and mildly hypertensive adults. *Journal of Nutrition* 140(2): 298–303.

Olalye MT, Rocha JB. (2007) Commonly used tropical medicinal plants exhibit distinct *in vitro* antioxidant activities against hepatotoxins in rat liver. *Experimental and Toxicologic Pathology* 58(6): 433–438.

Tseng TH, Kao ES, Chu CY, Chou FP, Lin Wu HW, Wang CJ. (1997) Protective effects of dried flower extracts of *Hibiscus sabdariffa* L. against oxidative stress in rat primary hepatocytes. *Food and Chemical Toxicology* 35(12): 1159–1164.

Wahabi HA, Alansary LA, Al-Sabban AH, Glasziuo P. (2010) The effectiveness of *Hibiscus sabdariffa* in the treatment of hypertension: a systematic review. *Phytomedicine* 17(2): 83–86.

Williamson EM, Driver S, Baxter K. (Eds.) (2013) *Stockley's Herbal Medicines Interactions*. 2nd Edition. Pharmaceutical Press, London, UK.

Rosemary　　　　　　　　　*Rosmarinus officinalis* L.

Family: Lamiaceae (Labiatae)

Drug name: Rosmarini folium; Rosmarini aetheroleum

Botanical drug used: Dried leaves; essential oil (obtained from steam distillation of the flowering aerial parts)

Indications/uses: In addition to its culinary uses, rosemary leaf has traditionally been used to treat a variety of conditions including dyspepsia, headache and nervous tension. Ancient Greeks used rosemary to improve memory. The German Commission E approved rosemary for internal use for dyspepsia and for external use in rheumatic disease and circulatory problems. The oil has a traditional use in promoting hair growth and is widely used in aromatherapy for its reputed stimulant properties.

Rosemary leaf is found in UK registered (THR) products in combination with centaury (*Centaurium erythraea* Rafn.) and lovage root (*Levisticum officinale* Koch.) for minor urinary complaints associated with cystitis in women. Registered products containing rosemary oil also contain other herbal ingredients, and are for external use only, for example, in combination with peppermint oil and eucalyptus oil as a decongestant and also to relieve muscular aches and pains.

Evidence: Clinical data to support traditional uses of rosemary are lacking. *In vitro* and animal studies indicate that rosemary has spasmolytic, antidepressant, antidiabetic, diuretic, antinociceptive, antioxidant, anti-inflammatory, anticancer and antimicrobial actions.

Safety: Rosemary herb is generally considered safe at recommended doses, although occasional allergic reactions have been reported. It should not be used during pregnancy in amounts greater than those used in food, and epileptic patients should not take high doses of rosemary due to the camphor content of the essential oil, which may induce convulsions. The oil should only be applied externally in a diluted form.

Main chemical compounds: The leaves contain up to 2.5% essential oil, the composition of which varies according to chemotype, and contains camphor (5–31%), 1,8-cineole (15–55%), α-pinene (9–26%), borneol (1.5–5.0%), camphene

---

Phytopharmacy: An evidence-based guide to herbal medicinal products, First Edition.
Sarah E. Edwards, Inês da Costa Rocha, Elizabeth M. Williamson and Michael Heinrich.
© 2015 John Wiley & Sons, Ltd. Published 2015 by John Wiley & Sons, Ltd.

(2.5–12.0%), β-pinene (2.0–9.0%), limonene (1.5–5.0%), verbenone (2.2–11.1%), β-caryophyllene (1.8–5.1%) and myrcene (0.9–4.5%). The leaves also contain flavonoids with a methylated aglycone (e.g. genkwanin and luteolin); phenolic acids, particularly rosmarinic, cholorogenic and caffeic; and terpenoids, including diterpenes (e.g. carnosol, carnosolic acid, rosmanol) and triterpenes (e.g. oleanolic and ursolic acids) (EMA 2010; Pharmaceutical Press Editorial Team 2013; WHO 2009; Williamson et al. 2013).

Clinical evidence:

Robust clinical data to support traditional medicinal uses of rosemary are lacking. Almost all trials using rosemary oil externally are using combination products for treating pain and muscle soreness. The aromatherapy use of rosemary oil is widely known but not well supported by clinical studies.

Effects on cognition: One short-term randomised, double-blind, controlled crossover study has investigated the acute effects of different doses of rosemary leaf powder on cognitive function in 28 older adults (mean age 75 years). The 'Cognitive Drug Research' computerised assessment system was used at intervals up to 6 hours following oral administration of four different doses of rosemary (750, 1500, 3000, 6000 mg) or placebo. Doses were counterbalanced, with a 7-day washout period between treatments, spread over 5 weeks. Results demonstrated a biphasic dose-dependent effect in speed of memory: the lowest dose (750 mg) had a significant beneficial effect compared with placebo ($p = 0.01$), while the highest dose (6000 mg) had a significant impairing effect ($p < 0.01$). Several doses of rosemary exhibited deleterious effects on other measures of cognitive performance, but these were inconsistent and not dose dependent (Pengelly et al. 2012). Given the small-scale study, and varied effects of different doses of rosemary, it is not possible to draw firm conclusions as to the efficacy of rosemary in memory enhancement, and further research is required.

Pre-clinical evidence and mechanisms of action:

A large number of studies have highlighted the diverse pharmacological actions of rosemary, a selection of which are summarised here.

Diuretic effects: Diuretic activity of an aqueous extract (8%) of rosemary was demonstrated in a rat model (Haloui et al. 2000).

Antispasmodic effects: A study demonstrated that ethanolic rosemary extract exhibited a significant and dose-dependent spasmolytic effect in isolated guinea pig ileum, mediated through dual blockade of muscarinic receptors and calcium channels. There was no evidence for the involvement of nicotinic receptors, prostaglandins or nitric oxide with this activity (Ventura-Martínez et al. 2011).

Anti-inflammatory and antinociceptive effects: Rosemary essential oil was shown to exhibit anti-inflammatory and peripheral antinociceptive activity in animal models (Takaki et al. 2008).

Antimicrobial activity: *In vitro* studies have demonstrated the antibacterial and antifungal properties of rosemary oil. Rosemary herb inhibits *Staphylococcus aureus* in meat, and is effective against a number of other bacteria. The constituents carnosol and ursolic acid inhibit a range of food-spoiling microbes, including *S. aureus*, *Escherichia coli* and *Lactobacillus brevis* (Pharmaceutical Press Editorial Team 2013).

Antiviral activity: Rosemary ethanolic extract and carnosolic acid have been shown to inhibit herpes simplex virus type 2 and human immunodeficiency virus type 1 (HIV-1), respectively (Pharmaceutical Press Editorial Team 2013).

Antidepressant properties: Chronic treatment with 10 mg/kg hydroalcoholic extract of rosemary has been shown to produce antidepressant effects in rodents, similar to those produced by fluoxetine (Prozac). The rosemary extract, like fluoxetine, was found to reduce the activity of acetylcholinesterase (AchE), thereby increasing the levels of the neurotransmitter acetylcholine (Machado et al. 2012). Polyphenol constituents of rosemary (luteolin, carnosic acid and rosmarinic acid) have also been shown to modulate dopaminergic, serotonergic and γ-aminobutyric acid (GABA)-ergic pathways (Sasaki et al. 2013).

Antidiabetic activity: An ethnobotanical survey in Morocco reported that rosemary is commonly used in the treatment of diabetes and hypertension (Tahraoui et al. 2007). Several studies have shown that orally administered rosemary extracts lower blood glucose in both healthy and diabetic animals (Gardner and McGuffin 2013). This is in contrast to an earlier study, where intramuscular administration of rosemary oil resulted in a hyperglycaemic effect in glucose-loaded rats and rabbits with alloxan-induced diabetes (Pharmaceutical Press Editorial Team 2013). Recently, rosemary extract has been shown to significantly increase glucose consumption in hepatocytes (HepG2 cells) and increase phosphorylation of $5'$ adenosine monophosphate-activated protein kinase (AMPK) and its substrate, acetyl-CoA carboxylase (ACC). Rosemary extract appears to regulate glucose and lipid metabolism by activating transcription factors through AMPK and peroxisome proliferator-activated receptor (PPAR) pathways (Tu et al. 2013).

Antioxidant and other effects: *In vivo* antioxidant activity of rosemary has been demonstrated in rats (Botsoglou et al. 2010). Other reported activities of rosemary extracts include anti-tumour, anti-hepatotoxic, and various other effects (Pharmaceutical Press Editorial Team 2013).

Interactions: Ingestion of a phenolic-rich extract of rosemary was found to significantly reduce absorption of non-haem iron in healthy women. While no other interactions have been reported in humans, rosemary has been shown to induce the drug-metabolising isoenzymes CYP1A1, CYP1A2 and CYP2B1/2 in rats. Rosemary may have a weak potential to interact with antidiabetics, as it has been shown to reduce serum glucose levels in animals. Inhibition of platelet aggregation has also been observed in animal and *in vitro* studies, so caution should be observed in patients on antiplatelet therapy (Williamson et al. 2013).

Contraindications: Hypersensitivity to rosemary. Safety of rosemary during pregnancy is uncertain. Older reports describe the use of rosemary (possibly the oil) as an abortifacient, although animal studies have not indicated such activity for aqueous extracts. Rosemary is also reported to be an emmenagogue (affects the menstrual cycle), thus during pregnancy and lactation amounts higher than those found in normal food intake should be avoided. Epileptics should avoid high doses of rosemary due to the potential of the constituent camphor to cause convulsions (EMA 2010; Gardner and McGuffin 2013; Pharmaceutical Press Editorial Team 2013).

Adverse effects: Allergic reactions including erythema and dermatitis, confirmed by patch test, have been reported after ingestion of rosemary or topical application of products containing rosemary. Photosensitivity has been associated with rosemary oil. Carnosol was identified as the major allergen in a case of occupational contact dermatitis (Gardner and McGuffin 2013; Pharmaceutical Press Editorial Team 2013).

Dosage: For products intended for internal use, follow manufacturers' instructions. Dried herb, for oral administration (adults): 2–4 g as an infusion three times daily; 4–6 g daily; for external use: 50 g for one bath. Liquid extract, for internal use: 2–4 ml (1:1 in 45% alcohol) three times daily (EMA 2010; Pharmaceutical Press Editorial Team 2013). Essential oil: for external use only, diluted to 2.5% with inert carrier oil.

General plant information: Native to the Mediterranean region, rosemary is widely cultivated for its culinary use. It is known as the 'herb of remembrance' and is mentioned in Anglo-Saxon herbals of the 11th century.

References

Botsoglou N, Taitzoglou I, Zervos I, Botsoglou E, Tsantarliotou M, Chatzopoulou PS. (2010) Potential of long-term dietary administration of rosemary in improving the antioxidant status of rat tissues following carbon tetrachloride intoxication. *Food and Chemical Toxicology* 48(3): 944–950.

EMA. (2010) Assessment report on *Rosmarinus officinalis* L., aetheroleum and *Rosmarinus officinalis* L., folium. European Medicines Agency http://www.ema.europa.eu/docs/en_GB/document_library/Herbal_-_HMPC_assessment_report/2011/02/WC500101693.pdf (accessed January 2014).

Gardner Z, McGuffin M. (Eds.) (2013) *American Herbal Product Association's Botanical Safety Handbook*. 2nd Edition. CRC Press, USA, 1072 pp.

Haloui M, Louedec L, Michel JB, Lyoussi B. (2000) Experimental diuretic effects of *Rosmarinus officinalis* and *Centaurium erythraea*. *Journal of Ethnopharmacology* 71(3): 465–472.

Machado DG, Cunha MP, Neis VB, Balen GO, Colla AR, Grando J, Brocardo PS, Bettio LE, Dalmarco JB, Rial D, Prediger RD, Pizzolatti MG, Rodrigues AL. (2012) *Rosmarinus officinalis* L. hydroalcoholic extract, similar to fluoxetine, reverses depressive-like behavior without altering learning deficit in olfactory bulbectomized mice. *Journal of Ethnopharmacology* 143(1): 158–169.

Pengelly A, Snow J, Mills SY, Scholey A, Wesnes K, Butler LR. (2012) Short-term study on the effects of rosemary on cognitive function in an elderly population. *Journal of Medicinal Food* 15(1): 10–17.

Pharmaceutical Press Editorial Team. (2013) *Herbal Medicines*. 4th Edition. Pharmaceutical Press, London, UK.

Sasaki K, El Omri A, Kondo S, Han J, Isoda H. (2013) *Rosmarinus officinalis* polyphenols produce anti-depressant like effect through monoaminergic and cholinergic functions modulation. *Behavioural Brain Research* 238: 86–94.

Tahraoui A, El-Hilaly J, Israili ZH, Lyoussi B. (2007) Ethnopharmacological survey of plants used in the traditional treatment of hypertension and diabetes in south-eastern Morocco (Errachidia province). *Journal of Ethnopharmacology* 110(1): 105–117.

Takaki I, Bersani-Amado LE, Vendruscolo A, Sartoretto SM, Diniz SP, Bersani-Amado CA, Cuman RK. (2008) Anti-inflammatory and antinociceptive effects of *Rosmarinus officinalis* L. essential oil in experimental animal models. *Journal of Medicinal Food* 11(4): 741–746.

Tu Z, Moss-Pierce T, Ford P, Jiang TA. (2013) Rosemary (*Rosmarinus officinalis* L.) extract regulates glucose and lipid metabolism by activating AMPK and PPAR pathways in HepG2 cells. *Journal of Agricultural and Food Chemistry* 61(11): 2803–2810.

Ventura-Martínez R, Rivero-Osorno O, Gómez C, González-Trujano ME. (2011) Spasmolytic activity of *Rosmarinus officinalis* L. involves calcium channels in the guinea pig ileum. *Journal of Ethnopharmacology* 137(3): 1528–1532.

WHO. (2009) *WHO Monographs on Selected Medicinal Plants*. Vol. 4. WHO, Geneva, 456 pp.

Williamson E, Driver S, Baxter K. (Eds.) (2013) *Stockley's Herbal Medicines Interactions*. 2nd Edition. Pharmaceutical Press, London, UK.

Sage   *Salvia officinalis* L.

Family: Lamiaceae (Labiatae)

Other common names: Common sage; garden sage

Drug name: Salviae officinalis folium

Botanical drug used: Leaves

Indications/uses: Colds, inflammation in the mouth and throat, sinusitis, chesty coughs and mild indigestion (including symptoms of heartburn and bloating). Sage is also used to reduce excess sweating and hot flushes associated with menopause. Traditionally, sage was used to improve memory, in the treatment of excessive lactation and to lower blood sugar levels in diabetes. Topically, it was used as a styptic (to stem bleeding) and to relieve minor skin inflammation. As a culinary herb, sage is widely used to add flavour.

Evidence: Several studies have demonstrated the antimicrobial, antidiabetic, anti-inflammatory and memory-enhancing activities of sage and its constituents, although clinical evidence is limited. A small clinical trial found fresh sage to be effective in reducing menopausal symptoms and some clinical studies indicate that *S. officinalis* has effectiveness in the treatment of cognitive impairment (see below).

Safety: Sage is safe to ingest at amounts normally found in food; however, high doses or prolonged use of extracts is not recommended due to the presence of the toxic compound thujone. Sage medicinal products should not be used during pregnancy or lactation.

Main chemical compounds: Volatile oil (1–2.8%), which contains as the major components α- and β-thujone (usually about 50%), with camphor, cineole, borneol and others. Sage also contains diterpenoids (including picrosalvin/carnosol), triterpenoids (including oleanolic acid and derivatives), phenolic acids (including caffeic, rosmarinic and salvianolic acids) and flavonoids (such as luteolin and salvigenin).

Note: The sub species *S. officinalis* subsp. *lavandulifolia* (Vahl) Gams from the western Mediterranean (earlier classified as a separate species *S. lavandulifolia* Vahl), is free of thujone, and, therefore, does not comply with the requirements of the European Pharmacopeia. It has, however, been tested in a range of clinical studies.

Phytopharmacy: An evidence-based guide to herbal medicinal products, First Edition.
Sarah E. Edwards, Inês da Costa Rocha, Elizabeth M. Williamson and Michael Heinrich.
© 2015 John Wiley & Sons, Ltd. Published 2015 by John Wiley & Sons, Ltd.

Clinical evidence:

The use of sage for colds, inflammation in the mouth and throat are based on empirical practice and not on clinical research.

Menopausal symptoms: Sage was found to be effective in reducing menopausal symptoms including hot flushes and excessive sweating in a small open multi-centred clinical trial (Bommer et al. 2011).

Memory enhancement: A systematic review of randomised controlled clinical studies demonstrated that *S. officinalis* had therapeutic effectiveness in the treatment of cognitive impairment in Alzheimer's disease, although further research is needed to test how it compares with existing pharmaceutical drugs and its impact in the control of cognitive deterioration (dos Santos-Neto et al. 2006).

Pre-clinical evidence and mechanisms of action:

Menopausal symptoms: An *in vitro* study suggests that oestrogenic flavonoids, especially luteolin, are responsible for the anti-flush effect of *S. officinalis* (Rahte et al. 2013).

Memory enhancement: Sage inhibits the enzyme acetylcholinesterase, which may explain its reported effects in enhancing memory. Since acetylcholine levels are diminished in Alzheimer's disease, this may also explain the mild, observed effects in treating this disease (Frances et al. 1999; Howes and Houghton 2012).

Antimicrobial effects: The essential oil of sage has antimicrobial (against bacteria and fungi) and antioxidant properties (Bozin et al. 2007). *In vitro* and *ex-vivo* studies have shown that aqueous extracts of sage leaves are also an effective antiviral at non-cytotoxic levels, including against herpes-simplex virus and HIV-1 (Geuenich et al. 2008; Nolkemper et al. 2006). Flavonoids and phenolic acids such as rosmarinic acid found in sage may be responsible for some of the biological activities, as these compounds have known antiviral, antibacterial, anti-inflammatory and antioxidant properties (Peterson and Simmonds 2003).

Antidiabetic effects: Studies in rats have demonstrated that ethanolic extracts of sage leaves show dose-dependent antidiabetic activity (Eidi and Eidi 2009).

Interactions: No drug interactions reported and although sage extracts have been shown *in vitro* to induce some CYP enzymes, it is considered to have a low potential for interaction (Williamson et al. 2013).

Contraindications: Sage should not be used during pregnancy and lactation, as traditionally, it is purported to have been used as an abortifacient and to affect the menstrual cycle (Pharmaceutical Press Editorial Team 2013).

Adverse effects: No adverse effects have been recorded. Sage *oil*, however, has high levels of thujone, a known pro-convulsant. A case of poisoning has been reported for an individual who took sage oil internally for acne (Pharmaceutical Press Editorial Team 2013).

Dosage: For herbal products, follow the manufacturers' instructions. Leaf: 1–4 g as an infusion 3 times per day; 4–6 g daily. Liquid extract: 1–4 ml (1:1 in 45% alcohol) 3 times daily. Gargles and rinses: 2.5 g/100 ml water (Pharmaceutical Press Editorial Team 2013).

General plant information: The original meaning of *Salvia* in Latin is 'the healing plant'. Sage has a long history of use in traditional European herbal medicine, and in the 17th century appears in the recipe for 'plague water', a remedy used against the plague, recorded by the diarist of the time, Samuel Pepys (Digby 1669).

References

Bommer S, Klein P, Suter A. (2011) First time proof of sage's tolerability and efficacy in menopausal women with hot flushes. *Advances in Therapy* 28(6): 490–500.

Bozin B, Mimica-Dukic M, Samojilik I, Jovin E. (2007) Antimicrobial and antioxidant properties of rosemary and sage (*Rosmarinus officinalis* L. and *Salvia officinalis* L., Lamiaceae) essential oils. *Journal of Agricultural and Food Chemistry* 55(19): 7879–7885.

Digby, K. (1669) *The Closet of Sir Kenelm Digby Knight, Opened.* H. Broome, London. http://www.foodsofengland.co.uk/plaguewater.htm (accessed 31 May 2013).

Eidi A, Eidi M. (2009) Antidiabetic effects of sage (*Salvia officinalis* L.) leaves in normal and streptozotocin-induced diabetic rats. *Diabetes & Metabolic Syndrome: Clinical Research and Reviews* 3: 40–44.

Frances P, Palmer A, Snape M, Wilcock GK. (1999) The cholinergic hypothesis of Alzheimer's disease: a review of progress. *Journal of Neurology, Neurosurgery and Psychiatry* 66(2): 137–147.

Geuenich S, Goffinet C, Venzke S, Nolkemper S, Baumann I, Plinkert P, Reichling J, Keppler OT. (2008) Aqueous extracts from peppermint, sage and lemon balm leaves display potent HIV-1 activity by increasing the virion density. *Retrovirology* 5: 27.

Howes MJR, Houghton PJ. (2012) Ethnobotanical treatment strategies against Alzheimer's disease. *Current Alzheimer Research* 9(1): 67–85.

Nolkemper S, Reichling J, Stintzing FC, Carle R, Schnitzler P. (2006) Antiviral effect of aqueous extracts from species of the Lamiaceae family against Herpes simplex virus type 1 and type 2 in vitro. *Planta Medica* 72(15): 1378–1382.

Peterson M, Simmonds MS. (2003) Rosmarinic acid. *Phytochemistry* 62(2): 121–125.

Pharmaceutical Press Editorial Team. (2013) *Herbal Medicines.* 4th Edition. Pharmaceutical Press, London, UK.

Rahte S, Evans R, Eugster PJ, Marcourt L, Wolfender JL, Kortenkamp A, Tasdemir D. (2013) *Salvia officinalis* for hot flushes: towards determination of mechanism of activity and active principles. *Planta Medica* 79(9): 753–760.

dos Santos-Neto LL, de Velena Toledo MA, Medeiros-Souza P. et al. (2006) The use of herbal medicine in Alzheimer's disease – a systematic review. *Evidence Based Complementary and Alternative Medicine* 3(4): 441–445.

Williamson EM, Driver S, Baxter K. (Eds.). (2013) *Stockley's Herbal Medicines Interactions.* 2nd Edition. Pharmaceutical Press, London, UK.

St. John's Wort　　　　　*Hypericum perforatum* L.

Family: Hypericaceae (Clusiaceae)

Other common names: Perforate St. John's wort

Drug names: Hyperici herba; Hyperici oleum

Botanical drug used: Aerial parts, including dried flowering tops; oil extract of herb

---

Indications/uses: St. John's wort extracts are used internally to treat low mood, depression, seasonal affective disorder, anxiety and insomnia (particularly if associated with menopause). Within the EU, products are licensed for treating mild-to-moderate forms of depression. Oil-based products are applied externally to treat minor wounds, bruises, burns and swellings.

Evidence: Considerable clinical data, including meta-analyses, have shown that St. John's wort is superior to placebo, and as effective as synthetic antidepressants in treating certain types of depression. There is some evidence for the external use.

Safety: St. John's wort is relatively safe, with the risks associated with its use considerably less than those of conventional antidepressants, although there is high potential for interactions with other medicines. Its use is not recommended in people who are taking immunosuppressants or cardiovascular drugs. With other medications, it is recommended that patients should only take preparations with a low hyperforin content, and even then under careful monitoring. Women who use St. John's wort preparations and are taking the contraceptive pill should use additional measures to avoid unwanted pregnancy.

Main chemical compounds: The main active components are the hypericins (naphthodianthrones), which include hypericin, isohypericin, pseudohypericin and protohypericin; and the hyperforins (prenylated phloroglucinols) including hyperforin, adhyperforin and their derivatives. Flavonoids including kaempferol, quercetin glycosides, hyperoside, biapigenin and amentoflavone, other polyphenolic constituents including caffeic and chlorogenic acids, and a volatile oil are also present (Butterweck 2009; Russo et al. 2014).

Much of the research, and most of the clinical studies, have been conducted using special extracts such as WS® 5570®, Li160® and Ze 117. These products are standardised to hypericins (approximately 0.3%) and sometimes to hyperforin and flavonoids (variable amounts). For details on compositions and individual trials, see Kasper et al. (2012), and for the importance of each type of constituent, see Butterweck (2009).

Preparations with low hyperforin content (see drug interaction section) are also available (e.g. Ze 117).

Clinical evidence:

Antidepressant activity: Many clinical studies have been carried out on St. John's wort (SJW), not all of good quality. The most recent Cochrane review evaluated 29 randomised double-blind trials (5489 patients with mild-to-moderately severe depression) comparing extracts of the herb with placebo or conventional antidepressants, administered between 4 to 12 weeks. Overall, SJW extracts were found to be superior to placebo, and with similar effectiveness to standard antidepressants, but with fewer side effects. SJW is often prescribed by physicians in German-speaking countries, where the study results were more favourable than from elsewhere (Linde et al. 2008). More recent reviews of randomised controlled clinical trials and other evidence (e.g. Gastpar 2013; Kasper et al. 2012; Russo et al. 2014) confirm the usefulness of SJW as an alternative to conventional antidepressants and conclude that it has a favourable safety profile, with adverse event rates similar to those of placebo and lower than that of synthetic antidepressants, but that potential drug interactions are a concern. Preparations containing low amounts of hyperforins (see drug interaction section) are available (e.g. Ze 117) and have been shown to retain antidepressant activity (Nahrstedt and Butterweck 2012), although the evidence for their efficacy is not as strong.

Skin healing effects: A hyperforin-rich cream was found to strengthen skin barrier function by reducing radical formation (78% compared to 45% placebo) and stabilising stratum corneum lipids during visible/near-infrared (VIS/NIR) irradiation over 4 weeks in human volunteers (Haag et al. 2014).

Hyperici oleum (HO), the oil extract of SJW, was tested for anti-inflammatory activity using a sodium lauryl sulphate test. Three oil-in-water creams containing 15% (w/v) of HO were formulated, and all demonstrated significant anti-inflammatory effects in an *in vivo* study. All possessed some antimicrobial activity, with the olive oil extract being the most potent (the others were palm and sunflower oils) (Arsić et al. 2011).

SJW extract appears to be well tolerated when applied externally: a bath oil containing a lipophilic extract was compared to sodium lauryl sulphate (SLS, 1%), and whereas the SLS caused a significant increase in skin erythema and transepidermal water loss when applied to the forearms of 18 volunteers, the SJW bath oil did not (Reuter et al. 2008).

Pre-clinical evidence and mechanisms of action:

Antidepressant activity: Very many *in vitro* and *in vivo* studies have investigated the mechanisms involved in the effects of SJW. It acts on dopaminergic, serotonergic and noradrenergic systems and the effects are complex and not fully known. The activity is also mediated by several of the different types of constituent, and particularly linked to the hypericin and hyperforin content. SJW extracts free from hyperforin and hypericin also exert antidepressant activity in animal behavioural models, supporting the hypothesis that flavonoids are part of the constituents responsible for the therapeutic efficacy. Hyperforin is known to contribute to the properties of SJW, confirming the hypothesis that the crude SJW extract contains several constituents with antidepressant activity (Butterweck et al. 2003). For further details on the complex pharmacology of SJW, see Butterweck (2009), Butterweck and Narstedt (2010), Russo et al. (2014).

Skin healing effects: Extracts from SJW have been shown to stimulate fibroblast migration and collagen synthesis *in vitro* (Dikmen et al. 2011) and aid excision wound healing *in vivo* (Süntar et al. 2010). They possess antibacterial and antiviral effects, including against mycobacteria, with hypericin and hyperforin both exhibiting activity (Avato et al. 2004; Mortensen et al. 2012; Saddiqe et al. 2010). Hypericin has also been reported to inhibit protein kinase C, to inhibit NF-kappaB activation (Bork et al. 1999) and inhibition of the release of arachidonic acid and leukotriene B4, which may account for some of the anti-inflammatory effects (WHO 2004).

Interactions: SJW has a high profile of interaction with many conventional drugs, as it affects the pharmacokinetics of many drugs by inducing cytochrome P450 (CYP) isozymes, such as CYP3A4, CYP2C19, CYP2C9, and the P-glycoprotein (P-gp) transporter. It inhibits the re-uptake of 5-hydroxytryptamine (5-HT, serotonin). There are conflicting reports of interaction with theophylline. The main components responsible for the interactions are the hyperforins (Nahrstedt and Butterweck 2010). *St. John's wort should therefore not be taken with other antidepressants, anticoagulants, oral contraceptives, immune suppressants, digoxin, opiates, antineoplastics, protease inhibitors, fexofenadine, statins, omeprazole and verapamil* (Russo et al. 2014; Thomas-Schoemann et al. 2014; Williamson et al. 2013). For example, the concomitant use of docetaxel and a specific SJW product in 10 cancer patients resulted in a decrease in the mean area under the docetaxel plasma concentration–time curve, but the maximum plasma concentration and elimination half-life of docetaxel were only non-significantly decreased (Goey et al. 2014). Patients should consult with a pharmacist or other healthcare professional before taking SJW with any other medicines. Products with low hyperforin content are available and have the potential to reduce the interactions, but the clinical efficacy (and indeed, any reduced interaction profile) is not well documented.

Contraindications: Internal use of SJW during pregnancy and breastfeeding should be avoided, as there is weak evidence from animal studies that it may cause uterine contractions and no safety data is available. Children should only use SJW under medical supervision due to lack of safety data (WHO 2004). *It is most important to ask the patient about concomitant medication, including the contraceptive pill* (see preceding text).

Adverse effects: Side effects of SJW are rare and mild and include minor gastrointestinal irritations, allergic reactions, tiredness and restlessness. Photosentisation in light sensitive individuals may occur. A single case of acute neuropathy after exposure to sunlight was reported in a patient taking the herb, and another case of reversible erythema after exposure to ultraviolet B was reported in a patient who had been taking the herb for 3 years (WHO 2004).

Dosage: Based on the existing evidence, preference should always be given to standardised products with clearly defined levels of hyperforins and hypericins up to a daily dose of 900 mg (equivalent to 0.2–2.7 mg total hypericin) (WHO 2004) and following manufacturers' instructions. If the herb is used in teas, a daily dosage of 2–4 g is suggested.

General plant information: SJW flowers around midsummer, coinciding with St. John's Day (24th June). Its leaves are well known to have characteristic oil glands, which appear as translucent dots when held up to the light.

References

St John's wort is very well investigated and has been reviewed many times. This is a selection of references and recent reviews to support the information included in this monograph.

Arsić I, Zugić A, Tadić V, Tasić-Kostov M, Mišić D, Primorac M, Runjaić-Antić D. (2011) Estimation of dermatological application of creams with St. John's Wort oil extracts. *Molecules* 17(1): 275–294.

Avato P, Raffo F, Guglielmi G, Vitali C, Rosato A. (2004) Extracts from St John's Wort and their antimicrobial activity. *Phytotherapy Research* 18(3): 230–232.

Bork P, Bacher S, Schmitz ML, Kaspers U, Heinrich M. (1999) Hypericin as an non-antioxidant inhibitor of NF-kappa B. *Planta Medica* 65(4): 297–300.

Butterweck. V. (2009) St. John's Wort: quality issues and active compounds. In: Cooper R, Kronenberg F. (Eds.) *Botanical Medicine: from Bench to Bedside.* Mary Ann Liebert Inc, USA. http://www.liebertpub.com/dcontent/files/samplechapters/Sample _BotanicalMedicineFromBenchtoBeds.pdf (accessed August 2013).

Butterweck V, Christoffel V, Nahrstedt A, Petereit F, Spengler B, Winterhoff H. (2003) Step by step removal of hyperforin and hypericin: activity profile of different Hypericum preparations in behavioral models. *Life Sciences* 73(5): 627–639.

Dikmen M, Oztürk Y, Sagratini G, Ricciutelli M, Vittori S, Maggi F. (2011) Evaluation of the wound healing potentials of two subspecies of *Hypericum perforatum* on cultured NIH3T3 fibroblasts. *Phytotherapy Research* 25(2): 208–214.

Gastpar M. (2013) Hypericum extract WS ® 5570 for depression – an overview. *International Journal of Psychiatry in Clinical Practice* 17(S1): 1–7.

Goey AKL, Meijerman I, Rosing H, Marchetti S, Mergui-Roelvink M, Keessen M, Burgers JA, Beijnen JH, Schellens JHM. (2014) The effect of St John's Wort on the pharmacokinetics of docetaxel. *Clinical Pharmacokinetics* 53(1): 103–110.

Haag SF, Tscherch K, Arndt S, Kleemann A, Gersonde I, Lademann J, Rohn S, Meinke MC. (2014) Enhancement of skin radical scavenging activity and stratum corneum lipids after the application of a hyperforin-rich cream. *European Journal of Pharmacy and Biopharmacy* 86(2): 227–233.

Kasper S, Caraci F, Forti B, Drago F, Aguglia E. (2012) Efficacy and tolerability of Hypericum extract for the treatment of mild to moderate depression. *European Neuropsychopharmacology* 20(11): 747–765.

Linde K, Berner MM, Kriston L. (2008) St. John's wort for major depression. *Cochrane Database of Systematic Reviews* 4: CD000448.

Mortensen T, Shen S, Shen F, Walsh MK, Sims RC, Miller CD. (2012) Investigating the effectiveness of St John's wort herb as an antimicrobial agent against mycobacteria. *Phytotherapy Research* 26(9): 1327–1333.

Nahrstedt A, Butterweck V. (2010) Lessons learned from herbal medicinal products: the example of St. John's Wort (perpendicular). *Journal of Natural Products* 73(5): 1015–1021.

Reuter J, Huyke C, Scheuvens H, Ploch M, Neumann K, Jakob T, Schempp CM. (2008). Skin tolerance of a new bath oil containing St. John's wort extract. *Skin Pharmacology and Physiology* 21(6): 306–11.

Russo E, Scicchitano F, Whalley BJ, Mazzitello C, Ciriaco M, Esposito S, Patanè M, Upton R, Pugliese M, Chimirri S, Mammì M, Palleria C, De Sarro G. (2014) *Hypericum perforatum*: pharmacokinetic, mechanism of action, tolerability, and clinical drug–drug interactions. *Phytotherapy Research* 28(5): 643–655.

Saddiqe Z, Naeem I, Maimoona A. (2010) A review of the antibacterial activity of *Hypericum perforatum* L. *Journal of Ethnopharmacology* 131(3): 511–521.

Süntar IP, Akkol EK, Yilmazer D, Baykal T, Kirmizibekmez H, Alper M, Yeşilada E. (2010) Investigations on the *in vivo* wound healing potential of *Hypericum perforatum* L. *Journal of Ethnopharmacology* 127(2): 468–477.

Thomas-Schoemann A, Blanchet B, Bardin C, Noé G, Boudou-Rouquette P, Vidal M, Goldwasser F. (2014) Drug interactions with solid tumour-targeted therapies (2014) *Critical Reviews in Oncology/Hematology* 89(1): 179–196.

WHO. (2004) *WHO Monographs on Selected Medicinal Plants.* Vol. 2. WHO, Geneva, 358 pp.

Williamson EM, Driver S, Baxter K. (Eds.) (2013) *Stockley's Herbal Medicines Interactions.* 2nd Edition. Pharmaceutical Press, London, UK.

Saw Palmetto *Serenoa repens* (W. Bartram) Small

Synonym: *Sabal serrulata* (Michx.) Schult.f.

Family: Arecaceae (Palmae)

Other common names: American dwarf palm; sabal

Drug name: Sabalis serrulatae fructus

Botanical drug used: Dried ripe fruit

Indications/uses: Saw palmetto extracts are used to relieve lower urinary tract symptoms such as dysuria, polyuria and urine retention, in men with a confirmed diagnosis of benign prostate hyperplasia (BPH). Prior to using saw palmetto, other serious urogenital conditions such as prostate cancer must have been ruled out. It has been advocated as a treatment for hair loss but there is no good evidence to support this.

Evidence: The clinical evidence is based on a large number of clinical studies, but remains contradictory for treating the symptoms of BPH.

Safety: There is only limited data available, but the constituents do not suggest any potential risks associated with the use of saw palmetto. The reported side effects are minor and mostly concerning the digestive system.

Main chemical compounds: Free fatty acids such as capric, caproic, caprylic, lauric, myristic, oleic, linoleic, linolenic, stearic and palmitic acids, and phytosterols such as β-sitosterol, stigmasterol, campesterol and β-sitosterol 3-O-D-glucoside. Flavonoids including rutin, kaempferol and isoquercetin, and carotenoids are also present (Booker et al. 2014; ESCOP 2003; Williamson et al. 2013).

Clinical evidence:

Benign prostrate hyperplasia: More than 90 clinical trials on the use of saw palmetto in BPH have been published but the overall evidence is conflicting. The most recent Cochrane Review concluded that 'at double and triple doses, [saw palmetto] did not improve urinary flow measures or prostate size in men with lower urinary tract symptoms consistent with BPH' (Tacklind et al. 2012). Earlier reviews by the Cochrane Collaboration (2000 and 2002) provided a more positive overall assessment but the shift was mainly due to a large placebo controlled clinical trial, carried

---

Phytopharmacy: An evidence-based guide to herbal medicinal products, First Edition.
Sarah E. Edwards, Inês da Costa Rocha, Elizabeth M. Williamson and Michael Heinrich.
© 2015 John Wiley & Sons, Ltd. Published 2015 by John Wiley & Sons, Ltd.

out in the United States, which had a negative outcome (Barry et al. 2011). In the same study, saw palmetto extract did not affect serum prostate specific antigen more than placebo, even at relatively high doses (Andriole et al. 2013).

Sexual dysfunction associated with BPH: An open multicentre clinical pilot trial has been conducted to investigate whether the saw palmetto preparation Prostasan® influenced BPH symptoms and sexual dysfunction (SDys). Eighty-two patients took part, each taking one capsule of 320 mg saw palmetto extract daily for 8-weeks. At the end of the treatment, the International Prostate Symptom Score was reduced from 14.4 ± 4.7 to 6.9 ± 5.2 ($p < 0.0001$); SDys measured with the brief Sexual Function Inventory improved from 22.4 ± 7.2 to 31.4 ± 9.2 ($p < 0.0001$), and the Urolife BPH QoL-9 sex total improved from 137.3 ± 47.9 to 195.0 ± 56.3 ($p < 0.0001$). Investigators' and patients' assessments confirmed the efficacy and treatment was well tolerated. Correlation analyses confirmed the relationship between improved BPH symptoms and reduced SDys (Suter et al. 2013).

Pre-clinical evidence and mechanisms of action: Anti-androgenic, anti-inflammatory and muscle relaxing and vasodilatory effects have been observed using lipophilic extracts of saw palmetto. In human foreskin fibroblasts, an extract inhibited the binding of [^{3}H]- dihydrotestosterone to both cytosolic and nuclear androgen receptors, and reduced conversion of testosterone into dihydrotestosterone by up to 50%. It also inhibited 5-alpha-reductase activity in rat ventral prostate by up to 90% (Sultan et al. 1984).

More recently, a saw palmetto extract enhanced erectile responses in rat and rabbit corpus cavernosum by inhibition of phosphodiesterase-5 activity and increasing inducible nitric oxide synthase (iNOS; Yang et al. 2013). It also inhibited contractions of the rat prostate gland, consistent with smooth muscle relaxant activity (Chua et al. 2014).

Interactions: None reported, and few expected, as saw palmetto does not affect most CYP enzymes. A single case report suggested that it may enhance anticoagulant therapy, but this has not been confirmed and the product cited contained other herbal ingredients (Agbabiaka et al. 2009; Williamson et al. 2013).

Adverse effects: The majority of adverse events reported are mild, infrequent and reversible, and include abdominal pain, diarrhoea, nausea and fatigue, headache, decreased libido and rhinitis (Agbabiaka et al. 2009).

Dosage: 160–320 mg of lipophilic extract. There is a lack of consistency between products in the amount of fatty acids present, and consequently in the daily (recommended) dose (Booker et al. 2014). Products containing the extract together with vitamins or other herbal extracts are also available.

General plant information: The plant is endemic to Florida (USA) and adjacent states, most notably Alabama and Georgia. It was used by Native Americans as a food and for its diuretic, sedative and aphrodisiac properties.

References

Agbabiaka TB, Pittler MH, Wider B, Ernst E. (2009) *Serenoa repens* (saw palmetto): a systematic review of adverse events. *Drug Safety* 32(8): 637–647.

Andriole GL, McCullum-Hill C, Sandhu GS, Crawford ED, Barry MJ, Cantor A; CAMUS Study Group. (2013) The effect of increasing doses of saw palmetto fruit extract on serum prostate specific antigen: analysis of the CAMUS randomized trial. *Journal of Urology* 189(2): 486–492.

Barry MJ, Meleth S, Lee JY, Kreder KJ, Avins AL, Nickel JC, Roehrborn CG, Crawford ED, Foster HE Jr, Kaplan SA, McCullough A, Andriole GL, Naslund MJ, Williams OD, Kusek JW, Meyers CM, Betz JM, Cantor A, McVary KT; CAMUS Study Group. (2011) Effect of increasing doses of saw palmetto extract on lower urinary tract symptoms: a randomized trial. *Journal of the American Medical Association (JAMA)* 306(12): 1344–1351.

Booker A, Suter A, Krnjic A, Strassel B, Zloh M, Said M, Heinrich M. (2014) A phytochemical comparison of saw palmetto products using gas chromatography and (1) H nuclear magnetic resonance spectroscopy metabolomic profiling. *Journal of Pharmacy and Pharmacology* 66(6): 811–822.

Chua T, Eise NT, Simpson JS, Ventura S. (2014) Pharmacological characterization and chemical fractionation of a liposterolic extract of saw palmetto (*Serenoa repens*): Effects on rat prostate contractility. *Journal of Ethnopharmacology* 152(2): 283–291.

ESCOP. (2003) *ESCOP Monographs: The Scientific Foundation for Herbal Medicinal Products*. 2nd Edition. Thieme, Exeter and London, UK.

Sultan C, Terraza A, Devillier C, Carilla E, Briley M, Loire C, et al. (1984) Inhibition of androgen metabolism and binding by a liposterolic extract of "*Serenoa repens* B" in human foreskin fibroblasts. *Journal of Steroid Biochemistry* 20(1): 515–519.

Suter A., Saller R, Riedi E, Heinrich M. (2013) Improving BPH symptoms and sexual dysfunctions with a saw palmetto preparation? Results from a pilot trial. *Phytotherapy Research* 27(2): 218–226.

Tacklind J, MacDonald R, Rutks I, Stanke JU, Wilt TJ. (2012) *Serenoa repens* for benign prostatic hyperplasia. *Cochrane Database of Systematic Reviews* 12: CD001423.

Williamson EM, Driver S, Baxter K. (Eds.) (2013) *Stockley's Herbal Medicines Interactions*. 2nd Edition. Pharmaceutical Press, London, UK.

Yang S, Chen C, Li Y, Ren Z, Zhang Y, Wu G, Wang H, Hu Z, Yao M. (2013) Saw palmetto extract enhances erectile responses by inhibition of phosphodiesterase 5 activity and increase inducible nitric oxide synthase messenger ribonucleic acid expression in rat and rabbit corpus cavernosum. *Urology* 81(6): 1380.e7–13.

Schisandra   *Schisandra chinensis* (Turcz.) Baill., *S. sphenanthera* Rehder & E.H.Wilson

Synonyms: *Sphaerostema japonicum* A.Gray (=*S. chinensis*); *Schisandra chinensis* var. *rubriflora* Franch. (=*S. sphenanthera*)

Family: Schisandraceae

Other common names: Five flavour fruit; magnolia vine; schizandra; Wuweizi. *S. chinensis*: Chinese magnolia vine; *S. sphenanthera:* Southern magnolia vine. The fruits of both *S. chinensis* and *S. sphenanthera* are known as Wuweizi, but since 2000 have been accepted by the Chinese Pharmacopoeia as two different herbal drugs, Bei-Wuweizi and Nan-Wuweizi, respectively.

Drug name: Schisandrae fructus

Botanical drug used: Dried ripe fruits

Indications/uses: Schisandra is considered an 'adaptogen' (i.e. it increases resistance to stress), and is used as a general tonic and to improve fatigue, for chronic coughs and asthma, gastritis, diabetes, urinary tract infections and psychosis. In China, extracts of *S. sphenathera* are also used as a liver protectant and to treat hepatitis, including after liver transplantation.

Evidence: Clinical data from controlled trials is lacking, although there have been some studies in humans that support the use of Schisandra for treating psychosis, gastritis, hepatitis and fatigue. Animal and *in vitro* studies also suggest that Schisandra improves liver function, stimulates liver cell growth and protects the liver from toxic damage, and may also have anticancer properties (see below).

Safety: Toxicity studies in rodents and pigs indicate that Schisandra is relatively safe for internal use. Paediatric use and use during pregnancy and lactation should be avoided due to lack of safety data. Schisandra interacts with certain drugs in a positive manner, increasing blood levels of, for example, ciclosporin and tacrolimus, drugs which are known to have variable or poor oral bioavailability. However, this type of therapy should only be managed by medical specialists.

Phytopharmacy: An evidence-based guide to herbal medicinal products, First Edition.
Sarah E. Edwards, Inês da Costa Rocha, Elizabeth M. Williamson and Michael Heinrich.
© 2015 John Wiley & Sons, Ltd. Published 2015 by John Wiley & Sons, Ltd.

Main chemical compounds: The major bioactive compounds in both species are lignans, the schisandrins, schisandrols and schisantherins (also known as gomisins and wuweizisus). The major components of *S. chinensis* are schisandrin, gomisin A, schisantherin B, γ-schisandrin and deoxyschisandrin, whereas the major constituents of *S. sphenanthera* are schisandrin A, schisandrin B, schisandrin C, schisandrol A, schisandrol B, schisantherin A and schisantherin B (Qin et al. 2014; Williamson et al. 2013). There are many others. Nortriterpenoids such as pre-schisanartanin and schindilactones have also been isolated from *S. chinensis* (Chan 2012; WHO 2007).

Clinical evidence:

Tonic and anti-depressive effects: Several Russian clinical studies have reported positive therapeutic effects of Schisandra preparations on psychogenic depression, or that associated with excessive fatigue and nervous exhaustion (Panossian and Wikman 2008).

Anti-ulcer effects: In an uncontrolled study, a Schisandra fruit tincture was used to treat patients with acute and chronic stomach and duodenal ulcers, with all subjects reporting a reduction in symptoms after just a few days, and ulcer healing in over 95% of subjects after 35 days of treatment (WHO 2007).

Enhancement of immunosuppressive therapy in liver transplant patients: Studies on *S. sphenanthera* extracts have shown that it can enhance the effects of tacrolimus, including in liver transplant patients (e.g. Jiang et al. 2010) by improving bioavailability.

Treatment of hepatitis: Clinical reports from China of more than 5,000 cases of hepatitis treated with Schisandra fruit preparations suggest beneficial effects measured by liver function tests; however, these reports are based on observational rather than placebo-controlled studies (WHO 2007).

Pre-clinical evidence and mechanisms of action:

Anti-stress and tonic effects: Anxiolytic properties have been shown in studies in mice (Chen et al. 2011). Reports indicate that schizandrin is the main active ingredient responsible for the effects on the CNS. In some *in vitro* studies, active ingredients including gomisin A,C, D and G and schizandrol B have been shown to inhibit acetylcholinesterase (which affects neurotransmission) in a dose-dependent manner, suggesting that the Schisandra lignans may improve cognitive impairment (Panossian and Wikman 2008).

Improvement in liver function: A number of studies in rodents have demonstrated hepatoprotective effects and improvement in liver function. These may be due to the lignans, which increase the antioxidative capability of liver cells by preventing over-production and accumulation of free radicals (Pu et al. 2012).

Anticancer properties: Potential anticancer effects of Schisandra have been highlighted by studies showing that *S. chinensis* lignans have anti-proliferative activity in human colorectal carcinoma and exhibit cytotoxicity against a colon cancer cell line (Chan 2012).

Other effects: Schisandra administration causes a reduction in serum glutamate-pyruvate transaminase levels, inhibition of platelet aggregation, calcium-antagonism, antioxidant, anti-tumour and antiviral effects, which have been attributed to the lignans (Chan 2012).

Interactions: The effects of Schisandra extract are variable and differ from the effects of some isolated constituents. For example, the extract greatly increases blood levels of tacrolimus and ciclosporin, immunosuppressant drugs used in organ transplant patients and reduces the side effects of tacrolimus (Jiang et al. 2010) whereas bifendate, a derivative of schisandrin C, can cause a *reduction* in ciclosporin blood levels. Schisandra modestly increases the bioavailability of the beta-blocker talinolol and the anticonvulsant midazolam, whereas bifendate has the opposite effect. Schisandra can significantly increase the blood concentration of paclitaxel in rats by inhibiting CYP3A-mediated metabolism and the P-gp-mediated efflux (Xue et al. 2013). In the case of warfarin, only experimental data are available but the effect was to increase clearance and reduce the half-life. Although no clinical interactions have been reported, co-administration with other prescription drugs that are metabolised by cytochrome P450 such as warfarin, protease inhibitors, oestrogen and progesterone combinations, and the herbal medicine St. John's wort, should only be carried out under expert supervision (WHO 2007; Williamson et al. 2013).

Contraindications: Due to lack of safety data, Schisandra should be avoided during pregnancy and lactation. For the same reason, it is not recommended for use in children.

Adverse effects: Minor side effects include heartburn, upset stomach, decreased appetite and stomach pain. Allergic skin reactions and urticaria have also been reported (WHO 2007).

Dosage: Average daily dose of dried fruit is 1.5–6 g per day (WHO 2007).

General plant information: *S. chinensis* is a climbing plant with a wide distribution including Russia, China, Korea and Japan. Its TCM name translated into English is 'five taste fruit', referring to the TCM theory of five tastes: sour, bitter, sweet, spicy and salty. It is believed that the fruit possesses all five essences of the elemental energies perceived in TCM: wood, fire, metal, earth and water. In China, Schisandra is considered to be a youth-preserving herb and beauty tonic (Chan 2012). Schisandra is recognised as an adaptogen in the official medicine of Russia where many pharmacological and clinical trials have been carried out since the 1960s (Panossian and Wikman 2008).

References

Chan SW. (2012) *Panax ginseng, Rhodiola rosa* and *Schisandra chinensis. International Journal of Food Sciences and Nutrition* 63(S1): 75–81.

Chen WW, He RR, Li YF, Li SB, Tsoi B, Kurihara H. (2011) Pharmacological studies on the anxiolytic effect of standardized Schisandra lignans extract on restraint-stressed mice. *Phytomedicine* 18(13): 1144–1147.

Jiang W, Wang X, Xu X, Kong L. (2010) Effect of *Schisandra sphenanthera* extract on the concentration of tacrolimus in the blood of liver transplant patients. *International Journal of Clinical Pharmacology and Therapeutics* 48(3): 224–229.

Panossian A, Wikman G. (2008) Pharmacology of *Schisandra chinensis* Bail.: an overview of Russian research. *Journal of Ethnopharmacology* 118(2): 183–212.

Pu HJ, Cao YF, He RR Zhao ZL, Song JH, Jiang B, Huang T, Tang SH, Lu JM, Kurihara H. (2012) Correlation between antistress and hepatoprotective effects of Schisandra lignans was related to its antioxidative actions in liver cells. *Evidence-Based Alternative and Complementary Medicine* 2012: 161062.

Qin XL, Chen X, Zhong GP, Fan XM, Wang Y, Xue XP, Wang Y, Huang M, Bi HC. (2014) Effect of Tacrolimus on the pharmacokinetics of bioactive lignans of Wuzhi tablet

Sea Buckthorn

Elaeagnus rhamnoides (L.) A.Nelson

Synonyms: *Hippophae littoralis* Salisb.; *H. rhamnoides* L.; and others

Family: Elaeagnaceae

Other common names: sallowthorn; sandthorn; seaberry; shaji

Drug name: Elaeagni fructus/Hippophae fructus

Botanical drug used: Ripe berries (pulp/peel) and/or the seeds

---

Indications/uses: Preparations of the fruit are commonly used as a source of vitamins and antioxidants. Therapeutic claims are mostly for chronic inflammatory conditions, including arthritis, gout and gastrointestinal ulcers. The seed oil is used for respiratory conditions, especially as an expectorant, but also for more serious conditions and is applied topically for inflammatory skin conditions including sunburn and minor wounds. It is an ingredient of many cosmetic products.

Evidence: In most cases, the evidence is anecdotal at best. Some clinical studies did not show superiority of the oil extract to placebo in atopic dermatitis. Berry preparations have not shown significant effects on biomarkers for cardiovascular health or on the risk or duration of the common cold.

Safety: Based on the long tradition of use as a food and knowledge of the phytochemistry, products complying with quality control guidelines appear to be very safe, but data are lacking both in terms of side effects and possible contraindications.

Main chemical compounds: The berries and seeds contain tocopherols, carotenoids, fatty acids and flavonoids (Bal et al. 2011). A range of acylated flavonoids, known as hippophins, are present in the seeds (Chen et al. 2013). The vitamin C content varies with origin and the variety used. The berries are nutritious, but astringent and acidic, so for food use, they may be frozen (bletted) and/or mixed with sweeter juices.

The seed oil contains mainly linoleic acid and α-linoleic acid (about 70% of the total fatty oil), while in the fruit oil (pulp and peel) the monounsaturated, omega-7,

---

Phytopharmacy: An evidence-based guide to herbal medicinal products, First Edition.
Sarah E. Edwards, Inês da Costa Rocha, Elizabeth M. Williamson and Michael Heinrich.
© 2015 John Wiley & Sons, Ltd. Published 2015 by John Wiley & Sons, Ltd.

palmitoleic acid predominates (approximately a third of the total fatty oil content). Both fruit and seeds contain small amounts of phytosterols (1–3%) and tocopherols/tocotrienols (vitamin E; 0.1–0.3%) (Bal et al. 2011; Christaki 2012; Pharmaceutical Press Editorial Team 2013).

Clinical evidence: Overall, clinical studies have shown limited or no significant effects for the prevention of cardiovascular diseases, or for immunomodulatory or hepatic applications, but some clinical data show encouraging results. Capsules of pulp oil were investigated for alleviating the symptoms of atopic dermatitis, and although some improvement was noted, no significant differences from placebo were observed (Yang et al. 1999).

Pre-clinical evidence and mechanisms of action: Some evidence for beneficial changes to plasma lipid levels (increase in HDL-cholerstol and decrease in LDL-cholesterol) were found in rabbits using a supercritical CO_2 extract (Basu et al. 2007). An aqueous extract of the seeds produced hypoglycaemic, hypotriglyceridaemic and antioxidant effects in streptozotocin-induce diabetic rats (Zhang et al. 2010). Substantial *in vitro* and *in vivo* evidence for antioxidant effects is available for fruit preparations (Pharmaceutical Press Editorial Team 2013).

Interactions: No data available.

Adverse effects: Based on the available data, sea buckthorn products are generally considered to be very safe. Although there is no evidence of toxicological risks in infants and breastfeeding mothers, caution is needed.

Dosage: For products, follow manufacturers' instructions. No dose range has been established for the medicinal use of the extracts, but the dose range for the entire fruit is reported to be 3–9 g (Pharmaceutical Press Editorial Team 2013).

General plant information: The species should not be confused with other 'buckthorns', for example, the common or purging buckthorn, *Rhamnus cathartica* L., or the alder buckthorn, *Frangula alnus* Mill. (syn. *Rhamnus frangula* L.), all Rhamnaceae, which are strong purgatives.

The English plant name makes reference to its habitat on sandy soils, especially along the coast but also on sandy river banks. Thanks to its robustness and a very fast growing basal shoot system the species is a popular landscaping and garden plant used in windbreaks and hedges, as well as in stabilising dunes, riverbanks and steep slopes.

References

Bal LM, Meda V, Naik SN, Satya S. (2011) Sea buckthorn berries: A potential source of valuable nutrients for nutraceuticals and cosmoceuticals. *Food Research International* 44(7): 1718–1727.

Basu M, Prasad R, Jayamurthy P, Pal K, Arumughan C, Sawhney RC. (2007) Anti-atherogenic effects of seabuck-thorn Hippophaea rhamnoides) seed oil. *Phytomedicine* 14(11): 770–777.

Chen C, Gao W, Ou-Yang DW, Kong DY. (2013). Three new flavonoids, hippophins K-M, from the seed residue of *Hippophae rhamnoides* subsp. *sinensis. Natural Product Research* 28(1): 24–29.

Christaki, E.. (2012) *Hippophae rhamnoides* L. (Sea Buckthorn): a potential source of nutraceuticals. *Food and Public Health* 2(3): 69–72.

Pharmaceutical Press Editorial Team. (2013) *Herbal Medicines*. 4th Edition. Pharmaceutical Press, London, UK.

Yang B, Kalimo KO, Mattila LM, Kallio SE, Katajisto JK, Peltola OJ, Kallio HP. (1999) Effects of dietary supplementation with sea buckthorn (*Hippophaë rhamnoides*) seed and pulp oils on atopic dermatitis. *Journal of Nutritional Biochemistry* 10(11): 622–630.

Zhang W, Zhao J, Wang J, Pang X, Zhuang X, Zhi X, Qu W. (2010) Hypoglycaemic effects of aqueous extract of sea buckthorn (*Hippophae rhamnoides* L.) seed residues in streptozotocin-induced diabetic rats. *Phytotherapy Research* 24(2): 228–232.

Senna   *Senna alexandrina* Mill.

Synonyms: *Cassia acutifolia* Delile; *C. angustifolia* Vahl; *Cassia senna* L.

Family: Fabaceae (Leguminosae)

Other common names: Senna pods; senna leaf; Alexandrian or Khartoum senna; Tinnevelly senna

Drug name: Sennae fructus; Sennae folium

Botanical drug used: Fruit (pods); leaf

Indications/uses: Senna products are used for the short-term relief of occasional constipation and are available both on prescription and over-the-counter. Dried senna pods, and the leaves, often as tea bags, are sometimes used to make an infusion (tea).

Evidence: There is very robust evidence supporting the use of *S. alexandrina* preparations as laxatives.

Safety: Overall, *S. alexandrina* is considered to be safe for short-term use as a laxative. However, senna products have been abused in the mistaken belief that they may help with weight loss. Prolonged use and high doses may cause severe adverse effects, including abdominal pain and water/electrolyte imbalance and it is usually recommended that they should not be taken for longer than 2 weeks. Fatalities have been recorded in young children overdosed with senna tea made from the pods.

Note: Patients who purchase senna products regularly should be counselled about the dose and frequency of use, and if excessive, or if misuse is suspected, should be referred to their GP. In extreme cases, albuminuria and haematuria may occur and potassium levels may need to be assessed.

Main chemical compounds: The main constituents of both the leaf and fruit are anthraquinone glycosides such as sennosides A, B, C and D and palmidin A, rhein-anthrone and aloe-emodin glycosides (Williamson et al. 2013). Other compounds present include kaempferol, mucilage polysaccharides, resin and calcium oxalate (Franz 1993). Alexandrian senna pods contain not less than 3.4%

Phytopharmacy: An evidence-based guide to herbal medicinal products, First Edition.
Sarah E. Edwards, Inês da Costa Rocha, Elizabeth M. Williamson and Michael Heinrich.
© 2015 John Wiley & Sons, Ltd. Published 2015 by John Wiley & Sons, Ltd.

of glycosides, calculated as sennoside B whereas Tinnevelly senna pods contain not less than 2.2%. Licensed preparations including tablets and granules are available, standardised to the total content of sennosides, expressed as sennoside B (usually 7.5 mg per tablet).

Clinical evidence: Several clinical studies have assessed the efficacy of senna preparations in the treatment of constipation and there is clear evidence of laxative effect, but not for bowel cleansing, and senna is not suitable for treating irritable bowel syndrome. Importantly, there have been reports of possible genotoxic or tumourigenic risk from anthranoid-containing laxatives and pharmacovigilance is necessary (see below; EMA 2007).

A Cochrane review on the use of laxatives in palliative care patients, who commonly need laxatives to counteract the effects of opioid analgesics, found no significant differences in effectiveness between lactulose and senna, or lactulose with senna, compared to magnesium hydroxide and liquid paraffin. However, in terms of stool frequency, in one study comparing lactulose and senna with co-danthramer (a synthetic anthranoid containing laxative), an advantage was found for lactulose and senna. When taken alone (senna or lactulose), there was no significant difference in terms of bowel function compared with the co-danthramer group. The authors concluded that no single laxative was superior to the rest, but that there is limited evidence on the use of laxatives in palliative care patients due to a lack of randomised clinical trials (Candy et al. 2011). However, senna has been used safely for many years and is less expensive than many other laxatives (Miles et al. 2006).

In the management of childhood constipation, a Cochrane review comparing senna with lactulose (21 participants under 15 years of age) concluded that there was no significant difference between the two agents in terms of passing stools. Minor adverse events were reported, mainly in the senna group, and included abdominal cramps and diarrhoea (Gordon et al. 2012).

Pre-clinical evidence and mechanisms of action: The sennosides are responsible for the laxative effect. They are not absorbed in the upper gut or split by digestive enzymes but are converted to the active metabolite, rhein-anthrone, by the bacteria present in the large intestine. The mechanism of action is by increasing gut motility and reducing colon transit time, thus inhibiting fluid re-absorption, resulting in defecation within 8–12 hours (EMA 2006).

Interactions: Care should be taken in patients taking cardiac glycosides, anti-arrhythmics or those inducing QT-prolongation, diuretics, adrenocorticosteroids or liquorice root, although no clinical evidence is available to draw any recommendations. Senna may reduce plasma levels of quinidine, by about 25% (EMA 2006; ESCOP 2003; Williamson et al. 2013).

Contraindications: The use of senna during pregnancy should be avoided, especially during the first trimester, and after that it should be used only if other measures (including dietary changes) fail and a stimulant laxative is required. It is not known to be harmful during breastfeeding but should be used sparingly and with caution. Senna containing preparations (such as 'Syrup of Figs' and senna syrup) are available for children under 12 years of age but these should only be used rarely, and usually in conjunction with a faecal softener or an osmotic laxative to prevent hard stools and painful motions. In children with chronic constipation, specialist advice is preferable to over-the-counter (OTC) purchase. Senna should not be used in any patient with intestinal obstruction or stenosis, appendicitis, inflammatory colon diseases (e.g. Crohn's disease, ulcerative colitis), abdominal pain of unknown origin, severe dehydration, or electrolyte depletion. It should not be taken in cases of faecal impaction without prior use of a faecal softener or an osmotic laxative, or undiagnosed, acute or persistent gastrointestinal complaints such as abdominal pain, nausea and vomiting except under medical advice. Long-term use should be avoided. If used in incontinent adults, pads should be changed more frequently to avoid skin irritation and patients should be warned that urine may be stained pink (BNF; EMA 2006; ESCOP 2003).

Adverse effects: Rare cases of hypersensitive reactions such as pruritus or rashes may occur. High doses may produce abdominal pain and spasm and passage of liquid stools, resulting in dehydration and loss of electrolytes. In such cases, the GP should be contacted. With chronic use and frequent OTC purchase, there may be a need to closely monitor self-medication, due to the potential for abuse in the mistaken belief that laxatives help with weight loss. Some studies have reported severe adverse effects such as liver damage (Posadzki et al. 2013; Teschke et al. 2012), but if used at recommended doses and for a short period of time no major adverse effect has been reported.

Routine purging, usually with senna or even more potent stimulant laxatives such as Cascara, has long been viewed as a means to 'cleanse' the body, but this concept is no longer considered valid and modern uses are limited to the treatment of chronic constipation. Fatalities have been recorded in young children overdosed with senna tea made from the pods.

Dosage: See manufacturers' instructions. The maximum daily dose of the anthraquinone glycosides is 30 mg, calculated as sennoside B (EMA 2006; Williamson et al. 2013). Senna pod tea is used more rarely and the dose depends on the type of senna; usually 6–12 pods per dose for an adult.

General plant information: *S. alexandrina* has a long tradition of use and has been an important item of trade, especially in Arabic medicine. The common name 'senna', which is given to the genus, is derived from Arabic 'saná' or 'sanná'. Note. The synonym genus name 'Cassia' also refers to an unrelated species, *Cinnamomum cassia* (L.) J.Presl, which is also known as 'Chinese Cassia' or 'Cassia bark', and is used as a spice and in traditional Chinese medicine. Senna is now a pan-tropical species but mostly grown in Southern India for commerce.

References

BNF (updated twice yearly), jointly published by the British Medical Association and the Royal Pharmaceutical Society: see www.BNF.org for the current edition.

Candy B, Jones L, Goodman ML, Drake R, Tookman A. (2011) Laxatives or methylnaltrexone for the management of constipation in palliative care patients. *Cochrane Database of Systematic Reviews* 1: CD003448.

EMA. (2006) Community herbal monograph on Senna leaf (Sennae folium). European Medicines Agency http://www.emea.europa.eu/docs/en_GB/document_library/Herbal_-_Community_herbal_monograph/2009/12/WC500018210.pdf.

EMA. (2007) Assessment report on *Cassia senna* L. and *Cassia angustifolia* Vahl, folium. European Medicines Agency http://www.ema.europa.eu/docs/en_GB/document_library/Herbal_-_HMPC_assessment_report/2009/12/WC500018219.pdf.

ESCOP. (2003) *ESCOP Monographs: The Scientific Foundation for Herbal Medicinal Products.* Thieme, Exeter and London, UK.

Franz G. (1993) The senna drug and its chemistry. *Pharmacology* 47(Suppl 1): 2–6.

Gordon M, Naidoo K, Akobeng AK, Thomas AG. (2012) Osmotic and stimulant laxatives for the management of childhood constipation. *Cochrane Database of Systematic Reviews* 7: CD009118.

Miles CL, Fellowes D, Goodman ML, Wilkinson S. (2006) Laxatives for the management of constipation in palliative care patients. *Cochrane Database of Systematic Reviews* 1: CD003448.

Posadzki P, Watson LK, Ernst E. (2013) Adverse effects of herbal medicines: an overview of systematic reviews. *Clinical Medicine* 13(1): 7–12.

Teschke R, Wolff A, Frenzel C, Schulze J, Eickhoff A. (2012) Herbal hepatotoxicity: a tabular compilation of reported cases. *Liver International* 32(10): 1543–1556.

Williamson EM, Driver S, Baxter K. (Eds.) (2013) *Stockley's Herbal Medicines Interactions.* 2nd Edition. Pharmaceutical Press, London, UK.

Shatavari　　　　　　　*Asparagus racemosus* Willd.

Synonyms: *Asparagopsis acerosa* Kunth.; *Asparagus dubius* Decne.; *Protasparagus racemosus* (Willd.) Oberm.; and others

Family: Asparagaceae (previously in Liliaceae)

Other common names: Indian asparagus; satavari

Botanical drug used: Roots

Indications/uses: Shatavari is an important Ayurvedic herb, used especially in women, as a galactagogue (to increase the production of breast milk), relieve pre-menstrual syndrome, increase fertility, reduce uterine bleeding and alleviate menopausal symptoms. It is considered a Rasayana (general tonic and adaptogen) and is also used for the prevention and treatment of gastrointestinal disturbances, including dyspepsia, diarrhoea and gastric ulcers, as an aphrodisiac, in nervous disorders including dementia and many others.

Evidence: Clinical data to support the traditional uses of shatavari are limited. Numerous pre-clinical studies have demonstrated a number of properties of *A. racemosus* extracts, including phytoestrogenic, antioxidant, antidiarrhoeal, anti-ulcer, diuretic, hepatoprotective, hypolipidaemic, antibacterial, immunomodulatory, anti-tumour and others.

Safety: Safety data is lacking. Although traditionally used as a galactagogue, until safety is confirmed it cannot be recommended during lactation or pregnancy. A study in rats showed teratogenic effects with high doses. It should not be used in children and adolescents under 18 years due to insufficient safety data.

Main chemical compounds: The major active constituents of shatavari are steroidal saponins, the shatavarins I-IV, shatavarosides A and B, immunoside and filiasparoside C, which are glycosides of sarsasapogenin. The roots have also been reported to contain an isoflavone, 8-methoxy-5,6,4'-trihydroxyisoflavone-7-O-β-D-glucopyranoside, and alkaloids including asparagamine, and racemosol, a 9,10-dihydrophenanthrene derivative. Flavonoids including kaempferol have been isolated from the woody portion of the tuberous roots (Alok et al. 2013; Bhutani et al. 2010; Bopana and Saxena 2007; Gautem et al. 2009; Sharma et al. 2009).

Phytopharmacy: An evidence-based guide to herbal medicinal products, First Edition.
Sarah E. Edwards, Inês da Costa Rocha, Elizabeth M. Williamson and Michael Heinrich.
© 2015 John Wiley & Sons, Ltd. Published 2015 by John Wiley & Sons, Ltd.

Clinical evidence: No recent clinical trials investigating *A. racemosus* for any indications were found. Few trials have been undertaken evaluating the effects of shatavari in humans, and the robustness of these is in some doubt. Most of the reported clinical studies concerned proprietary Ayurvedic products, and were uncontrolled and included polyherbal formulations. In one small cross-over study in eight normal healthy male volunteers, 2 g powdered root of shatavari was compared against metoclopramide (10 mg) in the reduction of gastric emptying time. No significant difference between the two treatments was found. In the treatment of diarrhoea, *A. racemosus* root powder was reported to be effective, but many details, including sample size etc., were not given (Bopana and Saxena 2007). The anti-ulcerogenic activity of *A. racemosus* root powder (12 g/day in four doses for average duration of 6 weeks) was also reported in another study in 32 patients with duodenal ulcers (Alok et al. 2013).

Two studies, cited in a 2007 review, investigated the galactagogue properties of products containing shatavari. While one proprietary product (tablets containing 40 mg *A. racemosus* extract per tablet) was found to increase milk secretion, another polyherbal product containing 15% shatavari extract was shown to be ineffective. Other studies that found effectiveness for shatavari-containing polyherbal products were for the following: regulating menstrual cycles in 20- to 45-year-old women with dysfunctional uterine bleeding, dysmenorrhoea, pre-menstrual syndrome and relief from post-menopausal symptoms (Bopana and Saxena 2007). These results should be interpreted with caution since the studies were not all of high quality and the products investigated contained other active herbal ingredients.

Pre-clinical evidence and mechanisms of action: The following effects have been observed for extracts of *A. racemosus*: stimulation of insulin secretion, which may explain the anti-hyperglycaemic activity (Hannan et al. 2007); inhibition of cholinesterase, which may increase *in vivo* cognitive effects (Ojhra et al. 2010); augmentation of Th1/Th2 response by the constituent immunoside, increasing IL-12 and IFN-γ synthesis and NO production in macrophages, which may contribute to immunomodulatory activity (Gautam et al. 2009; Sidiq et al. 2011); inhibition of oxytocin, resulting in uterine relaxation (Bopana and Saxena 2007); and inhibition of hepatic CYP2E1 activity, increasing removal of free radicals, and promoting hepatoprotective activity (Palanisamy and Manian 2012). Other activities which have been demonstrated *in vitro* and *in vivo* studies include anti-inflammatory, anti-urolithiatic, antibacterial, cardioprotective, diuretic, phytoestrogenic, hypolipidaemic, and anti-tumour effects (Acharya et al. 2013; Alok et al. 2013; Bhutani et al. 2010; Bopana and Saxena 2007; Hannan et al. 2007; Jagannath et al. 2012; Kumar et al. 2010; Mitra et al. 2012; Ojha et al. 2010; Williamson 2002).

Interactions: None known. However, *A. racemosus* extract has been shown to significantly inhibit cytochrome P450 isozyme CYP2E1 (Palanisamy and Manian 2012), which may inhibit the metabolism of paracetamol (acetaminophen), ethanol, theophylline, flurane anaesthetics and other drugs metabolised via CYP2E1.

Contraindications: Hypersensitivity to asparagus; patients with impaired renal function and heart disease. Shatavari should be avoided during pregnancy due to teratogenic effects observed in animals (Goel et al. 2006). Use during lactation or for children and adolescents under 18 years is not recommended due to lack of data.

Adverse effects: Allergic reactions to asparagus are known (Goel et al. 2006).

Dosage: Follow manufacturers' instructions for products. Recommended doses vary considerably but are usually up to 3 g/day, although Ayurvedic practitioners may use much higher doses, up to 20–30 g dried root powder per day (Williamson 2002).

General plant information: *A. racemosus* roots are used in traditional medicine systems throughout its range, including in India, tropical Africa, and northern Australia. Increase in demand, combined with habitat destruction, has led to a decline in its natural habitat, where it is considered endangered (Bopana and Saxena 2007).

References

Acharya SR, Acharya NS, Bhangale JO, Shah SK, Pandya SS. (2013) Antioxidant and hepatoprotective action of *Asparagus racemosus* Willd. root extracts. *Indian Journal of Experimental Biology* 50(11): 975–801.

Alok S, Jain SK, Verma A, Kumar M, Mahor A, Sabharwal M. (2013) Plant profile, phytochemistry and pharmacology of *Asparagus racemosus* (Shatavari): A review. *Asian Pacific Journal of Tropical Disease* 3(3): 242–251.

Bhutani KK, Paul AT, Fayad W, Linder S. (2010) Apoptosis inducing activity of steroidal constituents from *Solanum xanthocarpum* and *Asparagus racemosus*. *Phytomedicine* 17(10): 789–793.

Bopana N, Saxena S. (2007) *Asparagus racemosus* – ethnopharmacological evaluation and conservation needs. *Journal of Ethnopharmacology* 110(1): 1–15.

Gautam M, Saha S, Bani S, Kaul A, Mishra S, Patil D, Satti NK, Suri KA, Gairola S, Suresh K, Jadhav S, Qazi GN, Patwardhan B. (2009) Immunomodulatory activity of *Asparagus racemosus* on systemic Th1/Th2 immunity: implications for immunoadjuvant potential. *Journal of Ethnopharmacology* 121(2): 241–247.

Goel RK, Prabha T, Kumar MM, Dorababu M, Prakash, Singh G. (2006) Teratogenicity of *Asparagus racemosus* Willd. root, a herbal medicine. *Indian Journal of Experimental Biology* 44(7): 570–573.

Hannan JMA, Marenah L, Ali L, Rokeya B, Flatt PR, Abdel-Wahab YH. (2007) Insulin secretory actions of extracts of *Asparagus racemosus* root in perfused pancreas, isolated islets and clonal pancreatic beta-cells. *Journal of Endocrinology* 192(1): 159–168.

Jagannath N, Chikkannesetty SS, Govindadas D, Devasankaraiah G. (2012) Study of antiurolithiatic activity of *Asparagus racemosus* on albino rats. *Indian Journal of Pharmacology* 44(5): 576–579.

Kumar MC, Udupa AL, Sammodavardhana K, Rathnakar UP, Shvetha U, Kodancha GP. (2010) Acute toxicity an diuretic studies of the roots of *Asparagus racemosus* Willd in rats. *The West Indian Medical Journal* 59(1): 3–6.

Mitra SK, Prakash NS, Sundaram R. (2012) Shatavarins (containing Shatavarin IV) with anticancer activity from the roots of *Asparagus racemomsus*. *Indian Journal of Pharmacology* 44(6): 732–736.

Ojhra R, Sahy AN, Muruganandam AV, Singh GK, Krishnamurthy S. (2010) *Asparagus racemosus* enhances memory and protects against amnesia in rodent models. *Brain and Cognition* 74(1): 1–9.

Palanisamy N, Manian S. (2012) Protective effects of *Asparagus racemosus* on oxidative damage in isoniazid-induced hepatotoxic rats: an *in vivo* study. *Toxicology and Industrial Health* 28(3): 238–244.

Sharma U, Saini R, Kumar N, Singh B. (2009) Steroidal saponins from *Asparagus racemosus*. *Chemical and Pharmaceutical Bulletin* 57(8): 890–893.

Sidiq T, Khajuria A, Suden P, Singh S, Satti NK, Suri KA, Srinivas VK, Krishna E, Johri RK. (2011) A novel sarsasapogenin glycoside from *Asparagus racemosus* elicits protective immune responses against HBsAG. *Immunology Letters* 135(1–2): 129–135.

Williamson EM. (Ed.) (2002) *Major Herbs of Ayurveda*. Dabur Research Foundation, Churchill Livingstone Elsevier, UK.

Skullcap   *Scutellaria lateriflora* L.

Synonyms: *Cassida lateriflora* (L.) Moench

Family: Lamiaceae (Labiatae)

Other common names: American skullcap; helmet flower; hoodwort; quaker bonnet; scullcap.

Drug name: Scutellariae herba (Note: Scutellariae radix refers to the drug used in TCM, the root of a related species, *S. baicalensis* Georgi)

Botanical drug used: Dried aerial parts

Indications/uses: Traditionally, skullcap herb is used to relieve temporary symptoms of stress including mild anxiety and insomnia. It is a common ingredient in multi-herbal THR preparations used for the temporary relief of symptoms of mild anxiety. Baical skullcap (*S. baicalensis*) is very widely used in TCM to treat inflammation, virus infections and in the supportive treatment of cancer as part of several traditional formulae.

Evidence: Clinical data to support traditional uses are limited.

Safety: Safety and toxicity data are lacking. Adverse reports of liver damage in patients taking products purporting to contain skullcap have been reported. However, skullcap preparations have been known to be adulterated with other *Scutellaria* species, as well as potentially hepatotoxic germander species, *Teucrium spp.*, due to close morphological similarity. Use during pregnancy and lactation, or in children and adolescents under 18 years of age should be avoided.

Main chemical compounds: The major constituents of aerial parts of *S. lateriflora* are flavonoid glycosides, and their aglycones: predominantly baicalin (~5%), with dihydrobaicalin, baicalein, lateriflorin, lateriflorein, ikonnikoside I, scutellarin, oroxylin A, 2′-methoxy-chrysin-7-*O*-glucuronide, wogonin and wogonoside. The leaf contains significantly higher concentration of flavonoids (over 5%) than the roots (around 3%) or stems (around 2%) and flowering plants contain the highest amount of flavonoids. Phenolic acids found in skullcap herb include caffeic, cinnamic, *p*-coumaric and ferulic acids. The essential oil concentration and composition is uncertain due to conflicting data, and may be the result of misidentification

Phytopharmacy: An evidence-based guide to herbal medicinal products, First Edition.
Sarah E. Edwards, Inês da Costa Rocha, Elizabeth M. Williamson and Michael Heinrich.
© 2015 John Wiley & Sons, Ltd. Published 2015 by John Wiley & Sons, Ltd.

of plant species. Diterpenoids (neo-clerodanes) reported include scutelaterins A-C and scutecyprol A. Other constituents include potentially active constituents such as γ-aminobutyric acid (GABA) and melatonin, which have been found at low levels (Pharmaceutical Press Editorial Team 2013; Upton and Dayu 2012).

S. baicalensis roots contain similar flavonoids: baicalin, baicalein, oroxylin A, skullcap flavone I and II, tenaxin I, wogonin, wogonoside and many other hydroxylated and methoxylated flavones (Li et al. 2011; Williamson et al. 2013).

Clinical evidence: Clinical data for *S. lateriflora* are limited to two studies, which although both showing positive effects on mood, for ethical reasons used healthy subjects. It is therefore not possible to confirm if skullcap actually exerts anxiolytic effects in patients suffering from clinical depression and anxiety.

In one double-blind, placebo-controlled crossover study, the effects of various skullcap preparations on 19 healthy volunteers' energy, cognition and anxiety levels were assessed. Three active preparations were used: 350 mg freeze dried skullcap herb (Eclectic Institute, Oregon) and one or two 100 mg capsules of freeze dried skullcap extract (Phytos, California). Effects of treatment were evaluated over time (baseline and 30-minute intervals up to 120 minutes after administration). Results demonstrated little effect on cognition or energy, but a consistent reduction in anxiety (based on non-standardised subjective assessment scale), with a pronounced effect at 60 minutes after administration of the two capsules of extract (200 mg dose) (Wolfson and Hoffman 2003).

In another more recent placebo-controlled, double-blind crossover study assessing the effects of *S. lateriflora* on mood, 43 healthy subjects were randomly assigned to receive a sequence of three times daily *S. lateriflora* (350 mg) or placebo, each over 2 weeks. There was no significant difference between skullcap and placebo in this relatively non-anxious population, although a significant group effect suggested a carryover effect of skullcap. 'Total Mood Disturbance' (a standardised psychological score) was significantly decreased from pre-test scores with *S. lateriflora*, but not placebo. The authors concluded that *S. lateriflora* significantly enhanced global mood without a reduction in energy or cognition (Brock et al. 2013).

Pre-clinical evidence and mechanisms of action: A small number of *in vitro* and animal studies have been performed on skullcap, but evidence for effectiveness against anxiety is generally inconclusive. A greater number of published studies have been performed on root preparations of *S. baicalensis* (e.g. see Li et al. 2011), but it should not be assumed that it has the same therapeutic effects as *S. lateriflora* (Engels 2009). A study using both *in vitro* and *in vivo* models showed that *S. lateriflora* extracts and the constituents baicalin and baicelein showed promise as potential candidates against transmissible spongiform encephalopathies (TSE) and other neurodegenerative diseases such as Alzheimer's and Parkinson's disease. Beneficial effects of *S. lateriflora* against prion diseases can be explained by the anti-aggregatory and potential antioxidant effects of the constituents (Eiden et al. 2012). Another study showed that the methanolic extract of *S. lateriflora* inhibited alpha-glucosidase, suggesting a potential role as an antidiabetic agent (Kuroda et al. 2012).

Interactions: None reported (Williamson et al. 2013).

Contraindications: None documented. Skullcap should not be used during pregnancy, especially since traditionally it has been used to expel afterbirth.

Adverse effects: Several cases of liver damage have been reported in patients taking herbal preparations containing skullcap alone or in combination with other herbs, predominantly valerian. Whether these can be attributable to skullcap or another ingredient or adulterant, is uncertain (MacGregor et al. 1989).

Dosage: Oral administration (adults): (a) dried herb – 1–2 g as an infusion 3 times/day; (b) liquid extract – 2–4 ml (1:1 in 25% alcohol) 3 times/day; (c) tincture – 1–2 ml (1:5 in 45% alcohol) 3 times/day (Pharmaceutical Press Editorial Team 2013).

General plant information: Roots of *S. lateriflora* were used by Native Americans to treat menstrual disorders, diarrhoea, as a kidney medicine, as an emetic to expel afterbirth and to prevent smallpox (Moerman 1998). While some skullcap is wild harvested in North America, the majority of the world's market is supplied by small growers outside of North America. In 2001, it was estimated that 85% of the entire harvest of skullcap was from cultivated sources (Upton and Dayu 2012). European skullcap, *S. galericulata* L., has been used for similar purposes but has hardly been investigated. The roots of Baical or Chinese skullcap, *S. baicalensis* Georgi, are very widely used in TCM.

References

Brock C, Whitehouse J, Tewfik I, Towell T. (2013) American skullcap (*Scutellaria lateriflora*): a randomised, double-blind placebo-controlled crossover study of its effects on mood in healthy volunteers. *Phytotherapy Research* 28(5): 692–698.

Eiden M, Leidel F, Stohmeier B, Fast C, Groschup MH. (2012) A medicinal herb *Scutellaria lateriflora* inhibits PrP replication in vitro and delays the onset of prion disease in mice. *Frontiers in Psychiatry* 3: 9.

Engels G. (2009) Skullcap. *HerbalGram* 83: 1–2.

Kuroda M, Iwabuchi K, Mimaki Y. (2012) Chemical constituents of the aerial parts of *Scutellaria lateriflora* and their alph-glucosidase inhibitory activities. *Natural Product Communications* 7(4): 471–474.

Li C, Lin G, Zuo Z. (2011) Pharmacological effects and pharmacokinetics properties of Radix Scutellariae and its bioactive flavones. *Biopharmaceutics & Drug Disposition* 32(8): 427–445.

MacGregor FB, Abernethy VE, Dahabra S, Cobden I, Hayes PC. (1989) Hepatotoxicity of herbal remedies. *BMJ* 299(6708): 1156–1157.

Moerman DE. (1998) *Native American Ethnobotany*. Timber Press, Portland, OR, USA, 927 pp.

Pharmaceutical Press Editorial Team. (2013) *Herbal Medicines*. 4th Edition. Pharmaceutical Press, London, UK.

Williamson EM, Driver S, Baxter K. (Eds.) (2013) *Stockley's Herbal Medicines Interactions*. 2nd Edition. Pharmaceutical Press, London, UK.

Wolfson P, Hoffman DL. (2003) An investigation into the efficacy of *Scutellaria lateriflora* in healthy volunteers. *Alternative Therapies in Health and Medicine* 9(2): 74–78.

Upton R, Dayu RH. (2012) Skullcap: *Scutellaria lateriflora* L. an American nervine. *Journal of Herbal Medicine* 2: 76–96.

Slippery Elm   *Ulmus rubra* Muhl.

Synonyms: *Ulmus americana* var. *rubra* (Muhl.) Aiton; *U. fulva* Michx; *U. pendula* Willd.

Family: Ulmaceae

Other common names: Indian elm; moose elm; red elm; sweet elm

Drug name: Ulmi rubrae cortex

Botanical drug used: Inner bark

Indications/uses: Slippery elm bark is valued for its demulcent, emollient and nutritive properties, as a food for people recovering from a disease. Traditionally, it has been used to relieve the symptoms of indigestion, heartburn and flatulence, sore throats and coughs, to relieve inflammation of the urinary tract, colic, irritable bowel syndrome (IBS) and to treat diarrhoea. An important medicinal plant of Native Americans, a decoction of the bark was used as a laxative and to aid delivery in childbirth.

Evidence: Scientific research on medicinal properties of slippery elm is lacking.

Safety: Safety data for slippery elm is lacking, so not recommended during pregnancy and lactation and in children/ adolescents under 18 years of age.

Main chemical compounds: Slippery elm inner bark is predominantly composed of dietary fibre including celluloses, lignin, mucilage and gums. The mucilage consists of long-chain polysaccharides which combine with water to form a viscous, semi-solid mass, and is regarded as the main active constituent. The polysaccharides are composed of D-galactose, L-rhamnose and D-galacturonic acid and their methylated derivatives and are highly branched. Other constituents include phytosterols, starch, minerals and oleic and palmitic acids (Pengelly and Bennett 2011).

Clinical evidence:

No published clinical trials investigating slippery elm, other than in combination with other herbs, have been carried out.

Phytopharmacy: An evidence-based guide to herbal medicinal products, First Edition.
Sarah E. Edwards, Inês da Costa Rocha, Elizabeth M. Williamson and Michael Heinrich.
© 2015 John Wiley & Sons, Ltd. Published 2015 by John Wiley & Sons, Ltd.

Irritable bowel syndrome: An open-label, uncontrolled pilot study assessed the effects and tolerability of two herbal formulations containing slippery elm bark in 31 patients diagnosed with IBS. Of the two treatments, one significantly improved both bowel habit and IBS symptoms in the patients with constipation-predominant IBS. The other treatment was not effective in improving bowel habit in individuals with diarrhoea-predominating or alternating IBS, although it significantly improved a number of IBS symptoms (Hawrelak and Myers 2010). The constituents of the formulae were diarrhoea-IBS formula: powdered bilberry fruit, slippery elm bark, agrimony herb and cinnamon; constipation-IBS formula: slippery elm bark, lactulose, oat bran and liquorice root.

Sore throat and pharyngitis: A multi-centre, randomised double-blind placebo-controlled study was carried out to investigate a demulcent herbal tea formulation ('Throat Coat®', containing 8.5% *U. rubra* powder) for the symptomatic treatment of acute pharyngitis. Patients were given either Throat Coat ($n = 30$) or placebo ($n = 30$). Throat Coat was found to be significantly superior to placebo in providing rapid, temporary relief from sore throat pain in patients with pharyngitis (Brinckmann et al. 2003). As with the IBS study, it is not possible to attribute any perceived benefits of the treatment to a particular ingredient in the formulation.

Pre-clinical evidence and mechanisms of action:

Few studies on slippery elm have been reported.

Antioxidant activity in inflammatory bowel disease: An *in vitro* study investigated the antioxidant effects (in cell-free oxidant-generating systems and inflamed human colorectal biopsies) of six herbal therapies used to treat inflammatory bowel disease (IBD). Slippery elm, like the positive control 5-aminosalicylate (and most of the other herbal remedies tested), was found to have a dose-dependent antioxidant effect and was the most potent of the herbs tested in the biopsy study. The authors concluded that it was a promising candidate for formal evaluation of its therapeutic potential *in vivo* in patients with IBD and other chronic inflammatory conditions (Langmead et al. 2002). In another *in vitro* study, 28 herbs were screened for peroxynitrite scavenging activity. Slippery elm ranked fifth out of the 28 herbs, after witch hazel bark, rosemary, jasmine tea and sage (Choi et al. 2002).

Interactions: No case reports of interactions are available. However, the mucilage present in slippery elm may reduce absorption of medicines if taken concomitantly and it has been suggested that other medicines should be taken 1 hour prior to, or several hours after, slippery elm preparations (Gardner and McGuffin 2013).

Contraindications: Not recommended during pregnancy and lactation, in children and adolescents under 18 years of age, or in patients with liver disease, gallstones and other biliary disorders, due to lack of safety data. (MHRA 2013).

Adverse effects: No reports of side effects have been identified (Gardner and McGuffin 2013).

Dosage: As a food during recovery from an illness, slippery elm is made into a thin porridge or gruel. For manufactured herbal products follow manufacturers' dosage recommendations. No recommended dosage is available due to lack of scientific data.

General plant information: *U. rubra* is endemic to North America and cultivated there and in the British Isles (Morton 1990). In the United Kingdom, slippery elm bark can only be sold in premises which are registered pharmacies, under the supervision of a pharmacist. *U. rubra*, like other elms, is susceptible to Dutch elm disease, so larger trees are becoming increasingly rare, resulting in harvesters using greater quantities of smaller trees to obtain the same quantities of bark with consequent conservation issues (Pengelly and Bennett 2011).

References

Brinckmann J, Sigwart H, van Houten Taylor L. (2003) Safety and efficacy of a traditional herbal medicine (Throat Coat) in symptomatic temporary relief of pain in patients with acute pharyngitis: a multicenter, prospective, randomized, double-blinded, placebo-controlled study. *Journal of Alternative and Complementary Medicine* 9(2): 285–298.

Choi HR, Choi JS, Han YN, Base SJ, Chung HY. (2002) Peroxynitrate scavenging activity of herb extracts. *Phytotherapy Research* 16(4): 364–367.

Gardner Z, McGuffin M. (Eds.) (2013) *American Herbal Product Association's Botanical Safety Handbook*. 2nd Edition. CRC Press, USA, 1072 pp.

Hawrelak JA, Myers SP. (2010) Effects of two natural medicine formulations on irritable bowel syndrome symptoms: a pilot study. *Journal of Alternative and Complementary Medicine* 16(10): 1065–1071.

Langmead L, Dawson C, Hawkins C, Banna N, Loos S, Rampton DS. (2002) Antioxidant effects of herbal therapies used by patients with inflammatory bowel disease: an *in vitro* study. *Alimentary Pharmacology and Therapeutics* 16(2): 197–205.

MHRA. (2013) Summary of product characteristics, Potter's Slippery Elm Indigestion Relief. http://www.mhra.gov.uk/home/groups/spcpil/documents/spcpil/con1376454722191.pdf (accessed January 2014).

Morton JF. (1990) Mucilaginous plants and their uses in medicine. *Journal of Ethnopharmacology* 29(3): 245–266.

Pengelly A, Bennett K. (2011) Appalachian plant monographs: *Ulmus rubra* Muhl., Slippery Elm. http://www.frostburg.edu/fsu/assets/File/ACES/ulmus%20rubra%20-final.pdf (accessed 3 January 2014).

Spirulina

Arthrospira platensis Gomont, *A. maxima* Setchell & N.L.Gardner

Synonyms: *Spirulina platensis* (Gomont) Geitler (= *A. platensis*); *S. maxima* (Setchell & N.L.Gardner) Geitler (= *A. maxima*)

Family: Phormidiaceae

Botanical drug used: Biomass of cultured spirulina (a cyanobacterium/blue-green algae), as dried powder

---

Indications/uses: Used as a source of high quality protein and vitamins, spirulina is considered a 'nutraceutical' with diverse beneficial effects, including lowering of cholesterol, modulation of the immune system and suppression of cancer development and viral infection.

Evidence: Clinical data provides evidence of hypolipidemic and anti-inflammatory activity of spirulina. However, evidence is limited to small scale clinical trials, some of which are of poor quality. Numerous *in vitro* and *in vivo* animal studies have also demonstrated the immunomodulatory, antioxidant, anti-tumour, anti-ageing and radioprotective properties of spirulina.

Safety: Toxicological studies have confirmed the safety of spirulina for human consumption. The US FDA has designated spirulina produced by a few companies as 'generally regarded as safe' (GRAS). However, some spirulina products have been found to be contaminated with heavy metals or other species of cyanobacteria that produce microcystin and may pose a potential liver cancer risk.

Main chemical compounds: Spirulina contains high levels of proteins (55–70% dry weight), vitamins, minerals and essential fatty acids, including γ-linolenic acid (GLA) and β-carotene. Phycocyanin, a phycobiliprotein, is one of the major pigment constituents of spirulina (Habib et al. 2008; Soltani et al. 2012). Although spirulina has been promoted as a source of vitamin B_{12} (cobalamin) for use by vegans, pseudovitamin B_{12} is the predominant cobamide found in cyanobacteria and is not bioavailable to humans (Watanabe 2007).

Clinical evidence: A few preliminary clinical trials have investigated the functional effects of spirulina, but evidence from robust high quality trials is lacking.

Phytopharmacy: An evidence-based guide to herbal medicinal products, First Edition.
Sarah E. Edwards, Inês da Costa Rocha, Elizabeth M. Williamson and Michael Heinrich.
© 2015 John Wiley & Sons, Ltd. Published 2015 by John Wiley & Sons, Ltd.

Cumulative data from clinical trials demonstrate the hypolipidemic activity of spirulina, despite differences in study design and patient condition. Spirulina has been shown to increase levels of HDL-cholesterol and decrease LDL-cholesterol and reduce systolic and diastolic blood pressure (Deng and Chow 2010). In a study in Korean subjects, spirulina was found to reduce inflammatory cytokines IL-6 and TNF-α (Park et al. 2008). Spirulina has also been reported to activate the immune system in pan tobacco chewers (Hirahashi et al. 2002) and to significantly reduce symptoms associated with allergic rhinitis in a double-blind, placebo controlled study (Cingi et al. 2008).

Pre-clinical evidence and mechanisms of action: A number of *in vitro* and *in vivo* studies have reported the anti-inflammatory and antioxidant activities of spirulina. In one study with aged male rats, spirulina was shown to reverse age-related increase in pro-inflammatory cytokines in the cerebellum. A large number of animal studies have demonstrated the protective effect of spirulina against environmental toxicants, including heavy metals (Deng and Chow 2010). Immunomodulatory, antiviral and anticancer effects have also been demonstrated in rodents and *in vitro* studies (Hirahashi et al. 2002; Soltani et al. 2012).

The mechanisms of action behind the effects of spirulina observed in animal and human studies remain unclear, although phycocyanin is antioxidant and been shown to inhibit pro-inflammatory cytokines, while β-carotene also has antioxidant and anti-inflammatory properties (Deng and Chow 2010). Conversely, studies investigating the immunomodulatory properties of spirulina have been shown to enhance production of inflammatory mediators, including TNF-α, believed to be essential for successful control of infection (Soltani et al. 2012).

Interactions: None known.

Contraindications: People with the genetic condition phenylketonuria (PKU) cannot metabolise phenylalanine and thus should avoid using spirulina, as it is rich in this amino acid (Mazokopakis et al. 2008). Those with autoimmune diseases (such as multiple sclerosis, lupus and rheumatoid arthritis) should avoid using spirulina, as it may stimulate the immune system, although this is purely theoretical at present. Spirulina should be avoided by pregnant or breastfeeding women and in those under 18 years of age, given the risk of contamination with toxic substances and the overall lack of evidence.

Adverse effects: Rare cases of side effects have been reported in people taking 1 g of spirulina daily. These include headache, muscle pain, flushing of the face, sweating and difficulty concentrating. Liver damage and skin reactions have also been reported. One case of acute rhabdomyolysis was reported in a 28-year old man taking 3 g per day of a spirulina dietary supplement for 1 month (Mazokopakis et al. 2008).

Dosage: There is insufficient data to determine the appropriate dose of spirulina. Doses used in the few human trials that have been conducted have varied from 1 g to more than 5 g per day (Mazokopakis et al. 2008).

General plant information: Spirulina has long been used as a nutritional supplement by people living near to alkaline lakes where it naturally occurs, including the ancient Aztecs of Mesoamerica and the Kanembu people of Lake Chad region in Africa (Habib et al. 2008). More recently, it has been proposed as a potential primary food source to be cultivated on long-term space missions (Brown et al. 2006).

References

Brown I, Jones JA, Bayless D, Karakis S, Karpov L, McKay DS. (2006) Novel concept for LSS based on advanced microalgal biotechnologies. *Habitation* 10(3–4): 122. http://ntrs.nasa.gov/archive/nasa/casi.ntrs.nasa.gov/20080026048_2008025589.pdf (accessed 28 July 2013).

Cingi C, Conk-Dalay M, Cakli H, Bal C. (2008) The effects of spirulina on allergic rhinitis. *European Archives of Oto-Rhino-Laryngology* 265(10): 1219–1223.

Deng R, Chow T-J. (2010) Hypolipidemic, antioxidant, and anti-inflammatory activities of microalgae *Spirulina*. *Cardiovascular Therapeutics* 28(4): e33–e45.

Habib MAB, Parvin M, Huntington TC, Hasan MR. (2008) A review on culture, production and use of spirulina as food for humans and feeds for domestic animals and fish. *FAO Fisheries & Aquaculture Circular*. No. 1034. FAO, Rome.

Hirahashi T, Matsumoto M, Hazeki K, Saeki Y, Ui M, Seya T. (2002) Activation of the innate human immune system by Spirulina: augmentation of interferon production and NK cytotoxicity by oral administration of hot water extract of *Spirulina platensis*. *International Immunopharmacology* 2(4): 423–434.

Mazokopakis MM, Karefilakis CM, Tsartsalis AN, Milkas AN, Ganotakis ES. (2008) Acute rhabdomyolysis caused by Spirulina (*Arthrospira platensis*). *Phytomedicine* 15(6–7): 525–527.

Park HJ, Lee YJ, Ryu HK, Kim MH, Chung HW, Kim WY. (2008) A randomized double-blind, placebo-controlled study to establish the effects of spirulina in elderly Koreans. *Annals of Nutrition and Metabolism* 52(4): 322–328.

Soltani M, Khosravi A-R, Asadi F, Shokri H. (2012) Evaluation of protective effect of *Spirulina platensis* in Balb/C mice with candidiasis. *Journal de Mycologie Médicale* 22(4): 329–334.

Watanabe F. (2007) Vitamin B_{12} sources and bioavailability. *Experimental Biology and Medicine (Maywood)* 232(10): 1266–1274.

Squill   *Drimia maritima* (L.) Stearn

Synonyms: *Scilla maritima* L.; *Urginea maritima* (L.) Baker; *U. scilla* Steinh.; and others

Family: Asparagaceae (previously in Hyacinthaceae and Liliaceae)

Other common names: Maritime squill; red squill; sea onion; white squill

Drug name: Scillae bulbus

Botanical drug used: Bulb

Indications/uses: Squill is used for the relief of chesty coughs, as an expectorant.

Evidence: No clinical data available.

Safety: Squill should be avoided by patients with cardiac disorders, impaired hepatic or renal function, pregnancy and lactation, in high doses or for long periods. It is emetic and cardiotoxic in high doses.

Main chemical compounds: Contains cardiac glycosides, mostly based on scillarenin, the main compounds being scillaren A and proscillaridin A, with a number of minor glycosides. The bufadienolide proscillaridin A was identified as a cytotoxic compound in this species (El-Seedi et al. 2013). There is considerable variation between sources of the plant. Red squill contains, in addition, scilliroside and scillirubroside and a red pigment. Flavonoids such as quercetin, taxifolin and vitexin as well as phytosterols are also present (Williamson et al. 2013).

Clinical evidence: No clinical studies were found but squill has been used traditionally as an expectorant for chesty cough, chronic bronchitis, asthma and whooping cough. Squill oxymel, a compound of squill with honey, is an ingredient of Gee's linctus BP and other OTC cough preparations.

Pre-clinical evidence and mechanisms of action: No recent studies have been conducted. In the past, there was an interest in 'scillaren', a mixture of glycosides from squill, for its cardiac action, which is similar to digitalis but there is little information to support this therapeutic use (Gemmill 1974).

Phytopharmacy: An evidence-based guide to herbal medicinal products, First Edition.
Sarah E. Edwards, Inês da Costa Rocha, Elizabeth M. Williamson and Michael Heinrich.
© 2015 John Wiley & Sons, Ltd. Published 2015 by John Wiley & Sons, Ltd.

Interactions: It may have additive effects with cardiac glycosides such as digoxin or other cardioactive drugs and caution should be taken with thiazides, loop diuretics, carbenoxolone calcium and laxatives (NHS 2013; Williamson et al. 2013).

Contraindications: Squill is not recommended for patients with cardiac disorders or impaired hepatic or renal function. Squill has been reported to be an abortifacient and to affect the menstrual cycle. It must be avoided during pregnancy and lactation and is not recommended for children under 12 years old.

Adverse effects: Squill contains steroid cardiac glycosides with a narrow therapeutic range, thus therapeutic doses may induce side effects in susceptible individuals. These include gastric irritation or hypersensitivity reactions (urticaria/hives, flushing or dermatitis), salt and water retention, low potassium levels in blood and irregular pulse (NHS 2013).Having said that, Gee's Linctus is not considered to be particularly dangerous in recommended doses, and the cardiac glycosides are poorly absorbed. The emetic properties of squill may also prevent excessive absorption (Williamson et al. 2013).

Dosage: Dried bulb: 60–200 mg, as an infusion 3 times a day; squill liquid extract: (BPC 1973) 0.06–0.2 ml; squill tincture (BPC 1973): 0.3–2.0 ml; squill vinegar (BPC 1973): 0.6–2.0 ml (Pharmaceutical Press Editorial Team 2013).

General plant information: The plant is endemic to India, Africa and the Mediterranean region (Kameshwari et al. 2012) and has been used in medicine since the classic Greek period. Oxymel of Squill, used for coughs, was invented by Pythagoras (Gentry et al. 1987). Red squill was formerly employed as a rodenticide.

References

Gemmill CL. (1974) The pharmacology of squill. *Bulletin New York Academy of Medicine* 50(6): 747–750.

El-Seedi HR, Burman R, Mansour A, Turki, Z, Boulos L, Gullbo J, Göransson U. (2013) The traditional medical uses and cytotoxic activities of sixty-one Egyptian plants: Discovery of an active cardiac glycoside from *Urginea maritima*. *Journal of Ethnopharmacology* 145(3): 746–757.

Gentry HS, Verbiscar AJ, Banigan TF. (1987) Red Squill (*Urginea maritima*, Liliaceae). *Economic Botany* 41(2): 267–282.

Shiva Kameshwari MN, Bijul Lakshman, A, Paramasivam G. (2012) Biosystematics studies on medicinal plant *Urginea indica* Kunth. Liliaceae – A review. *International Journal of Pharmacy and Life Sciences* 3(1): 1394–1406.

NHS. (2013) Squill/Ipecacuanha/Licorice. http://www.nhs.uk/medicine-guides/pages /MedicineOverview.aspx?medicine=Squill/Ipecacuanha/Liquorice (retrieved January 2013).

Pharmaceutical Press Editorial Team. (2013) *Herbal Medicines*. 4th Edition. Pharmaceutical Press, London, UK.

Williamson EM, Driver S, Baxter K. (Eds.) (2013) *Stockley's Herbal Medicines Interactions*. 2nd Edition. Pharmaceutical Press, London, UK.

♦ ♦ ♦ ♦ ♦ ♦

Tea Tree (Oil)

Melaleuca alternifolia (Maiden & Betche) Cheel

Synonym: *M. linariifolia* var. *alternifolia* Maiden & Betche

Other common names: Australian tea tree; ti tree

Family: Myrtaceae

Drug name(s): Melaleucae alternifoliae aetherolium

Botanical drug used: Essential oil, extracted by steam distillation of the leaves and terminal branches (twigs)

Indications/uses: Tea tree oil is used topically as an antiseptic and to aid wound healing, and in products to treat acne, head lice and scabies. It is also used in inhalations to treat respiratory diseases (colds, bronchitis, asthma etc.).

Evidence: There is *in vitro* evidence to show that tea tree oil is effective against bacteria, fungi, viruses, human lice and protozoa. Other laboratory studies, including in animals, have demonstrated anti-inflammatory properties of diluted tea tree oil applied topically. There is some pre-clinical and clinical evidence indicating effectiveness of tea tree oil to treat acne, and fungal skin and nail infections but evidence for other indications, including control of MRSA, is limited.

Safety: Tea tree oil appears to be safe to use externally, with uncommon and minor adverse effects reported. However, it should not be used internally.

Main chemical compounds: There is variability in the chemical profile of tea tree oil and six chemotypes (varieties) of *M. alternifolia* are recognised, each with a distinct chemical composition. The most highly valued varieties contain at least 40% terpinen-4-ol. There is an international standard that stipulates ranges for 14 constituents of the oil, including terpinen-4-ol, terpinene, terpinoline, cineole, terpineol, cymene, pinene and limonene. The composition of the oil may alter considerably during storage, as its stability is affected by light, heat, air and moisture (Carson et al. 2006; Tisserand and Young 2014).

Clinical evidence: Clinical studies assessing the efficacy of tea tree oil in the treatment of acne have demonstrated that 5% tea tree oil reduces the number of

---

Phytopharmacy: An evidence-based guide to herbal medicinal products, First Edition.
Sarah E. Edwards, Inês da Costa Rocha, Elizabeth M. Williamson and Michael Heinrich.
© 2015 John Wiley & Sons, Ltd. Published 2015 by John Wiley & Sons, Ltd.

inflamed lesions on a par with benzoyl peroxide, but with significantly fewer side effects (Carson et al. 2006).

Pre-clinical evidence and mechanisms of action: Numerous laboratory studies have demonstrated the broad spectrum activity of tea tree oil against bacteria, including MRSA (methicillin resistant *Staphylcoccus aureus*), fungi, viruses and protozoa (Carson et al. 2006; Garozzo et al. 2011). *In vitro* and *in vivo* studies, including in humans, have demonstrated anti-inflammatory properties of tea tree oil. A study in mice demonstrated the effectiveness of diluted tea tree oil (10%) against skin tumours (Ireland et al. 2012).

In vivo studies have demonstrated that the main bioactive ingredient in tea tree oil, terpinen-4-ol, is effective against the human pathogenic yeast (that causes thrush), *Candida albicans* (Mondello et al. 2006) and human lice and their eggs (Priestley et al. 2006). Anecodotal *in vivo* evidence indicates that tea tree oil may be effective in treating vaginal infections caused by the protozoa *Trichomona vaginalis* (Carson et al. 2006) but safety has not been evaluated.

Tea tree oil has been shown to compromise the structural and functional integrity of bacterial membranes, which probably accounts for its antibacterial properties. It also affects the permeability of the yeast *C. albicans* cells (Carson et al. 2006). One of the main components of tea tree oil, terpinen-4-ol, inhibits growth of influenza virus by an interference with acidification of the intralysosomal compartment, thus potentially inhibiting viral uncoating (Garozzo et al. 2011).

Interactions: An *in vitro* study demonstrated a synergistic effect in combination with the antibiotic tobramycin against the bacteria *S. aureus* and *Escherichia coli* (D'Arrigo et al. 2010). However, whether this would apply in a clinical setting is uncertain.

Contraindications: Internal use of tea tree oil is not recommended, unless highly diluted in, for example, oral preparations such as throat pastilles. Topical use of tea tree oil should be avoided by those prone to contact dermatitis and its use should be discontinued if it causes skin irritation. There is limited data whether tea tree oil is safe to use during pregnancy and lactation, and for these patients the WHO advises that it should only be used under medical supervision. Diluted tea tree oil (5%) applied topically appears to be safe for use in children (Tisserand and Young 2014).

Adverse effects: Tea tree oil applied topically may cause drying of the skin. Cases of contact dermatitis have been associated with topical use of tea tree, especially at high concentrations, and if the oil becomes oxidised (Tisserand and Young 2014).

Dosage: Topical applications vary from 0.4–100% oil formulation in clinical trials, depending on type and location of skin disorder, but a maximum of 15% oil formulation has been suggested to avoid sensitisation. Internal use should be avoided, due to toxicity (Tisserand and Young 2014).

General plant information: Traditionally, Aboriginal peoples of Australia crushed the leaves to inhale the vapour to treat respiratory infections and used them topically to treat insect bites and skin infections. Today *M. alternifolia* is commercially cultivated, predominantly in northern New South Wales. The common name 'tea tree' is derived from the practice of early Australian settlers, who used boiled leaves of *Melaleuca* and related *Leptospermum* species as a tea substitute (Australian National Botanic Gardens 2012). It was once heralded as a 'medicine chest in a bottle', due to its antiseptic properties (Kew 2012).

References

Australian National Botanic Gardens. (2012) *Information about Australia's Flora: Teatrees – genus Leptospermum.* http://www.anbg.gov.au/leptospermum/ (accessed 14 December 2012).

Carson CF, Hammer A, Riley TV. (2006) *Melaleuca alternifolia* (tea tree) Oil: a review of antimicrobial and other medicinal properties. *Clinical Microbiology Reviews* 19(1): 50–62.

D'Arrigo M, Ginestra G, Mandalari G. et al. (2010) Synergism and postantibiotic effect of tobramycin and *Melaleuca alternifolia* (tea tree) oil against *Staphylococcus aureus* and *Escherichia coli. Phytomedicine* 17(5): 317–322.

Garozzo A, Timpanaro R, Stivala A., Bisignano G, Castro A. (2011) Activity of *Melaleuca alternifolia* (tea tree) oil on Influenza virus A/PR/8: Study on the mechanism of action. *Antiviral Research* 89(1): 83–88.

Ireland DJ, Greay SJ, Hooper CM, Kissick HT, Filion P, Riley TV, Beilharz MW. (2012) Topically applied *Melaleuca alternifolia* (tea tree) oil causes direct anti-cancer cytotoxicity in subcutaneous tumour bearing mice. *Journal of Dermatology* 67(2): 120–129.

Kew. (2012) *Melaleuca alternifolia.* http://www.kew.org/plants-fungi/Melaleuca-alternifolia .htm (accessed 14 December 2012).

Mondello F, De Bernardis F, Girolamo A, Cassone A, Salvatore G. (2006) *In vivo* activity of terpinen-4-ol, the main bioactive component of *Melaleuca alternifolia* Cheel (tea tree) oil against azole-susceptible and – resistant human pathogenic *Candida* species. *BMC Infectious Diseases* 6: 158.

Priestley CM, Burgess IF, Williamson EM. (2006) *In vitro* lethality of essential oil constituents towards the human louse, *Pediculus humanus,* and its eggs. *Fitoterapia* 77(4): 303–309.

Tisserand R, Young R. (2014) *Essential Oil Safety.* 2nd Edition. Churchill Livingstone Elsevier, UK, pp. 440–445.

Thyme	*Thymus vulgaris* L., *T. zygis* L.

Synonyms: *T. vulgaris* – *Origanum thymus* Kuntze; *T. collinus* Salisb.; *T. zygis* – *O. zygis* (L.) Kuntze; *T. angustifolius* Salisb.

Family: Lamiaceae (Labiatae)

Other common names: Common thyme; garden thyme; *T. zygis*: Spanish thyme

Drug name: Thymi herba; Thymi aetheroleum

Botanical drug used: Dried leaves and flowers; essential oil

Indications/uses: Traditionally, thyme is used in the treatment of upper respiratory tract infections, and as an expectorant for the relief of chesty coughs. Other recorded internal uses of the herb include in the treatment of colic, flatulence and intestinal worms. Thyme has also been used topically in the treatment of inflammatory skin disorders, minor wounds, rheumatism, arthritis and the common cold. UK registered products (THRs) containing thyme are only found in combination with other herbs and thyme oil is an ingredient of some multi-ingredient licensed products that are inhaled or applied topically. Thyme oil is also used as a disinfectant, and its main constituent, thymol, is used in dental products (mouthwashes and toothpaste), to prevent plaque and protect the oral cavity from bacterial decay.

Evidence: Clinical data are lacking, although there is some evidence to support the use of thyme in treatment of bronchial catarrh.

Safety: Thyme herb appears to be safe when used in usual quantities as a flavouring in food. Relatively higher doses used for therapeutic purposes are not recommended during pregnancy, and although no concerns have been identified, safety of thyme during lactation has not been established. Thyme oil should not be used internally, and only used topically if diluted in a suitable carrier oil.

Main chemical compounds: Flavonoids, including apigenin, eriodictyol, luteolin, naringenin and others are the main non-volatile constituents found in thyme herb. The volatile oil (about 2.5%), which varies in chemical composition according to chemotype, contains up to 70% thymol, with carvacrol, p-cymene, linalool, α-terpineol and thujan-4-ol (WHO 1999; Williamson et al. 2013).

Phytopharmacy: An evidence-based guide to herbal medicinal products, First Edition.
Sarah E. Edwards, Inês da Costa Rocha, Elizabeth M. Williamson and Michael Heinrich.
© 2015 John Wiley & Sons, Ltd. Published 2015 by John Wiley & Sons, Ltd.

Clinical evidence:

Respiratory conditions: There is some clinical evidence that thyme is effective in the treatment of bronchial catarrh, but, overall, well-designed clinical studies for thyme are lacking. In a randomised, controlled, double-blind trial involving 60 subjects with productive cough, treatment with syrup of thyme was compared to bromhexine over a 5-day period. Self-reported symptom relief was similar in both groups (Pharmaceutical Press Editorial Team 2013).

A combination product containing thyme herb with primrose root was investigated to assess efficacy and tolerability in another randomised, controlled, double-blind study involving 361 outpatients with acute bronchitis or >10 coughing fits per day. The thyme-primrose combination product was found to be superior to placebo in terms of efficacy, with a 50% reduction in coughing fits from baseline reached 2 days earlier than with placebo, and rapid improvement in acute bronchitis symptoms observed. Treatment was well-tolerated (Kemmerich 2007).

Pre-clinical evidence and mechanisms of action:

Respiratory conditions: Spasmolytic and antitussive activities of thyme have been attributed largely to the phenolic constituents of the volatile oil, thymol and carvacrol. These compounds have both been shown to prevent contractions in rodent ileum and trachea induced by histamine, acetylcholine and other reagents, but the concentrations of phenolics of aqueous extracts of thyme are too low to account for overall activity (Begrow et al. 2010; WHO 1999). *In vitro* studies suggest that flavonoids, notably luteolin, may also be responsible for the antispasmodic activity of thyme extracts, which appears to be due to synergistic effects between the phenolic constituents and flavone (Engelbertz et al. 2012).

Antimicrobial effects: Thyme essential oil was found to exhibit antibacterial activity (with MIC values ranging 0.1-4 v/v %) against clinical isolates of methicillin-resistant *Staphylococcus aureus* (MRSA), and other standard bacterial strains (both gram-negative and gram-positive) using disk diffusion and agar dilution methods (Tohidpour et al. 2010). Both *in vitro* and *in vivo* studies have also demonstrated antifungal activity of thyme oil and thymol, including against dermatomycetes and moulds from damp dwellings (de Lira Mota et al. 2012; Segvić Klarić et al. 2007; Soković et al. 2008).

Anti-inflammatory effects: Thyme oil, and its constituents thymol and carvacrol, have also been shown to have anti-inflammatory activity, attributable to the inhibition of inflammatory oedema and leukocyte migration (Fachini-Queroz et al. 2012). *In vitro* studies have shown that thyme essential oil inhibits platelet aggregation induced by ADP, arachidonic acid, collagen and thrombin, but had no effect on U46619 (a synthetic thromboxane A_2 receptor agonist)-induced aggregation (Okazaki et al. 2002).

Other effects: An *in vitro* study showed that an aqueous-ethanolic thyme extract bound competitively to oestradiol and progesterone receptors (Gardner and McGuffin 2013). Thymol, carvacrol and their derivatives, thymoquinone and thymohyrdoquinone, were found to be inhibitors of acetylcholinesterase (AChE), thus suggesting potential application in treatment for cognitive disorders (Jukic et al. 2007).

Interactions: No case reports are available. An *in vitro* study investigating the pharmacokinetics of an aqueous thyme extract found that it was a potent inhibitor of several cytochrome 450 isoenzymes, including CYP2C9, CYP2C19, CYP2D6, and CYP3A4, but the clinical relevance of these findings is uncertain (Williamson et al. 2013).

Contraindications: Thyme is reported to be an emmenagogue or abortifacient by several authors, and should not be used during pregnancy except under medical supervision. *In vitro* studies have shown that thyme has anticoagulant properties and may slow blood clotting (Gardner and McGuffin 2013), so should be avoided prior to surgery to reduce the risk of excessive bleeding. Thyme oil should not be used internally due to its toxicity, and only applied topically if diluted in a suitable carrier oil (Pharmaceutical Press Editorial Team 2013).

Adverse effects: Contact dermatitis and systemic allergy have been reported, sometimes with cross-reactions to rosemary or oregano (Gardner and McGuffin 2013). The major constituent of the essential oil, thymol, has documented toxic symptoms including nausea, vomiting, gastric pain, headache, dizziness, convulsions, coma, and cardiac and respiratory arrest (Pharmaceutical Press Editorial Team 2013).

Dosage: Oral administration (adults): (a) dried herb – 1–4 g as an infusion 3 times per day; (b) liquid extract of thyme (BPC 1949) 0.6–4.0 ml; (c) elixir of thyme (BPC 1949) 4–8 ml; (d) tincture 2–6 ml (1:5 in 45% alcohol) 3 times per day, 4 drops (Pharmaceutical Press Editorial Team 2013).

General plant information: Thyme is a popular culinary herb from the Mediterranean region. Medieval knights reputedly wore sprigs of thyme on their armour, as it was believed that the scent imbued them with strength during battle.

References

Begrow F, Engelbertz J, Feistel B, Lehnfeld R, Bauer K, Verspohl EJ. (2010) Impact of thymol in thyme extracts on their antispasmodic action and ciliary clearance. *Planta Medica* 76(4): 311–318.

Engelbertz J, Lechtenberg M, Studt L, Hensel A, Verspohl EJ. (2012) Bioassay guided fractionation of a thymol-deprived hydrophilic thyme extract and its antispasmodic effect. *Journal of Ethnopharmacology* 141(3): 848–853.

Fachini-Queiroz FC, Kummer R, Estevão-Silva CF, Carvalho MD, Cunha JM, Grespan R, Bersani-Amado CA, Cuman RK. (2012) Effects of thymol and carvacrol, constituents of *Thymus vulgaris* L. essential oil on the inflammatory response. *Evidence-Based Complementary and Alternative Medicine* 2012: 657026.

Gardner Z, McGuffin M. (Eds.) (2013) *American Herbal Product Association's Botanical Safety Handbook.* 2nd Edition. CRC Press, USA, 1072 pp.

Jukic M, Politeo O, Maksimovic M, Milos M, Milos M. (2007) *In vitro* acetylcholinesterase inhibitory properties of thymol, carvacrol and their derivatives thymoquinone and thymohydroquinone. *Phytotherapy Research* 21(3): 259–261.

Kemmerich B. (2007) Evaluation of efficacy and tolerability of a fixed combination of dry extracts of thyme herb and primrose root in adults suffering from acute bronchitis with productive cough. A prospective, double-blind, placebo-controlled multicentre clinical trial. *Arzneimittel-Forschung* 57(9): 607–615.

de Lira Mota KS, de Oliveira PF, de Oliveira WA, Lima IO, de Oliveira LE. (2012) Antifungal activity of *Thymus vulgaris* L. essential oil and its constituent phytochemicals against *Rhizopus oryzae*: interaction with ergosterol. *Molecules* 17(12): 14418–14433.

Okazaki K, Kawazoe K, Takaishi Y. (2002) Human platelet aggregation inhibitors from thyme (*Thymus vulgaris* L.). *Phytotherapy Research* 16(4): 398–399.

Pharmaceutical Press Editorial Team. (2013) *Herbal Medicines.* 4th Edition. Pharmaceutical Press, London, UK.

Segvić Klarić M, Kosalec I, Mastelić J, Piecková E, Pepeljnak S. (2007) Antifungal activity of thyme (*Thymus vulgaris* L.) essential oil and thymol against moulds from damp dwellings. *Letters in Applied Microbiology* 44(1): 36–42.

Soković M, Glamočlija J, Ćirić A, Kataranovski D, Marin PD, Vukojević J, Brkić D. (2008) Antifungal activity of the essential oil of *Thymus vulgaris* L. and thymol on experimentally induced dermatomoycoses. *Drug Development and Industrial Pharmacy* 34(12): 1388–1393.

Tohidpour A, Sattari M, Omidbaigi R, Yadegar A, Nazemi J. (2010) Antibacterial effect of essential oils from two medicinal plants against Methicillin-resistant *Staphylococcus aureus* (MRSA). *Phytomedicine* 17(2): 1420145.

WHO. (1999) *WHO Monographs on Selected Medicinal Plants.* Vol. 1. WHO, Geneva, 295 pp.

Williamson EM, Driver S, Baxter K. (Eds.) (2013) *Stockley's Herbal Medicines Interactions.* 2nd Edition. Pharmaceutical Press, London, UK.

Tongkat Ali   *Eurycoma longifolia* Jack

Synonyms: *E. latifolia* Ridl.; *E. merguensis* Planch.; *E. tavoyana* Wall.

Family: Simaroubaceae

Other common names: Long Jack; Malaysian ginseng; pasak bumi

Drug name: Eurycoma longifolia radix

Botanical drug used: Roots; less frequently, the leaves

Indications/uses: Preparations of Tongkat Ali are most commonly used as a tonic for restoring energy and vitality, including in sports, and for enhancing sexual performance and increasing libido. The herb has also been used traditionally to treat malaria, fevers, diabetes and fatigue, to strengthen bones and to reduce blood pressure.

Evidence: A few clinical studies on the tonic and aphrodisiac effects are available to support the use. However, due to the lack of data on safety and since there are no products on the market with an assured quality profile the use of Tongkat Ali can generally not be recommended. Recent studies have also focussed on its anticancer and antimalarial properties but there is no clinical evidence to support these uses as yet.

Safety: There is a lack of data on safety and side effects, but the most serious concerns about safety are regarding the poor quality of products sold as 'Tongkat Ali', which must be taken into account when recommending use of the herb. The reported cytotoxic effects should also be taken into account.

Main chemical compounds: The major constituents are the β-carboline alkaloids, such as canthin-6-one, and the quassinoids which include lonilactone, eurycomanone, 13α(21)-epoxyeurycomanone and eurycomanol, and the quassinoid diterpenoids eurycomalide A and B. Other compounds present include tirucallane-type triterpenes, squalene derivatives, biphenylneolignans, eurycolactone, laurycolactone, eurycomalactone and volatiles (curcumene, massoilactone) (Al-Salahi et al. 2013; Bhat and Karim 2010).

Clinical evidence:

Enhancement of sexual activity, fertility and well-being: A randomised, double-blind, placebo-controlled, parallel group study was carried out over

Phytopharmacy: An evidence-based guide to herbal medicinal products, First Edition.
Sarah E. Edwards, Inês da Costa Rocha, Elizabeth M. Williamson and Michael Heinrich.
© 2015 John Wiley & Sons, Ltd. Published 2015 by John Wiley & Sons, Ltd.

12 weeks to investigate the clinical effects of Tongkat Ali (TA) in men. A total of 109 men between 30 and 55 years of age were given either 300 mg of water extract of TA ('Physta') or placebo. Quality of life was assessed using the SF-36 questionnaire, and sexual well-being was measured using the International Index of Erectile Function (IIEF) and Sexual Health Questionnaires (SHQ). Seminal Fluid Analysis (SFA), fat mass and safety profiles were also evaluated. The EL group significantly improved in the domain 'Physical Functioning' of SF-36 and showed higher scores in the overall 'Erectile Function' domain in IIEF and 'Libido' (14% by week 12), SFA- with sperm motility at 44.4%, and semen volume at 18.2% at the end of treatment. Subjects with BMI $\geq$ 25 kg/m^2 significantly improved in loss of fat mass. All safety parameters were comparable to placebo (Ismail et al. 2012).

A standardised water extract of *E. longifolia* was also used to investigate its effects on male partners of sub-fertile couples with idiopathic infertility (75 patients, 100 mg twice a day for 9 months; 3 cycles of 3 months). The study showed that the extract improved sperm concentration, motility and morphologically normal sperm in men who suffered from infertility due to unknown cause. Further studies are needed to determine the potential use in the treatment of male infertility (Tambi and Imran 2010).

Physically active senior participants (25, from both sexes, aged between 57 and 72 years) were given a standardised TA water extract at a dose of 200 mg, twice a day for 5 consecutive weeks, and its effect on well-being assessed. There was an increase in serum testosterone levels (total and free) in both men and women as well as an increase of muscle mass (Henkel et al. 2013).

A placebo-controlled study assessing the effect of supplementation with TA on stress hormones and mood state in 63 subjects (32 men and 31 women) for 4 weeks found significant improvements in the TA group for tension (−11%), anger (−12%), and confusion (−15%). Stress hormone profiles (salivary cortisol and testosterone) were improved by TA supplementation, with reduced cortisol exposure (−16%) and increased testosterone status (+37%) (Talbott et al. 2013).

Even though some clinical data are available, there are no products on the market that comply with European quality requirements and, overall, the evidence is very limited.

Anticancer effects: No clinical studies available.

Anti-plasmodial effects: No clinical studies available.

Other effects: No clinical studies available.

Pre-clinical evidence and mechanisms of action:

Enhancement of well-being, fertility and sexual activity: Eurycomanone, the major quassinoid in *E. longifolia* root extract, has been shown to increase spermatogenesis by inhibiting the activity of phosphodiesterase and aromatase in steroidogenesis (Low et al. 2013). Extracts of the herb have also been shown to increase levels of testosterone in animals (Chan et al. 2009).

Anticancer effects: Eurycomanone (at concentrations of 5–20 μg/ml) exhibited significant anti-proliferative and anti-clonogenic cell growth effects on A549 lung cancer cells (Wong et al. 2012). The anti-angiogenic potential of a partially purified quassinoid-rich fraction (TAF273) of the root extract has recently been evaluated using *ex vivo* and *in vivo* angiogenesis models such as human umbilical vein endothelial cells (Al-Salahi et al. 2013).

Anti-plasmodial effects: The quassinoids eurycomanone, its derivatives and eurycomalactone, and the alkaloid 9-methoxycanthin-6-one, display anti-plasmodial activity (Chan et al. 2004).

Other effects: Recent studies in animals suggest an effect on osteoporosis via increasing hormone levels (Shuid et al. 2013). In an *in vivo* study with hyperglycaemic rats, the extract showed a blood glucose lowering effect although no effect was seen in normoglycaemic rats. The leaf and stem extracts of *E. longifolia* (but not the roots, the usual plant part used) have shown antimicrobial activity against several strains of gram-negative and gram-positive bacteria *in vitro* (Farouk and Benafri 2007; Solomon et al. 2013).

Interactions: There is one report showing that the bioavailability of propranolol (as a single dose) was significantly decreased when administered concomitantly with *E. longifolia* (200 mg of water-based extract) in healthy male subjects and caution is therefore advisable (Salman et al. 2010).

Contraindications: None known. Avoid during pregnancy and lactation, due to lack of data.

Adverse effects: None reported.

Dosage: There are insufficient data to recommend a suitable dose. For commercial products, follow the manufacturers' instructions.

General plant information: The plant is native to Southeast Asia and is found in Malaysia, Indonesia, Vietnam and the Philippines. Products are usually presented as raw crude powder or in capsules, alone or mixed with other 'aphrodisiac' herbs (Bhat and Karim 2010). The Health Sciences Authority of Singapore issued a warning in 2011 for adulterated Tongkat Ali 'herbal' medicinal products which actually contained tadalafil (MHRA 2011).

References

Al-Salahi OS, Kit-Lam C, Majid AM, Al-Suede FS, Mohammed Saghir SA, Abdullah WZ, Ahamed MB, Yusoff NM. (2013) Anti-angiogenic quassinoid-rich fraction from *Eurycoma longifolia* modulates endothelial cell function. *Microvascular Research* 90: 30–39.

Bhat R, Karim AA. (2010) Tongkat Ali (*Eurycoma longifolia* Jack): a review on its ethnobotany and pharmacological importance. *Fitoterapia* 81(7): 669–679.

Chan KL, Choo CY, Abdullah NR, Ismail Z. (2004) Antiplasmodial studies of *Eurycoma longifolia* Jack using the lactate dehydrogenase assay of *Plasmodium falciparum*. *Journal of Ethnopharmacology* 92(2–3): 223–7.

Chan KL, Low BS, Teh CH, Das PK. (2009) The effect of *Eurycoma longifolia* on sperm quality of male rats. *Natural Product Communications* 4(10): 1331–1336.

Farouk AE, Benafri A. (2007) Antibacterial activity of *Eurycoma longifolia* Jack. A Malaysian medicinal plant. *Saudi Medical Journal* 28(9): 1422–1424.

Henkel RR, Wang R, Bassett SH, Chen T, Liu N, Zhu Y, Mi T. (2013) Tongkat Ali as a potential herbal supplement for physically active male and female seniors-a pilot study. *Phytotherapy Research* 28(4): 544–550.

Ismail SB, Wan Mohammad WM, George A, Nik Hussain NH, Musthapa Kamal ZM, Liske E. (2012) Randomized clinical trial on the use of PHYSTA freeze-dried water extract of *Eurycoma longifolia* for the improvement of quality of life and sexual well-being in men. *Evidence-Based Complementary and Alternative Medicine* 2012: 429268.

Low BS, Choi SB, Wahab HA, Das PK, Chan KL. (2013) Eurycomanone, the major quassinoid in *Eurycoma longifolia* root extract increases spermatogenesis by inhibiting the activity of phosphodiesterase and aromatase in steroidogenesis. *Journal of Ethnopharmacology* 149(1): 201–207.

MHRA. (2011). Singapore Health Sciences Authority find 'tadalafil', a potent prescription medicine, in the capsule shells of a traditional herbal supplement for men. http://www .mhra.gov.uk/Safetyinformation/Generalsafetyinformationandadvice/Herbalmedicines /Herbalsafetyupdates/Allherbalsafetyupdates/CON051838.

Miyake K, Tezuka Y, Awale S, Li F, Kadota S. (2009) Quassinoids from *Eurycoma longifolia*. *Journal of Natural Products* 72(12): 2135–2140.

Salman SA, Amrah S, Wahab MS, Ismail Z, Ismail R, Yuen KH, Gan SH. (2010) Modification of propranolol's bioavailability by *Eurycoma longifolia* water-based extract. *Journal of Clinical Pharmacy and Therapeutics* 35(6): 691–696.

Shuid AN, El-arabi E, Effendy NM, Razak HS, Muhammad N, Mohamed N, Soelaiman IN. (2013) Eurycoma longifolia upregulates osteoprotegerin gene expression in androgen-deficient osteoporosis rat model. *BMC Complementary and Alternative Medicine* 12: 152.

Solomon MC, Erasmus N, Henkel RR. (2013) *In vivo* effects of *Eurycoma longifolia* Jack (Tongkat Ali) extract on reproductive functions in the rat. *Andrologia* 46(4): 339–348.

Talbott SM, Talbott JA, George A, Pugh M. (2013) Effect of Tongkat Ali on stress hormones and psychological mood state in moderately stressed subjects. *Journal of the International Society of Sports Nutrition* 10(1): 28.

Tambi MI, Imran MK. (2010) *Eurycoma longifolia* Jack in managing idiopathic male infertility. *Asian Journal of Andrology* 12(3): 376–380.

Wong PF, Cheong WF, Shu MH, Teh CH, Chan KL, AbuBakar S. (2012) Eurycomanone suppresses expression of lung cancer cell tumor markers, prohibitin, annexin 1 and endoplasmic reticulum protein 28. *Phytomedicine* 19(2): 138–144.

♠ ♠ ♠ ♠ ♠

Turmeric   *Curcuma longa* L.

Synonyms: *C. domestica* Valeton; and others

Family: Zingiberaceae

Other common names: Curcuma; haldi; Indian saffron

Drug name: Curcuma longae rhizoma

Botanical drug used: Rhizome (often erroneously called a 'root')

Indications/uses: Turmeric is mainly used to treat digestive, liver and skin disorders, and as an anti-inflammatory and immunomodulating agent in arthritis and psoriasis. It has many more traditional uses in TCM and Ayurveda, such as in treatment of menstrual problems, epilepsy, asthma, cough, haemorrhage, insect bites, kidney stones and to improve lactation. It is a well-known culinary spice and colourant.

Evidence: There is limited clinical evidence that turmeric (or its active constituent curcumin) may be effective in the treatment of gastrointestinal complaints, including dyspepsia and irritable bowel syndrome; that it reduces pain caused by osteoarthritis; reduces swelling in rheumatoid arthritis; and is effective in the treatment of psoriasis. However, robust clinical data for most indications are lacking. Numerous *in vitro* and animal studies have shown that turmeric has anti-inflammatory, antimicrobial, antioxidant and anticancer properties.

Safety: There is a lack of safety data for turmeric, although as it is eaten so widely it is likely to be safe if used at recommended doses by adults.

Main chemical compounds: The active compounds are the curcuminoids (3–5%), which are phenylpropanoids, predominantly curcumin (50–60%), with monodesmethoxycurcumin, dihydrocurcumin and many others; volatile oil (about 6%) consisting of mono- and sesquiterpenes including zingiberene, curcumene, α- and β-turmerene (ESCOP 2003; Williamson 2002). The products vary considerably in their chemical composition (Booker et al. 2014)

Clinical evidence: Despite the publication of several thousand research studies on turmeric or curcumin, and a few small clinical trials showing promise, robust scientific evidence for the effectiveness of turmeric in humans is lacking. The majority of trials have focused on the active constituent curcumin, although it has poor bioavailability due to its hydrophobicity. Low systemic bioavailability

---

Phytopharmacy: An evidence-based guide to herbal medicinal products, First Edition.
Sarah E. Edwards, Inês da Costa Rocha, Elizabeth M. Williamson and Michael Heinrich.
© 2015 John Wiley & Sons, Ltd. Published 2015 by John Wiley & Sons, Ltd.

after oral dosing may limit sufficient levels of curcumin for pharmacological effects reaching tissues outside of the gastrointestinal tract. Novel delivery methods are currently being developed, and a new highly bioavailable form of curcumin has entered phase 1 clinical trials (Kanai et al. 2013; Strimpakos and Sharma 2008; Thangapazham et al. 2013).

A number of studies have shown promising effects of curcumin in patients with various inflammatory diseases, including cancer, cardiovascular disease, arthritis, Crohn's disease, ulcerative colitis, peptic ulcer, gastric ulcer, irritable bowel disease, gastric inflammation, vitiligo, psoriasis, acute coronary syndrome, atherosclerosis, diabetes, renal conditions and β-thalassemia (Gupta et al. 2013; Heng et al. 2000; Strimpakos and Sharma 2008; Thangapazham et al. 2013).

Benefits of curcumin have also been demonstrated in healthy adults. A small controlled study in healthy adults (40–60 years), given either 80 mg/day curcumin (in a lipid-solubilised form to increase absorption) or placebo, demonstrated a number of potentially health-promoting effects. These included the statistically significant lowering of plasma triglyceride values, lowering of salivary amylase levels, raising of salivary radical scavenging capacities, raising of plasma catalase activities, lowering of plasma beta amyloid protein concentrations, lowering of plasma soluble intercellular adhesion molecule (sICAM) readings, increased plasma myeloperoxidase without increased C-reactive protein (CRP) levels, increased plasma nitric oxide and decreased plasma alanine amino transferase activities (DiSilvestro et al. 2012).

A few earlier clinical studies have investigated turmeric rhizome, rather than its constituent curcumin. Some of these suggest that turmeric may be effective in treating gastrointestinal conditions, including dyspepsia, irritable bowel syndrom, duodenal ulcer and cholecystitis (ESCOP 2003; Pharmaceutical Press Editorial Team 2013; WHO 2013). Another study has demonstrated that turmeric is as effective in reducing pain as ibuprofen in cases of osteoarthritis (Kuptniratsaikul et al. 2009). Potential antimutagenic and chemopreventive activities of turmeric have also been reported in humans. A study in 16 chronic smokers found that turmeric, given in doses of 1.5 g/day for 30 days, significantly reduced urinary excretion of mutagens, compared with 6 non-smokers (controls) (Polasa et al. 1992).

Pre-clinical evidence and mechanisms of action: Extensive *in vitro* and *in vivo* studies have demonstrated gastroprotective, anti-inflammatory, antioxidant, anti-arthritic, hypolipidaemic, hepatoprotective; antiviral, antibacterial, antifungal, anti-tumour and chemopreventive properties of turmeric and its active ingredients (ESCOP 2003; Funk et al. 2006; Jankasem et al. 2012; Pharmaceutical Press Editorial Team 2013; Strimpakos and Sharma 2008; Thangapazham et al. 2013).

Clearly, turmeric and its preparations act on multiple biological pathways. Curcumin and its derivatives are known antioxidants, and scavenge oxygen radicals that are implicated in the inflammatory process. Curcumin inhibits cycloxygenase-2 (COX-2), an enzyme associated with inflammation and tumourigenesis) and inhibits nuclear factor-kappa B (NF-κB). Curcuminoids have been shown to increase levels of CRP, a strong predictor and independent risk factor of atherosclerosis and cardiovascular disease (Sahebkar 2013). Curcumin modulates various signalling pathways including the mitogen-activated protein kinase pathway, regulating a wide range of transcription factors, growth factors and inflammatory cytokines. Evidence from a variety of cancer cell models suggests that curcumin has an anti-proliferative effect through induction of G2/M cell cycle arrest and apoptosis-like death (Colalto 2010; Gupta et al. 2013; Jackson et al. 2013; Thangapazham et al. 2013).

Evidence of curcumin toxicity has been demonstrated in animal models. Curcumin was shown to inhibit the proliferation of bovine aortic endothelial cells at concentrations relevant to the diet; concentrations as low as 0.1 μM affected DNA segregation and micronucleation (Jackson et al. 2013).

Interactions: Curcumin inhibits P-glycoprotein, and has been shown to enhance the effect of several drugs, often in a positive manner. Turmeric or curcumin also affects the absorption of some beta-blockers, increases the absorption of midazolam, but not the absorption of iron. Many Ayurvedic formulae of turmeric include pepper, as piperine enhances absorption (Williamson et al. 2013).

Contraindications: Obstruction of the biliary tract, cholangitis, gallstones, alcoholic liver disease. Although turmeric is used in Ayurveda to treat liver and biliary disease (Williamson 2002), these indications should only be treated by a medically qualified professional. A recent study found both advantageous and disadvantageous functions of curcumin on alcoholic liver injury in mice. Ethanol accelerates serum levels of serum aspartate aminotransferase (AST) and alanine aminotransferase (ALT), causing liver injury, production of tumour necrosis factor-alpha (TNF-α), transforming growth factor-beta (TGF-β) and NF-κB. Curcumin at a dose of 1×10^{-3} M actually accelerated liver injury with 5% ethanol, whereas at a lower dose (1×10^{-4} M) it did not lead to (and provided protection from) alcoholic liver injury, and may therefore have dual effects on alcoholic liver injury depending on its concentration (Zhao et al. 2012).

Adverse effects: Mild symptoms including dry mouth, flatulence and gastric irritation may also occur, but frequency is unknown (EMA 2009). Allergic dermatitis after topical application of turmeric and curcumin has been reported (ESCOP 2003).

There is a lack of data on the use of turmeric during pregnancy and lactation in humans, but a study in rats found no toxic effects (Ganiger et al. 2007) and it is likely to be safe in recommended doses.

A more important concern is the quality of turmeric-based food supplements. A product sold in Scandinavian countries was found to be contaminated with the drug nimesulide, which led to reports of liver damage, including one death (Stiefelhagen 2010). Consumers should be cautious about unregistered products sold with unsubstantiated claims, especially from unregulated sources over the Internet.

Dosage: Adults and elderly (oral administration): (a) Powdered herbal substance: 1.5–3.0 g daily; (b) comminuted herbal substance for tea preparation: 0.5–1 g, up to three times daily as an infusion; (c) tincture (1:10) 0.5–1 ml three times daily; (d) dry extract (13–25:1) 80–160 mg, divided in 2–5 partial doses daily; (e) dry extract (5.5–6.5:1): 100–200 mg, two times daily; (f) tincture (1:5): 10 ml once daily or 5 ml in 60 ml water three times daily (EMA 2009).

General plant information: Turmeric was at the centre of a so-called 'biopiracy' case, when in 1995 a US patent was issued for its use in wound healing to researchers at the University of Mississippi. After legal-wrangling, the patent was revoked, as evidence from ancient Sanskrit texts proved that the patent lacked novelty (Finetti 2011).

References

The literature on turmeric is immense. This is a selection of references in support of the information given above.

Booker A, Frommenwiler D, Johnston D, Umealajekwu Ch, Reich E, Heinrich M. (2014) Chemical variability along the value chains of turmeric: A comparison of Nuclear Magnetic Resonance (NMR) Spectroscopy and High Performance Thin Layer Chromatography (HPTLC). *Journal of Ethnopharmacology* 152(2): 292–301.

Colalto C. (2010) Herbal interactions on absorption of drugs: Mechanisms of action and clinical risk assessment. *Pharmacological Research* 62(3): 207–227.

DiSilvestro RA, Joseph E, Zhao S, Bomser J. (2012) Diverse effects of a low dose supplement of lipidated curcumin in healthy middle aged people. *Nutrition Journal* 11: 79.

EMA. (2009) Community herbal monograph on *Curcuma longa* L., rhizoma. European Medicines Agency. EMA/HMPC/456845/2008. http://www.ema.europa.eu/docs/en_GB/document_library/Herbal_-_Community_herbal_monograph/2010/02/WC500070703.pdf (accessed August 2014).

ESCOP. (2003) *ESCOP Monographs: The Scientific Foundation for Herbal Medicinal Products.* 2nd Edition. Thieme, Exeter and London, UK.

Finetti C. (2011) Traditional knowledge and the patent system: Two worlds apart? *World Patent Information* 33(1): 58–66.

Funk JL, Oyarzo JN, Frye JB, Chen G, Lantz RC, Jolad SD, Sólyom AM, Timmermann BN. (2006) Turmeric extracts containing curcuminoids prevent experimental rheumatoid arthritis. *Journal of Natural Products* 69(3): 351–355.

Ganiger S, Malleshappa HN, Krishnappa H, Rajashekhar G, Ramakrishna Rao V, Sullivan F. (2007) A two generation reproductive toxicity study with curcumin, turmeric yellow, in Wistar rats. *Food and Chemical Toxicology.* 45(1): 64–69.

Gupta SC, Patchva S, Aggarwal BB. (2013) Therapeutic roles of curcumin: lessons learned from clinical trials. *AAPS Journal* 15(1): 195–218.

Heng MC, Song MK, Harker J, Heng MK. (2000) Drug-induced suppression of phosphorylase kinase activity correlates with resolution of psoriasis as assessed by clinical, histological and immunohistochemical parameters. *British Journal of Dermatology* 143(5): 937–949.

Jackson SJ, Murphy LL, Venema RC, Singletary KW, Young AJ. (2013) Curcumin binds tubulin, induces mitotic catastrophe, and impedes normal endothelial cell proliferation. *Food and Chemical Toxicology* 60: 431–438.

Jankasem M, Withi-udomlert M, Gritsanapan W. (2012) Antifungal activity of turmeric creams at different concentrations. *Planta Medica* 78: PD56.

Kanai M, Otsuka Y, Otsuka K, Sato M, Nishimura T, Mori Y, Kawaguchu M, Hatan E, Kodama Y, Matsumoto S, Murakami Y, Imaizumi A, Chiba T, Nishihira J, Shibata H. (2013) A phase I study investigating the safety and pharmacokinetics of highly bioavailable curcumin (Theracurmin®) in cancer patients. *Cancer Chemotherapy and Pharmacology*. 71(6): 1521–1530.

Kuptniratsaikul V, Thanakhumtorn S, Chinswangwatanakul P, Wattanamongkonsil L, Thamlikitkul V. (2009) Efficacy and safety of *Curcuma domestica* extracts in patients with knee osteoarthritis. *Journal of Complementary and Alternative Medicine* 15(8): 891–897.

Pharmaceutical Press Editorial Team. (2013) *Herbal Medicines.* 4th Edition. Pharmaceutical Press, London, UK.

Polasa K, Raghuram TC, Krishna TP, Krishnaswamy K. (1992) Effect of turmeric on urinary mutagens. *Mutagenesis* 7(2): 107–109.

Sahebkar A. (2013) Are Curcuminoids Effective C-Reactive Protein-Lowering Agents in Clinical Practice? Evidence from a Meta-Analysis. *Phytotherapy Research* 28(5): 633–642.

Stiefelhagen P. (2010) "Doing something good" for the body? Definitely not! Liver damage caused by food supplements. *MMW Fortschritte der Medizin* 152(43): 21 [article in German].

Strimpakos AS, Sharma RA. (2008) Curcumin: preventive and therapeutic properties in laboratory studies and clinical trials. *Antioxidants and Redox Signalling* 10(3): 511–545.

Thangapazham RL, Sharad S, Maheshwari RK. (2013) Skin regenerative potentials of curcumin. *Biofactors* 39(1): 141–149.

Williamson EM, Driver S, Baxter K. (Eds.) (2013). *Stockley's Herbal Medicines Interactions.* 2nd Edition. Pharmaceutical Press, London, UK.

Williamson EM. (Ed.) (2002) *Major Herbs of Ayurveda.* Dabur Research Foundation. Churchill Livingstone Elsevier, UK.

Zhao HL, Song CH, Hee Chai OH. (2012) Negative Effects of Curcumin on Liver Injury Induced by Alcohol. *Phytotherapy Research* 26(12): 1857–1863.

Valerian   *Valeriana officinalis* L.

Family: Caprifoliaceae (Valerianaceae)

Other common names: All-heal; St. George's herb; setwall

Drug name: Valerianae radix

Botanical drug used: Roots, rhizomes and stolons

Indications/uses: Valerian is used as a mild sedative to treat symptoms of anxiety, stress and insomnia, including (and especially) during menopause.

Evidence: There is a large body of empirical and observational evidence but the results from controlled clinical trials investigating valerian for treating anxiety and sleep disorders are conflicting. Constituents from valerian have been demonstrated to be effective sedatives in animals (see the subsequent sections).

Safety: Short-term use (4–6 weeks) appears to be generally safe, but safety data for long-term use is lacking. Long-term use may result in benzodiazepine-like withdrawal symptoms.

Main chemical compounds: The chemical composition of *V. officinalis* varies considerably, depending on subspecies or variety, growing conditions and age or type of extract (WHO 1999). Significant constituents include the volatile oil, which contains bornyl acetate and bornyl isovalerate, β-carophyllene, valeranone, valerenic acid and other sesquiterpenoids and monoterpenes, iridoids (valepotriates), alkaloids, sterols and amino acids (Pharmaceutical Press Editorial Team 2013; WHO 1999).

Clinical evidence:

Sleeping disorders: Systematic reviews of clinical trials have been inconclusive as to the efficacy of valerian to treat either sleeping disorders or anxiety (Miyasaka et al. 2006; Nunes and Sousa 2011; Stevinson and Ernst 2000; Taibi et al. 2007). A recent study, in 100 women aged 50–60, found that a herbal preparation containing valerian with lemon balm produced a significant reduction in levels of sleep disorders compared to placebo (Taavoni et al. 2013).

Restless leg syndrome: A study by Cuellar and Ratcliffe (2009) suggested that the use of 800 mg of valerian for 8 weeks improved symptoms of restless leg syndrome

---

Phytopharmacy: An evidence-based guide to herbal medicinal products, First Edition.
Sarah E. Edwards, Inês da Costa Rocha, Elizabeth M. Williamson and Michael Heinrich.
© 2015 John Wiley & Sons, Ltd. Published 2015 by John Wiley & Sons, Ltd.

and decreased daytime sleepiness in patients who had reported an Epworth Sleepiness Scale (ESS) score of 10 or greater.

Menopausal symptoms: A double blind placebo-controlled clinical trial in 68 menopausal women with hot flushes investigated the effects of valerian capsules (255 mg 3 times a day for 8 weeks) on their severity and frequency as recorded through questionnaires, 2 weeks before, and four and eight weeks after the treatment. There was a statistical difference pre- and post-valerian treatment ($p < 0.001$) but not with placebo, regarding both severity and frequency 4 and 8 weeks after the treatment ($p < 0.001$) (Mirabi and Mojab 2013).

General: Variation in activity between commercially available valerian extracts has been demonstrated, irrespective of standardisation with valerenic acid (Ortiz et al. 2006) and may explain conflicting clinical trial data. Large variability in the pharmacokinetics of valerenic acid may contribute to the inconsistencies found in the effects of valerian (Anderson et al. 2010). Recent evidence suggests that the anxiolytic activity of the valerian extract is related to valerenic acid itself and not to derivatives such as acetoxy-valerenic acid (Felgentreff et al. 2012).

Pre-clinical evidence and mechanisms of action: Studies in rats have shown that valerian root extract, and its constituent valerenic acid, have potent anxiolytic effects (Murphy et al. 2010). A range of compounds, including the valerenic acids, amino acids and valepotriates, are thought to contribute to the effects either additively or synergistically (Houghton 1999). Valerian extracts modulate GABA, (Murphy et al. 2010) as does valerenic acid, but not acetoxy-valerenic acid (Felgentreff et al. 2012). Valepotriates have been implicated in a range of effects, but these are highly labile and decompose into valerenic acid very easily.

Interactions: Valerian extract has been shown to weakly inhibit some cytochrome P450 isoenzymes *in vitro,* but this is unlikely to be clinically relevant. Studies in mice indicate that valerian may increase sleeping time in response to barbiturates and alcohol. Case reports indicate there may be possible interaction with the antidiarrhoeal drug loperamide (Imodium) and the herbal medicines ginkgo and St. John's wort (Williamson et al. 2013). Valerian may potentiate the effects of anaesthetics that act on GABA receptors, thus valerian should be avoided pre-surgery to avoid a potential valerian-anaesthetic interaction (Yuan et al. 2004). There have been no clinical reports involving interactions with other sedatives.

Contraindications: Valerian is not recommended during pregnancy or lactation due to lack of safety data. Driving or operating machinery should be avoided when using valerian, due to its sedative effects.

Adverse effects: Few adverse effects for valerian have been reported. However, due to its GABA-receptor activity, it has been suggested that long-term use may lead to benzodiazepine-type withdrawal symptoms (Garges et al. 1998).

Dosage: Dried root and rhizome: 1–3 g as an infusion or decoction, 1–3 times per day. Tincture: 3–5 ml (1:5, 70% ethanol), once to several times daily. Extracts: amount equivalent to 2–3 g drug, once to several times daily; 2–6 ml of 1:2 liquid extract daily (Pharmaceutical Press Editorial Team 2013). External use: 100 g drug for a full bath (WHO 1999).

General plant information: Valerian was used to alleviate stress caused by air raids in the First World War (Grieve 1931). Indian valerian or 'Tagar' (*V. jatamansi*

Jones, synonym *V. wallichii* DC.), which is from the Himalayan region and used in Ayurvedic medicine, is sometimes a substitute for the European *V. officinalis*. Most of the material used pharmaceutically is derived from cultivation.

References

Anderson GD, Elmer GW, Taibi DM, Vitiello MV, Kantor E, Kalhorn TF, Howald WN, Barsness S, Landis CA. (2010) Pharmacokinetics of valerenic acid after single and multiple doses of valerian in older women. *Phytotherapy Research* 24(10): 1442–1446.

Cuellar NG, Ratcliffe SJ. (2009) Does valerian improve sleepiness and symptom severity in people with restless legs syndrome? *Alternative Therapies in Health and Medicine* 15(2): 22–8.

Felgentreff F, Becker A, Meier B, Brattström A. (2012) Valerian extract characterized by high valerenic acid and low acetoxy valerenic acid contents demonstrates anxiolytic activity. *Phytomedicine* 19(13): 1216–1222.

Garges HP, Varia I, Doraiswamy M. (1998) Cardiac complications and delirium associated with valerian root withdrawal. *JAMA* 280(18): 1566–1567.

Grieve M. (1931) *A Modern Herbal*. Harcourt, Brace & Company/(1971) Dover Publications, USA/(1995) online version Ed Greenwood, USA. http://botanical.com/botanical /mgmh/v/valeri01.html (accessed 10 May 2013).

Houghton PJ. (1999) The scientific basis for the reputed activity of valerian. *Journal of Pharmacy and Pharmacology* 51(5): 505–512.

Mirabi P, Mojab F. (2013) The effects of valerian root on hot flashes in menopausal women. *Iranian Journal of Pharmaceutical Research* 12(1): 217–222.

Miyasaka LS, Atallah ÁN, Soares B. (2006) Valerian for anxiety disorders. *Cochrane Database of Systematic Reviews* (4): CD004515.

Murphy K, Kubin ZJ, Shepherd JN, Ettinger RH. (2010) *Valeriana officinalis* root extracts have potent anxiolytic effects in laboratory rats. *Phytomedicine* 17(8–9): 674–678.

Nunes A, Sousa, M. (2011) Use of valerian in anxiety and sleep disorders: what is the best evidence? *Acta Médica Portuguesa* 24(Suppl. 4): 961–966.

Ortiz JG, Rassi N, Maldonado PM et al. (2006) Commercial valerian interactions with [3H]Flunitrazepam and [3H]MK-801 binding to rat synaptic membranes. *Phytotherapy Research* 20(9): 794–798.

Pharmaceutical Press Editorial Team. (2013) *Herbal Medicines*. 4th Edition. Pharmaceutical Press, London, UK.

Stevinson C, Ernst E. (2000) Valerian for insomnia: A systematic review of randomized clinical trials. *Sleep Medicine* 1(2): 91–99.

Taavoni S, Nazem Ekbatani N, Haghani H. (2013) Valerian/lemon balm use for sleep disorders during menopause. *Complementary Therapies in Clinical Practice* 19(4): 193–196.

Taibi DM, Landis CA, Petry H, Vitiello MV. (2007) A systematic review of valerian as a sleep aid: Safe but not effective. *Sleep Medicine Reviews* 11(3): 209–230.

WHO. (1999) *WHO Monographs on Selected Medicinal Plants*. Vol. 1. WHO, Geneva, 295 pp.

Williamson EM, Driver S, Baxter K. (Eds.) (2013) *Stockley's Herbal Medicines Interactions*. 2nd Edition. Pharmaceutical Press, London, UK.

Yuan CS, Mehendale S, Xiao Y, Aung HH, Xie JT, Ang-Lee MK. (2004) The gamma-aminobutyric acidergic effects of valerian and valerenic acid on rat brainstem neuronal activity. *Anesthesia and Analgesia* 98(2): 353–358.

Verbena *Verbena officinalis* L.

Family: Verbenaceae

Other common names: Vervain; European vervain

Drug name: Verbenae herba

Botanical drug used: Herb

Indications/uses: Verbena is used for the relief of nasal congestion and sinusitis (generally in combination products) and for the temporary relief of symptoms associated with stress, such as mild anxiety and to aid sleep.

Evidence: There is a lack of clinical studies on verbena preparations, although clinical evidence on its use as part of combination products (SinuComp®, Sinupret®) support its therapeutic use for sinusitis.

Safety: Verbena is considered to be safe when used at recommended doses.

Main chemical compounds: The main active constituents include the iridoid and secoiridoid glycosides verbenalin and hastatoside and verbenoside A and B; the phenylpropanoid glycosides such as verbascoside and flavonoids like luteolin, kaempferol, apigenin, quercetol and isorhamnetin (Duan et al. 2011; Schonbichler et al. 2013; Xu et al. 2010). A number of triterpenoids based on ursolic acid have been isolated (Shu et al. 2013).

Clinical evidence:

No clinical data are available for vervain and its preparations alone.

Sinusitis: A herbal combination product, Sinupret, containing gentian root, primrose flower, elderflower, sorrel herb and vervain herb, ratio 1:3:3:3:3) has been assessed in pre-clinical and clinical studies. A placebo-controlled, randomised double-blind clinical trial assessed the effect of this formulation in conjunction with antibiotics and nasal decongestants in acute sinusitis patients (160 subjects) and found Sinupret improved the symptoms of acute bacterial sinusitis (Neubauer and Marz 1994).

Phytopharmacy: An evidence-based guide to herbal medicinal products, First Edition.
Sarah E. Edwards, Inês da Costa Rocha, Elizabeth M. Williamson and Michael Heinrich.
© 2015 John Wiley & Sons, Ltd. Published 2015 by John Wiley & Sons, Ltd.

Pre-clinical evidence and mechanisms of action:

Antiviral effects: Two different formulations (dry extract and oral drops) of a combination product (described earlier) was tested *in vitro* for its antiviral potential against human pathogenic viruses known to cause infections of the upper respiratory tract. Both formulations showed a dose dependent (EC_{50} between 13.8 and 124.8 µg/ml) antiviral activity with a higher potency of the dry extract (Glatthaar-Saalmuller et al. 2011). Another study in mice also showed the antiviral potential of this formulation given as a prophylactic treatment on a respiratory tract infection model, the virus-induced mortality based Sendai virus (*Parainfluenza viridae*) infection. In general, Sinupret was more effective than the controls used (Schmolz et al. 2001).

Anti-inflammatory effects: A topical administration of a methanol extract of *V. officinalis* leaves (containing at least 3% of *V. officinalis*) has shown an anti-inflammatory and analgesic effect in a carrageenan-induced rat paw oedema and formalin test models (Calvo 2006). Three different extracts (petroleum ether, chloroform and methanol) of the aerial part of *V. officinalis* also showed anti-inflammatory action in a carrageenan paw oedema model, with the chloroform extract being the most active (Deepak and Handa 2000). Similar results were also observed in another study in rats on inflammatory effects in combination with protection of gastric damage and wound healing effect by different extracts of *V. officinalis* (Speroni et al. 2007).

Antioxidant and Antifungal Effects: A methanolic extract of the leaves showed antioxidant and antifungal activity *in vitro* (Casanova et al. 2008).

Cytotoxity: The triterpenoids have been shown to possess cytoxic effects against human hepatoma cell line Bel-7402 *in vitro* (Shu et al. 2013).

Interactions: Verbena may reduce iron absorption (Williamson et al. 2013).

Contraindications: Verbena is to not recommended during pregnancy due to lack of safety data.

Adverse effects: Allergic contact dermatitis has been reported (Del Pozo et al. 1994).

Dosage: Verbena should be used as recommended by the manufacturer.

General plant information: Verbena should not to be confused with lemon verbena (*Aloysia citrodora* Palau). Verbena is a traditional Chinese medicine (TCM) used for clearing away heat and detoxicating, promoting blood circulation and removing blood stasis, induce diuresis and excrete dampness (Duan et al. 2011). The name Vervain comes from the Celtic 'ferfaen' – to drive away a stone, referring to the traditional use of this herb for bladder problems and kidney stones.

References

Calvo MI. (2006) Anti-inflammatory and analgesic activity of the topical preparation of *Verbena officinalis* L. *Journal of Ethnopharmacology* 107(3): 380–382.

Casanova E, Garcia-Mina JM, Calvo MI. (2008) Antioxidant and antifungal activity of *Verbena officinalis* L. leaves. *Plant Foods for Human Nutrition* 63(3): 93–97.

Deepak M, Handa SS. (2000) Antiinflammatory activity and chemical composition of extracts of *Verbena* officinalis. *Phytotherapy Research* 14(6): 463–465.

Del Pozo MD, Gastaminza G, Navarro JA, Munoz D, Fernandez E, Fernandez de Corres L. (1994) Allergic contact dermatitis from *Verbena officinalis* L. *Contact Dermatitis* 31(3): 200–201.

Duan K, Yuan Z, Guo W, Meng Y, Cui Y, Kong D, Zhang L, Wang N. (2011) LC-MS/MS determination and pharmacokinetic study of five flavone components after solvent extraction/acid hydrolysis in rat plasma after oral administration of *Verbena officinalis* L. extract. *Journal of Ethnopharmacology* 135(2): 201–208.

Glatthaar-Saalmuller B, Rauchhaus U, Rode S, Haunschild J, Saalmuller A. (2011) Antiviral activity *in vitro* of two preparations of the herbal medicinal product Sinupret® against viruses causing respiratory infections. *Phytomedicine* 19(1): 1–7.

Neubauer N, Marz RW. (1994) Placebo-controlled, randomized double-blind clinical trial with Sinupret® sugar coated tablets on the basis of a therapy with antibiotics and decongestant nasal drops in acute sinusitis. *Phytomedicine* 1(3): 177–181.

Schmolz M, Ottendorfer D, Marz RW, Sieder C. (2001) Enhanced resistance to Sendai virus infection in DBA/2J mice with a botanical drug combination (Sinupret®). *International Immunopharmacology* 1(9–10): 1841–1848.

Schonbichler SA, Bittner LK, Pallua JD, Popp M, Abel G, Bonn GK, Huck CW. (2013) Simultaneous quantification of verbenalin and verbascoside in *Verbena officinalis* by ATR-IR and NIR spectroscopy. *Journal of Pharmaceutical and Biomedical Analysis* 84: 97–102.

Shu JC, Liu JQ, Chou GX. (2013) A new triterpenoid from *Verbena officinalis* L. *Natural Product Research* 27(14): 1293–1297.

Speroni E, Cervellati R, Costa S, Guerra MC, Utan A, Govoni P, Berger A, Muller A, Stuppner H. (2007) Effects of differential extraction of *Verbena officinalis* on rat models of inflammation, cicatrization and gastric damage. *Planta Medica* 73(3): 227–235.

Williamson EM, Driver S, Baxter K. (Eds.) (2013) *Stockley's Herbal Medicines Interactions.* 2nd Edition. Pharmaceutical Press, London, UK.

Xu W, Xin F, Sha Y, Fang J, Li YS. (2010) Two new secoiridoid glycosides from *Verbena officinalis*. *Journal of Asian Natural Products Research* 12(8): 649–653.

Wild Indigo   *Baptisia tinctoria* (L.) R.Br.

Synonyms: *Sophora tinctoria* L.

Family: Fabaceae (Leguminosae)

Other common names: False indigo; yellow broom

Drug name: Baptisie tinctoriae radix

Botanical drug used: Root

Indications/uses: Approved uses (THR) are for the relief of symptoms of the common cold, such as cough, catarrh, sore throat, runny or blocked nose.

Evidence: No clinical information is available to support the use of wild indigo alone, and there is only limited data to support its use as part of a fixed herbal combination.

Safety: Overall considered to be safe, but contains phytoestrogens (such as those found in red clover and soya, see pp. 311–313) which must be taken into consideration.

Main chemical compounds: Isoflavones including baptigenin, pseudobaptigenin, formononetin, genistein and biochanin A; quinolizidine alkaloids such as sparteine, cytisine (formerly known as baptitoxin), N-methylcytisine and anagyrine; polysaccharides, mainly arabinogalactans; coumarins (scopoletin); and glycoproteins, mainly arabinogalactan-proteins (APGs) such as baptisin and baptin (Classen et al. 2006; Williamson 2003).

Clinical evidence: No clinical studies are reported for *B. tinctoria* alone, but a randomised double-blind, placebo-controlled, clinical dose–response trial was found for a mixture of extracts containing *B. tinctoria* root, *Echinacea pallidal E. purpurea* root and *Thuja occidentalis* herb, for the treatment of patients with common cold. Patients (91 participants) received 9.6 mg or 19.2 mg of the fixed combination preparation for the treatment of upper respiratory tract infections (URIs), three times daily for 3–12 days. The study reported that the patients recovered faster (after 2 days), the effect was dose-dependent and there were no reported adverse effects (Naser et al. 2005). Similar effects were also observed in another multi-centre trial, including 239 participants with acute cold symptoms. In this study, a similar

---

Phytopharmacy: An evidence-based guide to herbal medicinal products, First Edition.
Sarah E. Edwards, Inês da Costa Rocha, Elizabeth M. Williamson and Michael Heinrich.
© 2015 John Wiley & Sons, Ltd. Published 2015 by John Wiley & Sons, Ltd.

formulation was used and patients received a similar dose three times a day for a period of 7–9 days (Henneicke-von Zepelin et al. 1999).

Pre-clinical evidence and mechanisms of action: Surprisingly little is known about the mechanism of action of wild indigo. An *in vitro* study investigating the immunological activities of *B. tinctoria* arabinogalactan-proteins in different test systems – proliferation of mouse lymphocytes, on nitrite- and IL6-production in alveolar mouse macrophage culture and IgM-production of mouse lymphocytes – showed that the AGPs had immunomodulatory effects in all three systems (Classen et al. 2006).

Interactions: None known.

Contraindications: Not recommended during pregnancy and lactation, as safety has not been established.

Adverse effects: Not known.

Dosage: Should be taken according to manufacturers' instructions.

General plant information: Wild indigo is native to North America. It is mainly used in combination products with other herbs, as described under clinical evidence. It should not be confused with *Indigofera tinctoria* L., 'true' indigo.

References

Classen B, Thude S, Blaschek W, Wack M, Bodinet C. (2006) Immunomodulatory effects of arabinogalactan-proteins from *Baptisia* and *Echinacea*. *Phytomedicine* 13(9–10): 688–694.

Henneicke-von Zepelin H, Hentschel C, Schnitker J, Kohnen R, Kohler G, Wustenberg P. (1999) Efficacy and safety of a fixed combination phytomedicine in the treatment of the common cold (acute viral respiratory tract infection): results of a randomised, double blind, placebo controlled, multicentre study. *Current Medical Research and Opinion* 15(3): 214–227.

Naser B, Lund B, Henneicke-von Zepelin HH, Kohler G, Lehmacher W, Scaglione F. (2005) A randomized, double-blind, placebo-controlled, clinical dose–response trial of an extract of *Baptisia, Echinacea* and *Thuja* for the treatment of patients with common cold. *Phytomedicine* 12(10): 715–722.

Williamson EM. (2003). *Potter's Herbal Cyclopaedia.* C W Daniels, Saffron Walden, UK.

Wild Lettuce   *Lactuca virosa* L.

Family: Asteraceae (Compositae)

Other common names: Bitter lettuce; lettuce opium; poisonous lettuce

Drug name: Lactucae virosae herba; Lactucarium (dried latex)

Botanical drug used: Flowering herb; dried latex

Indications/uses: Wild lettuce herb is used traditionally as a sedative in irritable cough and as a mild hypnotic in insomnia. It is rarely used alone but is common in products also containing valerian, hops and passion flower.

Evidence: Despite a long history of use, there is no clinical evidence to support the use of wild lettuce herb or latex as a sedative. The sesquiterpene lactones are mildly sedative and have analgesic properties.

Safety: Generally thought safe in normal doses, despite a long-standing reputation as a 'legal high' (drug of abuse) and previous use as a mild substitute for opium. No evidence of opioid-like effects has been found in either clinical or pre-clinical studies.

Main chemical compounds: Sesquiterpene lactones including lactucin and its derivatives lactucopicrin and 11β,13-dihydrolactucin (Wesołowska et al. 2006).

NB: wild lettuce is often reported to contain traces of the alkaloid hyoscyamine, to which the sedative action was attributed, but this has never been substantiated. Tropane alkaloids have never been reported as present in the Asteraceae family.

Clinical evidence: No evidence available for any indication.

Pre-clinical evidence and mechanisms of action:

Analgesic and anti-inflammatory effects: Lactucin, lactucopicrin and dihydrolactucin produced analgesic effects at doses of 15 and 30 mg/kg in the mouse hot plate test, similar to the effects of ibuprofen 30 mg/kg, and in the tail-flick test, similar to ibuprofen 60 mg/kg. Lactucopicrin appeared to be the most potent analgesic (Wesołowska et al. 2006).

Phytopharmacy: An evidence-based guide to herbal medicinal products, First Edition.
Sarah E. Edwards, Inês da Costa Rocha, Elizabeth M. Williamson and Michael Heinrich.
© 2015 John Wiley & Sons, Ltd. Published 2015 by John Wiley & Sons, Ltd.

Sedative effects: Lactucin and lactucopicrin, but not dihydrolactucin, showed sedative properties in the spontaneous locomotor activity test (Wesołowska et al. 2006).

Anti-cholinergic effects: Lactucopicrin showed a significant and dose-dependent inhibitory activity on acetyl cholinesterase (Rollinger et al. 2005).

Interactions: None reported.

Contraindications: May cause allergic reactions in sensitive individuals.

Adverse effects: Eight patients aged between 12 and 38 years were referred with various manifestations of wild lettuce toxicity; common presentations were mydriasis, dizziness and anxiety, urinary retention, decreased bowel sounds and sympathetic overactivity, suggesting an anticholinergic mechanism. However, no evidence of hyoscyamine was detected in blood, urine and gastric lavage fluid of the patients, all of whom recovered (Besharat et al. 2009). The patients had ingested large amounts of the fresh herb, but no reasons for doing so were offered.

Dosage: As part of a combination product, see manufacturers' instructions.

General plant information: The plant is pricklier than the edible lettuce, much more bitter in taste and reaches up to 2 m in height.

References

Besharat S, Besharat M, Jabbari A. (2009) Wild lettuce (*Lactuca virosa*) toxicity. *BMJ Case Reports* 2009. pii: bcr06.2008.0134.
Rollinger JM, Mock P, Zidorn C, Ellmerer EP, Langer T, Stuppner H. (2005) Application of the in combo screening approach for the discovery of non-alkaloid acetylcholinesterase inhibitors from *Cichorium intybus*. *Current Drug Discovery Technologies* 3(2): 185–193.
Wesołowska A, Nikiforuk A, Michalska K, Kisiel W, Chojnacka-Wójcik E. (2006) Analgesic and sedative activities of lactucin and some lactucin-like guaianolides in mice. *Journal of Ethnopharmacology* 107(2): 254–258.

Willow (Bark)
Salix alba L., *S. nigra* L., *S.* × *fragilis* L., *S. purpurea* L. and other *Salix spp.*

Family: Salicaceae

Other common names: *S. alba:* European willow, white willow; *S. nigra*: black willow, pussy willow; *S.* × *fragilis*: crack willow; *S. purpurea*: purple osier

Drug name: Salicis cortex

Botanical drug used: Bark

Indications/uses: The bark of various willow species has been used for treating a wide range of ailments, including fever, headache, influenza, rheumatism, gout and arthritis and is sold in products throughout Europe.

Evidence: There is moderate evidence for the use of willow bark extract in acute and chronic non-specific low-back pain and osteoarthritis.

Safety: Overall it is considered to be safe, although willow bark is not suitable for patients under 18 years and other groups of patients where aspirin is contraindicated: those allergic to salicylates, asthma and peptic ulcer, glucose-6-phosphate dehydrogenase deficiency, during the third trimester of pregnancy and when breastfeeding.

Main chemical compounds: The main constituents are the phenolic glycosides salicin, acetylsalicin, salicortin, salireposide, picein, triandrin. Esters of salicylic acid, salicyl alcohol, flavonoids including ampelopsin, taxifolin and derivatives, catechin and tannins are also present (Agnolet et al. 2012; EMA 2009; Williamson et al. 2013).

Clinical evidence: Several studies have shown the effectiveness of standardised willow bark extracts in arthritis and related inflammatory conditions. A two week, double-blind, randomised controlled trial assessing the effect of an extract (240 mg salicin/day) in patients with osteoarthritis showed a moderate analgesic effect compared to placebo (Schmid et al. 2001). Another trial demonstrated that the extract (standardised to daily dose of 240 mg salicin) reduced pain to the same degree as 12.5 mg/day rofecoxib (now withdrawn from sale due to cardiovascular toxicity) (Gagnier et al. 2006). A cohort study suggested that willow bark extract

Phytopharmacy: An evidence-based guide to herbal medicinal products, First Edition.
Sarah E. Edwards, Inês da Costa Rocha, Elizabeth M. Williamson and Michael Heinrich.
© 2015 John Wiley & Sons, Ltd. Published 2015 by John Wiley & Sons, Ltd.

(120–240 mg salicin) had a comparable effect to standard therapies (e.g. diclofenac and ibuprofen) with better tolerability and could be used for the treatment of osteoarthritis of the hip and knee (Beer and Wegener 2008).

A recent observational study over 6 months evaluated the dosage and safety of willow bark extract STW 33-I, (Proaktiv®), during long-term treatment with both mono- and combination therapy, in 436 patients with osteoarthritis and back pain. Co-medication with other NSAIDs and opioids was allowed. The results showed that the treatment was effective and well tolerated and the authors suggested that STW 33-I can be used in the long-term therapy of painful musculoskeletal disorders and can be combined with NSAIDs and opioids if necessary (Uehleke et al. 2013).

A Cochrane review assessing herbal medicines for the treatment of acute, sub-acute or chronic non-specific low-back pain in adults found moderate evidence that daily doses of *S. alba* extract, standardised to 120 or 240 mg salicin, were better than placebo for short-term pain relief. A systematic review of the effectiveness of willow bark extract in low back pain, with daily doses up to 240 mg salicin over periods of up to 6 weeks, also showed moderate evidence of effectiveness (Vlachojannis et al. 2009).

Pre-clinical evidence and mechanisms of action: After absorption, the phenolic glycoside salicin is metabolised into various salicylate derivatives. *In vitro* and *in vivo* studies have showed that willow bark extract inhibits cyclooxygenase-2 (COX-2), lipoxygenase, cytokine release, NF-kappaB activation, carrageenan-induced paw oedema, adjuvant-induced arthritis, heat-induced inflammation and dextran-induced hind paw oedema as well as acting as an antioxidant (Vlachojannis et al. 2011). However, the flavonoids and polyphenols present in willow bark provide a broader mechanism of action than aspirin and the extract does not damage the gastrointestinal mucosa to the same extent (Vlachojannis et al. 2011). Furthermore, a salicin-free willow bark extract induced antioxidant enzymes and prevented oxidative stress in human umbilical vein endothelial cells through activation of nuclear factor erythroid 2-related factor 2 (Nrf2), independent of salicin, providing a new potential explanation for the clinical usefulness of willow bark extract (Ishikado et al. 2013).

Interactions: It may increase the effects of anticoagulants such as warfarin and other coumarin derivatives, increasing the risk of bleeding (ESCOP 2003; Heck et al. 2000). Due to potential interaction with antiplatelet drugs, NSAIDs, and anticoagulants, caution is warranted (Williamson et al. 2013).

Contraindications: Willow bark extracts are contraindicated in children and adolescents less than 18 years, patients with asthma, active peptic ulcer disease, in cases of severe liver or renal dysfunction, coagulation disorders, gastric/duodenal ulcer and glucose-6-phosphate dehydrogenase deficiency and anyone allergic to salicylates. Willow bark extracts should be avoided during pregnancy and lactation (EMA 2009).

Adverse effects: Allergic reactions such as rashes, urticaria, asthma and gastrointestinal symptoms (nausea, vomiting, abdominal pain, diarrhoea, dyspepsia and heartburn) may occur (EMA 2009).

Dosage: Extracts are normally standardised to a minimum of 1.5% of total salicylic derivatives (expressed as salicin). For adults, the dose range for dry aqueous extract is 480–600 mg twice a day, for powdered herbal substance, it is 260–500 mg three times a day and as herbal tea, it is 1–3 g three to four times a day. Duration should be restricted to a maximum of 4 weeks (EMA 2009; Williamson et al. 2013).

General plant information: Willow preparations have been used throughout the world as antipyretics and analgesics. The origin of aspirin can be traced back to willow bark and its traditional uses, evolving from salicin to salicylic acid and then to the more effective and less toxic acetylsalicylic acid (Wick 2012).

References

Agnolet S, Wiese S, Verpoorte R, Staerk D. (2012) Comprehensive analysis of commercial willow bark extracts by new technology platform: combined use of metabolomics, high-performance liquid chromatography-solid-phase extraction-nuclear magnetic resonance spectroscopy and high-resolution radical scavenging assay. *Journal of Chromatography A* 1262: 130–137.

Beer AM, Wegener T. (2008) Willow bark extract (Salicis cortex) for gonarthrosis and coxarthrosis – Results of a cohort study with a control group. *Phytomedicine* 15(11): 907–913.

EMA. (2009) Community herbal monograph on Salix Cortex. European Medicines Agency http://www.ema.europa.eu/docs/en_GB/document_library/Herbal_-_Community_herbal _monograph/2009/12/WC500018256.pdf.

ESCOP. (2003) *ESCOP Monographs: The Scientific Foundation for Herbal Medicinal Products.* 2nd Edition. Thieme, Exeter and London, UK.

Gagnier JJ, van Tulder M, Berman B, Bombardier C. (2006) Herbal medicine for low back pain. *Cochrane Database of Systematic Reviews* (2): CD004504.

Heck AM, DeWitt BA, Lukes AL. (2000) Potential interactions between alternative therapies and warfarin. *American Journal of Health System Pharmacy* 57(13): 1221–1227; quiz 1228–1230.

Ishikado A, Sono Y, Matsumoto M, Robida-Stubbs S, Okuno A, Goto M, King GL, Keith Blackwell T, Makino T. (2013) Willow bark extract increases antioxidant enzymes and reduces oxidative stress through activation of Nrf2 in vascular endothelial cells and *Caenorhabditis elegans. Free Radical Biology and Medicine* 65: 1506–1515.

Schmid B, Ludtke R, Selbmann HK, Kotter I, Tschirdewahn B, Schaffner W, Heide L. (2001) Efficacy and tolerability of a standardized willow bark extract in patients with osteoarthritis: randomized placebo-controlled, double blind clinical trial. *Phytotherapy Research* 15(4): 344–350.

Uehleke B, Müller J, Stange R, Kelber O, Melzer J. (2013) Willow bark extract STW 33-I in the long-term treatment of outpatients with rheumatic pain mainly osteoarthritis or back pain. *Phytomedicine* 20(11): 980–984.

Vlachojannis JE, Cameron M, Chrubasik SA. (2009) Systematic review on the effectiveness of willow bark for musculoskeletal pain. *Phytotherapy Research* 23(7): 897–900.

Vlachojannis J, Magora F, Chrubasik S. (2011) Willow species and aspirin: different mechanism of actions. *Phytotherapy Research* 25(7): 1102–1104.

Wick JY. (2012) Aspirin: a history, a love story. *Consultant Pharmacist* 27(5): 322–329.

Williamson EM, Driver S, Baxter K. (Eds.) (2013) *Stockley's Herbal Medicines Interactions.* 2nd Edition. Pharmaceutical Press, London, UK.

Witch Hazel   *Hamamelis virginiana* L.

Family: Hamamelidaceae

Other common names: Hamamelis; snapping hazel; winterbloom

Drug name: Hamamelidis folium; Hamamelidis cortex; Hamamelidis ramunculus (twigs); Hamamelidis destillatum/ Aquae Hamamelidis (distillate)

Botanical drug used: Leaf; bark; twigs; steam distillate obtained from twigs

---

Indications/uses: Witch hazel is traditionally used externally to treat minor skin irritations, inflammation and dryness of skin. Witch hazel extract (leaf or bark) is considered a mild astringent, and has reported topical uses in treatment of haemorrhoids, varicose veins, bruises, minor abrasions and to reduce discomfort of the perineum after childbirth. Witch hazel water is also used for the temporary relief of eye discomfort due to dryness of the eye or exposure to wind and sun. Local anaesthetic and styptic properties have been attributed to it. In Germany, witch hazel leaf and bark tea preparations are approved for use as mouth gargles to treat inflammation of the gums and mucous membranes. Although historically witch hazel has been taken orally to treat diarrhoea, menorrhagia and dysmenorrhea, this is no longer recommended.

Evidence: There is some limited clinical evidence to support its traditional uses.

Safety: No significant problems have been reported for external use of witch hazel during pregnancy and lactation, although safety has not been conclusively established. Oral use is not recommended, as it may cause stomach irritation and liver damage.

Main chemical compounds: Tannins (10% bark, 3–10% leaf). Leaf tannins are a mixture of gallic acid (10%), hydrolysable hamamelitannin (1.5%) and condensed proanthocyanidins (88.5%). Bark tannins have a much higher hamamelitannin level (up to 65% of a hydroalcoholic extract). Witch hazel water (the steam distillate) does not contain tannins (Gardner and McGuffin 2013; WHO 2004). Hamamelis leaf should have minimum 3% tannin, expressed as pyrogallol, according to European and British Pharmacopoeias (BP 2012). Other constituents include small amounts of volatile oils (about 0.5%, e.g. hexenol, α- and β-ionones, eugenol, safrole and sesquiterpenes). The leaf contains flavonols (e.g. kaempferol, quercetin) and their glycosides (e.g. astragalin, quercitrin, afzelin and myricitrin). Resin, wax, saponins,

---

Phytopharmacy: An evidence-based guide to herbal medicinal products, First Edition.
Sarah E. Edwards, Inês da Costa Rocha, Elizabeth M. Williamson and Michael Heinrich.
© 2015 John Wiley & Sons, Ltd. Published 2015 by John Wiley & Sons, Ltd.

choline and free hamamelose are also found in witch hazel (Pharmaceutical Press Editorial Team 2013).

Clinical evidence:

Clinical studies supporting therapeutic uses of witch hazel are limited. A few studies investigating topical uses of witch hazel are highlighted here.

Skin inflammation: In one study involving 24 healthy volunteers, effectiveness in suppressing UV-induced skin erythema was found to be greater when low dose hamamelis distillate was administered using a vehicle of oil in water emulsion with phosphatidylcholine (PC), compared with hamamelis cream without PC. Although the study demonstrated anti-inflammatory activity of hamamelis distillate in a PC-containing vehicle, hydrocortisone 1% cream appeared to be superior in suppressing UV-induced erythema than the hamamelis preparations (Korting et al. 1993).

In a 2-week randomised, double-blind trial in 72 patients with atopic eczema, hamamelis distillate cream, reduced itching, scaling and erythema after 1 week. However, it was found to be no more effective than the base preparation (Korting et al. 1995).

An after-sun lotion containing 10% hamamelis distillate was assessed in 30 healthy volunteers, and found to suppress UVB-induced erythema between 20% and 27% within 48 h (Hughes-Formella et al. 1998). The same group compared the anti-inflammatory effectiveness (using a modified UV-erythema test) of different topical preparations, including three different lotions containing 10% hamamelis distillate, two vehicles, an antihistamine (dimethindene maleate) 0.1% gel, hydrocortisone 1% cream and hydrocortisone 0.25% lotion. All preparations exhibited anti-inflammatory effects, but the hydrocortisone preparations were most effective. One hamamelis lotion suppressed erythema significantly more than the vehicles at the UV dosage of 1.4 minimal erythema dose (MED) (Hughes-Formella et al. 2002).

In an observational study in 309 children aged 27 days to 11 years with minor skin injuries, 'nappy rash' or localised skin inflammation, treatment was either hamamelis ointment ($n = 231$) or dexpanthenol ointment ($n = 78$). Hamamelis ointment was found to be effective and well tolerated, with observed effects similar to dexpanthenol (Wolff and Kaiser 2007).

In a pilot study with healthy volunteers, a semi-solid formulation containing 1% *H. virginiana* procyanidin ('ProcyanoPlus') was found effective in reducing transepidermal water loss and erythema formation induced by sodium lauryl sulphate (Deters et al. 2001).

Haemorrhoids: There is some clinical evidence to support the use of witch hazel in rectal ointments to treat haemorrhoids. Studies found that hamamelis ointment significantly reduced pruritus, bleeding, burning sensation and pain associated with anorectal complaints (Pharmaceutical Press Editorial Team 2013; WHO 2004).

Perineal trauma: A Cochrane review reported a lack of effectiveness of witch hazel water in relieving pain caused by perineal trauma incurred during childbirth. No difference was found in the number of women reporting none or mild pain relief from treatment with ice packs compared to hamamelis water, with <5% incidence of perineal wound infection in both groups (East et al. 2012).

Pre-clinical evidence and mechanisms of action:

Anti-inflammatory and related effects: Antioxidant, anti-inflammatory and chemopreventive activities of witch hazel extracts have been reported. Topical dermal preparations of witch hazel have been shown to have a protective effect against oxygen radical damage in cultured murine dermal fibroblasts (Masaki et al. 1995). Hamamelitannin, which has been shown to have potent antioxidant activity, also inhibits TNFα-induced endothelial cell death *in vitro*. This may explain the traditional use of witch hazel as an anti-haemorrhagic, and its reputed use as a protective agent for UV radiation (Habtemariam 2002).

Antiviral effects: A witch hazel bark extract fraction, consisting mainly of proanthocyanadins, exhibited significant antiviral activity against *Herpes simplex* virus type 1, in addition to radical-scavenging properties, and strong anti-inflammatory effects (Erdelmeier et al. 1996). Weak antimicrobial activity has also been reported *in vitro* for a witch hazel topical preparation containing 90% distillate with 5% urea (Gloor et al. 2002).

Inhibition of mutagenesis and cancer cell proliferation: Constituent proanthocyanidins have been shown to act directly as desmutagens (i.e. they inactivate cancer-causing agents), with strong and dose-dependent inhibition of carcinogen-induced DNA damage demonstrated *in vitro* in human hepatoma cells (Hep G2). The degree of polymerisation of proanthocyanidins was found to influence the antimutagenic effect (Dauer et al. 1998; Dauer et al. 2003). *In vitro* studies have shown that witch hazel bark extracts, particularly polyphenolic fractions, inhibit cell proliferation in human colon cancer cell lines, with hamamelitannin being most effective (Lizárraga et al. 2008; Sánchez-Tena et al. 2012).

Interactions: None known.

Contraindications: None known.

Adverse effects: Allergic contact dermatitis has been reported rarely (EMA 2009; Gardner and McGuffin 2013). Safrole, a constituent of the volatile oil, is a known carcinogen, but is found in too small amounts to cause concern (Pharmaceutical Press Editorial Team 2013). Witch hazel is not recommended for oral use, as it may cause stomach irritation and liver damage (EMA 2009).

Dosage: *Hamamelis leaf* – external use: hamamelis ointment BPC 1973 (10% by weight of hamamelis liquid extract BPC 1973 in an ointment base); semi-solid or liquid preparations containing the equivalent 5–10% of leaf; for impregnated dressings and rinses, decoctions of 5–10 g of leaf to 250 ml of water, or 20 ml of tincture diluted to 100 ml. Local use for mouthwashes: as a decoction of 2–3 g in 150 ml, several times daily. Hamamelis suppositories BPC 1973 (200 mg of hamamelis dry extract BPC 1973 in a suitable base), or, more generally, suppositories containing the equivalent of 0.1–1 g of dried leaf.

Hamamelis bark – external use: semi-solid and liquid preparations containing the equivalent 5–10% of bark; as impregnated dressings, lotion or mouthwash, diluted tincture of hamamelis (1:10, 45% ethanol), or decoction of 5–10 g of bark to 250 ml of water.

Hamamelis water BPC 1973 – topical use: undiluted hamamelis water for applications to cuts, grazes, insect stings, other skin complaints and haemorrhoids, as a mouthwash, and in a saturated cotton wool swab as a nasal plug for nosebleeds, or to place over eyelids; for compresses, undiluted or diluted 1:3 in water; in semi-solid preparations, 20–30%. As an eyewash, use diluted hamamelis water, 10 drops to an eyebath half-filled with water (British Herbal Medicines Association, British Herbal Compendium Vol II 2006, in EMA 2009).

General plant information: Witch hazel is native to the eastern states of North America, and was used in the traditional medicine of Native Americans, including the Cherokee and Iroquois (Moerman 1998; USDA, NRCS 2014).

References

BP. (2012) *British Pharmacopoeia Volume IV.*

Dauer A, Metzner P, Schimmer O. (1998) Proanthocyanidins from the bark of *Hamamelis virginiana* exhibit antimutagenic properties against nitroaromatic compounds. *Planta Medica* 64(4): 324–327.

Dauer A, Hensel A, Lhoste E, Knasmüller S, Mersch-Sundermann V. (2003) Genotoxic and antigenotoxic effects of catechin and tannins from the bark of *Hamamelis virginiana* L. in metabolically competent, human hepatoma cells (Hep G2) using single cell gel electrophoresis. *Phytochemistry* 63(2): 199–207.

Deters A, Dauer A, Schnetz E, Fartasch M, Hensel A. (2001) High molecular compounds (polysaccharides and proanthocyanidins) from *Hamamelis virginiana* bark: influence on human skin keratinocyte proliferation and differentiation and influence on irritated skin. *Phytochemistry* 58: 949–958.

East CE, Begg L, Henshall NE, Marchant PR, Wallace K. (2012) Local cooling for relieving pain from perineal trauma sustained during childbirth. *Cochrane Database of Systematic Reviews* (5): CD006304.

EMA. (2009) Assessment report on *Hamamelis virginiana* L., cortex, *Hamamelis virginiana* L., folium, *Hamamelis virginiana* L., folium et cortex aut ramunculus destillatum. European Medicines Agency http://www.ema.europa.eu/docs/en_GB/document_library /Herbal_-_HMPC_assessment_report/2010/04/WC500089242.pdf (accessed 1 August 2013).

Erdelmeier CAJ, Cinatl Jr., J, Rabenau H, Doerr HW, Biber A, Kock E. (1996) Antiviral and antiphlogistic activities of *Hamamelis virginiana* bark. *Planta Medica* 62(3): 241–245.

Gardner Z, McGuffin M. (Eds.) (2013) *American Herbal Product Association's Botanical Safety Handbook.* 2nd Edition. CRC Press, USA, 1072 pp.

Gloor M, Reichling J, Wasik B, Holzgang HE. (2002) Antiseptic effect of a topical dermatological formulation that contains Hamamelis distillate and urea. *Forschende Komplementärmedizin und Klassische Naturheilkunde* 9(3): 153–159.

Habtemariam S. (2002) Hamamelitannin from *Hamamelis virginiana* inhibits the tumour necrosis factor-α (TNF)-induced endothelial cell death *in vitro*. *Toxicon* 40(1): 83–88.

Hughes-Formella BJ, Bohnsack K, Rippke F, Benner G, Rudolph M, Tausch I, Gassmueller J. (1998) Anti-inflammatory effect of hamamelis lotion in a UVB erythema test. *Dermatology* 196(3): 316–322.

Hughes-Formella BJ, Filbry A, Gassmueller J, Rippke F. (2002) Anti-inflammatory efficacy of topical preparations with 10% hamamelis distillate in a UV erythema test. *Skin Pharmacology and Applied Skin Physiology* 15(2): 125–132.

Korting HC, Schäfer-Korting M, Hart H, Laux P, Schmid M. (1993) Anti-inflammatory activity of hamamelis distillate applied topically to the skin. *Influence of vehicle and dose.* *European Journal of Clinical Pharmacology* 44(4): 315–318.

Korting HC, Schäfer-Korting M, Klövekorn W, Klövekorn G, Martin C, Laux P. (1995) Comparative efficacy of hamamelis distillate and hydrocortisone cream in atopic eczema. *European Journal of Clinical Pharmacology* 48(6): 461–465.

Lizárraga D, Tourino S, Reyes-Zurita FJ et al. (2008) Witch Hazel (*Hamamelis virginiana*) fractions and the importance of gallate moieties – electron transfer capacities in their anti-tumoral properties. *Journal of Agricultural and Food Chemistry* 56(24): 11675–11682.

Masaki H, Atsumi T, Sakurai H. (1995) Protective activity of hamamelitannin on cell damage of murine skin fibroblasts induced by UVB radiation. *Journal of Dermatological Science* 10(1): 25–34.

Moerman DE. (1998) *Native American Ethnobotany*. Timber Press, Portland, OR, USA, 927 pp.

Pharmaceutical Press Editorial Team (2013) *Herbal Medicines*. 4th Edition. Pharmaceutical Press, London, UK.

Sánchez-Tena S, Fernández-Cachón ML, Carreras A, Mateis-Martín ML, Costoya N, Moyer MP, Nuñez MJ, Torres JL, Cascante M. (2012) Hamamelitannin from witch hazel (*Hamamelis virginiana*) displays specific cytotoxic activity against colon cancer cells. *Journal of Natural Products* 75(1): 26–33.

USDA, NRCS. (2014) The PLANTS Database. http://plants.usda.gov (accessed 5 January 2014).

WHO. (2004) *WHO Monographs on Selected Medicinal Plants*. Vol. 2. WHO, Geneva, 358 pp.

Wolff HH, Kaiser M. (2007) Hamamelis in children with skin disorders and skin injuries: results of an observational study. *European Journal of Paediatrics* 166(9): 943–948.

Yohimbe

Pausinystalia johimbe (K.Schum.) Pierre

Synonyms: *Corynanthe johimbe* K.Schum.

Family: Rubiaceae

Other common names: Johimbe; yohimbebaum

Drug name: Yohimbe cortex

Botanical drug used: Bark

Indications/uses: Yohimbe is recommended for erectile dysfunction (ED) in males and loss of libido in females; it has also been advocated as an aid to weight loss and to enhance exercise tolerance and muscle development in bodybuilders.

Evidence: There is well-documented clinical evidence for the effects of isolated yohimbine in ED, but not for the bark extract or for any of the other indications.

Safety: Yohimbe bark is on the FDA list of dangerous supplements. Yohimbine, the main active compound, can cause insomnia, panic attacks, hallucinations, seizures, paralysis, tachycardia, hypertension and even heart attacks in doses above 40 mg. Levels of yohimbine in the bark extract are often very low, but are also variable and intractable priapism has been reported in a patient taking the bark extract.

Main chemical compounds: Indole alkaloids, the major active constituent being yohimbine, which accounts for 1–20% of the total alkaloid content. It also contains corynanthine, pseudoyohimine, α- and β-yohimbane and many others (Gardner and McGuffin 2013).

Clinical evidence: Yohimbine, the isolated alkaloid, has been shown to be effective in male ED but the levels present in yohimbe bark are often very low, so the extract may not be effective in treating ED (Gardner and McGuffin 2013). There is also little evidence for positive effects in bodybuilding or exercise tolerance (Cimolai and Cimolai 2011). A placebo-controlled study in professional soccer players showed that yohimbine (combined with resistance training) did not significantly alter body

---

Phytopharmacy: An evidence-based guide to herbal medicinal products, First Edition.
Sarah E. Edwards, Inês da Costa Rocha, Elizabeth M. Williamson and Michael Heinrich.
© 2015 John Wiley & Sons, Ltd. Published 2015 by John Wiley & Sons, Ltd.

or muscle mass or performance indicators, although fat mass was significantly lower in the yohimbine group (Ostojic 2006).

Pre-clinical evidence and mechanisms of action: Yohimbine is a selective alpha-2 adrenoceptor blocker (Verwaerde et al. 1997) and also a weak monoamine oxidase enzyme B inhibitor (Mazzio et al. 2013). Blockade of post-synaptic alpha-2 adrenoceptors causes weak corpus cavernosum smooth muscle relaxation, whereas blockade of both pre- and post-synaptic alpha-2 adrenoreceptors causes an increased release of noradrenaline and dopamine (Saenz De Tejada et al. 2000). Corynanthine is also an alpha-1 adrenoceptor blocker, so yohimbe bark extract in sufficient dosages may provide concomitant alpha-1 and alpha-2 adrenoceptors blockade and thus may better enhance erections than yohimbine alone (Gardner and McGuffin 2013).

Interactions: In a double-blind, placebo-controlled study in male patients with post-operative dental pain, the effect of pre-operative administration of yohimbine on analgesia produced by post-operative morphine was investigated. Yohimbine alone did not affect the pain, but the analgesic effect of morphine was significantly enhanced in the presence of yohimbine (Gear et al. 1995). The addition of yohimbine to fluoxetine or venlafaxine has been found to potentiate the antidepressant action of both of these drugs in mice (Dhir and Kulkarni 2007) and should be avoided. Yohimbe should not be taken concurrently with antihypertensive drugs and mono-amine oxidase inhibitors (MAOI).

Contraindications: Contraindications of yohimbe include liver or kidney disease, inflammation of the prostate, pregnancy, breast feeding, hypertension. Yohimbe bark should not be used in high doses and is not suitable for long-term use (Gardner and McGuffin 2013).

Adverse effects: Although yohimbine is generally thought to be safe in appropriate doses (Tam et al. 2001), overdoses can cause priapism and this has also been reported for the bark extract (Myers and Barrueto 2009). Depending on dosage, yohimbine can both lower and increase systemic blood pressure: small doses of yohimbine can increase blood pressure while large doses can cause dangerously low pressure and even death (Anderson et al. 2013). Other side effects include increased salivation, increased frequency of urination and anxiety (Gardner and McGuffin 2013).

Dosage: The dose of yohimbine used to treat ED is 15–30 mg/day, but since levels of yohimbine in the bark are variable, a safe dose cannot be given.

General plant information: Yohimbe is a tall evergreen tree native to the tropical rain forests of Cameroon, the Congo basin, Gabon and Nigeria. There are concerns that the tree may be endangered due to over-harvesting.

References

Anderson C, Anderson D, Harre N, Wade N. (2013) Case study: two fatal case reports of acute yohimbine intoxication. *Journal of Analytical Toxicology* 37(8): 611–614.

Cimolai N, Cimolai T. (2011) Yohimbine use for physical enhancement and its potential toxicity. *Journal of Dietary Supplements* 8(4): 346–354.

Dhir A, Kulkarni SK. (2007) Effect of addition of yohimbine (alpha-2-receptor antagonist) to the antidepressant activity of fluoxetine or venlafaxine in the mouse forced swim test. *Pharmacology* 80(4): 239–243.

Gardner Z, McGuffin M. (Eds.) (2013) *American Herbal Products Association's Botanical Safety Handbook*. 2nd Edition. CRC Press, USA, 1072 pp.

Gear RW, Gordon NC, Heller PH, Levine JD. (1995) Enhancement of morphine analgesia by the alpha 2-adrenergic antagonist yohimbine. *Neuroscience* 66(1): 5–8.

Mazzio E, Deiab S, Park K, Soliman KFA. (2013) High throughput screening to identify natural human monoamine oxidase B inhibitors. *Phytotherapy Research* 27(6): 818–828.

Myers A, Barrueto F Jr.. (2009) Refractory priapism associated with ingestion of yohimbe extract. *Journal of Medical Toxicology* 5(4): 223–225.

Ostojic SM. (2006) Yohimbine: the effects on body composition and exercise performance in soccer players. *Research in Sports Medicine* 14(4): 289–299.

Saenz De Tejada I, Kim NN, Goldstein I, Traish AM. (2000) Regulation of pre-synaptic alpha adrenergic activity in the corpus cavernosum. *International Journal of Impotence Research* 12(Suppl. 1): S20–S25.

Tam SW, Worcel M, Wyllie M. (2001) Yohimbine: a clinical review. *Pharmacology & Therapeutics* 91(3): 215–243.

Verwaerde P, Tran MA, Montastruc JL, Senard JM, Portolan G. (1997) Effects of yohimbine, an alpha 2-adrenoceptor antagonist, on experimental neurogenic orthostatic hypotension. *Fundamental and Clinical Pharmacology* 11(6): 567–575.

Index

Phytopharmacy: An evidence-based guide to herbal medicinal products, First Edition.
Sarah E. Edwards, Inês da Costa Rocha, Elizabeth M. Williamson and Michael Heinrich.
© 2015 John Wiley & Sons, Ltd. Published 2015 by John Wiley & Sons, Ltd.